ANALYSIS OF GROWTH FACTOR SIGNALING IN EMBRYOS

METHODS IN SIGNAL TRANSDUCTION SERIES

Joseph Eichberg, Jr., Series Editor

Published Titles

Lipid Second Messengers, Suzanne G. Laychock and Ronald P. Rubin

G Proteins: Techniques of Analysis, David R. Manning

Signaling Through Cell Adhesion Molecules, Jun-Lin Guan

G Protein-Coupled Receptors, Tatsuya Haga and Gabriel Berstein

Calcium Signaling, James W. Putney, Jr.

G Protein-Coupled Receptors: Structure, Function, and Ligand Screening, Tatsuya Haga and Shigeki Takeda

Calcium Signaling, Second Edition James W. Putney, Jr.

Analysis of Growth Factor Signaling in Embryos Malcolm Whitman and Amy K. Sater

ANALYSIS OF GROWTH FACTOR SIGNALING IN EMBRYOS

Edited by
Malcolm Whitman
Amy K. Sater

CRC Taylor & Francis
Taylor & Francis Group
Boca Raton London New York

CRC is an imprint of the Taylor & Francis Group,
an informa business

CRC Press
Taylor & Francis Group
6000 Broken Sound Parkway NW, Suite 300
Boca Raton, FL 33487-2742

Printed in the United States of America on acid-free paper
10 9 8 7 6 5 4 3 2 1

International Standard Book Number-10: 0-8493-3165-X (Hardcover)
International Standard Book Number-13: 978-0-8493-3165-7 (Hardcover)

Library of Congress Cataloging-in-Publication Data

Analysis of growth factor signaling in embryos / edited by Malcolm Whitman and Amy Sater.
p. cm. -- (Methods in signal transduction ; 8)
Includes bibliographical references and index.
ISBN 0-8493-3165-X (alk. paper)
1. Human embryo--Physiology. 2. Growth factors--Physiological effect. I. Whitman, Malcolm. II. Sater, Amy. III. Title. IV. Series.

QP277.A53 2006
612.6'46--dc22 2006044047

Visit the Taylor & Francis Web site at
http://www.taylorandfrancis.com

and the CRC Press Web site at
http://www.crcpress.com

Series Preface

The concept of signal transduction at the cellular level is now established as a cornerstone of the biological sciences. Cells sense and react to environmental cues by means of a vast panoply of signaling pathways and cascades. While the steady accretion of knowledge regarding signal transduction mechanisms is continuing to add layers of complexity, this greater depth of understanding has also provided remarkable insights into how healthy cells respond to extracellular and intracellular stimuli and how these responses can malfunction in many disease states.

Central to advances in unraveling signal transduction is the development of new methods and refinement of existing ones. Progress in the field relies upon an integrated approach that utilizes techniques drawn from cell and molecular biology, biochemistry, genetics, immunology and computational biology. The overall aim of this series is to collate and continually update the wealth of methodology now available for research into many aspects of signal transduction. Each volume is assembled by one or more editors who are leaders in their specialty. Their guiding principle is to recruit knowledgeable authors who will present procedures and protocols in a critical yet reader-friendly format. Our goal is to ensure that each volume will be of maximum practical value to a broad audience, including students, seasoned investigators and researchers who are new to the field.

Studies of growth factor signaling during embryonic development underlie a burgeoning field, whose aim is to address and explain a variety of dynamic processes of great complexity from the molecular to the systems level. The topics that make up the present volume illustrate the range of experimental and technical approaches now in use or under development for investigations of embryonic signal transduction. In many cases, to make progress in unraveling the intricate spatial and temporal signaling events occurring in embryos, existing methods used to study signaling in differentiated cells must either be modified or new techniques must be devised. Moreover, these approaches must be utilized in an interdisciplinary manner that can involve simultaneous applications of genetics, biochemistry, cell biology and systems biology. This volume is unique in that it brings together in one place descriptions of experimental strategies designed to elucidate growth factor signaling pathways in embryos.

Joseph Eichberg, Ph.D.
Advisory Editor for the Series

Preface

Analysis of growth factor signaling mechanisms in embryos is of critical importance, both because these signaling pathways regulate a broad range of developmental decisions, and because studies in embryos have elucidated key steps in growth factor signaling pathways. Although embryonic and differentiated cells share major features of growth factor signaling, experimental studies of growth factor signaling pathways in embryos differ from those in mammalian cultured cells in several respects. First, signaling is highly dynamic in embryos: The activity of any pathway is restricted in space and time. Thus, spatial and temporal factors must be considered in defining the experimental material (e.g., limiting analysis to a given tissue at a given stage of development). Second, because of these restrictions, the amount of material available for analysis is often very limited, and conventional biochemical assays must be scaled down or otherwise modified. Third, studies of embryonic cell signaling often assess a wide range of cellular responses to growth factors, some of which, such as specific cell behaviors, can only be evaluated *in vivo*. Finally, investigations using embryos can take advantage of gain and loss of function approaches using conventional or reverse genetics, which may already be linked to whole-embryo phenotypes.

Developmental biologists have been driven to investigate growth factor signaling in embryos in order to understand the regulatory mechanisms underlying a given developmental process. The availability of genome sequences, expressed sequence tag databases, and other genomic and bioinformatics resources create unparalleled opportunities for combining genetic, molecular, and biochemical analyses to understand the regulation and function of cell signaling pathways. As with any interdisciplinary undertaking, there is a danger of underestimating the complexities of signaling; thus, such opportunities demand a depth of understanding in technical methods as well as appropriate experimental design.

The goal of this book is to provide both methods and guidelines for experimental design in major aspects of cell signaling in vertebrate embryos. Section I focuses on specific signaling pathways or pathway components. In some instances, this section highlights the biochemistry of signaling pathways during development, which is often distinctive from that observed in cell culture systems. Section II discusses ionic regulatory mechanisms, while the two chapters in Section III examine ways of investigating gene regulation in response to extracellular signals. Section IV includes chapters that address emerging strategies that facilitate integrated analyses of cell signaling *in vivo* in embryonic systems. These chapters should help provide a foundation for further explorations of the cellular regulatory mechanisms governing vertebrate embryonic development.

Editors

Malcolm Whitman is an associate professor in the Department of Developmental Biology at the Harvard School of Dental Medicine and the Department of Cell Biology at Harvard Medical School. Dr. Whitman studied signal transduction as a graduate student in the laboratory of Lewis Cantley and *Xenopus* development as a postdoctoral fellow in the laboratory of Dr. Doug Melton. Dr. Whitman's laboratory studies the role of TGF-β superfamily signaling in early *Xenopus* embryogenesis.

Amy K. Sater is an associate professor in the Department of Biology and Biochemistry at the University of Houston. She earned her Ph.D. from the University of Texas at Austin, where she studied the induction of heart mesoderm in the laboratory of Antone Jacobson. Her postdoctoral work addressed mechanisms of ionic regulation during the specification of neural fate in the laboratories of Richard Steinhardt and Ray Keller at the University of California, Berkeley. Her laboratory investigates the mechanisms and functional significance of interactions between FGF and BMP signaling pathways in early vertebrate development.

List of Contributors

Dany S. Adams
The Forsyth Institute
Department of Developmental Biology
Harvard School of Dental Medicine
Boston, MA

Curtis Altmann
Department of Biomedical Sciences
FSU College of Medicine
Tallahassee, FL

Kevin Brown
Department of Molecular and Cellular Biology
Harvard University
Cambridge, MA

Daniel R. Buchholz
Laboratory of Growth and Development
National Institute of Child Health and Human Development,
National Institutes of Health
Bethesda, MD

Joanne Chan
Vascular Biology Program
Children's Hospital
Boston, MA

Jan L. Christian
Department of Cell and Developmental Biology
Oregon Health and Science University
Portland, OR

Wilson K. Clements
Department of Biology
Section of Cell and Developmental Biology
University of California at San Diego
La Jolla, CA

Gerald R. Crabtree
Department of Developmental Biology
Stanford University School of Medicine
Stanford, CA

Heithem M. El-Hodiri
Center for Molecular and Human Genetics
Columbus Children's Research Institute
Columbus, OH

Jason E. Gestwicki
Department of Developmental Biology
Stanford University School of Medicine
Stanford, CA

Raymond Habas
Department of Biochemistry
University of Medicine and Dentistry
New Jersey-Robert Wood Johnson School of Medicine
Piscataway, NJ

Xi He
Department of Neurology
Children's Hospital
Harvard Medical School
Boston, MA

Gregory R. Hoffman
Department of Cell Biology
Harvard Medical School
Boston, MA

David Kimelman
Department of Biochemistry
University of Washington
Seattle, WA

Michael Kühl
Department of Biochemistry
University of Ulm
Ulm, Germany

Ethan Lee
Department of Cell Biology
Vanderbilt University School of Medicine
Nashville, TN

Michael Levin
The Forsyth Institute
Department of Developmental Biology
Harvard School of Dental Medicine
Boston, MA

Karen J. Liu
Department of Developmental Biology
Stanford University School of Medicine
Stanford, CA

Malcolm Maden
MRC Centre for Developmental Neurobiology
King's College London
Guy's Campus, London

Randall T. Moon
Department of Pharmacology
University of Washington
Seattle, WA

Bindu Diana Paul
Laboratory of Growth and Development
National Institute of Child Health and Human Development
National Institutes of Health
Bethesda, MD

Thomas M. Roberts
Department of Cancer Biology
Dana Farber Cancer Institute
Boston, MA

Adrian Salic
Department of Cell Biology
Harvard Medical School
Boston, MA

Amy K. Sater (co-editor)
Department of Biology and Biochemistry
University of Houston
Houston, TX

Yun-Bo Shi
Laboratory of Growth and Development
National Institute of Child Health and Human Development
National Institutes of Health
Bethesda, MD

Diane C. Slusarski
Department of Biological Sciences
University of Iowa
Iowa City, IA

Shailaja Sopory
Department of Cell and Developmental Biology
Oregon Health and Science University School of Medicine
Portland, OR

Karen Symes
Department of Biochemistry
Boston University School of Medicine
Boston, MA

Malcolm Whitman (co-editor)
Harvard School of Dental Medicine
Department of Developmental Biology
Boston, MA

Contents

Section I

Signals and Pathways

1 Analysis of Canonical Wnt Signaling in *Xenopus* Embryos

Wilson K. Clements and David Kimelman

CONTENTS

I. INTRODUCTION

A. The Wnt Pathway

Wnt signaling plays a central role in development and disease (see References 1–5). Embryonic Wnt signals help to pattern the embryo at numerous points during development, including regulation of embryonic axis specification in anamniotic vertebrates, germ layer specification, neural patterning, and the formation of particular tissues and organs. In adult organisms, Wnt signaling is important for cell cycle regulation, the maintenance of stem cell populations, and tissue differentiation; whereas misregulation has been associated with numerous cancers (see References 1–5).

Wnt signaling is transduced through at least three different intracellular pathways. Activation of the canonical pathway involves stabilization of the multifunctional protein β-catenin and consequent activation of Wnt target genes. The non-canonical pathways are less well studied, but involve activation of either Jun N-terminal Kinase (JNK), Ca^{2+} signaling, or both (see Reference 11 and Chapters 2 and 6). Though many players in the non-canonical pathways are beginning to be elucidated, the

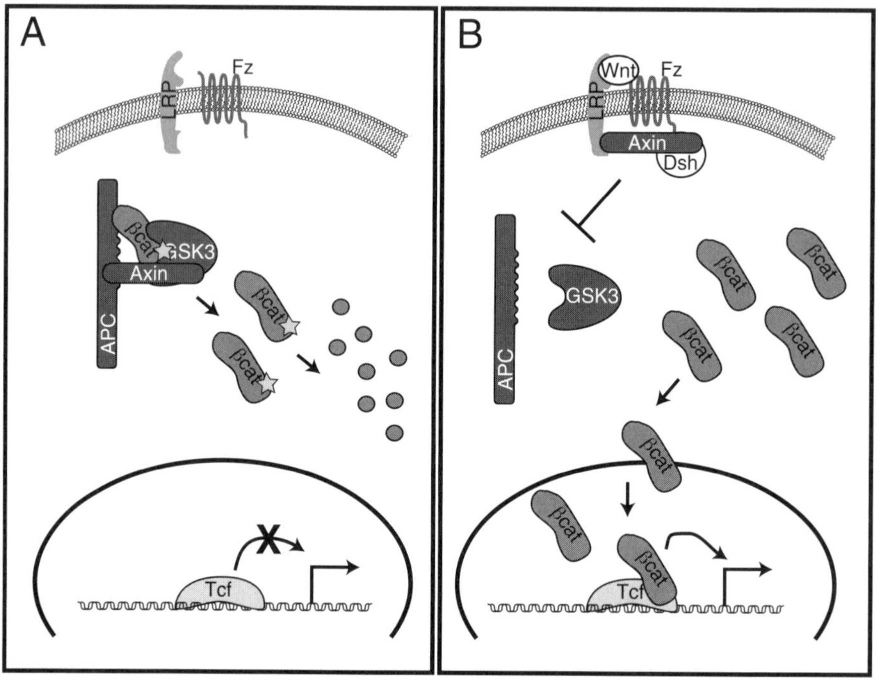

FIGURE 1.1 The canonical Wnt pathway. (a) When the Wnt pathway is inactive, the destruction complex, containing Axin, APC, and GSK-3, constitutively phosphorylates β-catenin. Phosphorylation of β-catenin (represented by stars) leads to degradation via the ubiquitin-proteasome path. Wnt target genes are repressed. (b) Binding of a Wnt ligand to its cognate Frizzled/LRP receptors leads to membrane localization of Axin and Dsh and inactivation of the destruction complex. β-catenin accumulates and enters the nucleus, where it activates, in cooperation with DNA-binding factors of the Tcf/Lef family and additional co-activators, the transcription of Wnt target genes.

specific pathways have yet to be fully described. In the past, the 19 known vertebrate Wnt ligands have generally been segregated into activators of either the canonical or the non-canonical pathways, although more recent evidence supports the idea that the specific intracellular pathway that is activated also depends on which receptors and co-receptors are present and what their particular post-translational state is.[12–25]

Canonical Wnt signaling (Figure 1.1) leads to the stabilization of a signaling pool of β-catenin, a molecule that, in addition to its role in the Wnt pathway, is important for cell-cell adhesion, linking cadherins to the actin cytoskeleton (see References 26–28). Thus, many cells not responding to a Wnt signal nevertheless require a basal level of non-signaling β-catenin, a fact that emphasizes the need for precise regulation of the molecules involved. Wnt pathway activation and consequent stabilization of the signaling β-catenin leads to transcription of Wnt target genes in cooperation with members of the Tcf/Lef family of DNA-binding proteins.[29–32] Under non-signaling conditions (Figure 1.1a), cytosolic levels of β-catenin are kept relatively low by the actions of a complex of proteins known as the destruction complex. This complex catalyzes the phosphorylation of β-catenin, which marks it for degradation by the ubiquitin-proteasome pathway (see References 3, 5, 33).

The core elements of the destruction complex are the scaffolding protein Axin, the large protein encoded by the *adenomatous polyposis coli* (APC) gene, and Glycogen Synthase Kinase-3 (GSK-3). Axin, the limiting component, functions to bring β-catenin together with GSK-3, the kinase that phosphorylates the critical N-terminal residues that direct β-catenin to the ubiquitin-proteasome pathway.[34–43] APC greatly improves Axin's ability to recruit β-catenin[36, 40, 44–46] and likely has a role in facilitating the release of phosphorylated β-catenin product, thus increasing the enzymatic efficiency of the destruction complex.[47–49] Additional proteins play auxiliary roles: For example, Casein Kinase 1α (CK1α) is required to provide priming phosphorylation of β-catenin, which allows its recognition and subsequent phosphorylation by GSK-3.[50–54] Another necessary component of the destruction complex is Protein Phosphatase 2A,[55–65] although its exact function remains unclear.

Binding of a Wnt ligand to receptors of the seven-transmembrane Frizzled family, in cooperation with receptors of the LDL Receptor-Related Proteins (LRP) family, leads to inhibition of the destruction complex. The mechanism of this inhibition is still relatively poorly understood, but it involves a pro-Wnt signaling protein, Dishevelled (Dsh) (see References 66 and 67). During destruction complex inhibition, both Dsh and the destruction complex scaffold Axin are recruited to the membrane (Figure 1.1b).[12–14, 16, 18–22, 68–73] The destruction complex is also inhibited by the *Xenopus* GSK3 Binding Protein (GBP)[74] and its mammalian orthologues in the frequently rearranged in advanced T-cell lymphomas (FRAT) family,[75, 76] which compete with Axin for binding to GSK-3.[77, 78] However, these proteins are not essential for destruction complex inhibition, because mice with targeted deletions of all three FRAT genes are viable, healthy, and fertile, and embryonic stem (ES) cells null for all three genes are able to respond to Wnt signals.[79] Inhibition of the destruction complex prevents phosphorylation and degradation of β-catenin, and thus promotes an increase in β-catenin levels. The stabilized β-catenin accumulates in the nucleus and activates Wnt-responsive target genes (Figure 1.1b).

A host of additional factors regulate canonical Wnt signaling (see References 3 and 5). These factors are both positive and negative modulators of the pathway and work both intra- and extracellularly. For example, a number of secreted factors have been shown to antagonize the Wnt pathway, including Wnt Inhibitory Factor-1 (WIF-1),[80] Secreted Frizzled-Related Proteins (SFRPs),[81–84] the multifunctional antagonist Cerberus,[85] and members of the Dickkopf family of proteins.[86–88] In contrast, heparin proteoglycan sulfate modifications to cell surface proteins appear to play a role in facilitating Wnt signaling[15] (see Reference 89). Intracellularly, the Wnt pathway is also modulated at every level. Numerous kinases modulate Wnt signaling.

Other types of post-translational modifications also play a role in the regulation of Wnt signaling, because members of the Tcf/Lef family have been shown to be SUMOylated.[90, 91] Intracellular proteins such as Inhibitor of Tcf and Catenin (ICAT)[92–95] and Chibby[96] can directly compete with β-catenin-binding proteins for binding to β-catenin. At the level of transcription, co-repressors (such as Groucho/TLE and CtBP)[97, 98] bind to Tcf, silencing Wnt target genes until β-catenin binds to Tcf, bringing along or recruiting co-activators (such as CBP/p300 and Brg-1)[99–101] and activating target gene transcription.

A further level of control over β-catenin is at the level of nuclear localization. The destruction complex proteins APC and Axin have been shown to have additional roles in limiting β-catenin's ability to signal, possibly by facilitating the constitutive export of low levels of β-catenin from the nucleus under non-signaling conditions.[102–106] In the presence of a Wnt signal, β-catenin is actively retained in the nucleus by the cooperative actions of the nuclear anchor protein Pygopus and the β-catenin-binding protein Legless.[107]

II. EXPERIMENTAL APPROACHES

Because of the numerous developmental and adult functions of the Wnt pathway and because of its roles in disease, the detailed molecular workings of the pathway are of great scientific, clinical, and therapeutic interest. One particularly valuable means of identifying the myriad molecular components of the pathway and understanding how they are integrated to turn on and off target gene expression and control cellular division, differentiation, and morphology has been through the investigation of early embryonic Wnt function.

Many of the established techniques presented in this chapter involve use of the frog *Xenopus laevis*, which has a number of advantages that make it an excellent tool for investigating the Wnt pathway. Synchronous eggs that develop rapidly *ex utero* are easy to obtain in large numbers. The large size of the embryos facilitates a variety of experimental manipulations. Importantly, the phenotypic and molecular effects of perturbing the Wnt pathway are well characterized, making it easy to analyze whether uncharacterized genes are involved in Wnt pathway regulation. One can then investigate their mechanism of action by asking how making molecular changes alters their involvement.

Because many of the techniques discussed here involve perturbing Wnt pathway regulation of the dorsal/ventral axis formation, a few preliminary remarks about this process are in order. In the early *Xenopus* embryo, fertilization leads to the formation of an array of "cortical" microtubules in the embryo just below the cell membrane (Figure 1.2) (see Reference 108). During the first cell cycle, the cell membrane and cytoplasm just below it are rotated relative to the inner "core" cytoplasm, so that the vegetal cortical cytoplasm is rotated approximately 30° toward the future dorsal side of the embryo. The cortical microtubules are arranged to provide a set of "tracks" along which factors can be moved toward this future dorsal side (Figure 1.2c, step 3). During normal development, a dorsalizing factor or factors (potentially including Dsh, GBP, kinesin, and *wnt11* RNA)[15, 109, 110] are transported along these tracks to the future dorsal side. These dorsalizing factors lead to activation of the Wnt pathway and stabilization of β-catenin on the future dorsal side of the embryo (Figure 1.2d, step 4). The stabilized β-catenin then activates genes involved in the formation of dorsoanterior structures. Ectopic activation of the Wnt pathway on the ventral side of the embryo through various experimental manipulations causes the formation of a second set of dorsoanterior structures, whereas

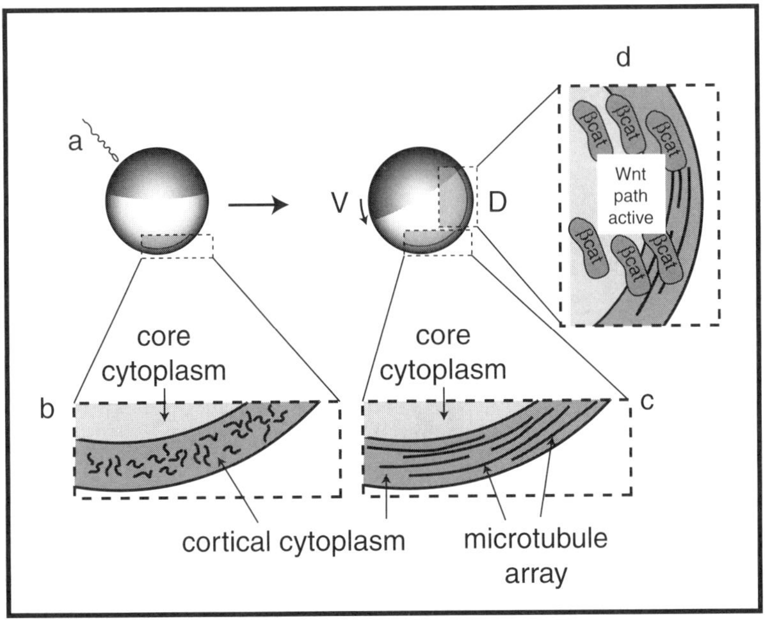

FIGURE 1.2 Cortical rotation leads to axis specification via Wnt pathway activation. (a) Sperm entry causes a rotation of the outer cortical cytoplasm (darker region) relative to the inner core cytoplasm (lighter region). Previously disorganized microtubules in the cortical cytoplasm (b) polymerize and become aligned in arrays (c) that provide a set of tracks along which dorsalizing factors are moved to the future dorsal side (d) of the embryo. The dorsalizing factors lead to inactivation of the destruction complex and stabilization of β-catenin, which activates the transcription of dorsal organizer genes, ultimately leading to the generation of dorsoanterior structures.

inhibition of the endogenous Wnt pathway on the dorsal side causes a loss of dorsoanterior structures.

A. Phenotype Assays

Because the Wnt pathway is active in establishing the dorsal/ventral axis in *Xenopus* embryos, a variety of overexpression assays can be used to address whether a suspected gene has Wnt regulatory effects (Figure 1.3). In these assays, the protein of interest is overexpressed in the developing embryo (typically from injected mRNA), and the phenotypic consequences on axis development are examined. Because other signaling pathways such as the Bone Morphogenetic Protein (BMP) and Nodal pathways contribute to axis specification, it is important to confirm that observed phenotypic effects are in fact due to perturbation of the Wnt pathway through additional molecular approaches (see Section B).

Because activation of the Wnt pathway specifies the future dorsal side of the embryo, overexpression of positive effectors on the opposite (ventral) side of the embryo can lead to ectopic activation of the pathway. Genes of interest are tested by microinjecting mRNA into the two ventral blastomeres of a four-cell embryo

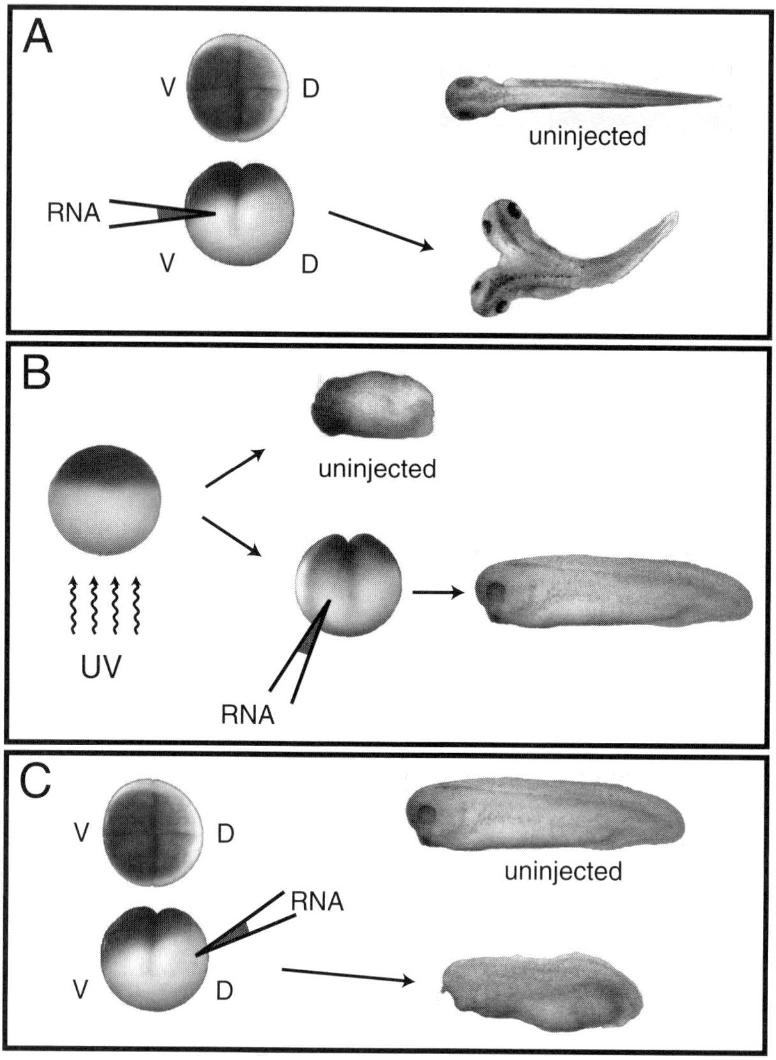

FIGURE 1.3 Phenotypic assays of Wnt pathway regulation in *Xenopus laevis*. (a) Overexpression of Wnt pathway-activating genes on the future ventral side of the embryo leads to axis twinning (bottom embryo). Genes to be tested are microinjected as mRNA into the future ventral side of the embryo at the four-cell stage. The ventral (V) side of four-cell embryos can be recognized because of its characteristically larger and darker blastomeres. (b) Overexpression of Wnt-pathway-activating genes in UV-irradiated embryos rescues axis formation. Embryos are irradiated with UV light early in the first cell cycle. At the two- or four-cell stage, they are microinjected with mRNA encoding the protein of interest. Wnt pathway-activating genes induce phenotypically normal (bottom) embryos; whereas uninjected irradiated embryos or irradiated embryos injected with mRNA that does not activate the Wnt pathway develop as ventralized embryos (top). (c) Overexpression of Wnt-pathway inhibiting genes on the future dorsal (D) side of the embryo leads to ventralization (bottom embryo). Four-cell embryos in (a) and (c) are depicted with animal (top) and lateral (bottom) views to show characteristic pigmentation.

(Figure 1.3a). The ventral blastomeres can be identified because of their characteristically darker pigmentation and larger size. Injection of mRNA encoding a Wnt pathway activator leads to a "twinned" embryo. This method has also been employed as a strategy to identify novel Wnt pathway targets (see, for example, Reference 111).

Another convenient way to test for positive regulators of the Wnt pathway in embryos is by rescuing the dorsal/ventral axis of embryos exposed to ultraviolet (UV) light during the first cell cycle (Figure 1.3b). UV light, at a low dose, inhibits the formation of the cortical array of microtubules necessary for transporting dorsalizing factors to the future dorsal side, thus inhibiting the activation of the Wnt pathway and the eventual specification of dorsoanterior structures (see Reference 108). In the UV rescue assay, embryos are first exposed to low-level UV illumination early in the first cell cycle. Then, at the two- or four-cell stage, two blastomeres are injected equatorially with mRNA encoding the protein to be tested. Proteins that activate the Wnt pathway are able to rescue axis formation, leading to wild-type embryonic development ("axis rescue"), whereas no treatment at all, or injection of mRNAs unrelated to axis formation, yields ventralized embryos. The advantage of this assay is that the degree of rescue can readily be quantified using the dorsoanterior index (DAI) of Kao and Elinson.[112] Quantification of the extent of rescue by scoring DAI allows a finer assessment of the level of axis induction; experimentally, it is much more difficult to assay the extent of a partially induced axis in a twinned embryo.

Proteins can also be tested for their ability to inhibit the Wnt pathway using a similar overexpression assay (Figure 1.3c). Here, mRNA encoding the protein of interest is injected equatorially into two dorsal blastomeres at the four-cell stage. Inhibition of the endogenous Wnt pathway characteristically leads to embryos that lack a head (see, for example, References 92 and 113). This assay is in general less useful than the activating assay, because there are many ways to produce this phenotype that do not involve regulation of the Wnt pathway. Moreover, this assay only works for intracellular regulators of the Wnt pathway, and only for some of these factors. Despite these limitations, it can be a useful assay in some instances.

B. Wnt Reporters and Target Genes

As noted in the previous section, multiple pathways contribute to dorsal/ventral axis specification. Determining whether a specific phenotypic response is a Wnt effect requires examination of molecular targets and transcriptional effects. A number of known genes are immediate early transcriptional targets of canonical Wnt signaling in the early *Xenopus* embryo. In addition, several reporter constructs exist that contain Tcf/Lef response elements. Looking for activation or inhibition of endogenous Wnt transcriptional targets, or activation of reporter constructs, can help strengthen the case that a protein of interest regulates Wnt/β-catenin signaling.

Siamois (*sia*), *xenopus nodal-related-3* (*xnr-3*), and *twin* (*xtwi*) have all been identified as immediate early gene targets of the Wnt pathway, activated by β-catenin in cooperation with Tcf/Lef family members.[32, 114–116] Proteins suspected of activating the Wnt pathway can be tested (Figure 1.4a) by injecting mRNA encoding them into the animal pole (presumptive ectoderm) of one- to two-cell embryos. Prior to the onset of gastrulation at approximately 5 hours post fertilization (hpf; Nieuwkoop and Faber stage 8),[117] the animal cap is removed by manual dissection and cultured until the point when sibling embryos begin gastrulation (10 hpf, stage 10.25).[117] Dissection and culture in isolation prevent inductive signals from reaching the otherwise naive ectoderm and thus allow testing of the inductive capabilities of the overexpressed protein(s) specifically. RNA is isolated from the harvested animal caps and processed by reverse transcriptase PCR (RT-PCR) for expression of the Wnt target genes *sia*, *xnr-3*, or *xtwi* (lists of successfully used primers can be found at http://www.xenbase.org/xmmr/Marker_pages/primers.html and in Reference 15) using either conventional gel electrophoresis or real-time PCR.[15, 118–121]

Another useful means of detecting whether proteins activate the Wnt pathway is by the use of reporter constructs containing Wnt-responsive promoters (Figure 1.4b). One common reporter utilizes a piece of the natural *siamois* promoter upstream of the luciferase gene. Because the *siamois* promoter contains three Tcf/Lef binding elements, it is activated by the Wnt pathway.[32, 122] Another common reporter construct is the TOPFLASH reporter, which contains three Tcf-binding DNA sequences multimerized upstream of a *c-Fos* basal promoter element.[123] In a typical assay (Figure 1.4b), the reporter plasmid is co-injected animally with mRNA encoding the protein to be tested into *Xenopus* embryos at the one- to two-cell stage. Animal caps may be dissected or, alternatively, whole embryos may be lysed and assayed for the expression of luciferase using a luminometer. This assay has also been widely used in cell culture.

The same approaches can be used to examine inhibitors of the Wnt pathway. In these experiments, RNA encoding a canonical Wnt (e.g., Wnt8) is injected along with mRNA encoding the protein of interest. The ectopic Wnt causes activation of the reporter construct unless the co-injected mRNA encodes a protein that inhibits the reporter or otherwise abrogates the signal transduction cascade.

C. Biochemical Studies in Egg/Embryo Lysate

One major challenge to studying the Wnt pathway has been determining the molecular mechanisms of various Wnt events. For example, although many of the protein constituents of the destruction complex have been identified, what their particular roles are within that complex has been difficult to ascertain. It would be helpful to be able to easily add and subtract components and to take kinetic measurements on interactions and downstream effects. A system utilizing *Xenopus* egg and embryo lysates was developed to provide a way of conveniently measuring kinetics and studying biochemical interactions *in vitro* (Figure 1.5).[34, 35] These kinds of measurements are particularly important considering that one aspect of pathway regulation seems to involve fine-tuning the relative levels of pathway constituents.

In vitro analysis has several advantages. First, it provides a way to measure concentrations and half-lives of proteins of interest. Second, it allows for easy addition and subtraction of proteins, a particularly valuable quality because *Xenopus* does not offer easy reverse genetic approaches to loss of function analysis and "knock-in" of altered genes. Finally, mutant or wild-type proteins can be added back in known amounts, and their effects determined. This has an advantage compared

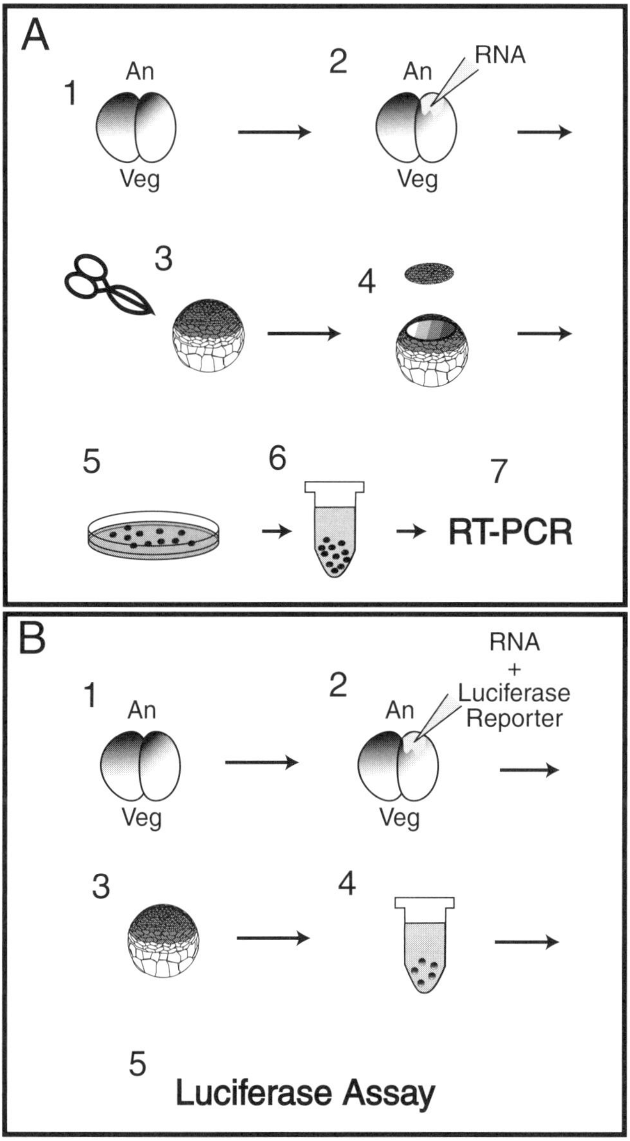

FIGURE 1.4 Wnt target gene and reporter assays. (a) Target gene assay. At the one- to two-cell stage (step 1), embryos are injected animally (An) with mRNA encoding the protein to be tested (step 2). At the late blastula stage (6–7 hpf; step 3), the animal caps are dissected (step 4) and cultured (step 5) until undissected sibling embryos reach the early gastrula stage (~10 hpf), when they are collected (step 6) and analyzed for expression of Wnt target genes by RT-PCR (step 7). (b) Reporter assay. At the one- to two-cell stage (step 1), embryos are co-injected animally with mRNA encoding the protein to be tested and a plasmid reporter containing a Wnt-responsive promoter upstream of the firefly luciferase gene (step 2; see text). The embryos are allowed to develop until the beginning of gastrulation, stage 10.25 (10 hpf; step 3) when they are collected (step 4) and assayed for luciferase activity (step 5).

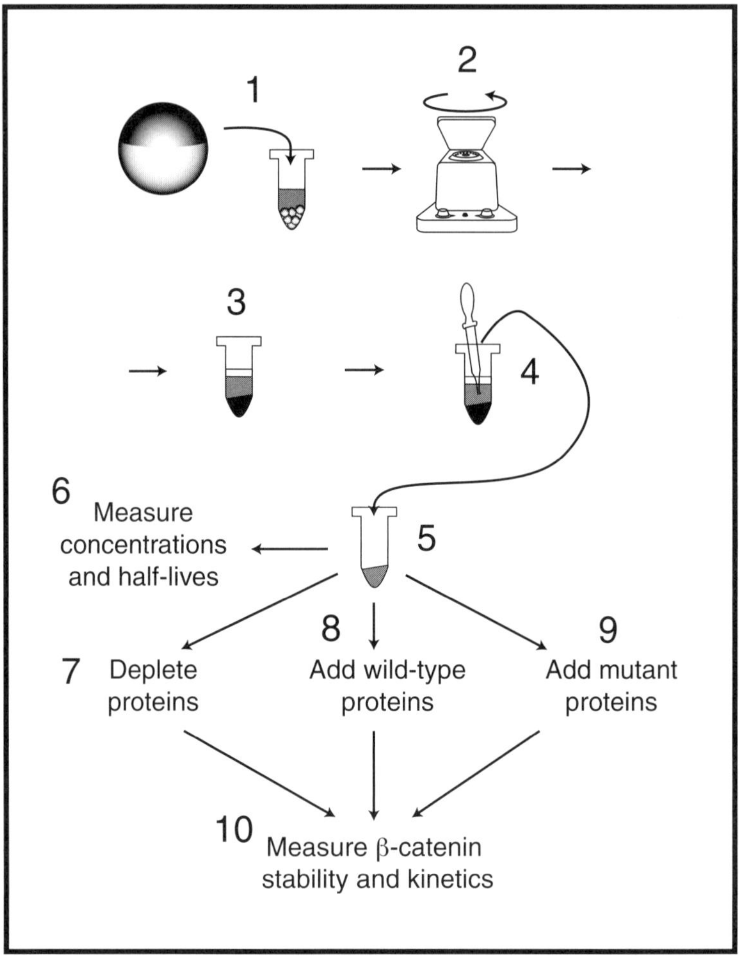

FIGURE 1.5 Biochemical studies in embryo/egg lysate. Eggs or embryos are collected (step 1) in protective buffer and spun (step 2) to separate the insoluble (bottom, black), cytosolic (middle, grey), and lipid (top, white) fractions (step 3). The cytosolic fraction is collected (steps 4–5), and protein concentrations and half-lives can be measured (step 6). The effects of protein depletion can be examined (step 7). Known amounts of wild-type protein can be added to lysate or added back to depleted lysate (step 8). Alternatively, mutant proteins can be added (step 9). Wnt effects are measured using readouts such as β-catenin stability (step 10).

to overexpression of mRNA in embryos because the amount of protein present is constantly changing as new protein is translated from the injected mRNA. Thus, *in vitro* studies can provide confirmation of suspected protein functions and give detailed insight into what components are scarce or abundant, what their relative half-lives are, and how protein interactions are controlled; they can also ultimately give insight into molecular mechanisms.

Lysates are relatively simply prepared (Figure 1.5). Eggs or embryos are pooled and placed in buffer. Centrifugation provides a cytosolic fraction that is removed and used for further studies. Protein concentrations can be measured immediately or over time by Western blotting and compared to known amounts of protein. mRNAs provided to the lysate can be transcribed to allow overexpression of proteins, or recombinant or *in vitro* translated proteins can be added in known amounts. Lysate proteins can be depleted using either antibodies or an affinity resin utilizing a known interacting protein or peptide. Depleted proteins can be replaced by mutant proteins or specific amounts of wild-type proteins to create the functional equivalent of a gene replacement analysis. Ultimately, effects on the Wnt pathway are measured. A common downstream Wnt pathway readout is the stabilization or degradation of β-catenin (see Section D); however, other methods, such as phosphorylation (e.g., of β-catenin, Axin, or APC) or ubiquitination state, could also be used.

There are a number of things to keep in mind when testing the function of proteins in lysate. Clearly, findings *in vitro* need to be confirmed *in vivo* to demonstrate that essential factors or properties are not significantly altered or lost. In addition, lysates are not able to provide information about regional effects. For example, although depletion of the pro-Wnt signaling protein Dsh from embryo lysate did not result in β-catenin stabilization,[34] it is nevertheless likely that Dsh plays a role in stabilizing β-catenin during dorsal/ventral axis specification in the embryo. However, this requirement most likely relies on a small, highly regionalized pool of Dsh confined to the developing dorsal side of the embryo,[110] and therefore depletion of Dsh from a whole embryo lysate does not produce a large enough effect on β-catenin levels to be measurable.

D. Glycogen Synthase Kinase-3 Assays

GSK-3 is involved in multiple cellular pathways in addition to playing a central role in regulation of the Wnt pathway (see References 33 and 124–126). GSK-3 was originally described as an enzyme regulating cellular use and storage of glucose.[127–129] Under conditions when Insulin is absent, GSK-3 constitutively phosphorylates Glycogen Synthase, inhibiting its ability to synthesize glycogen from glucose. In glucose-rich conditions, Insulin signaling activates Protein Kinase B (PKB, also known as Akt), which then phosphorylates and inactivates GSK-3, resulting in the high-level activation of Glycogen Synthase and the conversion of glucose into glycogen.[130, 131]

GSK-3 also regulates the Wnt pathway. As a member of the destruction complex, GSK-3 carries out the phosphorylation of β-catenin's N-terminal serine residues that permit recognition of β-catenin by the ubiquitin ligase complex, targeting β-catenin for degradation by the proteasome. Wnt signaling leads to decreased GSK-3 phosphorylation of β-catenin, though the mechanism of this decrease remains unclear.[33,124] Insulin signaling does not promote β-catenin stabilization,[132] and phosphorylation of β-catenin by GSK-3 can still be inhibited in cells containing a mutant form of GSK-3 that cannot be inhibited by the normal Insulin pathway.[133] Thus, an important unresolved problem is how the Wnt and Insulin signaling pathways can differentially regulate GSK-3.

It is not yet clear that the principal, or indeed even an essential, component of Wnt pathway activation is inhibition of GSK-3's overall kinase activity.[19, 33–35, 134]

GSK-3 depends on Axin and APC to efficiently phosphorylate β-catenin.[40–44, 46–49,135,136] One possibility is that Wnt signaling simply leads to a rearrangement or dissociation of the destruction complex that inhibits the catalytic efficiency of β-catenin phosphorylation or limits access to the substrate, without impairing the specific kinase activity of GSK-3 *per se*. For example, the GSK-3 Binding Protein is capable of promoting Wnt signaling but does not generally inhibit GSK-3 kinase activity.[74, 77] Nevertheless, several groups have observed a moderate decrease in GSK-3 specific activity in response to Wnt signaling.[132, 137–139] Because GSK-3 activity can also be regulated independently of Wnt signaling, it is important to determine whether observed changes in GSK-3 activity for a given set of cellular conditions is or is not a Wnt response by examining effects on β-catenin stabilization (see below) or looking at the activation or repression of Wnt-directed transcription (see Section B).

GSK-3 kinase activity can be tested *in vitro* (Figure 1.6a) using GSK-3 immunoprecipitated from embryos (or cultured cells) that have been subjected to a desired treatment (for example, overexpressed activators or inhibitors of the Wnt pathway). The embryos are lysed in a protective buffer, the cytosolic fraction retained, and then either endogenous or overexpressed tagged GSK-3 can be immunoprecipitated and incubated with any one of a variety of peptide substrates or even full-length proteins in the presence of [γ^{32}P]ATP. Peptide substrates derived from Glycogen Synthase,[140] eIF2b,[132] Myelin Basic Protein (MBP),[137] and cAMP Response Element Binding Protein (CREB)[77, 141] have all been used to assay immunoprecipitated GSK-3 kinase activity *in vitro*. A study has been done to investigate optimal peptide substrates.[142] Active GSK-3 incorporates ^{32}P, which can then be measured by autoradiography or liquid scintillation spectrometry (see, for example, References 139 and 141).

One means of assaying the Wnt-specific activity of GSK-3 is to assay the phosphorylation state of the particular N-terminal β-catenin serine/threonines targeted by GSK-3: Ser33, Ser37, and Thr41 (Figure 1.6b). A commercially available antibody directed against all three phospho-residues (Cell Signaling Technology, Beverly, MA) has been generated and used against cell culture-derived β-catenin,[51,52] and two other antibodies recognizing either Ser33/37 or Thr41/Ser45 phospho-β-catenin have been used against β-catenin from *Xenopus* lysate.[53]

More commonly, however, a convenient means of indirectly assaying GSK-3's Wnt-related activity is simply by examining the steady-state levels of β-catenin (Figure 1.6c), because active GSK-3 promotes β-catenin degradation. To assay β-catenin levels, it is possible to look at epitope-tagged, overexpressed β-catenin or endogenous β-catenin, but both strategies require care. If using overexpressed β-catenin, it is important to overexpress small amounts (< 100 pg mRNA in *Xenopus* embryos) to avoid overwhelming the destruction complex and allow observation of regulated degradation (see, for example, Reference 74). If examining endogenous β-catenin levels, it is important to isolate the cytosolic pool, because a large amount of β-catenin is found at the membrane in cell adhesion protein complexes and is therefore not subject to destruction complex-promoted degradation.

There are two strategies that have been used to isolate the cytosolic β-catenin: (1) Removal of the membrane-bound proteins (containing non-signaling

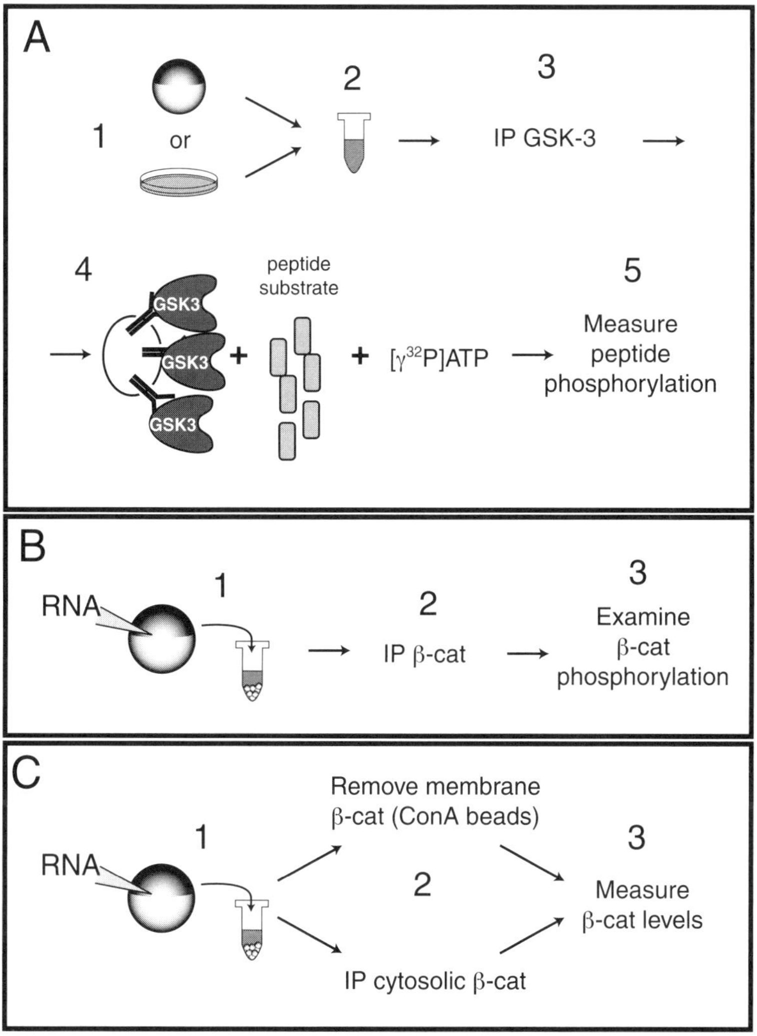

FIGURE 1.6 GSK-3 assays. (a) Measuring GSK-3 kinase activity. Cultured cells or embryos are subjected to the experimental treatment of choice (step 1), collected, and lysed, and the protein-containing fraction is collected (step 2). GSK-3 is immunoprecipitated (step 3), and the isolated GSK-3 is tested against a peptide or full-length protein substrate (step 4) to determine if phosphorylation is occurring (step 5; see text). (b) Testing whether GSK-3 is phosphorylating β-catenin. Embryos are injected at the one- to two-cell stage with mRNA encoding experimental proteins or epitope-tagged β-catenin (step 1). After 4 hours, ectopic (tagged) or endogenous β-catenin is immunoprecipitated (step 2) and its phosphorylation state is examined by Western blotting using antibodies specific for phosphorylated β-catenin (step 3; see text). (c) Testing GSK-3 kinase activity indirectly by examination of β-catenin stability. Injected embryos (step 1) are lysed and either the membrane-bound β-catenin is removed by incubating the cytosolic fraction with ConA beads, or the cytosolic fraction is immunoprecipitated directly by incubating with GST-cadherin or GST-Tcf (step 2). β-catenin stabilization is examined by Western blotting (step 3).

cadherin/β-catenin complexes) using Concanavalin A beads (ConA sepharose, Sigma-Aldrich, St. Louis, MO) (see, for example, Reference 143); the supernatant is retained because it contains the cytoplasmic proteins. (2) As an alternative, the cytosolic (signaling) pool of β-catenin can be affinity-purified directly using GST-Tcf or GST-cadherin (see, for example, References 95 and 144); the membrane pool does not bind to the column because it is already bound to endogenous cadherin. Levels of cytosolic β-catenin are then examined by Western blotting, with decreased β-catenin taken as evidence of an inhibition of GSK-3 activity.

It is not clear whether GSK-3-mediated phosphorylation causes other Wnt pathway proteins to be targeted for degradation. However, other Wnt pathway proteins are phosphorylation targets of GSK-3, including Axin[41,42,145,146] and APC.[40,48,49,59,135,147] The Wnt-specific regulation and consequences of phosphorylation by GSK-3 must be examined and evaluated for an individual protein.

To determine whether GSK-3 is capable of phosphorylating a particular protein (Figure 1.7), the first step is to examine the primary sequence of the protein to determine if it contains a GSK-3 phosphorylation consensus sequence: <u>S/T</u>XXXS/T(P), where S/T denotes a residue that could be either serine or threonine, an underline shows the GSK-3-phosphorylated residue, and (P) follows the site of priming phosphorylation by an ancillary kinase.[148, 149] Second, one would want to determine if the protein is in fact phosphorylated at all. An easy assay suggesting phosphorylation examines whether endogenous or overexpressed epitope-tagged protein exhibits multiple bands on a Western blot or increased mobility after phosphatase treatment (Figure 1.7b). More elaborate examinations are possible; for example, mass spectrophotometric phospho-peptide mapping.[150] Third, it should be determined whether GSK-3 is capable of phosphorylating the protein. Epitope-tagged overexpressed protein can be immunoprecipitated from embryos or cell culture and tested as a substrate with recombinant or immunoprecipitated GSK-3 (Figure 1.7c).

Note that it is important to test immunoprecipitated substrate protein and not recombinant protein, because GSK-3 requires that its substrates contain a priming phosphorylation and the priming kinase may not be known. If there is reason to think that the protein of interest is phosphorylated by GSK-3, the next step is to ask whether pharmacologic inhibition of GSK-3 relieves phosphorylation (Figure 1.7d). Lithium is a convenient and inexpensive inhibitor of GSK-3,[151, 152] but it has also been shown to inhibit adenylate cyclase and inositol monophosphatase, and activate JNK.[153–155] Thus, initial use of lithium as an inhibitor might be followed by any number of other more specific pharmacologic inhibitors such as SB 216763, SB 415286[156] (available from Sigma-Aldrich), or one of the GSK-3 inhibitors available from Calbiochem (La Jolla, CA).

As an alternative to pharmacological inhibition, some activities of GSK-3 can be blocked by the *Xenopus* GSK-3 binding protein or its mammalian homologues, the FRATs, if the phosphorylation of the substrate involves GSK-3 binding.[74–78] Because these proteins do not block all activities of GSK-3,[77] continued phosphorylation may not indicate that the protein/peptide in question is not a *bona fide* GSK-3 substrate. Determining whether *in vivo* phosphorylation has been blocked could be examined preliminarily by looking for increased electrophoretic mobility, but more rigorous

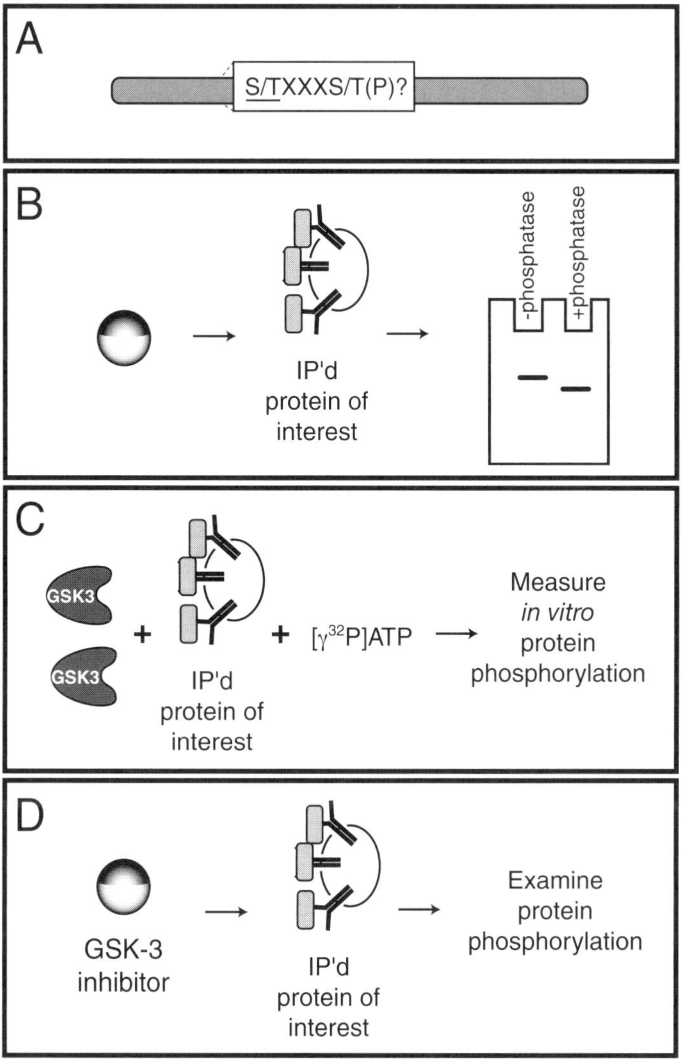

FIGURE 1.7 A strategy for determining if a protein is a GSK-3 kinase substrate. (a) Examine the primary sequence for GSK-3 phosphorylation consensus sequences: S/TXXXS/T(P), where S/T denotes a serine or threonine, X denotes any amino acid, and (P) indicates the site of priming phosphorylation. The underlined sequence represents the position normally phosphorylated by GSK-3. (b) Determine if the protein of interest is phosphorylated. Immunoprecipitated protein is examined for evidence of phosphorylation, for example, by looking for increased mobility after phosphatase treatment. (c) Determine if the protein of interest can be phosphorylated by GSK-3 *in vitro*. Recombinant or immunoprecipitated GSK-3 is incubated with immunoprecipitated protein and the protein is examined for phosphorylation. (d) Determine whether pharmacological inhibition of GSK-3 prevents phosphorylation of the protein. Embryos are incubated in a GSK-3 inhibitor, such as lithium or more specific pharmacological inhibitors. The protein is immunoprecipitated; its phosphorylation state is examined by metabolic labeling, gel shift, or phospho-specific antibodies and compared to proteins isolated from untreated embryos.

proof involves either *in vivo* labeling with [^{32}P]orthophosphate or a phospho-specific antibody to the putative substrate. The latter approach is preferable because it is much cleaner. The above steps should provide good insight into whether a protein of interest is in fact a substrate for GSK-3 kinase activity. Additional experiments might involve site-directed mutagenesis of putative target serines/threonines and co-immunoprecipitations of GSK-3 with the protein of interest (because GSK-3 often interacts stably with substrate proteins) (see, for example, References 74 and 157).

III. CONCLUSIONS

Xenopus embryos offer a powerful tool for investigation of the canonical Wnt pathway. A wealth of assays are available to find new genes involved, to examine known proteins for unknown roles, and to gain detailed insight into the molecular mechanisms that underlie both the on and off states of Wnt signal transduction. It is important to use an integrated approach to evaluating the role of a particular protein in the Wnt pathway. Phenotypic assays should be supported by evidence of Wnt transcriptional effects and regulation of β-catenin stability. One aspect of Wnt signal transduction involves regulation of the ability of GSK-3 to phosphorylate β-catenin and target it for degradation, but because many other pathways also regulate GSK-3 activity, it is important to determine whether kinase effects are Wnt effects.

ACKNOWLEDGMENTS

The authors wish to thank Dr. Carole Weaver for useful discussions in the preparation of the manuscript, Ujwal Pyati for critical comments, and Dr. Ethan Lee for providing his protocol for producing *Xenopus* embryo lysate. This work was supported by NIH grant HD27262 to D.K.

REFERENCES

1. Wang, J. and Wynshaw-Boris, A., The canonical Wnt pathway in early mammalian embryogenesis and stem cell maintenance/differentiation, *Curr Opin Genet Dev* 14 (5), 533–9, 2004.
2. Karim, R., Tse, G., Putti, T., Scolyer, R., and Lee, S., The significance of the Wnt pathway in the pathology of human cancers, *Pathology* 36 (2), 120–8, 2004.
3. Logan, C. Y. and Nusse, R., The Wnt signaling pathway in development and disease, *Annu Rev Cell Dev Biol* 20, 781–810, 2004.
4. van Es, J. H., Barker, N., and Clevers, H., You Wnt some, you lose some: oncogenes in the Wnt signaling pathway, *Curr Opin Genet Dev* 13 (1), 28–33, 2003.
5. Seidensticker, M. J. and Behrens, J., Biochemical interactions in the Wnt pathway, *Biochim Biophys Acta* 1495 (2), 168–82, 2000.
6. He, X. C., Zhang, J., Tong, W. G., Tawfik, O., Ross, J., Scoville, D. H., Tian, Q., Zeng, X., He, X., Wiedemann, L. M., Mishina, Y., and Li, L., BMP signaling inhibits intestinal stem cell self-renewal through suppression of Wnt-β-catenin signaling, *Nat Genet* 36 (10), 1117–21, 2004.

7. Liu, B. Y., McDermott, S. P., Khwaja, S. S., and Alexander, C. M., The transforming activity of Wnt effectors correlates with their ability to induce the accumulation of mammary progenitor cells, *Proc Natl Acad Sci U S A* 101 (12), 4158–63, 2004.
8. Rattis, F. M., Voermans, C., and Reya, T., Wnt signaling in the stem cell niche, *Curr Opin Hematol* 11 (2), 88–94, 2004.
9. Kielman, M. F., Rindapaa, M., Gaspar, C., van Poppel, N., Breukel, C., van Leeuwen, S., Taketo, M. M., Roberts, S., Smits, R., and Fodde, R., Apc modulates embryonic stem-cell differentiation by controlling the dosage of β-catenin signaling, *Nat Genet* 32 (4), 594–605, 2002.
10. Polakis, P., Wnt signaling and cancer, *Genes Dev* 14 (15), 1837–51, 2000.
11. Veeman, M. T., Axelrod, J. D., and Moon, R. T., A second canon. Functions and mechanisms of β-catenin-independent Wnt signaling, *Dev Cell* 5 (3), 367–77, 2003.
12. Cong, F., Schweizer, L., and Varmus, H., Wnt signals across the plasma membrane to activate the beta-catenin pathway by forming oligomers containing its receptors, Frizzled and LRP, *Development* 131 (20), 5103–15, 2004.
13. Schweizer, L. and Varmus, H., Wnt/Wingless signaling through β-catenin requires the function of both LRP/Arrow and Frizzled classes of receptors, *BMC Cell Biol* 4 (1), 4, 2003.
14. Mao, J., Wang, J., Liu, B., Pan, W., Farr, G. H., 3rd, Flynn, C., Yuan, H., Takada, S., Kimelman, D., Li, L., and Wu, D., Low-density lipoprotein receptor-related protein-5 binds to Axin and regulates the canonical Wnt signaling pathway, *Mol Cell* 7 (4), 801–9, 2001.
15. Tao, Q., Yokota, C., Puck, H., Kofron, M., Birsoy, B., Yan, D., Asashima, M., Wylie, C. C., Lin, X., and Heasman, J., Maternal Wnt11 activates the canonical Wnt signaling pathway required for axis formation in *Xenopus* embryos, *Cell* 120 (6), 857–71, 2005.
16. Liu, G., Bafico, A., Harris, V. K., and Aaronson, S. A., A novel mechanism for Wnt activation of canonical signaling through the LRP6 receptor, *Mol Cell Biol* 23 (16), 5825–35, 2003.
17. Bafico, A., Liu, G., Yaniv, A., Gazit, A., and Aaronson, S. A., Novel mechanism of Wnt signalling inhibition mediated by Dickkopf-1 interaction with LRP6/Arrow, *Nat Cell Biol* 3 (7), 683–6, 2001.
18. Liu, G., Bafico, A., and Aaronson, S. A., The mechanism of endogenous receptor activation functionally distinguishes prototype canonical and noncanonical Wnts, *Mol Cell Biol* 25 (9), 3475–82, 2005.
19. Tolwinski, N. S., Wehrli, M., Rives, A., Erdeniz, N., DiNardo, S., and Wieschaus, E., Wg/Wnt signal can be transmitted through arrow/LRP5,6 and Axin independently of Zw3/Gsk3β activity, *Dev Cell* 4 (3), 407–18, 2003.
20. Holmen, S. L., Salic, A., Zylstra, C. R., Kirschner, M. W., and Williams, B. O., A novel set of Wnt-Frizzled fusion proteins identifies receptor components that activate β-catenin-dependent signaling, *J Biol Chem* 277 (38), 34727–35, 2002.
21. Tamai, K., Semenov, M., Kato, Y., Spokony, R., Liu, C., Katsuyama, Y., Hess, F., Saint-Jeannet, J. P., and He, X., LDL-receptor-related proteins in Wnt signal transduction, *Nature* 407 (6803), 530–5, 2000.
22. Wehrli, M., Dougan, S. T., Caldwell, K., O'Keefe, L., Schwartz, S., Vaizel-Ohayon, D., Schejter, E., Tomlinson, A., and DiNardo, S., *Arrow* encodes an LDL-receptor-related protein essential for Wingless signalling, *Nature* 407 (6803), 527–30, 2000.
23. Park, M. and Moon, R. T., The planar cell-polarity *gene* stbm regulates cell behaviour and cell fate in vertebrate embryos, *Nat Cell Biol* 4 (1), 20–5, 2002.

24. Darken, R. S., Scola, A. M., Rakeman, A. S., Das, G., Mlodzik, M., and Wilson, P. A., The planar polarity gene *strabismus* regulates convergent extension movements in Xenopus, *Embo J* 21 (5), 976–85, 2002.
25. Goto, T. and Keller, R., The planar cell polarity gene *strabismus* regulates convergence and extension and neural fold closure in *Xenopus*, *Dev Biol* 247 (1), 165–81, 2002.
26. Gumbiner, B. M. and McCrea, P. D., Catenins as mediators of the cytoplasmic functions of cadherins, *J Cell Sci Suppl* 17, 155–58, 1993.
27. Gumbiner, B. M., Regulation of cadherin adhesive activity, *J Cell Biol* 148 (3), 399–404, 2000.
28. Gumbiner, B. M., Proteins associated with the cytoplasmic surface of adhesion molecules, *Neuron* 11 (4), 551–64, 1993.
29. van de Wetering, M., Cavallo, R., Dooijes, D., van Beest, M., van Es, J., Loureiro, J., Ypma, A., Hursh, D., Jones, T., Bejsovec, A., Peifer, M., Mortin, M., and Clevers, H., Armadillo co-activates transcription driven by the product of the *Drosophila* segment polarity gene *dTCF*, *Cell* 88, 789–99, 1997.
30. Behrens, J., von Kries, J. P., Kühl, M., Bruhn, L., Wedlich, D., Grosschedl, R., and Birchmeier, W., Functional interaction of β-catenin with the transcription factor LEF-1, *Nature* 382, 638–42, 1996.
31. Molenaar, M., van de Wetering, M., Oosterwegel, M., Peterson-Maduro, J., Godsave, S., Korinek, V., Roose, J., Destrée, O., and Clevers, H., XTcf-3 transcription factor mediates β-catenin-induced axis formation in *Xenopus* embryos, *Cell* 86, 391–99, 1996.
32. Brannon, M., Gomperts, M., Sumoy, L., Moon, R. T., and Kimelman, D., A β-catenin/XTcf-3 complex binds to the *siamois* promoter to regulate specification of the dorsal axis in *Xenopus*, *Genes Dev* 11, 2359–70, 1997.
33. Dominguez, I. and Green, J. B., Missing links in GSK3 regulation, *Dev Biol* 235 (2), 303–13, 2001.
34. Salic, A., Lee, E., Mayer, L., and Kirschner, M. W., Control of β-catenin stability: reconstitution of the cytoplasmic steps of the Wnt pathway in *Xenopus* egg extracts, *Mol Cell* 5, 523–32, 2000.
35. Lee, E., Salic, A., Kruger, R., Heinrich, R., and Kirschner, M. W., The roles of APC and Axin derived from experimental and theoretical analysis of the Wnt pathway, *PLoS Biol* 1 (1), E10, 2003.
36. Munemitsu, S., Albert, I., Souza, B., Rubinfeld, B., and Polakis, P., Regulation of intracellular β-catenin levels by adenomatous polyposis coli (APC) tumor-suppressor protein, *Proc Natl Acad Sci USA* 92, 3046–50, 1995.
37. Yost, C., Torres, M., Miller, J. R., Huang, E., Kimelman, D., and Moon, R. T., The axis-inducing activity, stability, and subcellular distribution of β-catenin is regulated in *Xenopus* embryos by glycogen synthase kinase 3, *Genes Dev* 10, 1443–54, 1996.
38. Aberle, H., Bauer, A., Stappert, J., Kispert, A., and Kemler, R., β-catenin is a target for the ubiquitin-proteasome pathway, *Embo J* 16 (13), 3797–804, 1997.
39. Orford, K., Crockett, C., Jensen, J. P., Weissman, A. M., and Byers, S. W., Serine phosphorylation-regulated ubiquitination and degradation of β-catenin, *J Biol Chem* 272 (40), 24735–8, 1997.
40. Hart, M. J., de los Santos, R., Albert, I. N., Rubinfeld, B., and Polakis, P., Downregulation of β-catenin by human Axin and its association with the APC tumor suppressor, β-catenin and GSK-3β, *Curr. Biol.* 8 (10), 573–81, 1998.
41. Ikeda, S., Kishida, S., Yamamoto, H., Murai, H., Koyama, S., and Kikuchi, A., Axin, a negative regulator of the Wnt signaling pathway, forms a complex with GSK-3β and β-catenin and promotes GSK-3β-dependent phosphorylation of β-catenin, *Embo J* 17, 1371–84, 1998.

42. Yamamoto, H., Kishida, S., Uochi, T., Ikeda, S., Koyama, S., Asashima, M., and Kikuchi, A., Axil, a member of the Axin family, interacts with both glycogen synthase kinase 3β and β-catenin and inhibits axis formation of *Xenopus* embryos, *Mol Cell Biol* 18 (5), 2867–75, 1998.
43. Behrens, J., Jerchow, B. A., Wurtele, M., Grimm, J., Asbrand, C., Wirtz, R., Kuhl, M., Wedlich, D., and Birchmeier, W., Functional interaction of an Axin homolog, Conductin, with β-catenin, APC, and GSK-3β, *Science* 280 (5363), 596–9, 1998.
44. Kawahara, K., Morishita, T., Nakamura, T., Hamada, F., Toyoshima, K., and Akiyama, T., Down-regulation of β-catenin by the colorectal tumor suppressor APC requires association with Axin and β-catenin, *J Biol Chem* 275 (12), 8369–74, 2000.
45. Kishida, S., Yamamoto, H., Ikeda, S., Kishida, M., Sakamoto, I., Koyama, S., and Kikuchi, A., Axin, a negative regulator of the Wnt signaling pathway, directly interacts with adenomatous polyposis coli and regulates the stabilization of β-catenin, *J Biol Chem* 273 (18), 10823–6, 1998.
46. Nakamura, T., Hamada, F., Ishidate, T., Anai, K., Kawahara, K., Toyoshima, K., and Akiyama, T., Axin, an inhibitor of the Wnt signalling pathway, interacts with β-catenin, GSK-3β and APC and reduces the β-catenin level, *Genes Cells* 3 (6), 395–403, 1998.
47. Xing, Y., Clements, W. K., Kimelman, D., and Xu, W., Crystal structure of a β-catenin/Axin complex suggests a mechanism for the β-catenin destruction complex, *Genes Dev* 17 (22), 2753–64, 2003.
48. Xing, Y., Clements, W. K., Le Trong, I., Hinds, T. R., Stenkamp, R., Kimelman, D., and Xu, W., Crystal structure of a β-catenin/APC complex reveals a critical role for APC phosphorylation in APC function, *Mol Cell* 15 (4), 523–33, 2004.
49. Ha, N. C., Tonozuka, T., Stamos, J. L., Choi, H. J., and Weis, W. I., Mechanism of phosphorylation-dependent binding of APC to β-catenin and its role in β-catenin degradation, *Mol Cell* 15 (4), 511–21, 2004.
50. Hagen, T., Di Daniel, E., Culbert, A. A., and Reith, A. D., Expression and characterization of GSK-3 mutants and their effect on β-catenin phosphorylation in intact cells, *J Biol Chem* 277 (26), 23330–5, 2002.
51. Hagen, T. and Vidal-Puig, A., Characterisation of the phosphorylation of β-catenin at the GSK-3 priming site Ser45, *Biochem Biophys Res Commun* 294 (2), 324–8, 2002.
52. Amit, S., Hatzubai, A., Birman, Y., Andersen, J. S., Ben-Shushan, E., Mann, M., Ben-Neriah, Y., and Alkalay, I., Axin-mediated CKI phosphorylation of β-catenin at Ser45: a molecular switch for the Wnt pathway, *Genes Dev* 16 (9), 1066–76, 2002.
53. Liu, C., Li, Y., Semenov, M., Han, C., Baeg, G. H., Tan, Y., Zhang, Z., Lin, X., and He, X., Control of β-catenin phosphorylation/degradation by a dual-kinase mechanism, *Cell* 108 (6), 837–47, 2002.
54. Yanagawa, S., Matsuda, Y., Lee, J. S., Matsubayashi, H., Sese, S., Kadowaki, T., and Ishimoto, A., Casein kinase I phosphorylates the Armadillo protein and induces its degradation in *Drosophila*, *Embo J* 21 (7), 1733–42, 2002.
55. Hsu, W., Zeng, L., and Costantini, F., Identification of a domain of Axin that binds to the serine/threonine protein phosphatase 2A and a self-binding domain, *J Biol Chem* 274 (6), 3439–45, 1999.
56. Fagotto, F., Jho, E., Zeng, L., Kurth, T., Joos, T., Kaufmann, C., and Costantini, F., Domains of Axin involved in protein-protein interactions, Wnt pathway inhibition, and intracellular localization, *J Cell Biol* 145 (4), 741–56, 1999.
57. Willert, K., Logan, C. Y., Arora, A., Fish, M., and Nusse, R., A *Drosophila* Axin homolog, Daxin, inhibits Wnt signaling, *Development* 126 (18), 4165–73, 1999.

58. Seeling, J. M., Miller, J. R., Gil, R., Moon, R. T., White, R., and Virshup, D. M., Regulation of β-catenin signaling by the B56 subunit of protein phosphatase 2A, *Science* 283 (5410), 2089–91, 1999.
59. Ikeda, S., Kishida, M., Matsuura, Y., Usui, H., and Kikuchi, A., GSK-3β-dependent phosphorylation of adenomatous polyposis coli gene product can be modulated by β-catenin and protein phosphatase 2A complexed with Axin, *Oncogene* 19 (4), 537–45, 2000.
60. Ratcliffe, M. J., Itoh, K., and Sokol, S. Y., A positive role for the PP2A catalytic subunit in Wnt signal transduction, *J Biol Chem* 275 (46), 35680–3, 2000.
61. Strovel, E. T., Wu, D., and Sussman, D. J., Protein phosphatase 2Cα dephosphorylates Axin and activates LEF-1-dependent transcription, *J Biol Chem* 275 (4), 2399–403, 2000.
62. Li, X., Yost, H. J., Virshup, D. M., and Seeling, J. M., Protein phosphatase 2A and its B56 regulatory subunit inhibit Wnt signaling in *Xenopus*, *Embo J* 20 (15), 4122–31, 2001.
63. Yamamoto, H., Hinoi, T., Michiue, T., Fukui, A., Usui, H., Janssens, V., Van Hoof, C., Goris, J., Asashima, M., and Kikuchi, A., Inhibition of the Wnt signaling pathway by the PR61 subunit of protein phosphatase 2A, *J Biol Chem* 276 (29), 26875–82, 2001.
64. Yang, J., Wu, J., Tan, C., and Klein, P. S., PP2A:B56ε is required for Wnt/beta-catenin signaling during embryonic development, *Development* 130 (23), 5569–78, 2003.
65. Bajpai, R., Makhijani, K., Rao, P. R., and Shashidhara, L. S., *Drosophila* Twins regulates Armadillo levels in response to Wg/Wnt signal, *Development* 131 (5), 1007–16, 2004.
66. Boutros, M. and Mlodzik, M., Dishevelled: at the crossroads of divergent intracellular signaling pathways, *Mech Dev* 83 (1–2), 27–37, 1999.
67. Wharton, K. A., Jr., Runnin' with the Dvl: proteins that associate with Dsh/Dvl and their significance to Wnt signal transduction, *Dev Biol* 253 (1), 1–17, 2003.
68. Cliffe, A., Hamada, F., and Bienz, M., A role of Dishevelled in relocating Axin to the plasma membrane during Wingless signaling, *Curr Biol* 13 (11), 960–6, 2003.
69. Hay, E., Faucheu, C., Suc-Royer, I., Touitou, R., Stiot, V., Vayssiere, B., Baron, R., Roman-Roman, S., and Rawadi, G., Interaction between LRP5 and Frat1 mediates the activation of the Wnt canonical pathway, *J Biol Chem* 280 (14), 13616–23, 2005.
70. Wong, H. C., Bourdelas, A., Krauss, A., Lee, H. J., Shao, Y., Wu, D., Mlodzik, M., Shi, D. L., and Zheng, J., Direct binding of the PDZ domain of Dishevelled to a conserved internal sequence in the C-terminal region of Frizzled, *Mol Cell* 12 (5), 1251–60, 2003.
71. Umbhauer, M., Djiane, A., Goisset, C., Penzo-Mendez, A., Riou, J. F., Boucaut, J. C., and Shi, D. L., The C-terminal cytoplasmic Lys-thr-X-X-X-Trp motif in Frizzled receptors mediates Wnt/β-catenin signalling, *Embo J* 19 (18), 4944–54, 2000.
72. Medina, A. and Steinbeisser, H., Interaction of Frizzled 7 and Dishevelled in *Xenopus*, *Dev Dyn* 218 (4), 671–80, 2000.
73. Yang-Snyder, J., Miller, J. R., Brown, J. D., Lai, C., and Moon, R. T., A *frizzled* homolog functions in a vertebrate *Wnt* signaling pathway, *Curr Biol* 6, 302–6, 1996.
74. Yost, C., Farr, G. H., III, Pierce, S. B., Ferkey, D. M., Chen, M. M., and Kimelman, D., GBP, an inhibitor of GSK-3, is implicated in *Xenopus* development and oncogenesis, *Cell* 93, 1031–41, 1998.

75. Jonkers, J., van Amerongen, R., van der Valk, M., Robanus-Maandag, E., Molenaar, M., Destree, O., and Berns, A., *In vivo* analysis of Frat1 deficiency suggests compensatory activity of Frat3, *Mech Dev* 88 (2), 183–94, 1999.
76. Jonkers, J., Korswagen, H. C., Acton, D., Breuer, M., and Berns, A., Activation of a novel proto-oncogene, *Frat1*, contributes to progression of mouse T-cell lymphomas, *Embo J* 16 (3), 441–50, 1997.
77. Farr, G. H., III, Ferkey, D. M., Yost, C., Pierce, S. B., Weaver, C., and Kimelman, D., Interaction among GSK-3, GBP, Axin, and APC in *Xenopus* axis specification, *J Cell Biol* 148 (4), 691–702, 2000.
78. Ferkey, D. M. and Kimelman, D., Glycogen synthase kinase-3 β mutagenesis identifies a common binding domain for GBP and Axin, *J Biol Chem* 277 (18), 16147–52, 2002.
79. van Amerongen, R., Nawijn, M., Franca-Koh, J., Zevenhoven, J., van der Gulden, H., Jonkers, J., and Berns, A., Frat is dispensable for canonical Wnt signaling in mammals, *Genes Dev* 19 (4), 425–30, 2005.
80. Hsieh, J. C., Kodjabachian, L., Rebbert, M. L., Rattner, A., Smallwood, P. M., Samos, C. H., Nusse, R., Dawid, I. B., and Nathans, J., A new secreted protein that binds to Wnt proteins and inhibits their activities, *Nature* 398 (6726), 431–6, 1999.
81. Shibata, M., Ono, H., Hikasa, H., Shinga, J., and Taira, M., *Xenopus crescent* encoding a Frizzled-like domain is expressed in the Spemann organizer and pronephros, *Mech Dev* 96 (2), 243–6, 2000.
82. Pera, E. M. and De Robertis, E. M., A direct screen for secreted proteins in *Xenopus* embryos identifies distinct activities for the Wnt antagonists Crescent and Frzb-1, *Mech Dev* 96 (2), 183–95, 2000.
83. Wang, S., Krinks, M., Lin, K., Luyten, F. P., and Moos, M., Jr., Frzb, a secreted protein expressed in the Spemann organizer, binds and inhibits Wnt-8, *Cell* 88 (6), 757–66, 1997.
84. Leyns, L., Bouwmeester, T., Kim, S. H., Piccolo, S., and De Robertis, E. M., Frzb-1 is a secreted antagonist of Wnt signaling expressed in the Spemann organizer, *Cell* 88 (6), 747–56, 1997.
85. Piccolo, S., Agius, E., Leyns, L., Bhattacharyya, S., Grunz, H., Bouwmeester, T., and De Robertis, E. M., The head inducer Cerberus is a multifunctional antagonist of Nodal, BMP and Wnt signals, *Nature* 397 (6721), 707–10, 1999.
86. Krupnik, V. E., Sharp, J. D., Jiang, C., Robison, K., Chickering, T. W., Amaravadi, L., Brown, D. E., Guyot, D., Mays, G., Leiby, K., Chang, B., Duong, T., Goodearl, A. D., Gearing, D. P., Sokol, S. Y., and McCarthy, S. A., Functional and structural diversity of the human Dickkopf gene family, *Gene* 238 (2), 301–13, 1999.
87. Glinka, A., Wu, W., Onichtchouk, D., Blumenstock, C., and Niehrs, C., Head induction by simultaneous repression of Bmp and Wnt signalling in *Xenopus*, *Nature* 389 (6650), 517–9, 1997.
88. Monaghan, A. P., Kioschis, P., Wu, W., Zuniga, A., Bock, D., Poustka, A., Delius, H., and Niehrs, C., Dickkopf genes are co-ordinately expressed in mesodermal lineages, *Mech Dev* 87 (1–2), 45–56, 1999.
89. Lin, X., Functions of heparan sulfate proteoglycans in cell signaling during development, *Development* 131 (24), 6009–21, 2004.
90. Yamamoto, H., Ihara, M., Matsuura, Y., and Kikuchi, A., Sumoylation is involved in β-catenin-dependent activation of Tcf-4, *Embo J* 22 (9), 2047–59, 2003.
91. Sachdev, S., Bruhn, L., Sieber, H., Pichler, A., Melchior, F., and Grosschedl, R., PIASy, a nuclear matrix-associated SUMO E3 ligase, represses LEF1 activity by sequestration into nuclear bodies, *Genes Dev* 15 (23), 3088–103, 2001.

92. Tago, K., Nakamura, T., Nishita, M., Hyodo, J., Nagai, S., Murata, Y., Adachi, S., Ohwada, S., Morishita, Y., Shibuya, H., and Akiyama, T., Inhibition of Wnt signaling by ICAT, a novel β-catenin-interacting protein, *Genes Dev* 14 (14), 1741–9, 2000.
93. Graham, T. A., Clements, W. K., Kimelman, D., and Xu, W., The crystal structure of the β-catenin/ICAT complex reveals the inhibitory mechanism of ICAT, *Mol Cell* 10 (3), 563–71, 2002.
94. Daniels, D. L. and Weis, W. I., ICAT inhibits β-catenin binding to Tcf/Lef-family transcription factors and the general coactivator p300 using independent structural modules, *Mol Cell* 10 (3), 573–84, 2002.
95. Gottardi, C. J. and Gumbiner, B. M., Role for ICAT in β-catenin-dependent nuclear signaling and cadherin functions, *Am J Physiol Cell Physiol* 286 (4), C747–56, 2004.
96. Takemaru, K., Yamaguchi, S., Lee, Y. S., Zhang, Y., Carthew, R. W., and Moon, R. T., Chibby, a nuclear β-catenin-associated antagonist of the Wnt/Wingless pathway, *Nature* 422 (6934), 905–9, 2003.
97. Cavallo, R. A., Cox, R. T., Moline, M. M., Roose, J., Polevoy, G. A., Clevers, H., Peifer, M., and Bejsovec, A., *Drosophila* Tcf and Groucho interact to repress Wingless signalling activity, *Nature* 395 (6702), 604–8, 1998.
98. Brannon, M., Brown, J. D., Bates, R., Kimelman, D., and Moon, R. T., XCtBP is a XTcf-3 co-repressor with roles throughout *Xenopus* development, *Development* 126, 3159–70, 1999.
99. Takemaru, K. I. and Moon, R. T., The transcriptional coactivator CBP interacts with β-catenin to activate gene expression, *J Cell Biol* 149 (2), 249–54, 2000.
100. Hecht, A., Vleminckx, K., Stemmler, M. P., van Roy, F., and Kemler, R., The p300/CBP acetyltransferases function as transcriptional coactivators of β-catenin in vertebrates, *Embo J* 19 (8), 1839–50, 2000.
101. Barker, N., Hurlstone, A., Musisi, H., Miles, A., Bienz, M., and Clevers, H., The chromatin remodelling factor Brg-1 interacts with beta-catenin to promote target gene activation, *Embo J* 20 (17), 4935–43, 2001.
102. Rosin-Arbesfeld, R., Cliffe, A., Brabletz, T., and Bienz, M., Nuclear export of the APC tumour suppressor controls β-catenin function in transcription, *Embo J* 22 (5), 1101–13, 2003.
103. Rosin-Arbesfeld, R., Townsley, F., and Bienz, M., The APC tumour suppressor has a nuclear export function, *Nature* 406 (6799), 1009–12, 2000.
104. Cong, F. and Varmus, H., Nuclear-cytoplasmic shuttling of Axin regulates subcellular localization of β-catenin, *Proc Natl Acad Sci U S A* 101 (9), 2882–7, 2004.
105. Wiechens, N., Heinle, K., Englmeier, L., Schohl, A., and Fagotto, F., Nucleo-cytoplasmic shuttling of Axin, a negative regulator of the Wnt-β-catenin pathway, *J Biol Chem* 279 (7), 5263–7, 2004.
106. Henderson, B. R., Nuclear-cytoplasmic shuttling of APC regulates β-catenin subcellular localization and turnover, *Nat Cell Biol* 2 (9), 653–60, 2000.
107. Townsley, F. M., Cliffe, A., and Bienz, M., Pygopus and Legless target Armadillo/β-catenin to the nucleus to enable its transcriptional co-activator function, *Nat Cell Biol* 6 (7), 626–33, 2004.
108. Weaver, C. and Kimelman, D., Move it or lose it: axis specification in *Xenopus*, *Development* 131 (15), 3491–9, 2004.
109. Weaver, C., Farr, G. H., III, Pan, W., Rowning, B. A., Wang, J., Mao, J., Wu, D., Li, L., Larabell, C. A., and Kimelman, D., GBP binds kinesin light chain and translocates during cortical rotation in *Xenopus* eggs, *Development* 130 (22), 5425–36, 2003.

110. Miller, J. R., Rowning, B. A., Larabell, C. A., Yang-Snyder, J. A., Bates, R. L., and Moon, R. T., Establishment of the dorsal-ventral axis in *Xenopus* embryos coincides with the dorsal enrichment of dishevelled that is dependent on cortical rotation, *J Cell Biol* 146 (2), 427–37, 1999.
111. Lemaire, P., Garrett, N., and Gurdon, J. B., Expression cloning of *Siamois*, a *Xenopus* homeobox gene expressed in dorsal-vegetal cells of blastulae and able to induce a complete secondary axis, *Cell* 81, 85–94, 1995.
112. Kao, K. R. and Elinson, R. P., The entire mesodermal mantle behaves as Spemann's organizer in dorsoanterior enhanced *Xenopus laevis* embryos, *Dev Biol* 127 (1), 64–77, 1988.
113. Dominguez, I., Mizuno, J., Wu, H., Song, D. H., Symes, K., and Seldin, D. C., Protein kinase CK2 is required for dorsal axis formation in *Xenopus* embryos, *Dev Biol* 274 (1), 110–24, 2004.
114. Brannon, M. and Kimelman, D., Activation of *Siamois* by the *Wnt* pathway, *Dev Biol* 180, 344–7, 1996.
115. McKendry, R., Hsu, S. C., Harland, R. M., and Grosschedl, R., LEF-1/TCF proteins mediate Wnt-inducible transcription from the *Xenopus* nodal-related 3 promoter, *Dev Biol* 192 (2), 420–31, 1997.
116. Laurent, M. N., Blitz, I. L., Hashimoto, C., Rothbächer, U., and Cho, K. W., The *Xenopus* homeobox gene *twin* mediates Wnt induction of *goosecoid* in establishment of Spemann's organizer, *Development* 124 (23), 4905–16, 1997.
117. Nieuwkoop, P. D. and Faber, J., *Normal table of Xenopus laevis,* North-Holland Publishing Company, Amsterdam, 1967.
118. Xanthos, J. B., Kofron, M., Tao, Q., Schaible, K., Wylie, C., and Heasman, J., The roles of three signaling pathways in the formation and function of the Spemann organizer, *Development* 129 (17), 4027–43, 2002.
119. Xanthos, J. B., Kofron, M., Wylie, C., and Heasman, J., Maternal VegT is the initiator of a molecular network specifying endoderm in *Xenopus laevis, Development* 128 (2), 167–80, 2001.
120. Heasman, J., Kofron, M., and Wylie, C., β-catenin signaling activity dissected in the early *Xenopus* embryo: a novel antisense approach, *Dev Biol* 222 (1), 124–34, 2000.
121. Kofron, M., Demel, T., Xanthos, J., Lohr, J., Sun, B., Sive, H., Osada, S., Wright, C., Wylie, C., and Heasman, J., Mesoderm induction in *Xenopus* is a zygotic event regulated by maternal VegT via TGFβ growth factors, *Development* 126 (24), 5759–70, 1999.
122. Fan, M. J., Gruning, W., Walz, G., and Sokol, S. Y., Wnt signaling and transcriptional control of Siamois in *Xenopus* embryos, *Proc Natl Acad Sci USA* 95 (10), 5626–31, 1998.
123. Korinek, V., Barker, N., Morin, P. J., van Wichen, D., de Weger, R., Kinzler, K. W., Vogelstein, B., and Clevers, H., Constitutive transcriptional activation by a β-catenin-Tcf complex in *APC-/-* colon carcinoma, *Science* 275 (5307), 1784–87, 1997.
124. Patel, S., Doble, B., and Woodgett, J. R., Glycogen synthase kinase-3 in insulin and Wnt signalling: a double-edged sword? *Biochem Soc Trans* 32 (Pt 5), 803–8, 2004.
125. Ferkey, D. M. and Kimelman, D., GSK-3: new thoughts on an old enzyme, *Dev Biol* 225 (2), 471–9, 2000.
126. Doble, B. W. and Woodgett, J. R., GSK-3: tricks of the trade for a multi-tasking kinase, *J Cell Sci* 116 (Pt 7), 1175–86, 2003.
127. Parker, P. J., Caudwell, F. B., and Cohen, P., Glycogen synthase from rabbit skeletal muscle; effect of insulin on the state of phosphorylation of the seven phosphoserine residues in vivo, *Eur J Biochem* 130 (1), 227–34, 1983.

128. Embi, N., Rylatt, D. B., and Cohen, P., Glycogen synthase kinase-3 from rabbit skeletal muscle. Separation from cyclic-AMP-dependent protein kinase and phosphorylase kinase, *Eur J Biochem* 107 (2), 519–27, 1980.
129. Woodgett, J. R. and Cohen, P., Multisite phosphorylation of glycogen synthase. Molecular basis for the substrate specificity of glycogen synthase kinase-3 and casein kinase-II (glycogen synthase kinase-5), *Biochim Biophys Acta* 788 (3), 339–47, 1984.
130. Welsh, G. I. and Proud, C. G., Glycogen synthase kinase-3 is rapidly inactivated in response to insulin and phosphorylates eukaryotic initiation factor eIF-2B, *Biochem J* 294 (Pt 3), 625–9, 1993.
131. Sutherland, C., Leighton, I. A., and Cohen, P., Inactivation of glycogen synthase kinase-3β by phosphorylation: new kinase connections in insulin and growth-factor signalling, *Biochem J* 296, 15–9, 1993.
132. Ding, V. W., Chen, R. H., and McCormick, F., Differential regulation of glycogen synthase kinase 3β by insulin and Wnt signaling, *J Biol Chem* 275 (42), 32475–81, 2000.
133. McManus, E. J., Sakamoto, K., Armit, L. J., Ronaldson, L., Shpiro, N., Marquez, R., and Alessi, D. R., Role that phosphorylation of GSK3 plays in insulin and Wnt signalling defined by knockin analysis, *Embo J*, 2005.
134. Tolwinski, N. S. and Wieschaus, E., Rethinking WNT signaling, *Trends Genet* 20 (4), 177–81, 2004.
135. Rubinfeld, B., Albert, I., Porfiri, E., Fiol, C., Munemitsu, S., and Polakis, P., Binding of GSK-3β to the APC-β-catenin complex and regulation of complex assembly, *Science* 272 (5264), 1023–6, 1996.
136. Kishida, M., Hino, S., Michiue, T., Yamamoto, H., Kishida, S., Fukui, A., Asashima, M., and Kikuchi, A., Synergistic activation of the Wnt signaling pathway by Dvl and casein kinase Iε, *J Biol Chem* 276 (35), 33147–55, 2001.
137. Itoh, K., Krupnik, V. E., and Sokol, S. Y., Axis determination in *Xenopus* involves biochemical interactions of Axin, glycogen synthase kinase 3 and β-catenin, *Curr Biol* 8 (10), 591–4, 1998.
138. Papkoff, J. and Aikawa, M., WNT-1 and HGF regulate GSK3 β activity and β-catenin signaling in mammary epithelial cells, *Biochem Biophys Res Commun* 247 (3), 851–8, 1998.
139. Dominguez, I. and Green, J. B., Dorsal downregulation of GSK-3β by a non-Wnt-like mechanism is an early molecular consequence of cortical rotation in early *Xenopus* embryos, *Development* 127 (4), 861–8, 2000.
140. Yuan, H., Mao, J., Li, L., and Wu, D., Suppression of glycogen synthase kinase activity is not sufficient for leukemia enhancer factor-1 activation, *J Biol Chem* 274 (43), 30419–23, 1999.
141. Wang, Q. M., Fiol, C. J., DePaoli-Roach, A. A., and Roach, P. J., Glycogen synthase kinase-3β is a dual specificity kinase differentially regulated by tyrosine and serine/threonine phosphorylation, *J Biol Chem* 269, 14566–74, 1994.
142. Welsh, G. I., Patel, J. C., and Proud, C. G., Peptide substrates suitable for assaying glycogen synthase kinase-3 in crude cell extracts, *Anal Biochem* 244 (1), 16–21, 1997.
143. Guger, K. A. and Gumbiner, B. M., A mode of regulation of β-catenin signaling activity in *Xenopus* embryos independent of its levels, *Dev Biol* 223 (2), 441–8, 2000.
144. Bafico, A., Gazit, A., Wu-Morgan, S. S., Yaniv, A., and Aaronson, S. A., Characterization of Wnt-1 and Wnt-2 induced growth alterations and signaling pathways in NIH3T3 fibroblasts, *Oncogene* 16 (21), 2819–25, 1998.
145. Willert, K., Shibamoto, S., and Nusse, R., Wnt-induced dephosphorylation of Axin releases β-catenin from the Axin complex, *Genes Dev* 13 (14), 1768–73, 1999.

146. Jho, E., Lomvardas, S., and Costantini, F., A GSK3β phosphorylation site in Axin modulates interaction with β-catenin and Tcf-mediated gene expression, *Biochem Biophys Res Commun* 266 (1), 28–35, 1999.
147. Easwaran, V., Song, V., Polakis, P., and Byers, S., The ubiquitin-proteasome pathway and serine kinase activity modulate adenomatous polyposis coli protein-mediated regulation of β-catenin-lymphocyte enhancer-binding factor signaling, *J Biol Chem* 274 (23), 16641–5, 1999.
148. Fiol, C. J., Mahrenholz, A. M., Wang, Y., Roeske, R. W., and Roach, P. J., Formation of protein kinase recognition sites by covalent modification of substrate. Molecular mechanisms for the synergistic action of casein kinase II and glycogen synthase kinase 3, *J Biol Chem* 262, 14042–8, 1987.
149. Fiol, C. J., Wang, A., Roeske, R. W., and Roach, P. J., Ordered multisite protein phosphorylation. Analysis of glycogen synthase kinase 3 action using model peptide substrates, *J Biol Chem* 265 (11), 6015–65, 1990.
150. Kintner, M. and Sherman, N. E., *Protein sequencing and identification using tandem mass spectrometry,* John Wiley & Sons, Inc., New York, 2000.
151. Stambolic, V., Ruel, L., and Woodgett, J. R., Lithium inhibits glycogen synthase kinase-3 activity and mimics wingless signalling in intact cells, *Curr. Biol.* 6, 1664–8, 1996.
152. Hedgepeth, C. M., Conrad, L. J., Zhang, J., Huang, H., Lee, V. M. Y., and Klein, P., Activation of the Wnt signaling pathway: A molecular mechanism for lithium action, *Dev Biol* 185, 82–91, 1997.
153. Marmol, F., Carbonell, L., Cuffi, M. L., and Forn, J., Demonstration of inhibition of cyclic AMP accumulation in brain by very low concentrations of lithium in the presence of alpha-adrenoceptor blockade, *Eur J Pharmacol* 226 (1), 93–6, 1992.
154. Masana, M. I., Bitran, J. A., Hsiao, J. K., and Potter, W. Z., In vivo evidence that lithium inactivates Gi modulation of adenylate cyclase in brain, *J Neurochem* 59 (1), 200–5, 1992.
155. Berridge, M. J., Downes, C. P., and Hanley, M. R., Neural and developmental actions of lithium: a unifying hypothesis, *Cell* 59 (3), 411–9, 1989.
156. Coghlan, M. P., Culbert, A. A., Cross, D. A., Corcoran, S. L., Yates, J. W., Pearce, N. J., Rausch, O. L., Murphy, G. J., Carter, P. S., Roxbee Cox, L., Mills, D., Brown, M. J., Haigh, D., Ward, R. W., Smith, D. G., Murray, K. J., Reith, A. D., and Holder, J. C., Selective small molecule inhibitors of glycogen synthase kinase-3 modulate glycogen metabolism and gene transcription, *Chem Biol* 7 (10), 793–803, 2000.
157. Zhou, B. P., Deng, J., Xia, W., Xu, J., Li, Y. M., Gunduz, M., and Hung, M. C., Dual regulation of Snail by GSK-3β-mediated phosphorylation in control of epithelial-mesenchymal transition, *Nat Cell Biol* 6 (10), 931–40, 2004.

2 How to Assay Non-Canonical Wnt Signaling: A Critical Analysis

Michael Kühl and Randall T. Moon

CONTENTS

I. INTRODUCTION

Wnt proteins constitute a family of secreted glycoproteins that interact with a Frizzled-mediated receptor complex to activate intracellular signaling.[1–3] Transmembrane proteins that are thought to be part of this receptor complex are Frizzleds, LRP-5/6, ryk, and Ror-2.[1, 4–6] Historically, Wnt proteins were subdivided into two functional classes depending on their biological activities. Members of the Wnt-1/wingless class transform C57mg cells and induce secondary axis formation in *Xenopus* embryos. Nowadays, we know that both biological effects are due to the activation of the so-called canonical Wnt/β-catenin signaling pathway. This signaling cascade involves stabilization of cytoplasmic β-catenin that can interact with transcription factors of the TCF/LEF family to regulate gene transcription.[1] Wnt ligands of the Wnt-5A class, however, are inactive in both assays described above, but activate β-catenin-independent, non-canonical Wnt signaling pathways.[2] Known mediators of non-canonical Wnt signaling include an intracellular release of calcium ions, activation of certain kinases like protein kinase C (PKC), calcium/calmodulin dependent kinase II (CamKII), jun-N-terminal kinase (JNK), rho kinase (ROK), protein kinase A (PKA), or the calcium-sensitive phosphatase calcineurin (CaCN), as well as GTPases of the rho/rac/cdc42 family.[2, 7–16] A recent publication indicates that vertebrate Dsh is a required part of non-canonical Wnt signaling as well as of the Wnt/β-catenin pathway.[17] Finally, there is strong evidence that both canonical

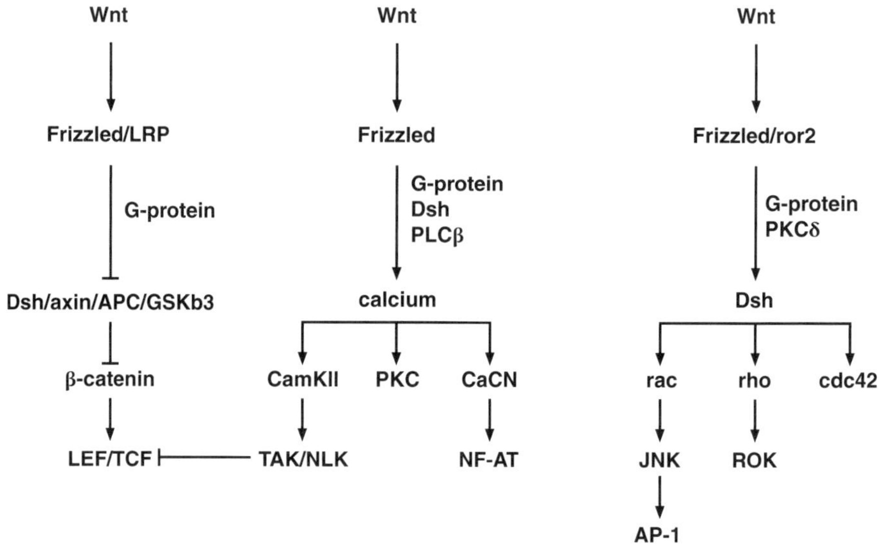

FIGURE 2.1 Outline of canonical and non-canonical Wnt pathways. Major properties of different Wnt signaling pathways are shown. Activation of the Wnt/β-catenin pathway involves inactivation of the Dsh/axin/APC/GSK-3β complex, which destabilizes β-catenin. Thus, β-catenin is stabilized upon a Wnt signal and can interact with LEF/TCF transcription factors. Non-canonical Wnt pathways are independent of β-catenin signaling and involve different intracellular mediators as indicated and discussed in the main text.

and non-canonical Wnt signaling pathways depend on the function of heterotrimeric G-proteins.[18–23] Figure 2.1 summarizes the canonical as well as non-canonical Wnt signaling pathways.

Non-canonical Wnt signaling has been shown to play important roles during embryonic development, i.e., to regulate cell adhesion,[24] cell migration,[25–28] cell polarity,[29] tissue separation,[30] dorso-ventral patterning,[8, 14, 31, 32] heart development,[11, 33–35] or eye development.[36, 37] Some of these functions may involve the inhibition of the β-catenin pathway.[2, 24, 38, 39] The enormous list of potential non-canonical Wnt signaling mediators raises the questions of how to prove that a given biological effect might be linked to non-canonical Wnt signaling and which of these mediators is involved. We here briefly describe biochemical assays to link intracellular mediators of non-canonical Wnt signaling to extracellular Wnt ligands and finally discuss which experiments need to be considered to link biological effects to non-canonical Wnt signaling. The bottom line is that although some assays exist, they are neither simple nor conclusive, and thus better assays are ultimately needed.

II. HOW TO ASSAY ACTIVATION OF NON-CANONICAL WNT SIGNALING

Experimental assays are required to analyze the regulation of β-catenin-independent mediators of Wnts in a given context. We here briefly discuss and reference these assays as they were published in various papers. The hallmark of the Wnt/calcium pathway is the release of intracellular calcium upon Wnt stimulation, which can be detected by calcium-sensitive fluorescent dyes that may require a computer-assisted calcium imaging setup.[21, 40] When performing calcium imaging studies, one should be aware that intracellular release of calcium upon receptor stimulation is a rapid response within seconds.[41] We refer to Chapter 8 by Diane Slusarski within this book for details.

This release of calcium is presaged by formation of inositoltrisphosphate (IP_3) and diacylglycerol (DAG) by phospholipase C. Calcium and DAG are activators of PKC, which can be monitored by a translocation of GFP-tagged PKC to the plasma membrane[7, 8] or phosphorylation of PKC by phospho-PKC-specific antibodies in Western Blots.[42] Finally, enzymatic activity assays using PKC-specific substrates can be performed to demonstrate activation of PKC.[7]

To monitor activation of CamKII, two independent methods are widely used. First, upon stimulation, CamKII becomes phosphorylated itself and becomes calcium independent. Therefore, Western Blot analysis can be used to show an increase in CamKII phosphorylation using a phospho-CamKII-specific antibody.[8, 43] Loading controls need to be included, and an optimal control will be CamKII itself. Thus, an activation of the enzyme will be indicated by an increase of the phosphorylated form and by an unchanged amount of total CamKII at the same time. Second, using enzymatic activity assays, the calcium-independent activity of CamKII can be determined using CamKII-specific substrates and inhibitors.[8, 17, 44] When performing CamKII activity assays, even the percentage of activated enzyme can be calculated if one measures the activity of CamKII after adding exogenous calcium (thus representing fully activated CamKII). Be aware that the fraction of active CamKII in correlation to total CamKII normally is 10% or less.

Similar methods can be applied to monitor activation of JNK. The detection of phosphorylated JNK or the phosphorylated JNK substrate c-jun are convenient read outs.[9, 11] Be aware to include a loading control to show equal amounts of JNK. Furthermore, a reporter gene assay can be used to analyze the activity of the JNK-regulated AP-1 transcription factor complex.[45–47]

For all of these kinases, phosphospecifc antibodies or enzymatic activity assays are commercially available that can be easily adapted to the experimental system used.

Finally, the involvement of G-proteins in non-canonical Wnt signaling pathways was demonstrated using pertussis toxin and antisense oligonucleotide approaches.[7, 8, 18, 19] Presumably, siRNA approaches are tenable as well.

Beside these biochemical assays, cellular read outs have been used to analyze non-canonical Wnt signaling. Wnt-5A has been shown to regulate cell adhesion, and overexpression of Wnt-5A in *Xenopus* embryos results in reduced cell mixing of lacZ expressing cells.[24, 25] Cell adhesion assays, however, are not easily performed and therefore cannot be recommended as a standard assay. As non-canonical Wnt signaling has been shown to regulate cell movements during gastrulation, phenotypic assays are often used in *Xenopus laevis* or zebrafish to monitor convergent extension movements. These include the analyses of *Xenopus* animal cap elongation after activin treatment or of convergence and extension movements in Keller explants.[48] These migration assays were recently adapted to eucaryotic cell culture lines using scratch assays. As an example of this assay, modulation of non-canonical Wnt signaling affects the ability of melanoma cells to close scratched defects in the monolayer.[49] Non-canonical signaling also regulates cell movement later in development, as evidenced by studies on neural crest migration.[28]

It needs to be noted, though, that the simple measurement of kinase activities or the analysis of cell movements does not provide a reliable read out for non-canonical Wnt signaling, as none of these mediators/biological effects are specific for non-canonical Wnt signaling. Just to name one example: PKC is integrated into a large number of different intracellular signaling cascades triggered by a variety of extracellular ligands. This issue will be discussed more intensively in Section IV of this chapter.

In many biological systems, activation of non-canonical signaling results in inhibition of canonical Wnt/β-catenin signaling.[24, 31, 32, 39] This raises the question of whether activation of non-canonical signaling always implicates inhibition of canonical signaling. If so, then routinously used Wnt/β-catenin assays might be a suitable tool to monitor the activity status of non-canonical Wnt pathways, e.g., measuring β-catenin stability or LEF/TCF mediated transcriptional activation. However, although many examples are known that demonstrate this correlation between active non-canonical and downregulated Wnt/β-catenin signaling, one should definitely not rely on these read outs to explore non-canonical signaling. For example, cell polarity in *Drosophila* is mediated by the Frizzled-dependent planar cell polarity pathway, which has, however, not been clearly linked to inhibition of β-catenin signaling.[29] Furthermore, one should keep in mind that β-catenin is regulated also by Wnt-independent mechanisms, e.g., cadherins or integrin-linked kinase.[24, 50, 51]

III. DISCUSSION OF BIOCHEMICALS ASPECTS OF NON-CANONICAL WNT SIGNALING

One important issue when studying signal transduction pathways is the question of the time course of activation. Normally, triggering intracellular signaling pathways results in a rapid cellular response. Calcium is released within seconds from intracellular stores after receptor stimulation. Subsequently, enzymes are activated within minutes. Newly formed transcripts can be detected within hours, and even newly synthesized proteins are detectable soon after stimulation. Many phenotypic assays, however, are performed after hours or days (e.g., assays for convergent extension movements in frogs), raising the question of whether the observed effect is of a primary or secondary nature. One therefore should prefer short-term experiments using purified Wnt proteins (Wnt-3A and Wnt-5A are commercially available) or at least conditioned media instead of overexpressing Wnt ligands by plasmid transfection. An additional valuable tool, that of chimeric receptors, has been used in various publications.[8, 19, 37, 52] These are inducible Frizzled receptors consisting of extracellular and transmembrane domains of the β_2-adrenergic receptor, and intracellular domains of Frizzled receptors that can be stimulated by adrenergic agonist or antagonist to stimulate intracellular signaling. Cell lines stably transfected with the receptors will respond in a rapid manner to receptor stimulation, allowing a detailed biochemical and inexpensive analysis of canonical and non-canonical signaling pathways.

IV. FINAL REMARKS

Considering the fact that all currently known mediators of non-canonical Wnt signaling are not exclusively involved in Wnt signaling, an important question needs to be addressed: What does one require, even in a preliminary manner, to link a biological effect to non-canonical Wnt signaling? Obviously, this involves a series of experiments: First, an observed phenotype needs to be linked to a specific Wnt ligand; for example, loss of function of a Wnt results in a distinct phenotype. Second, this effect needs to be independent of β-catenin-mediated transcription. Therefore, overexpression of β-catenin should be insufficient to rescue the observed Wnt loss of function phenotype. Third, the phenotype should be Dishevelled dependent, as Dishevelled is an integral component of non-canonical Wnt signaling. More precisely, a deletion mutant of Dishevelled, dshΔDIX, which activates non-canonical but not canonical Wnt signaling, should be able to rescue the phenotype.

Fourth, having established that a phenotype is Wnt dependent, but β-catenin independent, one should begin efforts to determine which of the above-mentioned mediators are involved. This will require three lines of experiments. The loss of function of the mediator should phenocopy the Wnt loss of function effect. Inhibitors for some of the non-canonical Wnt mediators are available and have been used successfully.[3, 11, 37, 34] The gain of function of the mediator should rescue the Wnt loss of function, and finally, the Wnt ligand should regulate the activity of the mediator, as discussed before. Examples where such criteria have been employed include the function of non-canonical Wnt signaling during convergent extension movement as well as heart, eye, and dentritic development.[11, 12, 26, 34, 36, 37, 53]

REFERENCES

1. Logan, C.Y. and Nusse, R., The Wnt signaling pathway in development and disease, *Annu Rev Cell Dev Biol*, 20, 781, 2004.
2. Veeman, M.T., Axelrod, J.D., and Moon, R.T., A second canon: Functions and mechanisms of β-catenin-independent Wnt signaling, *Dev Cell*, 5, 367, 2003.
3. Kühl, M., The Wnt/calcium pathway: biochemical mediators, tools and future requirements, *Front Biosci.,* 9, 967, 2004.
4. Hikasa, H. et al., The *Xenopus* receptor tyrosine kinase Xror2 modulates morphogenetic movements of the axial mesoderm and neuroectoderm via Wnt signaling, *Development*, 129, 5227, 2002.
5. Oishi, I. et al., The receptor tyrosine kinase Ror2 is involved in non-canonical Wnt5a/JNK signalling pathway, *Genes Cells*, 8, 645, 2003.
6. Lu, W. et al., Mammalian Ryk is a Wnt coreceptor required for stimulation of neurite outgrowth, *Cell*, 119, 97, 2004.
7. Sheldahl, L.C. et al., Protein kinase C is differentially stimulated by Wnt and Frizzled homologs in a G-protein-dependent manner, *Current Biology*, 9, 695, 1999.
8. Kühl, M. et al., Ca^{2+}/Calmodulin-dependent protein kinase II is stimulated by Wnt and Frizzled homologs and promotes ventral cell fates in *Xenopus*, *J Biol Chem*, 275, 12701, 2000.
9. Boutros, M. et al., Dishevelled activates JNK and discriminates between JNK pathways in planar polarity and wingless signaling, *Cell*, 94, 109, 1998.

10. Cai, Y. et al., Phosphorylation of Pax2 by the c-Jun N-terminal kinase and enhanced Pax2-dependent transcription activation, *J Biol Chem,* 277, 1217, 2002.
11. Pandur, P. et al., Wnt-11 activation of a non-canonical Wnt signaling pathway is required for cardiogenesis, *Nature*, 418, 636, 2002.
12. Yamanaka, H. et al., JNK functions in the non-canonical Wnt pathway to regulate convergent extension movements in vertebrates, *EMBO Rep.,* 3, 69, 2002.
13. Chen, A.E., Ginty, D.D., and Fan, C.M., Protein kinase A signaling via CREB controls myogenesis induced by Wnt proteins, *Nature*, 433, 317, 2005.
14. Saneyoshi, T. et al., The Wnt/calcium pathway activates NF-AT and promotes ventral cell fate in *Xenopus* embryos, *Nature*, 417, 295, 2002.
15. Habas, R., Dawid, I.B., and He, X., Coactivation of Rac and Rho by Wnt/Frizzled signaling is required for vertebrate gastrulation, *Genes Dev.*, 17, 295, 2003.
16. Habas, R., Kato, Y., and He, X., Wnt/Frizzled activation of Rho regulates vertebrate gastrulation and requires a novel Formin homology proteins Daam1, *Cell*, 107, 843, 2001.
17. Sheldahl, L.C. et al., Dishevelled activates Ca^{2+} flux, PKC, and CamKII in vertebrate embryos, *J Cell Biol*, 161, 769, 2003.
18. Liu, T. et al., G protein signaling from activated rat frizzled-1 to the beta-catenin-Lef-Tcf pathway, *Science*, 292, 1718, 2001.
19. Liu, X. et al., Activation of a frizzled-2/beta-adrenergic receptor chimera promotes Wnt signaling and differentiation of mouse F9 teratocarcinoma cells via Galphao and Galphat, *Proc Natl Acad Sci USA*, 96, 14383, 1999.
20 . Wu, C. et al., RGS proteins inhibit XWnt-8 signaling in *Xenopus* embryonic development, *Development,* 127, 2773, 2000.
21. Slusarski, D.C., Corces, V.G., and Moon, R.T., Interaction of Wnt and a Frizzled homologue triggers G-protein-linked phosphatidylinositol signaling, *Nature*, 390, 410, 1997.
22. Penzo-Mendez, A. et al., Activation of Gbetagamma signaling downstream of Wnt-11/Xfz7 regulates cdc42 activity during *Xenopus* gastrulation, *Dev Biol*, 257, 302, 2003.
23. Katanaev, V.L. et al., Trimeric G protein-dependent frizzled signaling in *Drosophila*, *Cell,* 120, 111, 2005.
24. Torres, M.A. et al., Activities of the Wnt-1 class of secreted signaling factors are antagonized by the Wnt-5A class and by a dominant negative cadherin in early *Xenopus* development, *J Cell Biol*, 133, 1123, 1996.
25. Moon, R.T. et al., Dissecting Wnt signalling pathways and Wnt-sensitive developmental processes through transient misexpression analyses in embryos of *Xenopus laevis*, *Dev Suppl,* 85, 1993.
26. Heisenberg, C.P. et al., Silberblick/Wnt11 mediates convergent extension movements during zebrafish gastrulation, *Nature*, 405, 76, 2000.
27. Wallingford, J.B. et al., Dishevelled controls cell polarity during *Xenopus* gastrulation, *Nature*, 405, 81, 2000.
28. DeCalisto, J. et al., Essential role of non-canonical Wnt signaling in neural crest migration, *Development*, 132, 2587, 2005.
29. Strutt, D., Frizzled signalling and cell polarisation in *Drosophila* and vertebrates, *Development,* 130, 4501, 2003
30. Winklbauer, R. et al., Frizzled-7 signaling controls tissue separation during *Xenopus* gastrulation, *Nature*, 413, 856, 2001.
31. Westfall, T.A. et al., Wnt-5/pipetail functions in vertebrate axis formation as a negative regulator of Wnt/beta-catenin activity, *J Cell Biol*, 162, 889, 2003.

32. Westfall, T.A., Hjertos, B., and Slusarski, D.C., Requirement for intracellular calcium modulation in zebrafish dorsal-ventral patterning, *Dev Biol*, 259, 380, 2003.
33. Eisenberg, C.A. and Eisenberg, L.M., Wnt11 promotes cardiac tissue formation of early mesoderm, *Dev. Dynam.*, 216, 45, 1999.
34. Garriock, R.J. et al., Wnt11-R, a protein closely related to mammalian Wnt11, is required for heart morphogenesis in *Xenopus*. *Dev Biol*, 279, 179, 2005.
35. Terami, H. et al., Wnt11 facilitates embryonic stem cell differentiation to Nkx2.5-positive cardiomyocytes, *Biochem Biophys Res Commun*, 325, 968, 2004.
36. Rasmussen, J.T. et al., Regulation of eye development by frizzled signaling in *Xenopus*, *Proc Natl Acad Sci USA*, 98, 3861, 2001.
37. Maurus, D. et al., Non-canonical Wnt-4 signaling and EAF2 are required for eye development in *Xenopus laevis*, *EMBO J.*, 24, 1181, 2005.
38. Maye, P., Multiple mechanisms for Wnt11-mediated repression of the canonical Wnt signaling pathway, *J Biol Chem*, 279, 24659, 2004.
39. Topol, L. et al., Wnt-5a inhibits the canonical Wnt pathway by promoting GSK-3 independent beta-catenin degradation, *J Cell Biol*, 162, 899, 2003.
40. Slusarski, D.C. et al., Modulation of embryonic intracellular Ca^{2+} signaling by Wnt-5A, *Developmental Biology*, 182, 114, 1997.
41. Clapham, D.E., Calcium signaling, *Cell*, 80, 259, 1995.
42. Koyanagi, M. et al., Non-canonical Wnt signaling enhances differentiation of human circulating progenitor cells to cardiomyogenic cells, *J Biol Chem*, 280, 16838, 2005.
43. Robitaille, J. et al., Mutant frizzled-4 disrupts retinal angiogenesis in familial exudative vitreoretinopathy, *Nat Genet*, 32, 326, 2002.
44. Ishitani, T. et al., The TAK1-NLK mitogen-activated protein kinase cascade functions in the Wnt-5a/Ca^{2+} pathway to antagonize Wnt/beta-catenin signaling, *Mol Cell Biol*, 23, 2003.
45. Park, M. et al., The planar cell polarity gene stbm regulates cell behaviour and cell fate in vertebrate embryos, *Nat Cell Biol*, 4, 20, 2002.
46. Le Floch, N. et al., The proinvasive activity of Wnt-2 is mediated through a noncanonical Wnt pathway coupled to GSK-3beta and c-jun/AP-1 signaling, *FASEB J*, 19, 144, 2005.
47. Lyu, J. and Joo, C.K., Wnt-7a upregulates matrix metalloproteinase-12 expression and promotes cell proliferation in corneal epithelial cells during wound healing, *J Biol Chem*, 280, 21653, 2005.
48. Kühl, M. et al., Antagonistic regulation of convergent extension movements in *Xenopus* by Wnt/beta-catenin and Wnt/Ca^{2+} signaling, *Mech Dev*, 106, 61, 2001.
49 Weeraratna, A.T. et al., Wnt5a signaling directly affects cell motility and invasion of metastatic melanoma, *Cancer Cell*, 1, 279, 2002.
50. Finnemann, S. et al., Cadherin transfection of *Xenopus* XTC cells downregulates expression of substrate adhesion molecules, *Mol Cell Biol*, 15, 5082, 1995.
51. Novak, A. et al., Cell adhesion and the integrin-linked kinase regulate the LEF-1 and beta-catenin signaling pathways, *Proc Natl Aca. Sci USA*, 95, 4374, 1998.
52. DeCostanzo, A.J. et al., The Frizzled-1/beta$_2$-adrenergic receptor chimera: pharmacological properties of a unique G protein-linked receptor, *Naunyn-Schmiedeberg's Arch Pharmacol*, 365, 341, 2002.
53. Rosso, S.B. et al., Wnt signaling through Dishevelled, Rac and JNK regulates dendritic development, *Nature Neurosci*, 8, 34, 2005.

3 Regulation of TGF-β Family Activity by Proprotein Processing

Shailaja Sopory and Jan L. Christian

CONTENTS

I. INTRODUCTION

A large number of proteins, including growth factors, hormones, extracellular matrix proteins, cell surface receptors, bacterial toxins, and viral surface glycoproteins, are synthesized as inactive proproteins that need to be cleaved to gain physiological activity. These precursors are proteolytically activated by a novel family of enzymes, called proprotein convertases (PCs) or subtilisin-like proprotein convertases (SPCs).

Among the many known PC substrates are cell-to-cell signaling molecules that play essential roles during embryonic development. Consistent with this, targeted mutation of genes encoding PC family members can lead to embryonic lethality or severe birth defects due to loss of growth factor activity.[1] PCs do not merely provide an "on/off" switch for growth factors, however, because proprotein maturation has also been shown to provide temporal, spatial, or subcellular regulation of their activity. Thus, in analyzing the embryonic functions of growth factors that are generated from inactive precursor proteins, it is critical to take into consideration when and how these factors are proteolytically activated. In this chapter, we give a few examples of how proteolytic activation can regulate the activity of signaling molecules, a brief overview of the PC family, and a description of various *in vitro* and *in vivo* approaches that have been useful for studying the role of proteolysis in regulating growth factor activity during vertebrate embryonic development. For the purpose of our discussion, we shall focus on proteolytic activation of transforming growth factor-β (TGF-β) superfamily members and the candidate PCs that are responsible for their cleavage.

II. REGULATION OF TGF-β FAMILY ACTIVITY BY PROPROTEIN PROCESSING

The TGF-β superfamily is composed of a diverse group of proteins that share sequence similarity and a conserved cystine knot structure.[2] Prominent members include TGF-β1-3, nodal, activin, inhibin, Mullerian inhibiting substance, and bone morphogenetic proteins (BMPs). Many TGF-β family proteins signal in the early embryo and are essential for normal development. Moreover, several members of this family, most notably activin, nodal, and selected BMPs, have been shown to establish a morphogenetic gradient within the embryo and can induce distinct cell fates in a concentration-dependent manner. Consequently, proper spatial and temporal activation of these proteins by PCs is vital.

Processing of BMP4, one of the best characterized BMPs, provides an example of how proteolytic activation can provide both spatial and temporal regulation of ligand activity. BMP4 is required for the specification or patterning of many organ systems throughout embryogenesis.[3] In the frog embryo, endogenous BMP transcripts are present maternally,[4,5] and yet the intracellular BMP signaling pathway cannot be activated by endogenous or ectopic BMPs until the mid-blastula stage.[6,7] Interestingly, ectopically expressed BMP-4 accumulates in precursor form in early *Xenopus* embryos but appears not to be cleaved until the mid-blastula transition,[8] despite the fact that all candidate convertases are present in oocytes.[9,10] Thus, the timing of activation of the intracellular BMP-signaling pathway may be regulated by temporally restricted proteolytic activation of the precursor protein.

In addition to providing temporal regulation of BMP4 activity, cleavage of the BMP4 precursor can spatially regulate the range over which the mature ligand can signal. ProBMP4 is sequentially cleaved at two sites within the prodomain: initially at a site adjacent to the mature domain and then at an upstream site within the prodomain.[8, 11] Differential cleavage at the second site regulates the activity and signaling range of mature BMP4 by directing endosomal trafficking of cleaved BMP4 to either degradatory (if the upstream site is not cleaved) or secretory/recycling (if the upstream site is cleaved) pathways.[8, 12] We have hypothesized that the upstream site of proBMP4 is selectively cleaved in only a subset of tissues, thereby providing a mechanism for tissue-specific regulation of the range of action of BMP4. Consistent with this hypothesis, mice carrying a targeted mutation that prevents cleavage of the upstream site have phenotypic abnormalities in only a subset of tissues that are known to be sensitive to BMP4 dosage.[13] Consequently, differential expression of various PCs, or other factors that regulate cleavage at the upstream site, may affect the signaling range of BMP4.

Proteolytic processing has also been shown to play an important role in regulating the signaling range of nodal, a gene that is required for mesoderm induction and patterning.[14] For most TGF-β family members, the prodomain functions as an essential intracellular chaperone to direct dimerization and folding.[15] Nodal appears to be an exception to this rule because a prodomain deleted form of nodal is active when expressed in zebrafish embryos.[16] Removal of the nodal prodomain generates a mature ligand that is more active in autocrine signaling but is less stable, and thus signals at shorter range than does the same ligand generated from the native precursor. Because the nodal precursor is cleaved in the extracellular space, following secretion,[17] these results suggest that the prodomain may enable uncleaved nodal precursor to travel to distal cells, where it is then processed and can activate the intracellular signal transduction cascade.

III. THE PROPROTEIN CONVERTASE FAMILY

A. Domain Structure of PC Family Members

PCs are a family of calcium-dependent serine proteases with catalytic domains that are similar to that of bacterial subtilisin.[1, 18] PCs have been highly conserved throughout evolution, with seven known family members in mammals: furin (also known as SPC1), PC2 (also known as SPC2), PC1 (also known as PC3 or SPC3), PACE4 (also known as SPC4), PC4 (also known as SPC5), PC6 (also known as SPC6 or PC5), and PC7 (also known as LPC, PC8, or SPC7).

Features of the primary structure of PCs are remarkably conserved. All PC family members contain an N-terminal signal sequence for admission into the secretary pathway, followed by a prodomain, a catalytic domain, a domain unique to PCs called the "Homo B" or "P" domain, and a carboxy (C)-terminal domain of varying composition (illustrated in Figure 3.1). The prodomain is required for proper folding

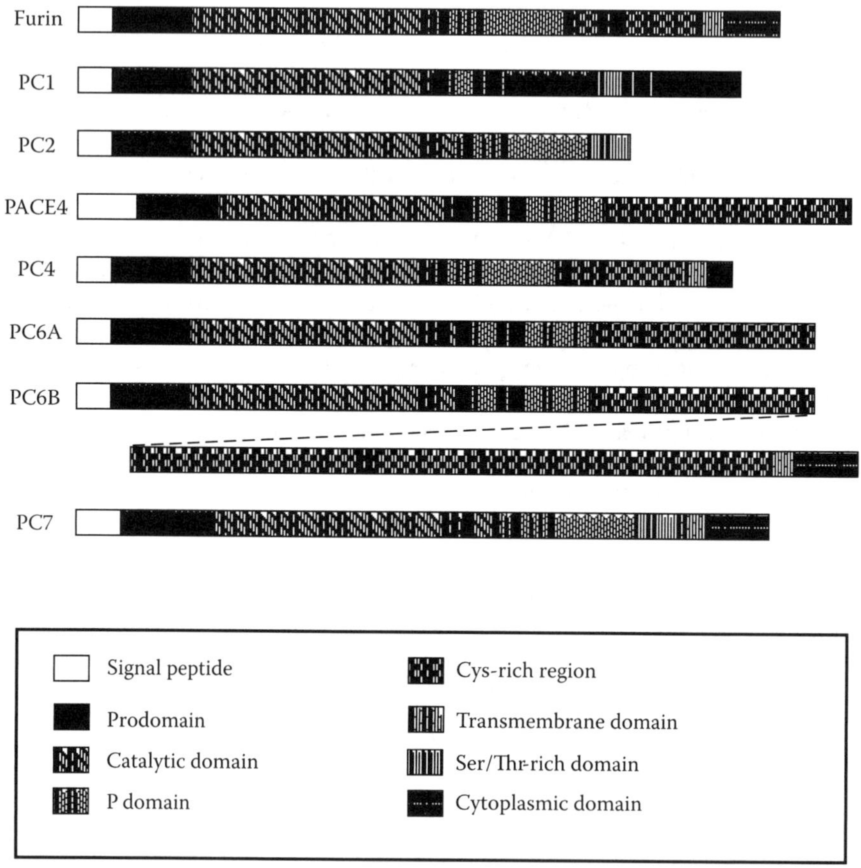

FIGURE 3.1 Domain structure of mammalian PC family members.

and maintains the precursor in an inactive state until it is autocatalytically removed.[19] With the exception of PC2, cleavage of the prodomain occurs in the ER and is required for transport beyond this compartment. Following cleavage, the propeptide remains associated with the enzyme, masking the catalytic cleft, until the complex traffics to a trans-Golgi network (TGN)/endosomal compartment, where a second internal cleavage releases the active enzyme.[1]

The catalytic domain contains a characteristic catalytic triad of Asp, His, and Ser residues, and a negatively charged substrate-binding pocket that contributes to selectivity for substrates containing dibasic or multibasic amino acid motifs. Substrate selectivity is further influenced by the P domain, which structurally stabilizes the catalytic domain.[20, 21] Finally, the C-terminal domain is dispensable for catalytic activity and is the most divergent among PCs. Furin, PC4, PACE4, PC6, and PC7 all contain a cysteine-rich (Cys-rich) domain with a variable number of repeated motifs, each ~50 amino acids in length and containing a well-conserved Cys topography.[10, 22] The function of this domain is unknown. Although mouse PC4 is reported to lack this domain, isoforms of human and *Xenopus* PC4 that contain a conserved Cys-rich domain as well as a putative transmembrane domain and short cytoplasmic tail have been identified.[10] Furin, PC7, and one of the two characterized splice isoforms of PC6 (PC6B) also include a transmembrane domain near the C-terminus and a short cytoplasmic tail. The cytoplasmic tails of furin and PC6B contain sorting signals for trafficking between the TGN, endosomes, and the cell surface.[18, 23, 24]

Proprotein processing occurs in both the constitutive and regulated secretory pathways, with different PCs playing a predominant role in protein maturation in these two pathways. PC1 and PC2 are expressed exclusively in neuroendocrine tissues, where they are localized to dense core secretory granules of the regulated pathway and are released upon stimulation by a secretagogue. They are responsible for processing of neuroendocrine hormones like POMC, proinsulin, and proglucagon.[25] PC4 also functions in the regulated secretory pathway and is expressed exclusively in reproductive tissues.[26, 27]

By contrast, the active forms of furin, PACE4, PC6B, and PC7 are broadly expressed and are localized to the TGN and secretory vesicles of the constitutive pathway. This location provides access to many precursor proteins, including TGF-β family members, which move to the cell surface via constitutive secretory vesicles.[28, 29] Secreted or shed forms of furin, PACE4, and PC6B have also been identified,[23, 30–32] consistent with the finding that some TGF-β precursor proteins are cleaved in the extracellular space.[17, 33] Although the soluble splice isoform of PC6 (PC6A) has been shown to sort to the regulated secretory pathway in cultured neuroendocrine cells,[23] it is likely to function in the constitutive pathway as well, given its broad distribution outside of neuroendocrine tissues.[34] In summary, furin, PACE4, PC6, and PC7 are all good candidates for endogenous TGF-β family convertases based on their broad expression patterns and localization within the constitutive secretory pathway. By contrast, PC1, PC2, and PC4 are unlikely to function in this capacity, based on their tissue-restricted expression patterns and localization to regulated secretory granules, and thus will not be discussed further.

B. Substrate Selectivity of PCs

Considerable effort has been invested in determining the sequence-specific requirements for PC cleavage.[25] PCs recognize specific residues around the scissile bond by accommodating individual amino acid side chains in their corresponding substrate binding site. According to the nomenclature used to identify residues surrounding the cleavage site, successive residues beginning immediately N-terminal to the cleavage site are designated P_1, P_2, and so on, whereas those beginning immediately C-terminal to the cleavage site are designated P_1', P_2', and so on.

Furin preferentially cleaves after the motif Arg-X-Arg/Lys-Arg (where X is any amino acid), which is often referred to as an optimal furin consensus motif.[35, 36] A basic residue in the P_2 position is not essential, however, because furin will also recognize the minimal consensus motif, -Arg-X-X-Arg-, and in rare cases the presence of Arg or Lys residues at the P_2 and P_6 position can substitute for the P_4 Arg.[25, 37] The crystal structure of the furin catalytic and P domains supports the stringent selectivity for substrates with basic residues in the P_1, P_4, or P_6 positions.[38]

The cleavage specificity of PACE4, PC6, and PC7 is less well studied than that of furin, but homology modeling suggests that the substrate binding region of all PCs contains a large number of negatively charged residues that enable the protease to recognize and cleave following multiple basic amino acids.[39] PACE4 and PC6 are more similar to furin than is PC7 and contain a greater negative charge density within the active site cleft, suggesting that they may prefer additional basic residues near the cleavage site. Preliminary characterization of the cleavage specificity of PACE4 and PC7 indicates a requirement for Arg at the P_1 and P_4 positions for both enzymes[25] and a further requirement for a basic residue at the P_2 position for PC7.[40] By contrast, PC6 may not have a strict requirement for Arg at P_4 and may not distinguish between Arg and Lys at P_1.

Although PCs will readily cleave -Arg-X-X-Arg- motifs in peptide substrates *in vitro*, not all such sites are accessible in native proteins. This relatively simple motif is found in many proteins that are not PC substrates, and *bona fide* PC substrates often harbor multiple -Arg-X-X-Arg- motifs but are selectively cleaved at a single site. Thus, PC substrates cannot be defined by the mere presence of a consensus cleavage motif in the absence of biochemical analysis. Methods for identifying substrates and studying the functional significance of putative cleavage sites are discussed in the second half of this chapter.

C. Developmental Expression Patterns of Proprotein Convertases

Furin, PACE4, PC6, and PC7 are all broadly expressed during vertebrate embryogenesis. With the exception of PC7, which is ubiquitously expressed,[41] individual PCs exhibit tissue-specific expression patterns that change dynamically throughout development. The major embryonic sites of expression of furin, PACE4, and PC6 are summarized in the following paragraphs.

The human furin gene is transcribed from at least three distinct promoters, yielding multiple mRNA isoforms.[28] To date, only a single transmembrane domain containing protein isoform has been described, although this protein can be cleaved to generate a soluble protein that is released into the extracellular space.[30] In mice, furin transcripts are first detected at embryonic day (E) 5.5 in the extraembryonic ectoderm and in proximal portions of the epiblast and visceral endoderm.[17] This pattern persists until E7.5, at which time transcripts are also detected in the extraembryonic and precardiac mesoderm and in the node.[42] By E8–8.5, furin is highly expressed in the anterior intestinal portal, allantois, primitive heart, lateral plate mesoderm, notochordal plate, and foregut endoderm. Expression of furin is not detected in the nervous system at any point during embryogenesis, but outside of this organ it is fairly ubiquitously expressed by E13.5.[42, 43] In *Xenopus,* furin is ubiquitously expressed throughout gastrulation, but transcripts become restricted primarily to nonneural ectoderm and nonaxial mesoderm by the neurula stage.[10, 11] Later in development, furin is most highly expressed in the developing kidney, eye, otic vesicle, and pharyngeal arches.[10]

The primary PACE4 transcript undergoes extensive alternative splicing. Currently, six different splice isoforms have been reported for human PACE4 (PACE4A,

-B, -C, -CS, -D, and -E).[44, 45] PACE4A and PACE4E encode full-length, functional enzymes that differ only in the presence of a 75 amino acid insertion near the C-terminus of PACE4E. This novel sequence includes a hydrophobic cluster that may increase intracellular retention.[45] By contrast, no activity has been attributed to PACE4B, which lacks the P domain; PACE4C and PACE4CS, which are truncated near the end of the P domain; or PACE4D, which lacks the signal peptide, propeptide, and cysteine-rich region.[31] These inactive isoforms have been hypothesized to act as dominant negative proteins that titrate the level of active PACE4A, although this has not been tested.[31]

Developmental expression patterns of the different PACE4 splice isoforms have not been reported, but embryonic expression of murine PACE4, detected with a probe that recognizes all transcripts, overlaps extensively with that of furin.[41] PACE4 transcripts are first detected in the extraembryonic ectoderm at E5.5, but become more broadly distributed as development proceeds with high levels of expression observed in the gut, heart, neural tube, limb buds, and developing bone. Relative to the mouse, expression of *Xenopus* PACE4 is both temporally and spatially restricted. Maternal PACE4 transcripts are localized to the vegetal hemisphere of the oocyte and persist in the prospective endoderm and mesoderm until the end of gastrulation. After this time, weak expression of PACE4 can be detected in the notochord, the brain, and a subset of endodermal cells that may represent primordial germ cells.[46]

Murine PC6 also undergoes alternative splicing to generate multiple isoforms, two of which (PC6A and PC6B) have been well characterized.[47] PC6A is a soluble isoform and contains sorting information in its unique carboxy terminus that directs trafficking to regulated secretory granules in neuroendocrine cells.[23] By contrast, membrane-bound PC6B localizes to the TGN[23] or to a post-TGN compartment within the constitutive pathway.[24] Like furin, PC6B can undergo carboxy terminal processing to release a soluble isoform that may be involved in processing extracellular substrates.[23] Splice isoforms encoding PC6A and PC6B have also been identified in *Xenopus*, together with four C-terminally truncated, and presumably secreted, versions of PC6.[10]

Quantitatively, PC6A is more broadly and more abundantly expressed in both embryonic and adult tissues than is PC6B.[47–49] *In situ* hybridization analysis, using a probe that detects all isoforms, reveals that mouse PC6 is expressed in the extraembryonic endoderm, the amnion, and the distal primitive streak by E7.5.[41] Later in development, prominent expression is observed in the somites, limb buds, and vertebral and facial cartilage. In *Xenopus,* PC6A is highly expressed in the developing nervous system, notochord, otic vesicle, eye, pharyngeal arches, and kidney primordia.[10] PC6B is much more weakly expressed in a subset of these tissues.

Comparison of the expression patterns of PCs with those of potential substrates can guide identification of candidate convertases, but several caveats should be kept in mind when doing so. First, because PCs are synthesized as inactive precursors that must be autocatalytically cleaved and sorted to the correct processing compartment before they can function, the spatial and temporal expression pattern of a PC at the RNA level does not strictly reflect that of the active enzyme. Second, because secreted or shed forms of PCs have been shown to cleave some substrates in the extracellular space, the site of synthesis is not necessarily the site of action of a given PC. Thus, one can neither rule in nor rule out a given PC as the convertase responsible for cleaving a substrate of interest based on comparison of RNA expression patterns alone.

In the following section, we provide an overview of several alternative or complimentary approaches to studying the embryonic functions of PCs using *in vitro* or *in vivo* approaches in mouse or *Xenopus*. Other animal model systems, such as chick, *C. elegans*, and *Drosophila*, provide a unique set of advantages and disadvantages for studying PC function but are beyond the scope of this chapter.

IV. ANALYSIS OF PROPROTEIN PROCESSING *IN VITRO*

A. *In Vitro* Cleavage of Fluorogenic Peptides

A simple and rapid approach to studying enzyme substrate specificity, which precludes the need to generate recombinant substrates, involves the use of fluorogenic peptides. The basic principle underlying cleavage of fluorogenic peptides is that the intact substrate is poorly fluorescent, whereas the free fluorophore, when liberated by cleavage, is highly fluorescent. Several substrates, in which the fluorophore is directly linked to the bond undergoing cleavage, have been utilized to assay PC selectivity.[50, 51] This technique enables rapid spectroscopic analysis of cleavage of a large number of sequences by different PCs. An obvious disadvantage is that most fluorophores are bulky aromatic groups, unlike the amino acid residues that naturally occur downstream of the cleavage site, and this could prevent optimal interaction of the enzyme with the substrate. The use of internally quenched fluorogenic substrates overcomes this drawback.[51] These substrates consist of a peptide chain corresponding to a region of the protein encompassing the PC cleavage site with a fluorescent donor on one end and an acceptor (quencher) on the other.

In the uncleaved state, the quencher reduces the fluorescent intensity of the fluorophore, even when the two groups are separated by several amino acids. Cleavage leads to an increase in fluorescence that is proportional to the amount of cleaved substrate generated. The design of intramolecularly quenched substrates should take the following into consideration: The fluorophore should not be bulky, which might interfere with interaction between the enzyme and substrate; the quencher should have low absorbance in the wavelength suitable for excitation; and the whole molecule should be resistant to other enzymes and sufficiently hydrophilic to be soluble in water.[51] An example of a donor-acceptor pair is 2-aminobenzoic acid as the fluorescent donor and nitrotyrosine as the acceptor. These peptides offer a number of advantages in terms of high sensitivity and throughput. They allow for rapid characterization of multiple sequences and enable one to determine kinetic parameters and substrate specificity for different proteases.

The obvious concern with using fluorogenic peptides to assay PC selectivity is physiological relevance. This technique can be used to identify consensus cleavage motifs for a particular protease, and based on this information, one can identify potential substrates and cleavage sites, but the presence of a perfect consensus cleavage motif in a protein does not necessarily mean that this site is recognized *in vivo*. Many proteins, including members of the TGF-β family, contain multiple -Arg-X-X-Arg- motifs, some or all of which are not utilized *in vivo*, and some presumed substrates are cleaved by PCs at nonconsensus motifs. Therefore, although analysis of cleavage of fluorogenic peptides can be used to identify potential substrates for a specific PC, other factors, such as protein conformation, will influence which sites are cleaved *in vivo*.

B. *In Vitro* Cleavage of Recombinant Proteins

Another useful approach to test whether a substrate can be cleaved by a given PC is to ask whether full-length recombinant proteins can be cleaved *in vitro*. Radiolabeled substrates can be generated either *in vitro*, for example, using rabbit reticulocyte lysates, or *in vivo*, in cultured cells or *Xenopus* oocytes. Substrates are then partially purified by immunoprecipitation, incubated *in vitro* with a particular recombinant PC and cleavage products analyzed by electrophoresis and autoradiography. Cleavage of *in vitro* synthesized proteins is a fast and easy way to look for sites in candidate substrates that can be cleaved by specific PCs. It is important to keep in mind, however, that *in vitro* synthesized protein will not have undergone post-translational modifications, such as glycosylation, and may not be correctly folded or dimerized. Proteins expressed in oocytes have the advantage of undergoing post-translational modifications and folding, and therefore may provide more biologically relevant information. In the case of TGF-β family members, however, it is important to recognize that dimerization and folding are rate limiting, and thus the majority of TGF-β precursor protein isolated from oocytes is likely to be in monomer form. Therefore, cleavage information obtained from these experiments, though informative, may not fully reflect the *in vivo* situation.

In vitro cleavage reactions can be used to identify or rule out potential PCs as endogenous convertases for a particular substrate. In general, if a PC fails to cleave a substrate *in vitro*, it is unlikely to do so *in vivo*. The converse, however, is not necessarily true because substrate selectivity *in vitro* is much broader than that *in vivo*. Despite these limitations, anecdotal evidence suggests that recombinant PCs will accurately identify and selectively cleave only those -Arg-X-X-Arg- motifs that are normally used *in vivo*, even when the substrate has been generated in rabbit reticulocyte lysates. Thus, whereas its usefulness as a tool to match a single PC with a substrate is limited, *in vitro* cleavage reactions can be used to determine whether a protein is a PC substrate and which potential cleavage sites are utilized *in vivo*. A detailed protocol for synthesis of radiolabeled substrates and *in vitro* cleavage is provided later in Section VII.A of this chapter.

V. ANALYSIS OF PROPROTEIN PROCESSING *IN VIVO*

A. Analysis in Cultured Cells

Cell culture provides a convenient way to study proteolytic processing by co-expressing the substrate of interest with specific PC family members. Because PCs are themselves generated as inactive precursors, an important consideration in choosing a cell line for analysis is whether autocatalytic activation occurs efficiently in a given cell type. A wide variety of cell lines have been utilized to study the biosynthesis of various PCs, and literature is available regarding which cell lines optimally process a particular family member. PACE4, for example, is efficiently processed and secreted from HEK293, CHO, and LoVo cells, but not from COS-1 or AtT-20 neuroendocrine cells.[52] Furin, PC6, and PC7 have all been successfully expressed in BSC-40 cells.[35, 53, 54] Regardless of the cell line that is chosen, a critical control in transfection experiments is to verify by Western blot that the mature, active form of a PC is generated in order to rule out the trivial possibility that failure to detect substrate processing is due to inefficient PC maturation.

In addition to overexpressing candidate PCs along with potential substrates, it is possible to use knowledge of which endogenous PCs are enriched in, or absent from, different cell lines to restrict candidate convertases for a particular substrate. Furin-deficient cell lines like LoVo, a human colon carcinoma cell line, and RPE 40, a mutant cell line derived from CHO cells, are available[55] and have been used to study the necessity or redundancy of furin in processing numerous substrates.[56]

Processing of substrates can also be studied in cultured cells that have been depleted of PC activity using PC-selective inhibitors or antisense knock downs. Treatment of vascular smooth muscle cells with the pan-PC inhibitor Decanoyl-Arg-Val-Lys-Arg-chloromethylketone (dec-RVKR-CMK), for example, led to loss of cleavage of endogenous α_v integrin, and the specific protease involved was demonstrated to be PC6 using specific antisense oligonucleotides.[57] Antisense oligonucleotides have also been used to demonstrate that furin, PACE4, and PC7 function redundantly to process proalbumin in HepG2 cells.[1]

A final way in which cultured cells can be used to study proprotein processing is as a tool to verify that putative cleavage sites, identified on the basis of consensus cleavage motifs or the size of proteolytic fragments generated by *in vitro* cleavage, are used *in vivo*. Specifically, amino acid substitutions that disrupt putative PC recognition motifs can be introduced into substrates, and their effect on processing analyzed in transiently transfected cells.[56, 58]

Analysis of proprotein processing in cultured cells is limited by the fact that proteins are highly overexpressed in a heterologous cell type, and the results may not reflect cleavage *in vivo*. In mammalian COS cells, for example, processing of BMP4 is dramatically increased by co-expression of furin or PACE4.[34] Analysis of mice lacking both of these proteases, however, suggests that furin and PACE4 are not required for cleavage of BMP4 in the early embryo, despite being co-expressed in the same subset of cells.[17] Furthermore, we have observed that mutations that disrupt folding or cleavage of proBMP4 in *Xenopus* embryos do not necessarily do so in cultured cells, as will be described in Section V.E.3 of this chapter. Therefore, although analysis of precursor cleavage in cultured cells can serve as a starting point for further studies, the results often need to be validated in a whole animal model.

B. Mouse Knock Out Models

Targeted disruption of specific PC genes in mice has provided valuable information concerning the role of these proteases in regulating various developmental processes, although embryonic lethality combined with complex phenotypes and redundancy can make results difficult to interpret. This is particularly true in the case of the broadly expressed PCs, furin and PACE4, as detailed below.

Furin knock out mice die early in embryogenesis due to a variety of morphological abnormalities. Mutant embryos do not undergo axial rotation and develop abnormal yolk sac vasculature, fail to undergo chorioallantoic fusion, and exhibit severe ventral closure defects, leading to cardiac insufficiency and lethality by E10.5–11.5.[42] Similar phenotypes are observed in mice deficient for specific TGF-β or BMP ligands, consistent with the possibility that furin is required for their proteolytic activation.[42] Chimeric analysis has shown that furin is required within epiblast derivatives, including the primitive heart, gut, and extraembryonic mesoderm, for normal patterning of these organs.[59] More recent studies highlight an important role for extraembryonic furin in patterning the epiblast, as well.[17, 59] An interferon-inducible, conditional furin knock out mouse has been generated that will provide an important tool for analysis of tissue- and stage-specific embryonic functions of furin. This mouse has been used in initial studies to show that furin is not required for normal liver function in the adult.[60]

Loss of function of PACE4 leads to prenatal death in 25% of homozygous mutant embryos, most likely due to cardiac malformations.[59] PACE4-deficient embryos also show complex craniofacial abnormalities, anterior truncations, or defects in left/right patterning. All of these processes are regulated to some extent by BMPs and nodal, suggesting that PACE4 is required for processing of one or both of these precursor proteins. Consistent with this possibility, expression of putative nodal target genes is downregulated in PACE4 mutant embryos.

As described earlier, the patterns of expression of furin and PACE4 overlap extensively in early embryos, making it difficult to determine whether the dissimilar phenotypes observed in mice mutant for each gene reflects unique substrate specificity, or functional redundancy and partial compensation by the retained PC. To answer this question, mice lacking both furin and PACE4 were generated.[17] The double mutants have a more severe phenotype than that observed in either single mutant and essentially phenocopy embryos lacking nodal. Biochemical evidence that PACE4 and furin function redundantly to cleave nodal was provided by the demonstration that embryonic stem cells mutant for both enzymes are unable to process ectopically expressed nodal. Interestingly, expression of furin and PACE4 is restricted to extraembryonic ectoderm in early embryos, whereas nodal is expressed many cell diameters away, in the epiblast. This suggests that cleavage of nodal occurs outside of the cell and requires secreted or shed forms of furin and PACE4, rather than occurring within the biosynthetic pathway, as has commonly been assumed.

Many endogenous PC substrates, including nodal, cannot be detected at the protein level in early embryos, and this has hampered definitive identification of substrates in mouse knock out models. Despite this limitation, mouse mutants have provided valuable information about the developmental roles of PCs. These studies have also shown that individual PCs function redundantly in some cases, but in other cases, they display a unique substrate selectivity that is not determined by the presence or absence of a simple consensus cleavage motif. Future tissue-specific PC knock outs or targeted knock ins of mutations that alter or ablate cleavage sites will improve our understanding of how proprotein processing regulates the embryonic functions of TGF-β family members.

C. Analysis in *Xenopus*

Targeted mutagenesis is not feasible in *Xenopus*, but several alternative approaches to knocking down PC function are available. *Xenopus* has been used extensively as a model organism to understand various aspects of early embryogenesis, including dorsal-ventral patterning, mesoderm induction, and specification of neural fate, and members of the TGF-β family are known to play critical roles in each of these processes. Thus, it is often possible to predict which TGF-β ligands are candidate substrates for specific PCs based on analysis of early patterning defects resulting from loss of PC function.[11] This knowledge can then be combined with biochemical analysis to test whether cleavage of candidate substrates is impaired in PC-deficient embryos or oocytes as described in the following section.

1. Inhibition of Proprotein Convertase Activity in Embryos

One approach to knocking down endogenous PC activity in *Xenopus* embryos involves the use of chemical- or protein-based PC inhibitors. Dec-RVKR-CMK, for example, is a potent inhibitor of all PC family members that is more commonly applied to cultured cells[54, 61] but could potentially be microinjected into *Xenopus* embryos or added to the culture media of embryonic explants. Protein-based inhibitors have also been developed, the most potent and best characterized of these being α1-PDX. α1-PDX is a genetically engineered mutant form of the naturally occurring serine protease inhibitor, α1–antitrypsin (α1–AT), which contains the minimal furin consensus motif -Arg-Ile-Pro-Arg- in its reactive site and serves as a potent suicide substrate.[62] Microinjection of RNA encoding α1-PDX into *Xenopus* embryos phenocopies the effects of blocking endogenous BMP4 activity, suggesting that an α1-PDX-sensitive protease is required to proteolytically activate BMP4. This was further substantiated by *in vitro* biochemical analysis of BMP4 cleavage in the presence of purified PCs.[11]

One drawback to using PC inhibitors *in vivo* is that they either are global inhibitors of all PCs or, at best, are selective for specific PCs but will inhibit more broadly at higher doses. α1-PDX, for example, is highly selective for furin and PC6, but can also inhibit PACE4 and PC7.[53] Efforts to develop specific inhibitors of individual PCs by introducing amino acid substitutions into the reactive site loop of α1-AT, or using unbiased screens of chemical libraries, are ongoing.[63] Recently, a naturally occurring furin-directed serpin has been identified in *Drosophila* that may provide improved selectivity, although this has not yet been tested.[64, 65] In summary, whereas currently available inhibitors can be used to determine whether the protease responsible for processing a given substrate belongs to the PC family, careful titration of dosage is required to determine which family member is involved.

Another option for knocking down PC function in *Xenopus* embryos involves the use of antisense morpholino oligonucleotides. Morpholinos bind to the 5′ untranslated region of endogenous mRNAs and prevent the ribosomal initiation complex from binding to the cognate mRNA, thus blocking translation through steric interference. Morpholinos can also be used to block splicing of pre-mRNAs, although this option requires knowledge of genomic sequence, which is currently limited in *Xenopus*. One advantage to the use of antisense morpholinos in *Xenopus* is the ability to target specific cell lineages by selectively injecting only a subset of blastomeres of the cleaving embryo. This reduces phenotypic complexity and enables one to focus on specific developmental events in which TGF-β family members are known to participate. A disadvantage to the use of morpholinos is that they do not deplete maternally stored proteins. *Xenopus* furin, PACE4, PC6, and PC7 are all maternally expressed and are relatively stable proteins, which may preclude analysis of early embryonic functions following morpholino knockdown.

2. Host Transfer Technique for Studying Maternal PC Function

The host transfer technique provides a means to study early embryonic functions of maternally expressed genes and can be used to overcome problems associated with persistence of stored proteins. This technique involves depleting *Xenopus* oocytes of specific maternal RNAs, reimplanting the oocytes into host females, and fertilizing the spawned eggs in order to study the effect this depletion has on development of the embryo.[66] Briefly, oocytes are manually defolliculated and injected with phosphorothioate modified antisense oligodeoxynucleotides, which lead to rapid degradation of the targeted RNA, or with antisense morpholinos, which block translation. Oocytes are then cultured for up to 3 days to allow for turnover of maternal stores of the targeted protein before they are induced to undergo maturation. Antisense injected and control oocytes are stained with different vital dyes to distinguish them from the host eggs and transferred into the coelomic cavity of a female frog that has been hormonally stimulated to lay eggs, and the spawned eggs are then fertilized *in vitro*. A detailed protocol for this technique can be found at http://www.xenbase.org/methods/anti-methods/ and Reference 67.

Recently, the host transfer technique has been used to study the role of PACE4 in early *Xenopus* development.[46] Based on previous studies showing that mouse nodal is a substrate of mammalian PACE4,[17] this group focused initially on studying the potential involvement of PACE4 in cleavage of *Xenopus* nodal-related (Xnr) proteins. PACE4-depleted embryos were analyzed for loss of Xnr activity and target gene expression. Embryos were also injected with RNA encoding epitope-tagged versions of various candidate substrates, followed by Western blot analysis to determine whether proprotein processing was affected.[46] These studies showed that PACE4 is required for mesoderm induction, consistent with the known role for Xnrs in this process.[68, 69] Surprisingly, biochemical analysis demonstrated that PACE4 is essential for cleavage of only a subset of Xnrs, despite the presence of an optimal furin consensus motif in all family members. This provides some of the best evidence that individual PCs show substrate selectivity *in vivo* that is not determined solely by the amino acid sequence surrounding the cleavage site or by tissue-restricted co-expression of substrate and protease.

Depletion of maternal PCs in *Xenopus* presents many of the same challenges as mouse knock outs including the potential for early lethality, complex phenotypes, or lack of phenotype due to redundancy. In addition, the host transfer technique is fairly labor intensive, and given the current lack of antibodies that recognize *Xenopus* PCs, it may not be possible to determine whether the absence of a phenotype is due to redundancy as opposed to insufficient depletion of maternal protein. This procedure has significant advantages as well, however, such as the ability to test for redundancy by targeting multiple PCs in a single set of embryos by simply injecting several antisense oligonucleotides. In addition, it is relatively easy to analyze cleavage of candidate substrates in PC-depleted *Xenopus* embryos by ectopically expressing epitope-tagged substrates, and in many cases these can be targeted to the specific cell lineages where they are normally expressed.

3. Biochemical Analysis of Proprotein Processing in Oocytes

Xenopus oocytes provide a sensitive system with which to study cleavage of TGF-β precursor proteins by endogenous PCs. Oocytes are quiescent and tolerate much higher levels of ectopic TGF-β proteins than do developing embryos. This enables one to detect even unstable cleavage products without risking lethality. One can thus deplete endogenous PCs from oocytes using antisense oligonucleotides or morpholinos, as described above, followed by biochemical analysis of substrate cleavage without going through the host transfer procedure. One caveat to using oocytes for this purpose is that they represent only a single cell type at one developmental stage, whereas it is possible that a given substrate is cleaved by more than one PC along the secretory pathway or by different PCs in a tissue-specific manner.

The ability to culture oocytes for several days makes it possible to do many of the cell biological manipulations that are routinely employed in cultured cells. For example, it is possible to use pulse chase analysis to study the rate and order of processing, in cases where more than one site is cleaved, or the effect of cleavage on ligand stability. It is also possible to chemically modulate subcellular trafficking, or to block lysosomal or proteosomal function in oocytes in order to study the effect of processing on endocytic recycling or degradation of the ligand. A protocol for pulse chase analysis of *in vivo* cleavage in *Xenopus* oocytes is provided at the end of this chapter in Section VII.B.

The oocyte system has recently been used to study the cell biology of proBMP4 processing. As described earlier, proBMP4 is sequentially cleaved at two sites within the prodomain.[11] Functional analysis in *Xenopus* embryos had shown that failure to cleave at the upstream site severely decreased the activity and signaling range of mature BMP4,[8] but the mechanism by which this occurred was unknown. Biochemical analysis in *Xenopus* oocytes revealed that the pro- and mature domains of BMP4 remain noncovalently associated following the first cleavage, generating a complex that is targeted for rapid lysosomal degradation unless the upstream site is cleaved. Cleavage at the upstream site liberates mature BMP4 from the prodomain, thereby stabilizing the protein. These studies also showed that cleavage at the second, but not the first, site is enhanced at reduced pH, consistent with the possibility that the two cleavages occur in distinct subcellular compartments, and they identified a critical histidine residue that may function as a pH sensor.[12]

An important advantage to using *Xenopus* oocytes to study the cell biology of TGF-β precursor processing is that, in many cases, it appears to more closely recapitulate processing in embryos than does analysis in cultured cells. We have found, for example, that mutations that disrupt cleavage at the upstream site of proBMP4 lead to rapid degradation of the mature ligand in both *Xenopus* embryos and oocytes but have no effect on ligand half-life in at least five mammalian cell lines that have been tested (unpublished data). Furthermore, deletion mutant forms of proBMP4 that fail to fold properly, and thus do not traffick out of the ER, are not cleaved in *Xenopus* oocytes or embryos but are efficiently cleaved in cultured cells. These differences may be due to the very high levels of expression achieved in transiently transfected cultured cells or to the absence of molecular chaperones or other critical cofactors that are expressed, and regulate precursor trafficking or processing, in embryonic cells but not in stably transformed cell types.

VI. SUMMARY

In this chapter, we have provided an overview of how proprotein maturation affects growth factor activity and have described several *in vitro* and *in vivo* approaches to studying this process. Different model systems each have their own unique advantages or disadvantages, and as with all questions in science, analysis of proprotein maturation is best tackled using a combination of complimentary methods. In the following section, we provide optimized protocols for two methods that are not routinely used in the development community.

VII. PROTOCOLS

A. *In Vitro* Cleavage of Radiolabeled Substrates

1. Anesthetize a mature female frog by submerging in 0.02% tricaine (ethyl 3-aminobenzoate methanesulfonate salt, Sigma A5040) for approximately 20 min until unconscious.
2. Place the frog belly side up on bench paper with the plastic side up and loosely cover the frog with paper towels soaked in tricaine to prevent the skin from drying. Make an ~1 cm incision through the muscle and skin layer on the side of the abdomen, avoiding the blood vessel along the ventral midline, and remove a small piece of ovarian tissue using forceps and scissors. After examining under the dissecting microscope to ensure that the oocytes are healthy, remove enough oocytes for the experiment. Suture the incision (the muscle wall and the skin must be sutured separately) and allow the frog to recover in water at room temperature.
3. Using a sharp pair of forceps, tear the ovarian tissue into small pieces and transfer to a 50 ml polyproylene tube containing 20 ml OR-2 (82.5 mM NaCl, 2 mM KCl, 1 mM $MgCl_2$, 5 mM Hepes pH 7.5). Add Liberase Blendzyme 3 (Roche 1814176) to a final concentration of 0.28 U/ml (0.07

mg/ml) and incubate at 25°C for 1–3 h with gentle shaking to remove follicle cells (see Notes).

4. Wash the oocytes in two to three changes of OR-2, transfer to a petri dish (see Notes), and remove immature and dying oocytes.
5. Incubate overnight at 18°C in ND-96 (96 mM NaCl, 2 mM KCl, 1.8 mM $CaCl_2$, 1 mM $MgCl_2$, 5 mM Hepes pH 7.5) containing 1.5% horse serum, 2.5 mM sodium pyruvate, and 50 μg/ml gentamycin (ND-96+). The next day, select healthy, stage VI oocytes for injection.[72]
6. Orient oocytes with the pigmented animal pole up. Aiming the needle near the equator, inject oocytes with *in vitro* synthesized capped RNAs encoding the substrate (0.5–50 ng) and α1-PDX (5 ng, to inhibit endogenous PC activity), along with 700 nCi of ^{35}[S]Met/Cys (Easy Tag Express Protein labeling mix, Perkin Elmer, NEG 772, see Notes) in a volume of 10–50 nl.
7. Transfer injected oocytes into 12-well dishes (~10–15 oocytes per well) and incubate in ND-96+ at 18°C with gentle shaking. The incubation time necessary to achieve maximal protein expression can vary from 20–72 h and must be determined empirically for each substrate.
8. Harvest oocytes (6–8 oocytes per sample, see Notes) by resuspending in 30 μl of homogenization buffer (0.25 M sucrose, 50 mM Tris-Cl, pH 7.5, 5 mM potassium acetate, 5 mM magnesium acetate, 1 mM DTT (fresh) and protease inhibitors (Roche, 11836153001). Repeatedly force the oocytes in and out of the pipette tip until they are completely resuspended. Add 100 μl SDS solution (1% SDS, 0.1 M Tris-HCl, pH 8.0) and incubate for 30 min at 37°C.
9. Add 500 μL of TxSWB (1% Trition × 100, 0.1 M NaCl, 0.1 M Tris-HCl, pH 8.0, 10 mM EDTA) and incubate on ice for 45 min.
10. Microfuge for 10 min at 12,000 rpm at 4°C and transfer the supernatant to a new tube (see Notes).
11. Add the appropriate antibody (see Notes) to the supernatant at a concentration of 1–10 μg/ml (determined empirically) and incubate for 8–12 h at 4°C, with gentle shaking.
12. Add Protein A/ProteinG beads (20 μl of a 1:1, bead:buffer, suspension) and continue incubation for another 2–4 h to pull down the antigen-antibody complex. One can also add the beads along with the antibody and let the incubation go overnight if nonspecific binding is not a problem.
13. Wash the beads three times with TxSWB containing 1 mM PMSF and then twice with the same buffer lacking detergent.
14. Resuspend the beads in a known volume of TxSWB lacking detergent, mix well, and count a small aliquot (e.g., 5 μL out of 500 μl, see Notes).
15. After determining the amount of radioactivity incorporated, take the appropriate volume (at least 3000 counts per time point, see Notes), spin the beads, and remove the buffer. Resuspend the bead slurry containing the bound substrate in 100 μl of cleavage buffer (100 mM Hepes, pH 7.5 containing 0.5% Triton X-100 and 1 mM $CaCl_2$).

16. Add recombinant PC (2–5 nM) to start the reaction (see Notes). Immediately remove a 20 μl aliquot to a tube containing denaturing buffer (0.5% SDS, 1% β-mercaptoethanol) and heat at 90°C for 10 min to stop the reaction. This is the 0 min time point. Incubate the remaining reaction at 25–30°C and remove aliquots at various time intervals (minutes to hours, determined empirically), stopping the reaction as described above. Depending on the amount of radioactive protein obtained, one can do more time points by increasing the reaction volume accordingly.
17. Resolve the cleavage products by polyacrylamide gel electrophoresis and visualize by autoradiography.

Notes

1. Enzymatic defolliculation of oocytes can compromise oocyte viability and impair translational efficiency.[72] We have not found this to be a problem for the current application, particularly if partially purified preparations of collagenase such as Blendzyme (Roche) are used and incubation time is kept to a minimum. If necessary, oocytes can be defolliculated manually or injected with follicle cells intact.
2. Defolliculated oocytes should be cultured in glass or tissue culture treated polystyrene dishes or multiwell plates to prevent sticking.
3. ^{35}[S]Met/Cys is dried under vacuum, resuspended in DEPC treated water at a concentration of 110 μCi/μl, and stored at –80°C.
4. Most TGF-β family precursor proteins are retained intracellularly until they are cleaved, and thus radiolabeled substrate, consisting predominantly of unfolded monomers as discussed earlier, is recovered from oocyte lysates. Some proproteins (e.g., nodal) are secreted in precursor form once folding is complete, and it may be possible to selectively purify dimerized precursor protein from the culture media in these cases.
5. Recombinant substrates can also be generated by translating RNA *in vitro* in the presence of ^{35}[S]Met/Cys using commercially prepared rabbit reticulocyte lysates. In this case, the translation reactions are diluted to 500 μl with TxSWB supplemented with protease inhibitors and substrate is immunoprecipitated following steps 11–14 of the protocol.
6. If antibodies specific for the substrate of interest are not available, one can introduce epitope tags, for which commercial antibodies are available, into expression constructs. Tagged proteins should be assayed *in vivo* to ensure that the epitope does not interfere with proper folding, cleavage, or activity of the mature protein. In general, epitope tags inserted into the prodomain are better tolerated than those in the mature domain.
7. The counts obtained will vary for different proteins, depending on the number of Cys/Met residues in the molecule. A range of 30,000–200,000 cpm/reaction is recommended. Inability to detect counts could be due to inefficient protein synthesis (possibly due to degradation of the RNA used for injection or poor quality oocytes) or inefficient immunoprecipitation.

8. When comparing the cleavage of wild-type and mutant proteins, or when comparing cleavage by multiple PCs, it is important to take equal numbers of counts for each cleavage reaction.
9. The pH of the buffer used for *in vitro* cleavage reactions may vary depending on the pH optima of the enzyme of interest. Most PCs are active between a pH range of 6–8, with a calcium requirement of 0.2–2 mM, but PC1 has a narrow acidic pH optima (pH 5.5–6) and a higher calcium requirement (4 mM).
10. Recombinant furin is commercially available (R&D Systems, New England Biolabs) but other PCs must be generated *de novo*. High-level expression of PCs has been achieved using standard techniques such as transient or stable transfection of cultured cells,[63] or infection with recombinant vaccinia virus,[53] or baculovirus.[70, 71] For *in vitro* cleavage reactions, PCs can be concentrated by ultrafiltration from the media of expressing cells, and no further purification is required. When comparing cleavage of a substrate by different PCs, the concentration of each PC can be determined as described in Reference 11.

B. Pulse Chase Analysis of Cleavage in *Xenopus* Oocytes

1. Isolate oocytes as described in protocol 1.
2. Inject oocytes with 0.5–8 ng *in vitro* synthesized capped RNA encoding the substrate of interest along with 700 nCi of [^{35}S]Met/Cys (see Notes).
3. Transfer injected oocytes into a 12- or 24-well dish (~10–15 oocytes per well, one well per time point) and culture in a minimal volume (~500 μl) of ND-96+ for 3 h.
4. After 3 h, add Met/Cys to a final concentration of 5 mM to initiate the chase.
5. Harvest the oocytes and collect the conditioned media at increasing time intervals (see Notes).
6. Immunoprecipitate protein from the clarified lysates and from the culture media, as described in protocol 1 (see Notes).
7. Resolve the cleavage products by polyacrylamide gel electrophoresis under reducing or nonreducing conditions and visualize by autoradiography (see Notes).

Notes

1. For pulse chase analysis, inject the minimal amount of RNA necessary to visualize the precursor at the earliest time point. This enhances the ability to see the appearance of cleavage products, and corresponding disappearance of precursor, over a reasonable period of time.
2. The length of time required for complete cleavage must be determined empirically for each substrate. In general, time points separated by 4–8 h over several days are appropriate.

3. Whereas most TGF-β precursor proteins are retained in the oocyte, cleaved prodomain and mature domain fragments accumulate predominantly or exclusively in the culture media, and thus it is important to analyze both fractions.
4. It may be necessary to deglycosylate samples with PNGaseF prior to electrophoresis in order to visualize or quantitate cleavage products.

ACKNOWLEDGMENTS

We thank Catherine Degnin, Francois Jean, Gary Thomas, and Bill Skach for assistance in developing the methodology used for analysis of cleavage, and Devorah Goldman and Sylvia Nelsen for helpful comments on the manuscript. The authors were supported by grants from the NIH (RO1HD37976 and RO1HD42598).

REFERENCES

1. Taylor, N. A., Van De Ven, W. J., Creemers, J. W., Curbing activation: proprotein convertases in homeostasis and pathology. *Faseb*, 17, 1215–27, 2003.
2. Shi, Y., Massague, J., Mechanisms of TGF-beta signaling from cell membrane to the nucleus. *Cell*, 113, 685–700, 2003.
3. Hogan, B. L., Bone morphogenetic proteins: multifunctional regulators of vertebrate development. *Genes Dev*, 10, 1580–94, 1996.
4. Hemmati-Brivanlou, A., Thomsen, G. H., Ventral mesodermal patterning in Xenopus embryos: expression patterns and activities of BMP-2 and BMP-4. *Dev Gene*, 17, 78–89, 1995.
5. Clement, J. H., Fettes, P., Knochel, S., Lef, J., Knochel, W., Bone morphogenetic protein 2 in the early development of Xenopus laevis. *Mech Dev*, 52, 357–70, 1995.
6. Faure, S., Lee, M. A., Keller, T., ten Dijke, P., Whitman, M., Endogenous patterns of TGFbeta superfamily signaling during early Xenopus development. *Development*, 127, 2917–31, 2000.
7. Kurata, T., Nakabayashi, J., Yamamoto, T., Mochii, M., Ueno, N., Visualization of endogenous BMP signaling during Xenopus development. *Differentiation*, 67, 33–30, 2000.
8. Cui, Y., Hackenmiller, R., Berg, L., Jean, F., Nakayama, T., Thomas, G., Christian, J. L., The activity and signaling range of mature BMP-4 is regulated by sequential cleavage at two sites within the prodomain of the precursor. *Genes Dev*, 15, 2797–802, 2001.
9. Nelsen, S., Unpublished data.
10. Nelsen, S., Berg, L., Wong, C., Christian, J. L., Proprotein convertase genes in Xenopus development. *Dev Dyn,* 233, 1038–44, 2005.
11. Cui, Y., Jean, F., Thomas, G., Christian, J. L., BMP-4 is proteolytically activated by furin and/or PC6 during vertebrate embryonic development. *Embo J*, 17, 4735–43, 1998.
12. Degnin, C., Jean, F., Thomas, G., Christian, J. L., Cleavages within the prodomain direct intracellular trafficking and degradation of mature BMP-4. *Mol Biol Cell,* 11, 5012–20, 2004.
13. Goldman, D., Hackenmiller, R., Nakayama, T., Sopory, S., Wong, C., Kulessa, H., and Christian, J. L, Mutation of an upstream cleavage site in the BMP4 prodomain leads to tissue-specific loss of activity. *Development,* 133, 1933–42.

14. Whitman, M., Nodal signaling in early vertebrate embryos: themes and variations. *Dev Cell,* 1, 605–17, 2001.
15. Gray, A. M., Mason, A. J., Requirement for activin A and transforming growth factor-beta 1 pro-regions in homodimer assembly. *Science*, 247, 1328–30, 1990.
16. Le Good, J. A., Joubin, K., Giraldez, A. J., Ben-Haim, N., Beck, S., Chen, Y., Schier, A. F., Constam, D. B., Nodal stability determines signaling range. *Curr Biol*, 15, 31–36, 2005.
17. Beck, S., Le Good, J. A., Guzman, M., Ben Haim, N., Roy, K., Beermann, F., Constam, D. B., Extraembryonic proteases regulate Nodal signalling during gastrulation. *Nat Cell Biol*, 4, 981–85, 2002.
18. Thomas, G., Furin at the cutting edge: from protein traffic to embryogenesis and disease. *Nat Rev Mol Cell Biol*, 3, 753–66, 2002.
19. Anderson, E. D., Molloy, S. S., Jean, F., Fei, H., Shimamura, S., Thomas, G., The ordered and compartment-specific autoproteolytic removal of the furin intramolecular chaperone is required for enzyme activation. *J Biol Chem*, 277, 12879–90, 2002.
20. Lipkind, G. M., Zhou, A., Steiner, D. F., A model for the structure of the P domains in the subtilisin-like prohormone convertases. *Proc Natl Acad Sci U S A*, 95, 7310–15, 1998.
21. Zhou, A., Martin, S., Lipkind, G., LaMendola, J., Steiner, D. F., Regulatory roles of the P domain of the subtilisin-like prohormone convertases. *J Biol Chem*, 273, (18), 11107–14, 1998.
22. Roebroek, A. J., Creemers, J. W., Pauli, I. G., Kurzik-Dumke, U., Rentrop, M., Gateff, E. A., Leunissen, J. A., Van de Ven, W. J., Cloning and functional expression of Dfurin2, a subtilisin-like proprotein processing enzyme of Drosophila melanogaster with multiple repeats of a cysteine motif. *J Biol Chem*, 267, 17208–15, 1992.
23. De Bie, I., Marcinkiewicz, M., Malide, D., Lazure, C., Nakayama, K., Bendayan, M., Seidah, N. G., The isoforms of proprotein convertase PC5 are sorted to different subcellular compartments. *J Cell Biol*, 135, 1261–75, 1996.
24. Xiang, Y., Molloy, S. S., Thomas, L., Thomas, G., The PC6B cytoplasmic domain contains two acidic clusters that direct sorting to distinct trans-Golgi network/endosomal compartments. *Mol Biol Cell*, 11, 1257–73, 2000.
25. Rockwell, N. C., Krysan, D. J., Komiyama, T., Fuller, R. S., Precursor processing by kex2/furin proteases. *Chem Rev*, 102, 4525–48, 2002.
26. Seidah, N. G., Day, R., Hamelin, J., Gaspar, A., Collard, M. W., Chretien, M., Testicular expression of PC4 in the rat: molecular diversity of a novel germ cell-specific Kex2/subtilisin-like proprotein convertase. *Mol Endocrinol*, 6, 1559–70, 1992.
27. Nakayama, K., Kim, W. S., Torii, S., Hosaka, M., Nakagawa, T., Ikemizu, J., Baba, T., Murakami, K., Identification of the fourth member of the mammalian endoprotease family homologous to the yeast Kex2 protease. Its testis-specific expression. *J Biol Chem*, 267, 5897–900, 1992.
28. Denault, J. B., Leduc, R., Furin/PACE/SPC1: a convertase involved in exocytic and endocytic processing of precursor proteins. *FEBS Lett*, 379, 113–16, 1996.
29. Zhou, A., Webb, G., Zhu, X., Steiner, D. F., Proteolytic processing in the secretory pathway. *J Biol Chem*, 274, 20745–48, 1999.
30. Vidricaire, G., Denault, J. B., Leduc, R., Characterization of a secreted form of human furin endoprotease. *Biochem Biophys Res Commun*, 195, 1011–18, 1993.
31. Zhong, M., Benjannet, S., Lazure, C., Munzer, S., Seidah, N. G., Functional analysis of human PACE4-A and PACE4-C isoforms: identification of a new PACE4-CS isoform. *FEBS Lett*, 396, 31–36, 1996.

32. Tsuji, A., Sakurai, K., Kiyokage, E., Yamazaki, T., Koide, S., Toida, K., Ishimura, K., Matsuda, Y., Secretory proprotein convertases PACE4 and PC6A are heparin-binding proteins which are localized in the extracellular matrix. Potential role of PACE4 in the activation of proproteins in the extracellular matrix. *Biochim Biophys Acta*, 1645, 95–104, 2003.
33. Leitlein, J., Aulwurm, S., Waltereit, R., Naumann, U., Wagenknecht, B., Garten, W., Weller, M., Platten, M., Processing of immunosuppressive pro-TGF-beta 1,2 by human glioblastoma cells involves cytoplasmic and secreted furin-like proteases. *J Immunol*, 166, 7238–43, 2001.
34. Constam, D. B., Robertson, E. J., Regulation of bone morphogenetic protein activity by pro domains and proprotein convertases. *J Cell Biol*, 144, 139–49, 1999.
35. Molloy, S. S., Bresnahan, P. A., Leppla, S. H., Klimpel, K. R., Thomas, G., Human furin is a calcium-dependent serine endoprotease that recognizes the sequence Arg-X-X-Arg and efficiently cleaves anthrax toxin protective antigen. *J Biol Chem*, 267, 16396–402, 1992.
36. Walker, J. A., Molloy, S. S., Thomas, G., Sakaguchi, T., Yoshida, T., Chambers, T. M., Kawaoka, Y., Sequence specificity of furin, a proprotein-processing endoprotease, for the hemagglutinin of a virulent avian influenza virus. *J Virol*, 68, 1213–18, 1994.
37. Krysan, D. J., Rockwell, N. C., Fuller, R. S., Quantitative characterization of furin specificity. Energetics of substrate discrimination using an internally consistent set of hexapeptidyl methylcoumarinamides. *J Biol Chem*, 274, 23229–34, 1999.
38. Henrich, S., Cameron, A., Bourenkov, G. P., Kiefersauer, R., Huber, R., Lindberg, I., Bode, W., Than, M. E., The crystal structure of the proprotein processing proteinase furin explains its stringent specificity. *Nat Struct Biol*, 10, 520–26, 2003.
39. Henrich, S., Lindberg, I., Bode, W., Than, M. E., Proprotein convertase models based on the crystal structures of furin and kexin: explanation of their specificity. *J Mol Biol*, 345, 211–27, 2005.
40. van de Loo, J. W., Creemers, J. W., Bright, N. A., Young, B. D., Roebroek, A. J., Van de Ven, W. J., Biosynthesis, distinct post-translational modifications, and functional characterization of lymphoma proprotein convertase. *J Biol Chem*, 272, 27116–23, 1997.
41. Constam, D. B., Calfon, M., Robertson, E. J., SPC4, SPC6, and the novel protease SPC7 are coexpressed with bone morphogenetic proteins at distinct sites during embryogenesis. *J Cell Biol*, 134, 181–91, 1996.
42. Roebroek, A. J., Umans, L., Pauli, I. G., Robertson, E. J., van Leuven, F., Van de Ven, W. J., Constam, D. B., Failure of ventral closure and axial rotation in embryos lacking the proprotein convertase Furin. *Development*, 125, 4863–76, 1998.
43. Zheng, M., Streck, R. D., Scott, R. E., Seidah, N. G., Pintar, J. E., The developmental expression in rat of proteases furin, PC1, PC2, and carboxypeptidase E: implications for early maturation of proteolytic processing capacity. *J Neurosci*, 14, 4656–73, 1994.
44. Tsuji, A., Mori, K., Hine, C., Tamai, Y., Nagamune, H., Matsuda, Y., The tissue distribution of mRNAs for the PACE4 isoforms, kexin-like processing protease: PACE4C and PACE4D mRNAs are major transcripts among PACE4 isoforms. *Biochem Biophys Res Commun*, 202, 1215–21, 1994.
45. Mori, K., Kii, S., Tsuji, A., Nagahama, M., Imamaki, A., Hayashi, K., Akamatsu, T., Nagamune, H., Matsuda, Y., A novel human PACE4 isoform, PACE4E is an active processing protease containing a hydrophobic cluster at the carboxy terminus. *J Biochem (Tokyo)*, 121, 941–48, 1997.

46. Birsoy, B., Berg, L., Williams, P. H., Smith, J. C., Wylie, C. C., Christian, J. L., Heasman, J., XPACE4 is a localized pro-protein convertase required for mesoderm induction and the cleavage of specific TGF-β proteins in Xenopus development. *Development*, 132, 591–602, 2005.
47. Nakagawa, T., Murakami, K., Nakayama, K., Identification of an isoform with an extremely large Cys-rich region of PC6, a Kex2-like processing endoprotease. *FEBS Lett*, 327, 165–71, 1993.
48. Lusson, J., Vieau, D., Hamelin, J., Day, R., Chretien, M., Seidah, N. G., cDNA structure of the mouse and rat subtilisin/kexin-like PC5: a candidate proprotein convertase expressed in endocrine and nonendocrine cells. *Proc Natl Acad Sci U S A*, 90, 6691–95, 1993.
49. Seidah, N. G., Chretien, M., Day, R., The family of subtilisin/kexin like pro-protein and pro-hormone convertases: divergent or shared functions. *Biochimie*, 76, 197–209, 1994.
50. Jean, F., Boudreault, A., Basak, A., Seidah, N. G., Lazure, C., Fluorescent peptidyl substrates as an aid in studying the substrate specificity of human prohormone convertase PC1 and human furin and designing a potent irreversible inhibitor. *J Biol Chem*, 270, 19225–31, 1995.
51. Yaron, A., Carmel, A., Katchalski-Katzir, E., Intramolecularly quenched fluorogenic substrates for hydrolytic enzymes. *Anal Biochem*, 95, 228–35, 1979.
52. Mains, R. E., Berard, C. A., Denault, J. B., Zhou, A., Johnson, R. C., Leduc, R., PACE4: a subtilisin-like endoprotease with unique properties. *Biochem J*, 321, 587–93, 1997.
53. Jean, F., Stella, K., Thomas, L., Liu, G., Xiang, Y., Reason, A. J., Thomas, G., Alpha1-Antitrypsin Portland, a bioengineered serpin highly selective for furin: application as an antipathogenic agent. *Proc Natl Acad Sci U S A*, 95, 7293–98, 1998.
54. Basak, A., Zhong, M., Munzer, J. S., Chretien, M., Seidah, N. G., Implication of the proprotein convertases furin, PC5 and PC7 in the cleavage of surface glycoproteins of Hong Kong, Ebola and respiratory syncytial viruses: a comparative analysis with fluorogenic peptides. *Biochem J*, 353, 537–45, 2001.
55. Nakayama, K., Furin: a mammalian subtilisin/Kex2p-like endoprotease involved in processing of a wide variety of precursor proteins. *Biochem. J.*, 327, 625–35, 1997.
56. Logeat, F., Bessia, C., Brou, C., LeBail, O., Jarriault, S., Seidah, N. G., Israel, A., The Notch1 receptor is cleaved constitutively by a furin-like convertase. *Proc Natl Acad Sci U S A*, 95, 8108–12, 1998.
57. Stawowy, P., Kallisch, H., Veinot, J. P., Kilimnik, A., Prichett, W., Goetze, S., Seidah, N. G., Chretien, M., Fleck, E., Graf, K., Endoproteolytic activation of alpha(v) integrin by proprotein convertase PC5 is required for vascular smooth muscle cell adhesion to vitronectin and integrin-dependent signaling. *Circulation*, 109, 770–76, 2004.
58. Siegfried, G., Basak, A., Cromlish, J. A., Benjannet, S., Marcinkiewicz, J., Chretien, M., Seidah, N. G., Khatib, A. M., The secretory proprotein convertases furin, PC5, and PC7 activate VEGF-C to induce tumorigenesis. *J Clin Invest*, 111, 1723–32, 2003.
59. Constam, D. B., Robertson, E. J., Tissue-specific requirements for the proprotein convertase furin/SPC1 during embryonic turning and heart looping. *Development*, 127, 245–54, 2000.
60. Roebroek, A. J., Taylor, N. A., Louagie, E., Pauli, I., Smeijers, L., Snellinx, A., Lauwers, A., Van de Ven, W. J., Hartmann, D., Creemers, J. W., Limited redundancy of the proprotein convertase furin in mouse liver. *J Biol Chem*, 279, 53442–50, 2004.

61. Nejjari, M., Berthet, V., Rigot, V., Laforest, S., Jacquier, M. F., Seidah, N. G., Remy, L., Bruyneel, E., Scoazec, J. Y., Marvaldi, J., Luis, J., Inhibition of proprotein convertases enhances cell migration and metastases development of human colon carcinoma cells in a rat model. *Am J Pathol*, 164, 1925–33, 2004.
62. Anderson, E. D., Thomas, L., Hayflick, J. S., Thomas, G., Inhibition of HIV-1 gp160-dependent membrane fusion by a furin-directed alpha 1-antitrypsin variant. *J Biol Chem*, 268, 24887–91, 1993.
63. Tsuji, A., Ikoma, T., Hashimoto, E., Matsuda, Y., Development of selectivity of alpha1-antitrypsin variant by mutagenesis in its reactive site loop against proprotein convertase. A crucial role of the P4 arginine in PACE4 inhibition. *Protein Eng*, 15, 123–30, 2002.
64. Osterwalder, T., Kuhnen, A., Leiserson, W. M., Kim, Y. S., Keshishian, H., Drosophila serpin 4 functions as a neuroserpin-like inhibitor of subtilisin-like proprotein convertases. *J Neurosci*, 24, 5482–91, 2004.
65. Richer, M. J., Keays, C. A., Waterhouse, J., Minhas, J., Hashimoto, C., Jean, F., The Spn4 gene of Drosophila encodes a potent furin-directed secretory pathway serpin. *Proc Natl Acad Sci U S A*, 101, 10560–65, 2004.
66. Heasman, J., Holwill, S., Wylie, C. C., Fertilization of cultured Xenopus oocytes and use in studies of maternally inherited molecules. *Methods Cell Biol*, 36, 213–30, 1991.
67. Zuck, M., Wylie, C., Heasman, J., *Maternal mRNAs in* Xenopus *embryos: an antisense approach*. Oxford University Press: Oxford, UK, 1998, pp. 341–54.
68. Jones, C. M., Kuehn, M. R., Hogan, B. L. M., Smith, J. C., Wright, C. V. E., Nodal-related signals induce axial mesoderm and dorsalize mesoderm during gastrulation. *Development*, 121, 3651–62, 1995.
69. Osada, S. I., Wright, C. V., Xenopus nodal-related signaling is essential for mesendodermal patterning during early embryogenesis. *Development*, 126, 3229–40, 1999.
70. Komiyama, T., Fuller, R. S., Engineered eglin c variants inhibit yeast and human proprotein processing proteases, Kex2 and furin. *Biochemistry*, 39, 15156–65, 2000.
71. Cain, B. M., Vishnuvardhan, D., Wang, W., Foulon, T., Cadel, S., Cohen, P., Beinfeld, M. C., Production, purification, and characterization of recombinant prohormone convertase 5 from baculovirus-infected insect cells. *Protein Expr Purif*, 24, 227–33, 2002.
72. Smith, L. D., Xu, W. Varnold, R. L. Oogenesis and oocyte isolation, *Methods Cell Biol* 36, 213–30, 1991.

4 Analysis of MAP Kinase Pathways in Vertebrate Development

Amy K. Sater and Heithem M. El-Hodiri

CONTENTS

I. INTRODUCTION

MAP kinase (MAPK) cascades have been implicated in a wide range of cellular and developmental regulatory processes, including cell proliferation, differentiation, apoptosis, stress response, cytokine response, and planar cell polarity. The canonical version of the cascade is that a MAP kinase kinase kinase (MAP3K) phosphorylates and activates a dual-specificity MAP kinase kinase (MAP2K), which in turn phosphorylates and activates a MAPK family member (1). Although the Extracellular signal Regulated Kinase (Erk) MAPK pathway is the best known, Jun-N-terminal Kinase (JNK), p38 kinase, and, most recently, Nemo-Like Kinase (NLK) pathways have also received considerable attention. Generally, MAPK family members have a wide range of substrates, including transcription factors, signaling pathway components, other kinases, and cytoskeletal regulators. In many cases, the efficiency, substrate specificity, or intracellular localization are regulated by the formation of ternary complexes, consisting of the MAP3K, MAP2K, and MAPK, bound to a scaffold protein (2, 3). The scaffold protein brings the kinase components into proximity to generate a MAP kinase module, which may increase the efficiency of signaling while reducing the signal amplification that would result from a conventional series of bimolecular interactions. Scaffolds may also direct the subcellular localization of the kinase module, thus restricting the range of available targets.

Historically, Erk MAPKs have been thought of as primary effectors of Receptor Tyrosine Kinase (RTK) signaling, whereas JNKs and p38s are associated with inflammation and response to cellular stress. It is now clear, however, that each of the major MAPK pathways can mediate responses to a wide range of stimuli. MAPKs can be activated by G Protein Coupled Receptors (GPCRs), Transforming Growth Factor-β (TGF-β) receptors, Ephrin/Eph interactions, integrins, noncanonical wnt signals, and elevated intracellular calcium. Thus, these evolutionarily ancient kinase modules have been incorporated into many regulatory circuits.

In embryos, MAPKs act at the interface between specific developmental decisions, often elicited by growth factors or other external signals, and more fundamental cellular activities such as mitosis, motility, or programmed cell death (Table 4.1). Through their role in the planar cell polarity system, JNKs may contribute to the coordination of cell movement during gastrulation. Both JNKs and p38Ks have been implicated in apoptosis in multiple embryonic tissues. Conversely, Erk MAPKs may be associated with the protection from apoptosis conferred by Fibroblast Growth Factor (FGF) in other embryonic cell types. The p38Ks play poorly understood roles in the early embryonic patterning of several vertebrate and invertebrate embryos; in some cases, they are required for the establishment of early asymmetries. Erk MAPKs are required for the specification of mesoderm and neural ectoderm, as well as for subsequent patterning around the midbrain-hindbrain boundary, among many examples. Given that erk MAPKs and p38Ks are major effectors of FGF and Bone Morphogenetic Protein-4 (BMP4) signals, respectively, it is likely that future studies will reveal specific roles for these MAPK subfamilies in mediating many of the cell fate decisions elicited by these growth factors.

Although MAP kinase pathways were discovered in part via genetic studies in yeast, *Drosophila*, and *C. elegans*, they have largely eluded genetic analysis in

Table 4.1
Comparison of MAPK Subfamilies

	Erk MAPK	JNK	p38	NLK
Key activators/ stimuli in embryos	FGFs, other growth factors acting through RTKs, integrins	Noncanonical wnt signals (via Rac), Ephrins	BMP4	Noncanonical wnt signals
MAP3Ks	Raf1, A-Raf, B-Raf, mos	MEKK1,2 Mixed Lineage Kinase (MLK)[141]	MEKK2,3, TAK1	TAK1
MAP2Ks	MEK1, 2	MKK4,7	MKK3,7	HIPK2
MAPKs	erk MAPK1, 2	JNK 1,2	p38α, p38β, p38γ, p38δ	NLK
Key substrates	Ets TFs, Elk1, RSK, MKP1, ATF2, Irx2[142]	Jun/AP1, Elk1	ATF2, RSKB	TCF4, Lef1, Stat3, Myb
Negative regulators	MKP3, MKP4 (MKP1)	MKP1,2,5,7	MKP1,2,5,7	Unknown
Pharmacological inhibitors	PD098059, U0126	SP600125	SB203580, SB202190, BIRB796	Unknown

Note: Only major examples are listed in some categories. See text for additional references. PGFs, Peptide Growth Factors; RTKs, Receptor Tyrosine Kinases; TFs, transcription factors; RSK, Ribosomal S6 Kinase.

vertebrates for several reasons. First, each of the three major types of MAPKs exists as multiple isoforms in vertebrates, and there is considerable redundancy between isoforms in each group. Second, MAPKs carry out multiple functions, including those essential for cell survival as well as those mediating specific developmental decisions. Third, conventional approaches to functional analysis in embryos have been relatively uninformative. For example, the major types of MAPKs are expressed throughout most if not all cell types, and activity is regulated by phosphorylation. Thus, *in situ* hybridization patterns reflecting expression of specific MAPKs or other components of the central kinase cascades provide little information regarding function or activity. In contrast, some of the noncatalytic modulators show more specific patterns of expression, which may suggest distinct functions for the associated pathways. Finally, the same MAPK may be regulated by multiple signals simultaneously; for example, the basal level of erk MAPK activity resulting from sustained integrin signaling may be unrelated to the elevated level triggered by stage-specific FGF signals.

In part for these reasons, our understanding of MAPK regulation in a wide range of mammalian cultured cell types has outstripped our understanding of MAPK functions in vertebrate embryos. Studies in mammalian cultured cells have revealed highly complex and possibly cell-type-specific regulatory circuitries for the primary types of MAPKs. A major challenge for developmental biologists investigating MAPK signaling is to identify which regulatory interactions identified in cell culture systems actually correspond to functionally significant interactions in developmental processes.

The goal of this review is to introduce the landscape of MAP kinase regulation and approaches for experimental investigation, rather than to provide a comprehensive analysis of the major pathways. In the first section, each of the major MAP kinase subfamilies and their associated cascades will be described, as well as several examples illustrating their roles in specific developmental processes. The second section discusses scaffolds and modulatory interactions, using examples relating to the erk MAPKs. The third section provides strategies for the analysis and manipulation of pathway activity and the identification of candidate substrate proteins.

II. ERK MAP KINASES

Although best known for their roles in the mitogenic response, erk MAPKs are essential regulators for multiple developmental processes in vertebrate embryos. Members of this subfamily are activated by dual phosphorylation at threonine and tyrosine within the TEY motif located within the activation or "T" loop. Mammals have two isoforms of erk MAPK, at 42 and 44 kD, whereas a single isoform is present in *Xenopus*. This subfamily also includes MAPK5 (BMK, 4). Erk MAPKs are activated by the MAP2Ks MAP or ERK Kinase (MEK)1 and MEK2, which in turn can be activated by the MAP3Ks Raf1, A-Raf, B-Raf, or mos. The basic mechanisms underlying this pathway have been extensively reviewed elsewhere (5, 6).

Activation of erk MAPK is required during *Xenopus* oocyte maturation, and the oocyte has provided an outstanding model system for the quantitative analysis of flux through the signaling cascade (7, 8). During oocyte maturation, erk MAPK activity increases rapidly and dramatically via zero-order kinetics, essentially switching from an off state to an on state. In other cell types, however, low basal levels of erk MAPK activity undergo moderate increases in response to mitogenic or other signals, and the level, duration, and intracellular localization of the elevated MAPK activity are all subject to regulation (9).

Most commonly, the erk MAPK cascade is activated in response to activation of ras, which leads to recruitment and activation of Raf1. Whereas ras-dependent activation of the erk MAPK pathway is a well-understood response to FGFs and other peptide growth factors that bind receptor tyrosine kinases, TGF-β signaling can also lead to erk MAPK activation via Ras (see Reference 10 for review). Basal levels of erk MAPK activity are maintained through integrin signaling, again via Ras (11). Activation of the pathway can also be sustained via activation of protein kinase C, which can promote activation of Raf. Activation of erk MAPKs by protein kinase C (PKC) isoforms has not yet been explicitly demonstrated in vertebrate embryonic tissues. However, numerous examples can be found among mammalian cell lines, including human intestinal epithelial cells (12) and colon cancer cells (13). In the former, PKCα promotes Ras-dependent erk MAPK activation, whereas activation of PKCδ can lead to erk MAPK activation independent of Ras (14). PKC-delta has also been implicated in the p38 MAPK-dependent antagonism of erk MAPKs during avian chondrogenesis, as discussed below. Finally, calmodulin-dependent kinases (CaMs) have been shown to mediate erk MAPK activation following depolarization of neuroblastoma cells; this interaction also participates in neurite outgrowth (15).

The erk MAPK cascade is a primary effector pathway for FGF signaling, and erk MAPK has been frequently implicated in FGF-dependent developmental processes. However, FGF signals are also mediated by PI-3 kinase, phospholipaseC-gamma (PLCγ), and other Ras-dependent pathways, so the involvement of FGFs does not necessarily implicate erk MAPK (16, 17). Erk MAPK activation is required for the specification of mesoderm in *Xenopus* (18–22), where it participates in a positive feedback loop with FGF and the transcription factor brachyury (Xbra), expressed as an immediate-early response to FGF (23, 24). Neural specification also requires erk MAPK (25), which downregulates BMP4 signals via inhibitory phosphorylation of the BMP4 effector Smad1 (26, 27). FGF8 is a major signal produced by the isthmic organizer (28), and erk MAPK is critical for cerebellar development and cell survival (29). In addition, recent studies suggest that erk MAPK is needed for the development of the retina and lens (30), the sclerotome (31), and cardiovascular system (32). Finally, cyclic activation of erk MAPK within the presomitic mesoderm is required for segmentation (33). Moreover, in numerous instances, erk MAPK activity is required for cell survival during embryonic development (e.g., Reference 34), as well as in differentiated cell types (e.g., Reference 35).

III. JUN-N-TERMINAL KINASES

JNKs were initially associated with stress response and inflammation, but are now known to have specific functions in embryonic development. Members of this family carry a TPY motif in the activation loop and are activated by MKKs 4 and 7. Three JNK genes have been identified in mammalian genomes, and *Drosophila* and *C. elegans* have a single JNK. The role of JNKs in the planar cell polarity pathway is discussed in Chapter 2 by Kühl and Moon, and in Chapter 8 by Slusarski; it will not be considered further here. However, JNKs have also been implicated in other developmental processes in vertebrates, most notably in the regulation of apoptosis in the developing nervous system (36, 37) and other embryonic tissues (38).

JNKs, together with p38 MAPKs, are required for cavity formation in the preimplantation mouse embryo, and JNK functions at this stage include both positive and negative regulation of specific target genes (4–10). In *Xenopus*, JNK levels are relatively high in the mature oocyte and remain high throughout cleavage stages (39); because zygotic transcription is almost entirely inactive during this period, the authors suggest that JNKs may serve functions independent of transcriptional regulation. JNKs have also been implicated in inside-out EphrinB1 signaling (40, 41), although the functional significance of JNK signaling in EphrinB1-mediated processes such as skeletal patterning (42) or neural crest cell differentiation (43) has not been elucidated.

IV. P38 KINASES

Although they were originally identified as stress-activated kinases associated with inflammation, members of the p38 kinase subfamily regulate multiple events in embryonic development, in part through their role as BMP4 effector kinases. Mammalian genomes include alpha, beta, gamma, and delta forms of p38 MAPK; alpha (MPK2) and gamma (Xp38 gamma/SAPK3) (44) have been identified in both *Xenopus* and zebrafish. Members of this family have a TGY motif in the activation loop, and p38 MAPK cascades can involve the MAP3Ks MEKKS 2 and 3 or TGF-β-Activated Kinase (TAK1), and the MAP2Ks MKKs 3 and 7. MKK4 can phosphorylate p38 subfamily members *in vitro*, but the significance of MKK4 as a p38 activator *in vivo* is unclear. In *Xenopus* oocytes and *in vitro*, MKK6 activates p38α more efficiently at low levels of activity; at higher levels of pathway stimulation, MKK6 can phosphorylate both p38α and p38γ (45). These findings suggest that alterations in signal intensity may result in the activation of different p38 MAPKs, yielding different ratios of activated p38 MAPKs under low versus high signaling conditions. Key substrates for p38 MAPKs include the transcription factors ATF-2 (46), MEF-2 (47), SAP1-A, and Elk-1 (48), as well as other kinases, such as Ribosomal S6 Kinase-B (RSK-B), Mitogen and Stress-activated Kinase-1 (MSK1), and MAPKAP K2/3 (reviewed in Reference 49).

Recent findings indicate that p38 activity is required during cleavage stages in zebrafish, sea urchin, and mouse embryos. Asymmetric activation of p38 in zebrafish embryos is necessary for proper cleavage, and the region of activation coincides with the area in which *dharma bozozok* is expressed (50). In sea urchin embryos, p38 MAPK is required for the initiation of nodal expression (51), whereas in the preimplantation mouse embryo, the activation of a p38 MAPK is associated with the mobilization of the actin cytoskeleton (52). Additional expression profiling studies suggest that p38 MAPK mediates alterations in gene activity that are required for cavity formation within the blastocyst, as well as other early developmental events (53).

The p38 MAPK pathway has also been implicated in differentiation and remodeling of several embryonic tissues. In chick ciliary ganglion neurons, the p38 pathway restricts trafficking of specific subtypes of K+ channels to the plasma membrane, a key step in terminal neuronal differentiation, via its effects on the actin cytoskeleton (54). BMP5-dependent p38 activity is required for the expression of several target genes in the interdigital region of the chick limb (55). The p38 MAPKs serve as critical regulators of apoptosis in several tissues or cell types, including cardiomyocytes (56), embryonic neurons (57), and others.

The function of p38 MAPK during mammalian chondrogenesis in culture has been investigated more extensively (see Reference 58 for review). The TGF-β family member Growth/Differentiation Factor 5 (GDF5) induces chondrogenesis in mammalian ATDC5 cells, accompanied by activation of p38 MAPKs, and chondrogenesis is inhibited by the p38 MAPK inhibitor SB 202190 (59). A second p38 MAPK inhibitor, SB 203580, partially inhibited chondrogenesis in primary cultures of mesenchymal cells (60), transcriptional activation of the cyclin-dependent kinase inhibitor p21 is required for the initiation of chondrogenesis, and p38 MAPK is required for this step (61). Erk MAPKs and p38 MAPKs act antagonistically in primary cultures (60), in ATDC5 cells (59), and in prechondrogenic cell lines, such as mesenchymal progenitor cells obtained from adult bone (62). Moreover, EGF inhibits chondrogenesis, presumably via inhibition of p38 MAPKs and simultaneous activation of erk MAPKs (63). Interestingly, inhibition of p38 MAPKα or p38 MAPKβ in primary cultures of chick limb mesenchyme promotes myogenesis and myofibrillar differentiation (64).

V. THE TAK1/NEMO-LIKE KINASE PATHWAY

Nemo and the corresponding Nemo-Like Kinase orthologues are divergent members of the MAPK superfamily, characterized by an activation loop motif of TQY in *Drosophila* and vertebrate orthologues and TCY in LIT-1, the *C. elegans* orthologue. Additional characteristics of NLKs include poly-His and poly-Ala tracts in the N-terminal region and a primarily nuclear localization (65). The *Drosophila nemo* gene undergoes differential splicing, resulting in eight distinct isoforms (66). Vertebrate genomes have a single NLK gene, with a single primary isoform (67).

Genetic analyses indicate that NLKs play diverse roles in development of *Drosophila* and *C. elegans*. In *Xenopus,* NLK is both maternally and zygotically expressed throughout the embryo (68). Although relatively few specific substrates for NLK phosphorylation have been identified, NLKs have been shown to carry out several developmentally significant regulatory events. First, NLKs phosphorylate TCF4 and LEF1, reducing their affinity for DNA (69) and targeting them for nuclear export (70). This relationship was first demonstrated using genetic analyses in *C. elegans,* where the phosphorylation of the TCF orthologue POP-1 by LIT-1 leads to nuclear export of POP-1, resulting in a disruption of lineage specification (71, 72). Subsequent studies have verified this interaction in mammalian cells (73, 74). In each case, activation of NLK is dependent upon the activation of TAK1 by noncanonical wnt signals, resulting in inhibitory crosstalk from the noncanonical wnts to the canonical wnt pathway.

A. TGF-β-Activated Kinase-1

TAK1 was first identified as a MAP kinase kinase kinase (MAP 3K7) that served as an effector of TGF-β signals (75). It has subsequently been shown to transduce signals from the BMP-4 receptor complex, with which it interacts via TAB1 (76). It has been shown to activate the MAPKK family members MKK3 and MKK7, leading to the activation of JNKs (77), NLK (78), or p38 kinases (79), rather than erk MAPKs.

Although the relationship between TAK1 and NLK has been well established, the intermediate protein kinase has eluded identification until recently. TAK1 signals to NLK via Homeodomain-Interacting Protein Kinase-2 (HIPK2) in hematopoietic cells (78), and HIPK2 and NLK have been shown to mediate the phosphorylation of c-Myb (80) and several other transcription factors (81), either inactivating them or targeting them for ubiquitin-mediated proteolysis. Other direct activators of NLK have not been identified. In *Xenopus*, NLK has been implicated in neural specification (68), and the TAK1-NLK-STAT3 pathway contributes to the induction of mesoderm by nodal signals in *Xenopus* embryos (82). Moreover, recent work has shown that activation of NLK by TAK1 in response to Interleukin-1 (Il-1) involves the formation of a ternary complex mediated by STAT3, which is also a substrate of NLK (83).

VI. SCAFFOLDING

Scaffolding of MAP kinase pathway components via the formation of ternary complexes is thought to provide specificity and modulation for MAP kinase cascades (see Reference 84 for review). Although the assembly and function of scaffolding complexes have been extensively studied in yeast and mammalian cells in culture, the functional significance of scaffolding in developmental systems has not been extensively investigated. A common characteristic of scaffolding proteins is that expression at levels above the optimum can lead to a "titrating-out" of proteins that must interact via the scaffold complex, through the accumulation of partial complexes that include the scaffolding protein plus either of the two binding partners (85). Thus, genetic analyses of such proteins often show that gain-of-function phenotypes resemble phenotypes resulting from loss of function. Scaffold proteins have been identified for the JNK cascade (86, 87), and STAT3 serves as a scaffold for the TAK1/NLK pathway, as noted earlier (83). Two particularly important examples of scaffolding proteins are Kinase Suppressor of Ras (KSR) (88) and β-arrestin (89).

Kinase Suppressor of Ras was originally identified in *Drosophila* and *C. elegans*. KSR binds MEK, and Ras activation leads to interactions with Raf and Erk MAPK. Formation of this complex is thought to facilitate the phosphorylation of MEK by Raf (see Reference 88 for review). Two KSR genes have been identified in vertebrate genomes, and although KSR2 Expressed Sequence Tags (ESTs) have been detected from *Xenopus* oocytes and gastrulae, it is unclear whether KSRs are expressed in early embryos of mouse or zebrafish. KSR is negatively regulated by phosphorylation on Ser392 and Thr274, which restrict the duration of erk MAPK activation (90). Because one of these sites is a consensus MAPK phosphorylation sequence, phosphorylation at this site may represent a mechanism of negative feedback. KSR is also negatively regulated by a Ras-responsive E3 ubiquitin ligase known as Impedes Mitogenic Signal Propagation (IMP), which binds KSR and restricts interactions between Raf and MEK (91). Ras activation leads to inhibition of IMP via autoubiquitination. IMP is thought to regulate the level of MAPK response to a given mitogenic signal; it does not restrict the duration of MAPK activation.

Arrestins bind to phospho-amino acids in the C-terminal cytoplasmic domain of seven membrane spanning receptors, recruiting Assembly Particle-2 (AP-2) to initiate clathrin-dependent endocytosis (see Reference 89 for review). Originally identified as mediators of β-adrenergic receptor desensitization, arrestins regulate endocytosis of many types of G protein-coupled receptors, including rhodopsins (92) and Frizzled-4 (93), as well as a TGF-β receptor III (TGFBRIII) (94). More recently, they have been shown to act as a scaffold for all three types of MAP kinase cascades, although their role as a scaffold for the erk MAP kinase cascade has been examined most extensively. Endocytosis of the receptor/arrestin/MAPK complex moves the active MAPK scaffold complex to the interior of the cytoplasm. Four arrestins have been identified in vertebrate genomes; of these, two are expressed solely in retinal photoreceptors, while the remaining two, referred to as β-arrestins 1 and 2, are widely expressed throughout differentiated tissues. National Center for Biotechnology Information (NCBI) UniGene data indicate that β-arrestin 2 is also expressed at embryonic stages in *Xenopus*, zebrafish, and mice. β-arrestin 2-dependent erk MAPK scaffold complexes are retained in the cytoplasm, which directs erk MAPK activity toward cytoplasmic substrates and reduces the level of erk MAPK-inducible transcription. The net result is prolonged cytoplasmic activation of erk MAPK (95). It has yet to be determined whether β-arrestin 2 plays a significant role in the modulation of MAP kinase activity in embryos.

Space limitations preclude any significant discussion of other scaffold proteins that direct JNK or p38 activity. Scaffold proteins associated with these pathways have been identified, however: JNK-Interacting Protein (JIP-1) (86) facilitates JNK signals, and two other isoforms, JIP-2, and JIP-4, mediate activation of p38 MAPKs (84). In addition, the JNK scaffold Plenty of SH3s (POSH) (87) has recently been implicated in amphibian anterior neural development and associated apoptosis (96).

VII. NEGATIVE REGULATION

A. MAP Kinase Phosphatases

These dual specificity phosphatases (DUSPs) are among the chief negative regulators of MAPKs (reviewed in Reference 97). Mammalian genomes contain 10 MKPs, and 6 have been identified in *Drosophila*. There are three classes of MKPs, characterized by differences in gene structure. Members of all classes have two CDC25 Homology (CH2) domains in the N-terminal domain; this region also includes a docking site incorporating several basic amino acids, which largely dictates the substrate preference of the phosphatase. The catalytic site is located near the C-terminus. MKPs in the first group, which includes MKP1 (DUSP1) and MKP2 (DUSP4), as well as DUSP2 and DUSP5, are localized primarily in the nucleus; MKPs1 and 2, along with DUSP2, have the broadest substrate specificity, acting preferentially on JNKs and p38s, but also able to dephosphorylate erk MAPKs. The second group includes MKP3, also known as PYST1 (DUSP6) and PYST2 (DUSP7), which are both cytosolic phosphatases. Additional members such as MKP4 (DUSP9) and MKP5 (DUSP10), are found in both the nucleus and cytoplasm. MKP3, MKP4, and PYST2, as well as DUSP5, are specific for erk MAPKs. MKP5 and the third group, the cytosolic MKP7 and the ubiquitous DUSP8, specifically dephosphorylate JNKs and p38s.

Many of the MKPs so far investigated are under either transcriptional or post-translational control by MAP kinase family members. The best-studied MKP, MKP1, is phosphorylated by erk MAPK, which inhibits proteolysis of the phosphatase, prolonging activity. It is also transcriptionally activated by both erk MAPK and p38 signals; regulation of MKP transcription is a key mechanism for inhibitory cross-regulation between MAPK pathways (98, 99).

The functional significance *in vivo* for most specific MKPs has not been evaluated. Transgenic mice carrying a null mutation in MKP1 have no discernible phenotype, presumably because of extensive redundancy. MKP3, however, shows a distinct pattern of expression in vertebrate embryos (100); it is activated in the embryonic mouse brain in response to FGF8 signals originating in the isthmic organizer (101), as well in the neural plate (102) and the limb bud (103). These and other findings suggest that activation of MKP3 expression functions as a negative modulator of erk MAP kinase following activation of the latter by FGF signals to control the level or spatial extent of erk MAPK activation.

B. Protein Phosphatases

Phosphoprotein Phosphatase (PP) 1 and PP2A have both been shown to dephosphorylate MAPKs and their upstream regulators (e.g., MEK1). These protein phosphatases act on MAP kinases localized in the cytoplasm without inhibiting nuclear MAP kinases. PP2A is particularly important in the regulation of Erk MAPK; it has also been implicated in the downregulation of JNK activity in monocytes (104) and p38 activity in aortic endothelial cells (105) and platelets (106). Specific roles for PP1 and PP2A in the regulation of the non-Erk MAPKs have not been extensively studied across a range of cell types, so it is unclear whether these phosphatases are significant modulators of activity in general.

C. Noncatalytic Modulators: Sprouty, Spred, and Sef

Several noncatalytic inhibitory modulators of the erk MAPK pathway have been identified in recent years, as part of negative feedback loops that restrict intracellular signaling in response to FGF (100). Sprouty, Sprouty-Related EVH Domain (Spred), and Sef are transcriptionally activated by FGF effector pathways, similar to MKP3, as discussed earlier. The best-known, Sprouty, was first identified in *Drosophila* via a mutation that expanded the spatial extent of FGF signaling during tracheal development (107). The four mammalian orthologues have received considerable attention from biochemical and genetic studies (reviewed in Reference 108). Sprouty family members can interact directly with several components of the RTK pathway, including the FGFR adapter FRS2, Grb2, Raf1, B-Raf, and Shp2. The best-studied family members, Sprouty1 and Sprouty2, have been shown to antagonize Receptor Tyrosine Kinase signaling at multiple points, such as binding to the Grb2/SOS complex (109), inhibition of Raf1 activation by RAS (110), both of which attenuate erk MAPK signaling, and inhibiting release of intracellular calcium (111), presumably via inhibition of PLC-γ activation.

Spry1 and Spry2 undergo complex post-translational modification, including lipid modifications altering intracellular localization (112), tyrosine phosphorylation at multiple sites (113), inhibitory nitrosylation (114), and ubiquitylation resulting from interaction with the E3 ubiquitin ligase c-CBL (113). Interaction with c-CBL is one of several events controlled by phosphorylation and is critical for the enhancements of EGF Receptor (EGFR) stability by Spry (115). Interestingly, *Xenopus* Spry2 diverges from the mammalian Spry2, in that xSpry2 does not disrupt erk

MAPK activity but, rather, inhibits calcium signaling (111) and PKCδ (116). Specific mechanisms underlying Spry function have not yet been thoroughly evaluated in other nonmammalian vertebrate systems.

Expression patterns and loss-of-function analyses suggest that Sprys are most often associated with modulation of FGF signaling, although Sprys affect PDGF and EGF signals, as well. In embryos, *Sprys* are expressed in areas of high FGF expression, such as mesoderm in *Xenopus* gastrulae (111, 116), the zebrafish midbrain-hindbrain boundary (117), and the developing kidney (118, 119). *Sprouty2* null mice exhibit defects in ear development (120) and abnormalities in the enteric nervous system (121). The restricted phenotypic abnormalities obtained in these studies suggest considerable functional redundancy among *Sprouty* family members.

The related but more recently identified Spred is also associated with inhibition of erk MAPK signals (122). Spred family members share a conserved domain with Sprouty in addition to the EVH domain associated with the actin cytoskeletal regulator Ena/VASP. Two family members have been identified in mammalian (123) and *Xenopus* (116) genomes. Both Spreds have been shown to inhibit hematopoiesis (123, 124). *Spreds* are expressed in the embryonic brain, heart, epidermis, and lung (125), and *Spred* and *Sprouty* family members can be expressed in distinct regions within the same tissue, e.g., the embryonic lung (125). Sprouty and Spred functions, however, are nonoverlapping in *Xenopus* mesoderm: Xspry2 inhibits calcium-dependent signals required for cell movement during gastrulation, whereas Spred inhibits erk MAPK activity required for mesodermal commitment (116). As the authors suggest, this provides a means to direct FGF signaling along specific avenues of development.

Sef represents the third group of modulators, and it has the most specific effects on erk MAPK signaling. Well conserved throughout vertebrates, Sef is a transmembrane protein that blocks tyrosine phosphorylation of FGFR1 and FRS2 (126, 127). It also interacts directly with active MEK, inhibiting the release of phosphorylated erk MAPK and preventing the accumulation of activated erk MAPK in the nucleus (128). Sef is localized primarily in the Golgi; upon mitogenic stimulation, some fraction of the pool relocalizes to the plasma membrane, and the activated MEK/erk MAPK complex is associated with both the plasma membrane and the Golgi Sef pools (128). When *Sef* is expressed, mitogenic stimulation leads to phosphorylation of known cytoplasmic substrates of erk MAPK, while phosphorylation of the nuclear substrate Elk-1 is reduced/limited. Rather than inhibiting MEK or erk MAPK activity, Sef is thus thought to direct activated erk MAPK toward cytoplasmic targets. Sef has additional functions as well: It activates JNK via a TAK1/MKK4 pathway in PC12 cells, leading to apoptosis (129). *Sef* null mice are viable (117).

VIII. MANIPULATION OF ACTIVITY STATES

A. Gene-Based Approaches

Experimental manipulation of MAPK pathways can be carried out via multiple strategies, and careful consideration of the entire pathway and comparisons of embryonic expression patterns for key pathway components should identify the most effective and feasible approach. Many MAPK pathway components are expressed in oocytes, and the presence of maternal stores of mRNA or protein may complicate the use of forward or reverse genetic approaches. For example, in some cases, mutations must be maternally as well as zygotically expressed in zebrafish embryos to block gene function in early development or during gastrulation. The use of antisense morpholino oligonucleotides (MO) may be problematic if substantial amounts of the protein of interest are maternally expressed, as is the case for many components of MAPK cascades. For *Xenopus*, host-transfer techniques that allow introduction of the antisense oligonucleotide into immature oocytes may resolve this problem. Redundancy between different isoforms is a more difficult issue; such situations may call for pharmacological inhibitors or the use of dominant negative mutations in specific pathway components. The critical step for any genetic or pharmacological strategy is to monitor the phosphorylation state of the target MAPK itself, as well as the function under investigation.

B. Pharmacological Inhibitors

Numerous pharmacological inhibitors have been identified that act on specific pathway components; however, few have been tested in embryos or embryonic explants. The best known of these are PD 098059 and U0126, which inhibit phosphorylation of MEK1 and MEK2, and thus can be used to inhibit the erk MAPK pathway. Although both have been effective in *Xenopus* embryos, it has been suggested that U0126 may have higher affinity to the *Xenopus* MEK1 (J. Maller, personal communication). Neither compound has any effect on phosphorylated MEK1, however, so embryonic tissues must be pretreated with either compound before the introduction of growth factors or other activators of the erk MAPK pathway.

Both p38α and p38β can be inhibited by SB compounds, such as SB 203580. In mammalian cultured cells, this compound inhibits p38α and p38β at relatively low concentrations (e.g., 1 μM); however, p38γ is unaffected. At 10-fold higher concentrations, it inhibits JNKs as well as p38α and p38β. The effectiveness of this compound requires the presence of a threonine at a specific position within the ATP-binding site, and in JNKs and p38γ, a methionine replaces the threonine at this site (130). At 10 μM, SB 203580 has also been shown to inhibit the activity of TGF-βRI both *in vitro* and *in vivo* (131). Thus, this compound should be used with caution when investigating TGF-β R pathways; a control experiment monitoring levels of the appropriate phospho-Smad should establish whether TGF-β receptor activity is affected. A recently identified compound, BIRB796, has been shown to inhibit all p38 isoforms, although p38α and p38β are more sensitive than p38γ (132). Selective JNK inhibitors include the anthrapyrazolone strila inhibitor SP 600125 (133) and a peptide based on the JNK-binding domain of JIP-1, a JNK scaffold (134).

PD 098059, U0126, and SB 203580 can be used reliably on embryonic explants; however, their effectiveness in whole embryos can be variable, depending upon the degree of penetration. In some instances, they can be introduced via microinjection.

IX. ASSESSMENT OF STATES OF ACTIVATION

A. Western Blots with Phospho-Specific Antibodies

The simplest way to determine the level of activity for a specific MAPK is via immunoblots probed with antibodies directed against the phospho-specific form of the kinase. Duplicate blots are probed with antibodies against the kinase. As long as the detection signal is kept within the linear range, the ratio of the phospho-specific antibody signal to the total kinase-antibody signal can be used as an indicator of relative activity. This use of phospho-specific antibodies is discussed elsewhere in this volume (see Chapter 7 by Whitman).

B. Whole-Mount Immunohistochemistry with Phospho-Specific Antibodies

Phospho-specific antibodies against erk MAPK have often been used in whole-mount immunohistochemistry (see Chapter 7). The monoclonal anti-diphospho MAPK initially generated by Seger and colleagues (135) and commercially available from Sigma-Aldrich has been particularly useful in studies of vertebrate embryos, as well as in *Drosophila*. The chief caveat regarding such studies is that in whole-mounts, the antibody will detect only relatively high levels of activity, and it is more effective in detecting diphospho MAPK in the nucleus than in the cytoplasm. For example, in whole-mount immunohistochemistry of *Xenopus* gastrulae, diphospho-MAPK is detectable in the dorsal and lateral marginal zones; no diphospho-MAPK is visible in the dorsal ectoderm (McClaskey and Sater, unpublished observations). Both MAPK activity assays and immunoblots probed with anti-diphospho-MAPK antibodies, however, indicate that a substantial proportion of erk MAPK is phosphorylated in the dorsal ectoderm.

X. ACTIVITY ASSAYS

Although immunoblotting using phospho-specific antibodies is usually the easiest way to evaluate activity states of MAPK family members, direct measurements of activity are preferable under some conditions. For example, a suitable phospho-specific antibody may not be available with sufficient cross-reactivity for the species under investigation, or there may be multiple subfamily members mediating the phosphorylation of a specific substrate, or the expected change in activity may be relatively small, so that comparisons of activity, rather than phosphorylation, may provide a more accurate view.

Studies of specific MAPKs in mammalian cultured cells often use immune complex kinase assays to measure activity. Here, the specific MAPK is immunoprecipitated and incubated in the presence of γ-^{32}P ATP and a substrate peptide, usually derived from myelin basic protein or EGF Receptor. These assays require large amounts (400–600 μg protein) of homogeneous starting material, hardly a problem for *in vitro* systems, but often prohibitive for embryos. They also depend upon an antibody suitable for immunoprecipitation, which can be problematic for nonmammalian embryos, particularly when investigating kinases other than erk MAPKs.

We have developed an assay for erk MAPK activity in *Xenopus* embryonic tissues that can be performed with very small quantities of material (25, 136). It assesses MAPK activity in tissue lysates via ^{32}P-phosphorylation of a substrate peptide derived from *Xenopus* nuclear factor7 (Xnf7) (137, 138). The entire kinase reaction is then separated on a tris/tricine gel (139), which allows resolution of both erk MAPK and the 2.8 kD peptide. The gel is then cut; the lower part with the phosphorylated peptide is then autoradiographed. The upper part is used for immunoblotting, using an antibody directed against erk MAPK, and visualized by chemiluminescent detection. When both the autoradiographic and chemiluminescent exposures are kept within the linear range, they can be quantified by scanning densitometry or other image analysis methods. Relative specific activity is reflected by the ratio of the phosphorylated substrate peptide to the total erk MAPK. With appropriate controls, this method could be adapted to other MAPK family members, using substrate peptides shown to be highly specific for individual MAPKs.

XI. ERK MAP KINASE ASSAY

A. Substrate Peptide

Our substrate peptide was based on the erk MAPK phosphorylation site of Xnf7; serines located outside the phosphorylation site have been changed to alanines. The sequence is as follows:

AKKAACAPV**T**PVEKKARPLEKC. The consensus phosphorylation motif is underlined, and the phosphorylated threonine is shown in bold. Previous studies have shown that either immunodepletion of erk MAPK from mature oocyte extracts (137) or treatment of ectoderm cells with PD 098059 before lysis (Uzgare and Sater, unpublished results) eliminates 90% of the phosphorylation at this site (for additional discussion, see Section XII of this chapter). The advantage of this substrate over other commercially available substrates, such as peptides based on the EGFR or Myelin Basic Protein, is that it is large enough to resolve on a high-percentage acrylamide gel, as described herein. A similar approach could be used to develop an assay for other MAPK family members. For instance, a peptide based on the NLK phosphorylation sites of LEF-1 might be used as a substrate for NLK activity assays. This sequence (150 - AVHPL**T**PLITYSDEHF**S**PGSHP) includes both phosphorylation sites (69), as well as four other potential phosphorylation sites (dotted underline). For an optimal substrate peptide, these sites would be converted to alanines and phenylalanine, to minimize the likelihood of phosphorylation by other kinases.

B. Kinase Reaction

1. Use approximately 100 µg of tissue. We have found that this is equivalent to approximately 15 animal caps, or ~12 isolates of midgastrula (st 11) neural ectoderm. As a preliminary step, protein content of tissues should be measured using a Bradford assay or other standard protein assay.

2. Lyse tissues on ice in 20 µl Kinase Buffer (KB: 20 mM Hepes, pH 7.5; 40 mM MgCl2; 20 mM EGTA; 1 mM DTT; 80 mM β-glycerophosphate; 50 mM NaF; 1 mM sodium orthovanadate; plus either a commercially available protease inhibitor cocktail at the concentration recommended by the manufacturer, or the following protease inhibitors: 3 µg/ml leupeptin, 1 µM microcystin, 20 µg/ml aprotinin, 2 mM PMSF). KB is stored as aliquots at –20°C; if PMSF is used, it is added immediately before use.
3. Centrifuge at 14,000 rpm for 20 min. at 4°C.
4. Prepare the following reaction mix on ice:

   5 µl supernatant
   10.5 µl KB
   1 µl Xnf7 substrate peptide (2 µg/µl)
   1 µl 1 mM ATP
   2.5 µCi γ-^{32}P – ATP (30 Ci/mmol) (1 µCi/µl)

   ---
   20 µl total volume

   Note: Duplicate reactions should be run for each sample.
5. Incubate at 24°C for 10 min.
6. Stop reaction by adding 20 µl 2X SDS sample buffer (100 mM tris, pH 6.8, 5% SDS, 4% β-mercaptoethanol, 24% glycerol, 0.12% Bromophenol Blue) and boil 4 min. Samples can now be stored at –20°C or –80°C, or loaded directly onto tris/tricine gels.

C. Tris/Tricine Gel Electrophoresis

This gel system is a modification of that developed by Shagger and von Jagow (Reference 139). It uses a 4.5%/10%/17% SDS-polyacrylamide gel, tris/tricine buffer system, and a significantly longer run time to separate polypeptides across a wide molecular weight range, allowing resolution of peptides as low as 2 kD.

1. Resolving gel: 17% acrylamide (30 ml total volume):
   10 ml 3X gel buffer (GB: 3 M tris HCl, pH 8.45; 0.3% SDS)
   16.8 ml 30% acrylamide (standard 30: 0.8 acrylamide:bis-acrylamide)
   0.1 g bis-acrylamide (final bis-acrylamide concentration equals that obtained by using a 30:1.2 ratio of acrylamide to bis-acrylamide)
   3.2 ml glycerol
   150 µl 10% ammonium persulfate (APS)
   15 µl tetraethylmethylenediamine (TEMED)
   Overlay with distilled water; allow 20 min. for polymerization.
2. Spacer gel: 10% acrylamide (10 ml total volume):
   3.3 ml 3X gel buffer
   3.3 ml 30% acrylamide
   3.3 ml distilled water
   100 µl 10% APS

10 μl TEMED
Overlay with isobutanol; allow 15 min. for polymerization.

3. Stacker gel: 4.5% acrylamide (12.5 ml total volume):
   3.1 ml 3X gel buffer
   1.6 ml 30% acrylamide
   7.8 ml distilled water
   100 μl 10% APS
   10 μl TEMED
   Allow 20 min. for polymerization.
4. Cathode buffer (for upper buffer reservoir):
   0.1 M tris
   0.1 M tricine
   0.1% SDS
   The resulting solution should be at pH 8.25 without adjustment. Store at room temperature and use within 1 month. This buffer cannot be used for multiple runs.
5. Anode buffer (for lower buffer reservoir):
   0.2 M tris HCl, pH 8.9
   This solution can be prepared as 5X stock (i.e., 1 M tris). Do not use for multiple runs.

Running conditions: Gels should be run at constant voltage in two phases. Run at 50 V overnight, then increase the voltage to 120 V for 9 hours. The use of prestained molecular weight markers will make it easier to carry out subsequent steps.

Cut gel to separate upper and lower halves; cutting between 20 and 30 kD provides large enough pieces. Dry the lower half by preferred methods and autoradiograph. Transfer the upper half to nitrocellulose or Hybond and probe with antibodies directed against MAPK.

XII. ASSESSMENT OF POTENTIAL TARGET PROTEINS

A third arena involves determining whether a protein of interest is a direct target of MAPKs. Members of the MAPK family are proline-directed kinases; for example, erk MAPK recognizes a minimal sequence of S/T-P and an optimal sequence of P-X-S/T-P. However, proline-directed kinases abound, and potential phosphorylation sites are not necessarily accessible for phosphorylation. In some instances, binding at a secondary docking site referred to as the common docking domain (CD) is essential for phosphorylation (see Reference 140 for review). Thus, demonstrating that a given phosphorylation site is a target for a specific kinase under physiological conditions requires several steps.

First, it is necessary to determine whether the protein of interest shows a difference in phosphorylation under conditions in which the MAPK family member is activated (or inactivated). As long as there are not a large number of phosphorylation sites, there are (at least) two ways to do this. First, one can look for a shift in electrophoretic mobility. In our experience, this is considerably easier to see on a 2-D Western blot than an SDS-PAGE blot, because the addition of one or more phosphates has a greater impact on charge than on mass. We have found that 100 μg of protein can easily be detected on 2-D gels, and it may be possible to use less on Western blots. For instance, tissues can be prepared under conditions in which kinase activity is increased or inhibited as described earlier, lysed in kinase buffer, separated by 2-D electrophoresis using standard methods, and blotted. If an antibody against the protein of interest is available, then one can evaluate the phosphorylation state of the endogenous protein.

As an alternative, one can express an epitope-tagged protein and probe the blot using an antibody directed against the epitope. Phosphorylation can be detected as an acidic shift in the isoelectric point of the protein; often, multiple closely spaced spots will appear as a stutter along a horizontal line, reflecting multiple phosphorylation states of a single protein. To confirm that the stutter is due to phosphorylation, one should run a second identical set of samples in which endogenous phosphatases are not inhibited; we have found that omission of orthovanadate from the kinase buffer is sufficient, although one can also add phosphatases (e.g., calf intestinal phosphatase) to the lysate. The presence of active phosphatases should lead to a collapse of the stutter line into a single spot at the most basic position. By comparing 2-D blot patterns from samples prepared under conditions in which a MAPK is highly active or inactive, one can establish whether activation of the pathway leads to a change in the phosphorylation state of the protein. This may be a simpler approach if the amount of starting material is limiting or immunoprecipitation is not feasible for other reasons.

If sufficient starting material is available, a second approach is to add γ-^{32}P-ATP to tissue lysates to determine whether the protein can be phosphorylated by endogenous kinases; the protein can then be immunoprecipitated, and phosphorylated protein can be visualized by autoradiography.

If these preliminary studies suggest that the protein is phosphorylated by a MAPK, two additional experiments are particularly useful in extending these findings and in evaluating the functional significance of the putative phosphorylation. First, one can immunodeplete the lysate of the specific MAPK(s) by carrying out serial immunoprecipitations and retaining the supernatant, as described in Reference 137. One can then add a kinase substrate (a synthetic peptide, or a bacterially expressed or *in vitro*-translated protein) and determine whether immunodepletion of the kinase prevents phosphorylation of the protein. In the case of Xnf7, immunodepletion of erk MAPK from mature oocyte extracts inhibits approximately 90% of the phosphorylation; the remainder is apparently due to cdc2p34 (137). Second, one can investigate the effects of phosphorylation on protein function or localization by disrupting the phosphorylation sites via site-directed mutagenesis. As is well-known, mutations of serines or threonines to alanines (or tyrosines to phenylalanines) mimic the nonphosphorylated state, whereas mutations of Ser or Thr residues to aspartates or glutamates mimic phosphorylated residues. These experiments should clarify the relative importance of potential MAPK phosphorylation in the regulation of a putative substrate.

XIII. SUMMARY

Recent and ongoing studies have greatly expanded our understanding of the multifaceted roles played by MAPK family members in vertebrate development. Moreover, complex regulatory circuitries operating at several levels control the amount, duration, and subcellular localization of MAPK activity. An integration of relatively simple biochemical analyses of kinase activity states with cellular, transcriptional, or developmental outputs will provide a clearer and more detailed picture of the means by which MAPKs contribute to the unfolding of developmental processes.

ACKNOWLEDGMENTS

The authors wish to thank J. Akif Uzman and Malcolm Whitman for comments on the manuscript, and members of the staff of the National Center for Biotechnology Information for the development and annotation of UniGene sets for MAP kinase family members. Studies in the lab of A.K.S. were supported by the National Science Foundation.

REFERENCES

1. Pearson G, Robinson F, Beers Gibson T, Xu BE, Karandikar M, Berman K, Cobb MH. (2001). Mitogen-activated protein (MAP) kinase pathways: regulation and physiological functions. Endocr Rev. 22(2):153–83.
2. Kolch W, Calder M, Gilbert D. (2005). When kinases meet mathematics: the systems biology of MAPK signalling. FEBS Lett. 579(8):1891–5.
3. Kolch W. (2005). Coordinating ERK/MAPK signalling through scaffolds and inhibitors. Nat Rev Mol Cell Biol. 6(11):827–37.
4. Hii CS, Anson DS, Costabile M, Mukaro V, Dunning K, Ferrante A. (2004). Characterization of the MEK5-ERK5 module in human neutrophils and its relationship to ERK1/ERK2 in the chemotactic response. J Biol Chem. 279(48):49825–34
5. Kyosseva SV. (2004). Mitogen-activated protein kinase signaling. Int Rev Neurobiol. 59:201–20.
6. Roux PP, Blenis J. (2004). ERK and p38 MAPK-activated protein kinases: a family of protein kinases with diverse biological functions. Microbiol Mol Biol Rev. 68(2):320–44.
7. Ferrell JE Jr. (1999). Xenopus oocyte maturation: new lessons from a good egg. Bioessays. 21(10):833–42.
8. Ferrell JE Jr. (2002). Self-perpetuating states in signal transduction: positive feedback, double-negative feedback and bistability. Curr Opin Cell Biol. 14(2):140–8.
9. Pouyssegur J, Volmat V, Lenormand P. (2002). Fidelity and spatio-temporal control in MAP kinase (ERKs) signalling. Biochem Pharmacol. 64(5–6):755–63.
10. Mulder KM. (2000). Role of Ras and Mapks in TGFbeta signaling. Cytokine Growth Factor Rev. 11(1–2):23–35.
11. Howe AK, Aplin AE, Juliano RL. (2002). Anchorage-dependent ERK signaling—mechanisms and consequences. Curr Opin Genet Dev. 12(1):30–5.

12. Clark JA, Black AR, Leontieva OV, Frey MR, Pysz MA, Kunneva L, Woloszynska-Read A, Roy D, Black JD. (2004). Involvement of the ERK signaling cascade in protein kinase C-mediated cell cycle arrest in intestinal epithelial cells. J Biol Chem. 279(10):9233–47.
13. Wang X, Wang Q, Hu W, Evers BM. (2004). Regulation of phorbol ester-mediated TRAF1 induction in human colon cancer cells through a PKC/RAF/ERK/NF-kappaB-dependent pathway. Oncogene. 2004 Mar 11;23(10):1885–95.
14. Paruchuri S, Hallberg B, Juhas M, Larsson C, Sjolander A. (2002). Leukotriene D(4) activates MAPK through a Ras-independent but PKCepsilon-dependent pathway in intestinal epithelial cells. J Cell Sci. 115(Pt 9):1883–93.
15. Schmitt JM, Wayman GA, Nozaki N, Soderling TR. (2004). Calcium activation of ERK mediated by calmodulin kinase I. J Biol Chem. 279(23):24064–72.
16. Carballada R, Yasuo H, Lemaire P. (2001). Phosphatidylinositol-3 kinase acts in parallel to the ERK MAP kinase in the FGF pathway during Xenopus mesoderm induction. Development. 128(1):35–44.
17. Thisse B, Thisse C. (2005). Functions and regulations of fibroblast growth factor signaling during embryonic development. Dev Biol. 287(2):390–402.
18. Hartley RS, Lewellyn AL, Maller JL. (1994). MAP kinase is activated during mesoderm induction in Xenopus laevis. Dev Biol. 163(2):521–4.
19. LaBonne C, Burke B, Whitman M. (1995). Role of MAP kinase in mesoderm induction and axial patterning during Xenopus development. Development. 121(5):1475–86.
20. Gotoh Y, Masuyama N, Suzuki A, Ueno N, Nishida E. (1995). Involvement of the MAP kinase cascade in Xenopus mesoderm induction. EMBO J. 14(11):2491–8.
21. Umbhauer M, Marshall CJ, Mason CS, Old RW, Smith JC. (1995). Mesoderm induction in Xenopus caused by activation of MAP kinase. Nature. 376(6535):58–62.
22. LaBonne C, Whitman M. (1997). Localization of MAP kinase activity in early Xenopus embryos: implications for endogenous FGF signaling. Dev Biol. 183(1):9–20.
23. Isaacs HV, Pownall ME, Slack JM. (1994). eFGF regulates Xbra expression during Xenopus gastrulation. EMBO J. 13(19):4469–81.
24. Diaz J, Martinez-Mekler G. (2005). Interaction of the IP3-Ca2+ and MAPK signaling systems in the Xenopus blastomere: a possible frequency encoding mechanism for the control of the Xbra gene expression. Bull Math Biol. 67(3):433–65.
25. Uzgare AR, Uzman JA, El-Hodiri HM, Sater AK. (1998). Mitogen-activated protein kinase and neural specification in Xenopus. Proc Natl Acad Sci U S A. 95(25):14833–8.
26. Sater AK, El-Hodiri HM, Goswami M, Alexander TB, Al-Sheikh O, Etkin LD, Akif Uzman J. (2003). Evidence for antagonism of BMP-4 signals by MAP kinase during Xenopus axis determination and neural specification. Differentiation. 71(7):434–44.
27. Pera EM, Ikeda A, Eivers E, De Robertis EM. (2003). Integration of IGF, FGF, and anti-BMP signals via Smad1 phosphorylation in neural induction. Genes Dev. 17(24):3023–8.
28. Riou JF, Delarue M, Mendez AP, Boucaut JC. (1998). Role of fibroblast growth factor during early midbrain development in Xenopus. Mech Dev. 78(1–2):3–15.
29. Chi CL, Martinez S, Wurst W, Martin GR. (2003). The isthmic organizer signal FGF8 is required for cell survival in the prospective midbrain and cerebellum. Development. 130(12):2633–44.

30. Gotoh N, Ito M, Yamamoto S, Yoshino I, Song N, Wang Y, Lax I, Schlessinger J, Shibuya M, Lang RA. (2004). Tyrosine phosphorylation sites on FRS2alpha responsible for Shp2 recruitment are critical for induction of lens and retina. Proc Natl Acad Sci U S A. 101:17144–9.
31. Smith TG, Sweetman D, Patterson M, Keyse SM, Munsterberg A. (2005). Feedback interactions between MKP3 and ERK MAP kinase control scleraxis expression and the specification of rib progenitors in the developing chick somite. Development. 132(6):1305–14.
32. Liberatore CM, Yutzey KE. (2004). MAP kinase activation in avian cardiovascular development. Dev Dyn. 230:773–80.
33. Delfini MC, Dubrulle J, Malapert P, Chal J, Pourquie O. (2005). Control of the segmentation process by graded MAPK/ERK activation in the chick embryo. Proc Natl Acad Sci U S A. 102:11343–8.
34. Hood JD, Frausto R, Kiosses WB, Schwartz MA, Cheresh DA. (2003). Differential alphav integrin-mediated Ras-ERK signaling during two pathways of angiogenesis. J Cell Biol. 162:933–43.
35. Zhu Y, Culmsee C, Klumpp S, Krieglstein J. (2004). Neuroprotection by transforming growth factor-beta1 involves activation of nuclear factor-kappaB through phosphatidylinositol-3-OH kinase/Akt and mitogen-activated protein kinase-extracellular-signal regulated kinase1,2 signaling pathways. Neuroscience 123(4):897–906.
36. Schuster N, Dunker N, Krieglstein K. (2002). Transforming growth factor-beta induced cell death in the developing chick retina is mediated via activation of c-jun N-terminal kinase and downregulation of the anti-apoptotic protein Bcl-X(L). Neurosci Lett. 330:239–42.
37. Willaime-Morawek S, Brami-Cherrier K, Mariani J, Caboche J, Brugg B. (2003). C-Jun N-terminal kinases/c-Jun and p38 pathways cooperate in ceramide-induced neuronal apoptosis. Neuroscience. 119:387–97.
38. Lisovsky M, Itoh K, Sokol SY (2002). Frizzled receptors activate a novel JNK-dependent pathway that may lead to apoptosis. Curr Biol. 12:53–8.
39. Bagowski CP, Xiong W, Ferrell JE Jr. (2001). c-Jun N-terminal kinase activation in Xenopus laevis eggs and embryos. A possible non-genomic role for the JNK signaling pathway. J Biol Chem. 276:1459–65.
40. Huynh-Do U, Vindis C, Liu H, Cerretti DP, McGrew JT, Enriquez M, Chen J, Daniel TO. (2002). Ephrin-B1 transduces signals to activate integrin-mediated migration, attachment and angiogenesis. J Cell Sci. 115:3073–81.
41. Xu Z, Lai KO, Zhou HM, Lin SC, Ip NY. (2003). Ephrin-B1 reverse signaling activates JNK through a novel mechanism that is independent of tyrosine phosphorylation. J Biol Chem. 278(27):24767–75.
42. Compagni A, Logan M, Klein R, Adams RH. (2003). Control of skeletal patterning by ephrinB1-EphB interactions. Dev Cell. 5:217–30.
43. Davy A, Aubin J, Soriano P. (2004). Ephrin-B1 forward and reverse signaling are required during mouse development. Genes Dev. 18:572–83.
44. Perdiguero E, Pillaire MJ, Bodart JF, Hennersdorf F, Frodin M, Duesbery NS, Alonso G, Nebreda AR. (2003). Xp38gamma/SAPK3 promotes meiotic G(2)/M transition in Xenopus oocytes and activates Cdc25C. EMBO J. 22(21):5746–56.
45. Alonso G, Ambrosino C, Jones M, Nebreda AR (2000). Differential activation of p38 mitogen-activated protein kinase isoforms depending on signal strength. J Biol Chem. 275(51):40641–8.
46. Morton S, Davis RJ, Cohen P. (2004). Signalling pathways involved in multisite phosphorylation of the transcription factor ATF-2. FEBS Lett. 572(1–3):177–83.

47. Yang SH, Galanis A, Sharrocks AD. (1999). Targeting of p38 mitogen-activated protein kinases to MEF2 transcription factors. Mol Cell Biol. 19(6):4028–38.
48. Yang SH, Whitmarsh AJ, Davis RJ, Sharrocks AD. (1998). Differential targeting of MAP kinases to the ETS-domain transcription factor Elk-1. EMBO J. 17(6): 1740–9.
49. Shi Y, Gaestel M. (2002). In the cellular garden of forking paths: how p38 MAPKs signal for downstream assistance. Biol Chem. 383(10):1519–36.
50. Fujii R, Yamashita S, Hibi M, Hirano T. (2000). Asymmetric p38 activation in zebrafish: its possible role in symmetric and synchronous cleavage. J Cell Biol. 150(6):1335–48.
51. Bradham CA, McClay DR. (2005). p38 MAPK is essential for secondary axis specification and patterning in sea urchin embryos. Development. 133, 21–32.
52. Paliga AJ, Natale DR, Watson AJ. (2005). p38 mitogen-activated protein kinase (MAPK) first regulates filamentous actin at the 8-16-cell stage during preimplantation development. Biol Cell. 97(8):629–40.
53. Maekawa M, Yamamoto T, Tanoue T, Yuasa Y, Chisaka O, Nishida E. (2005). Requirement of the MAP kinase signaling pathways for mouse preimplantation development. Development. 132(8):1773–83.
54. Chae KS, Dryer SE. The p38 mitogen-activated protein kinase pathway negatively regulates Ca2+-activated K+ channel trafficking in developing parasympathetic neurons. J Neurochem. 94(2):367–79.
55. Zuzarte-Luis V, Montero JA, Rodriguez-Leon J, Merino R, Rodriguez-Rey JC, Hurle JM. (2004). A new role for BMP5 during limb development acting through the synergic activation of Smad and MAPK pathways. Dev Biol. 272(1):39–52.
56. Porras A, Zuluaga S, Black E, Valladares A, Alvarez AM, Ambrosino C, Benito M, Nebreda AR. (2004). P38 alpha mitogen-activated protein kinase sensitizes cells to apoptosis induced by different stimuli. Mol Biol Cell. 15(2):922–33.
57. Horstmann S, Kahle PJ, Borasio GD. (1998). Inhibitors of p38 mitogen-activated protein kinase promote neuronal survival in vitro. J Neurosci Res. 52(4):483–90.
58. Stanton LA, Underhill TM, Beier F. (2003). MAP kinases in chondrocyte differentiation. Dev Biol. 263(2):165–75.
59. Nakamura K, Shirai T, Morishita S, Uchida S, Saeki-Miura K, Makishima F. (1999). p38 mitogen-activated protein kinase functionally contributes to chondrogenesis induced by growth/differentiation factor-5 in ATDC5 cells. Exp Cell Res. 250(2):351–63.
60. Oh CD, Chang SH, Yoon YM, Lee SJ, Lee YS, Kang SS, Chun JS. (2000). Opposing role of mitogen-activated protein kinase subtypes, erk-1/2 and p38, in the regulation of chondrogenesis of mesenchymes. J Biol Chem. 275(8):5613–9.
61. Nakajima M, Negishi Y, Tanaka H, Kawashima K. (2004). p21(Cip-1/SDI-1/WAF-1) expression via the mitogen-activated protein kinase signaling pathway in insulin-induced chondrogenic differentiation of ATDC5 cells. Biochem Biophys Res Commun. 320(4):1069–75.
62. Tuli R, Tuli S, Nandi S, Huang X, Manner PA, Hozack WJ, Danielson KG, Hall DJ, Tuan RS. (2003). Transforming growth factor-beta-mediated chondrogenesis of human mesenchymal progenitor cells involves N-cadherin and mitogen-activated protein kinase and Wnt signaling cross-talk. J Biol Chem. 278(42):41227–36.
63. Yoon YM, Oh CD, Kim DY, Lee YS, Park JW, Huh TL, Kang SS, Chun JS. (2000). Epidermal growth factor negatively regulates chondrogenesis of mesenchymal cells by modulating the protein kinase C-alpha, Erk-1, and p38 MAPK signaling pathways J Biol Chem. 275(16):12353–9.

64. Weston AD, Sampaio AV, Ridgeway AG, Underhill TM. (2003). Inhibition of p38 MAPK signaling promotes late stages of myogenesis. J Cell Sci. 116(Pt 14):2885–93.
65. Brott BK, Pinsky BA, Erikson RL. (1998). Nlk is a murine protein kinase related to Erk/MAP kinases and localized in the nucleus. Proc Natl Acad Sci U S A. 95(3):963–8.
66. Drysdale RA, Crosby MA, and The Flybase Consortium. (2005). Flybase: genes and gene models. Nucleic Acids Research. 33:D390–D395. http://flybase.org.
67. Kortenjann M, Wehrle C, Nehls MC, Boehm T. (2001). Only one nemo-like kinase gene homologue in invertebrate and mammalian genomes. Gene. 278(1–2):161–165.
68. Hyodo-Miura J, Urushiyama S, Nagai S, Nishita M, Ueno N, Shibuya H. (2002). Involvement of NLK and Sox11 in neural induction in Xenopus development. Genes Cells. 7(5):487–96.
69. Ishitani T, Ninomiya-Tsuji J, Matsumoto K. (2003). Regulation of lymphoid enhancer factor 1/T-cell factor by mitogen-activated protein kinase-related Nemo-like kinase-dependent phosphorylation in Wnt/beta-catenin signaling. Mol Cell Biol. 23(4):1379–89.
70. Lo M-C, Gay F, Odom R, Shi Y, Lin R. (2004). Phosphorylation by the b-catenin/MAPK complex promotes 14-3-3-mediated nuclear export of TCF/POP-1 in signal-responsive cells in C. elegans. Cell. 117:95–106.
71. Ishitani T, Ninomiya-Tsuji J, Nagai S, Nishita M, Meneghini M, Barker N, Waterman M, Bowerman B, Clevers H, Shibuya H, Matsumoto K. (1999). The TAK1-NLK-MAPK-related pathway antagonizes signalling between beta-catenin and transcription factor TCF. Nature. 399(6738):798–802.
72. Meneghini MD, Ishitani T, Carter JC, Hisamoto N, Ninomiya-Tsuji J, Thorpe CJ, Hamill DR, Matsumoto K, Bowerman B. (1999). MAP kinase and Wnt pathways converge to downregulate an HMG-domain repressor in Caenorhabditis elegans. Nature. 399(6738):793–7.
73. Ishitani T, Kishida S, Hyodo-Miura J, Ueno N, Yasuda J, Waterman M, Shibuya H, Moon RT, Ninomiya-Tsuji J, Matsumoto K. (2003). The TAK1-NLK mitogen-activated protein kinase cascade functions in the Wnt-5a/Ca(2+) pathway to antagonize Wnt/beta-catenin signaling. Mol Cell Biol. 23(1):131–9.
74. Smit L, Baas A, Kuipers J, Korswagen H, van de Wetering M, Clevers H. (2004). Wnt activates the Tak1/Nemo-like kinase pathway. J. Biol. Chem. 279(17): 17232–40.
75. Yamaguchi K, Shirakabe K, Shibuya H, Irie K, Oishi I, Ueno N, Taniguchi T, Nishida E, Matsumoto K. (1995). Identification of a member of the MAPKKK family as a potential mediator of TGF-beta signal transduction. Science. 270:2008–11.
76. Yamaguchi K, Nagai S, Ninomiya-Tsuji J, Nishita M, Tamai K, Irie K, Ueno N, Nishida E, Shibuya H, Matsumoto K. (1995). XIAP, a cellular member of the inhibitor of apoptosis family, links the receptors to TAB1-TAK1 in the BMP signaling pathway. EMBO J. 18:179–87.
77. Shirakabe K, Yamaguchi K, Shibuya H, Irie K, Matsuda S, Moriguchi T, Gotoh Y, Matsumoto K, Nishida E. (1997). TAK1 mediates the ceramide signaling to stress-activated protein kinase/c-Jun N-terminal kinase. J Biol Chem. 272:8141–4.
78. Kanei-Ishii C, Ninomiya-Tsuji J, Tanikawa J, Nomura T, Ishitani T, Kishida S, Kokura K, Kurahashi T, Ichikawa-Iwata E, Kim Y, Matsumoto K, Ishii S. (2004). Wnt-1 signal induces phosphorylation and degradation of c-Myb protein via TAK1, HIPK2, and NLK. Genes Dev. 18(7):816–29.

79. Hanafusa H, Ninomiya-Tsuji J, Masuyama N, Nishita M, Fujisawa J, Shibuya H, Matsumoto K, Nishida E. (1999). Involvement of the p38 mitogen-activated protein kinase pathway in transforming growth factor-beta-induced gene expression. J Biol Chem. 274:27161–7.
80. Kurahashi T, Nomura T, Kanei-Ishii C, Shinkai Y, Ishii S. (2005). The Wnt-NLK signaling pathway inhibits A-Myb activity by inhibiting the association with coactivator CBP and methylating histone H3. Mol Biol Cell. 16:4705–13.
81. Yasuda J, Yokoo H, Yamada T, Kitabayashi I, Sekiya T, Ichikawa H. (2004). Nemo-like kinase suppresses a wide range of transcription factors, including nuclear factor-kappaB. Cancer Sci. 95(1):52–7.
82. Ohkawara B, Shirakabe K, Hyodo-Miura J, Matsuo R, Ueno N, Matsumoto K, Shibuya H. (2004). Role of the TAK1-NLK-STAT3 pathway in TGF-beta-mediated mesoderm induction. Genes Dev. 18:381–6.
83. Kojima H, Sasaki T, Ishitani T, Iemura S, Zhao H, Kaneko S, Kunimoto H, Natsume T, Matsumoto K, Nakajima K. (2005). STAT3 regulates Nemo-like kinase by mediating its interaction with IL-6-stimulated TGFbeta-activated kinase 1 for STAT3 Ser-727 phosphorylation. Proc Natl Acad Sci U S A. 102(12):4524–9.
84. Morrison DK, Davis RJ. (2003). Regulation of MAP kinase signaling modules by scaffold proteins in mammals. Annu Rev Cell Dev Biol. 19:91–118.
85. Levchenko A, Bruck J, Sternberg PW. (2000). Scaffold proteins may biphasically affect the levels of mitogen-activated protein kinase signaling and reduce its threshold properties. Proc Natl Acad Sci U S A. 97(11):5818–23.
86. Yasuda J, Whitmarsh AJ, Cavanagh J, Sharma M, Davis RJ. (1999). The JIP group of mitogen-activated protein kinase scaffold proteins. Mol Cell Biol. 19(10):7245–54.
87. Xu Z, Kukekov NV, Greene LA. (2003). POSH acts as a scaffold for a multiprotein complex that mediates JNK activation in apoptosis. EMBO J. 22(2):252–61.
88. Raabe T, Rapp UR. (2002). KSR: a regulator and scaffold protein of the MAPK pathway. Sci STKE. 2002(136):PE28.
89. Lefkowitz RJ, Whalen EJ. (2004). Beta-arrestins: traffic cops of cell signaling. Curr Opin Cell Biol. (2):162–8.
90. Razidlo GL, Kortum RL, Haferbier JL, Lewis RE. (2004). Phosphorylation regulates KSR1 stability, ERK activation, and cell proliferation. J Biol Chem. 279(46):47808–14.
91. Matheny SA, Chen C, Kortum RL, Razidlo GL, Lewis RE, White MA. (2004). Ras regulates assembly of mitogenic signalling complexes through the effector protein IMP. Nature. 427(6971):256–60.
92. Satoh AK, Ready DF. (2005). Beta-arrestin mediates light-dependent rhodopsin endocytosis and cell survival. Curr Biol. 15:1722–33.
93. Chen W, ten Berge D, Brown J, Ahn S, Hu LA, Miller WE, Caron MG, Barak LS, Nusse R, Lefkowitz RJ. (2003). Dishevelled 2 recruits beta-arrestin 2 to mediate Wnt5A-stimulated endocytosis of Frizzled 4. Science. 301(5638):1391–4.
94. Chen W, Kirkbride KC, How T, Nelson CD, Mo J, Frederick JP, Wang XF, Lefkowitz RJ, Blobe GC. (2003). Beta-arrestin 2 mediates endocytosis of type III TGF-beta receptor and down-regulation of its signaling. Science. 301(5638):1394–7.
95. Luttrell LM. (2005). Composition and function of g protein-coupled receptor signalsomes controlling mitogen-activated protein kinase activity. J Mol Neurosci. 26(2–3):253–64.
96. Kim GH, Park E, Han JK. (2005). The assembly of POSH-JNK regulates Xenopus anterior neural development. Dev Biol. 286(1):256–69.

97. Theodosiou A, Ashworth A. (2002). MAP kinase phosphatases. Genome Biol.3(7):REVIEWS3009.
98. Li J, Gorospe M, Hutter D, Barnes J, Keyse SM, Liu Y. (2001). Transcriptional induction of MKP-1 in response to stress is associated with histone H3 phosphorylation-acetylation. Mol Cell Biol. 21(23):8213–24.
99. Laderoute KR, Mendonca HL, Calaoagan JM, Knapp AM, Giaccia AJ, Stork PJ. (1999). Mitogen-activated protein kinase phosphatase-1 (MKP-1) expression is induced by low oxygen conditions found in solid tumor microenvironments. A candidate MKP for the inactivation of hypoxia-inducible stress-activated protein kinase/c-Jun N-terminal protein kinase activity. J Biol Chem. 274(18):12890–7.
100. Tsang M, Dawid IB. (2004). Promotion and attenuation of FGF signaling through the Ras-MAPK pathway. Sci STKE. (228):pe17.
101. Echevarria D, Martinez S, Marques S, Lucas-Teixeira V, Belo JA. (2005). Mkp3 is a negative feedback modulator of Fgf8 signaling in the mammalian isthmic organizer. Dev Biol. 277(1):114–28.
102. Eblaghie MC, Lunn JS, Dickinson RJ, Munsterberg AE, Sanz-Ezquerro JJ, Farrell ER, Mathers J, Keyse SM, Storey K, Tickle C. (2003). Negative feedback regulation of FGF signaling levels by Pyst1/MKP3 in chick embryos. Curr Biol. 13(12): 1009–18.
103. Kawakami Y, Rodriguez-Leon J, Koth CM, Buscher D, Itoh T, Raya A, Ng JK, Esteban CR, Takahashi S, Henrique D, Schwarz MF, Asahara H, Izpisua Belmonte JC. (2003). MKP3 mediates the cellular response to FGF8 signalling in the vertebrate limb. Nat Cell Biol. 5(6):513–9.
104. Shanley TP, Vasi N, Denenberg A, Wong HR. (2001). The serine/threonine phosphatase, PP2A: endogenous regulator of inflammatory cell signaling. J Immunol. 166(2):966–72.
105. Lee T, Kim SJ, Sumpio BE. (2003). Role of PP2A in the regulation of p38 MAPK activation in bovine aortic endothelial cells exposed to cyclic strain. J Cell Physiol. 194(3):349–55.
106. Sundaresan P, Farndale RW. (2002). P38 mitogen-activated protein kinase dephosphorylation is regulated by protein phosphatase 2A in human platelets activated by collagen. FEBS Lett. 528(1–3):139–44.
107. Hacohen N, Kramer S, Sutherland D, Hiromi Y, Krasnow MA. (1998). Sprouty encodes a novel antagonist of FGF signaling that patterns apical branching of the Drosophila airways. Cell 92(2):253–63.
108. Mason JM, Morrison DJ, Albert Basson M, Licht JD. (2006). Sprouty proteins: multifaceted negative-feedback regulators of receptor tyrosine kinase signaling. Trends Cell Biol. 16(1):45–54.
109. Hanafusa H, Torii S, Yasunaga T, Nishida E. (2002). Sprouty1 and Sprouty2 provide a control mechanism for the Ras/MAPK signalling pathway. Nat Cell Biol. 4(11):850–8.
110. Yusoff P, Lao DH, Ong SH, Wong ES, Lim J, Lo TL, Leong HF, Fong CW, Guy GR. (2002). Sprouty2 inhibits the Ras/MAP kinase pathway by inhibiting the activation of Raf. J Biol Chem. 277(5):3195–201.
111. Nutt SL, Dingwell KS, Holt CE, Amaya E. (2001). Xenopus Sprouty2 inhibits FGF-mediated gastrulation movements but does not affect mesoderm induction and patterning. Genes Dev. 15(9):1152–66.
112. Impagnatiello MA, Weitzer S, Gannon G, Compagni A, Cotten M, Christofori G. (2001). Mammalian sprouty-1 and -2 are membrane-anchored phosphoprotein inhibitors of growth factor signaling in endothelial cells. J Cell Biol. 152(5):1087–98.

113. Mason JM, Morrison DJ, Bassit B, Dimri M, Band H, Licht JD, Gross I. (2004). Phosphorylation of Sprouty proteins regulates their ability to inhibit growth factor signaling: a dual feedback loop. Mol Biol Cell. 15(5):2176–88.
114. Wu X, Alexander PB, He Y, Kikkawa M, Vogel PD, McKnight SL. (2005). Mammalian sprouty proteins assemble into large monodisperse particles having the properties of intracellular nanobatteries. Proc Natl Acad Sci U S A. 102(39):14058–62.
115. Hall AB, Jura N, DaSilva J, Jang YJ, Gong D, Bar-Sagi D. (2003). hSpry2 is targeted to the ubiquitin-dependent proteasome pathway by c-Cbl. Curr Biol. 13(4):308–14.
116. Sivak JM, Petersen LF, Amaya E. (2005). FGF signal interpretation is directed by Sprouty and Spred proteins during mesoderm formation. Dev Cell. 8(5):689–701.
117. Lin W, Jing N, Basson MA, Dierich A, Licht J, Ang SL. (2005). Synergistic activity of Sef and Sprouty proteins in regulating the expression of Gbx2 in the mid-hindbrain region. Genesis. 41(3):110–5.
118. Gross I, Morrison DJ, Hyink DP, Georgas K, English MA, Mericskay M, Hosono S, Sassoon D, Wilson PD, Little M, Licht JD. (2003). The receptor tyrosine kinase regulator Sprouty1 is a target of the tumor suppressor WT1 and important for kidney development. J Biol Chem. 278(42):41420–30.
119. Chi L, Zhang S, Lin Y, Prunskaite-Hyyrylainen R, Vuolteenaho R, Itaranta P, Vainio S. (2004). Sprouty proteins regulate ureteric branching by coordinating reciprocal epithelial Wnt11, mesenchymal Gdnf and stromal Fgf7 signalling during kidney development. Development. 131(14):3345–56.
120. Shim K, Minowada G, Coling DE, Martin GR. (2005). Sprouty2, a mouse deafness gene, regulates cell fate decisions in the auditory sensory epithelium by antagonizing FGF signaling. Dev Cell. 8(4):553–64.
121. Taketomi T, Yoshiga D, Taniguchi K, Kobayashi T, Nonami A, Kato R, Sasaki M, Sasaki A, Ishibashi H, Moriyama M, Nakamura K, Nishimura J, Yoshimura A. (2005). Loss of mammalian Sprouty2 leads to enteric neuronal hyperplasia and esophageal achalasia. Nat Neurosci. 8(7):855–7.
122. Wakioka T, Sasaki A, Kato R, Shouda T, Matsumoto A, Miyoshi K, Tsuneoka M, Komiya S, Baron R, Yoshimura A. (2001). Spred is a Sprouty-related suppressor of Ras signalling. Nature. 412(6847):647–51.
123. Nobuhisa I, Kato R, Inoue H, Takizawa M, Okita K, Yoshimura A, Taga T. (2004). Spred-2 suppresses aorta-gonad-mesonephros hematopoiesis by inhibiting MAP kinase activation. J Exp Med. 199(5):737–42
124. Nonami A, Kato R, Taniguchi K, Yoshiga D, Taketomi T, Fukuyama S, Harada M, Sasaki A, Yoshimura A. (2004). Spred-1 negatively regulates interleukin-3-mediated ERK/mitogen-activated protein (MAP) kinase activation in hematopoietic cells. J Biol Chem. 279(50):52543–51.
125. Hashimoto S, Nakano H, Singh G, Katyal S. (2002). Expression of Spred and Sprouty in developing rat lung. Gene Expr Patterns. 2(3-4):347–53.
126. Kovalenko D, Yang X, Nadeau RJ, Harkins LK, Friesel R. (2003). Sef inhibits fibroblast growth factor signaling by inhibiting FGFR1 tyrosine phosphorylation and subsequent ERK activation. J Biol Chem. 278(16):14087–91.
127. Furthauer M, Lin W, Ang SL, Thisse B, Thisse C. (2002). Sef is a feedback-induced antagonist of Ras/MAPK-mediated FGF signalling. Nat Cell Biol. 4(2):170–4.
128. Torii S, Kusakabe M, Yamamoto T, Maekawa M, Nishida E. (2004). Sef is a spatial regulator for Ras/MAP kinase signaling. Dev Cell. 7(1):33–44.
129. Yang X, Kovalenko D, Nadeau RJ, Harkins LK, Mitchell J, Zubanova O, Chen PY, Friesel R. (2004). Sef interacts with TAK1 and mediates JNK activation and apoptosis. J Biol Chem. 279(37):38099–102.

130. Eyers PA, Craxton M, Morrice N, Cohen P, Goedert M. (1998). Conversion of SB 203580-insensitive MAP kinase family members to drug-sensitive forms by a single amino-acid substitution. Chem Biol. 5:321–8.
131. Yakymovych I, Engstrom U, Grimsby S, Heldin CH, Souchelnytskyi S. (2002). Inhibition of transforming growth factor-beta signaling by low molecular weight compounds interfering with ATP- or substrate-binding sites of the TGF-beta type I receptor kinase. Biochemistry 41(36):11000–7.
132. Kuma Y, Sabio G, Bain J, Shpiro N, Marquez R, Cuenda A. (2005). BIRB796 inhibits all p38 MAPK isoforms in vitro and in vivo. J Biol Chem. 280(20):19472–9.
133. Bennett BL, Sasaki DT, Murray BW, O'Leary EC, Sakata ST, Xu W, Leisten JC, Motiwala A, Pierce S, Satoh Y, Bhagwat SS, Manning AM, Anderson DW. (2001). SP600125, an anthrapyrazolone inhibitor of Jun N-terminal kinase. Proc Natl Acad Sci U S A. 98(24):13681–6.
134. Barr RK, Boehm I, Attwood PV, Watt PM, Bogoyevitch MA. (2004). The critical features and the mechanism of inhibition of a kinase interaction motif-based peptide inhibitor of JNK. J Biol Chem. 279:36327–38.
135. Yung Y, Dolginov Y, Yao Z, Rubinfeld H, Michael D, Hanoch T, Roubini E, Lando Z, Zharhary D, Seger R. (1997). Detection of ERK activation by a novel monoclonal antibody. FEBS Lett. 408(3):292–6.
136. Goswami M, Uzgare AR, Sater AK. (2003). Regulation of MAP kinase by the BMP-4/TAK1 pathway in Xenopus ectoderm. Dev Biol. 236:259–70.
137. El-Hodiri HM, Che S, Nelman-Gonzalez M, Kuang J, Etkin LD. (1997). Mitogen-activated protein kinase and cyclin B/Cdc2 phosphorylate Xenopus nuclear factor 7 (xnf7) in extracts from mature oocytes. Implications for regulation of xnf7 subcellular localization. J Biol Chem. 272(33):20463–70.
138. El-Hodiri HM, Shou W, Etkin LD. (1997). xnf7 functions in dorsal-ventral patterning of the Xenopus embryo. Dev Biol. 190(1):1–17.
139. Schagger H, von Jagow G. (1987). Tricine-sodium dodecyl sulfate-polyacrylamide gel electrophoresis for the separation of proteins in the range from 1 to 100 kDa. Anal Biochem. 166(2):368–79.
140. Tanoue T, Nishida E. (2003). Molecular recognitions in the MAP kinase cascades. Cell Signal. 15(5):455–62.
141. Brancho D, Ventura JJ, Jaeschke A, Doran B, Flavell RA, Davis RJ. (2005). Role of MLK3 in the regulation of mitogen-activated protein kinase signaling cascades. Mol Cell Biol. 25(9):3670–81.
142. Matsumoto K, Nishihara S, Kamimura M, Shiraishi T, Otoguro T, Uehara M, Maeda Y, Ogura K, Lumsden A, Ogura T. (2004). The prepattern transcription factor Irx2, a target of the FGF8/MAP kinase cascade, is involved in cerebellum formation. Nat Neurosci. 7(6):605–12.

5 Analysis of Retinoid Signaling in Embryos

Malcolm Maden

CONTENTS

I. INTRODUCTION

In common with other signalling pathways that operate in developing embryos, there are three striking features of the retinoid signalling pathway. First, it is utilized not just once, but again and again at different times throughout development; there is therefore a strong temporal control of the pathway. Second, it is involved in not just one but several developing systems of the embryo; there is therefore a tissue-specific or spatial control of the pathway. Third, it is the precise level of retinoid signalling that is crucial to these embryonic systems because development can be disrupted either by raising or by lowering the signalling levels.

Because of this situation, if the retinoid signalling pathway is eliminated or disrupted from the beginning of development, then many organ systems will be affected, which inevitably results in the early death of the embryos. Far better, then, to perform temporally or spatially controlled disruptions so that effects on the particular organ system of choice can be investigated.

II. THE RETINOID SIGNALLING PATHWAY

The biologically active component of the retinoid pathway is all-*trans*-retinoic acid (tRA) because this is the compound that activates the retinoic acid receptors (see page 89) and is present in embryos. Its isoform, 9-*cis*-retinoic acid, also activates the receptors and is present in adult mouse liver (Heyman et al., 1992), but has not been detected in embryos and does not rescue retinoic acid-deficient mice (Mic et al., 2003), so its biological significance is limited. There are also other active retinoic acids including didehydroretinoic acid, which is found especially in avian embryos (Maden et al., 1998b), and 4-*oxo*-retinoic acid, which is normally considered to be a catabolic breakdown product of all-*trans*-retinoic acid, but can transactivate the retinoic acid receptors (Pijnappel et al., 1993). These compounds are of low molecular weight (around 300 Da), are lipophilic and can rapidly diffuse across a field of cells in an embryo to affect cells at a distance. They are synthesized from all-*trans*-retinol (otherwise known as vitamin A), and collectively, the molecules that are derived from vitamin A metabolism are known as retinoids.

Retinoids are obtained from the diet in animal sources such as retinyl esters or in plant sources such as β-carotene, and following digestion and transport to the liver they are stored there as retinyl esters (Napoli, 1994). Upon demand, retinyl esters are converted to retinol in the liver hepatocytes; retinol is released into the bloodstream and transported to the cells that require it bound to retinol binding protein (RBP). Retinol enters the cells and is bound in the cytoplasm by cellular retinol binding protein (CRBP). All-*trans*-retinoic acid is then generated in the cytoplasm as a result of two oxidative reactions. In the first, retinol is oxidized to retinaldehyde by the alcohol dehydrogenases (ADHs) or short chain dehydrogenases (SDRs), and in the second, retinaldehyde is oxidized to tRA by the aldehyde dehydrogenases (ALDHs) (Duester, 2000). Having been synthesized, tRA is bound in the cytoplasm by cellular retinoic acid binding protein (CRABP) so that neither retinol, nor retinaldehyde, nor tRA are ever free in the cytoplasm, but bound to specific proteins (retinaldehyde also binds to CRBP). The function of CRBP is to present retinol to the two classes of enzymes described above, and thus it is involved in the metabolism of retinol to tRA (Napoli, 1996). CRABP is thought to deliver tRA to the nucleus and interact with the nuclear transcription complex (Delva et al., 1999; Dong et al., 1999). There are several genes for each of these proteins and enzymes: There are four CRBPs, two CRABPs, five ADHs, five SDRs, and four RALDHs (Duester,

2001). The *Raldhs*, in particular, have a tissue-specific distribution, and the expression of *Raldh1*, *Raldh2*, and *Raldh3* in the embryo is an excellent example of this where there is virtually no overlap in distribution (Blentic et al., 2003).

There is a further class of enzyme involved in tRA metabolism, and this is a cytochrome P450 class called Cyp26. Again, there are several members, in this case three, namely Cyp26A1, Cyp26B1, and Cyp26C1, and these are involved in the catabolism of tRA to inactive products. These enzymes generate 4-*oxo*-RA, 4-OH-RA, 18-OH-RA, and 5,6-*epoxy*-RA (Fujii et al., 1997; White et al., 1996; White et al., 2000a), and again they have a tissue-specific distribution in the embryo (Reijntjes et al., 2004). Although this is a catabolic and, thus, inactivating process, several of these Cyp products are biologically active. For example, 4-*oxo*-RA respecifies the head-to-tail axis of the *Xenopus* embryo (Pijnappel et al., 1993), the overexpression of *Cyp26A1* in embryonal carcinoma cells induces neuronal differentiation (Sonneveld et al., 1999), and 4-*oxo*-RA, 4-OH-RA, and 5,6-*epoxy*-RA completely rescue the retinoid-deficient quail embryo (Kostetskii et al., 1998; Reijntjes et al., 2005).

tRA acts in the nuclei of cells of the embryo or adult to establish, change, or maintain patterns of gene activity. To perform this function, it binds to ligand activated nuclear transcription factors, of which there are two classes, the retinoic acid receptors (RARs) and the retinoid X receptors (RXRs). There are three RAR genes, α, β, and γ, each of which has multiple isoforms (Kastner et al., 1994a), and three RXRs, α, β, and γ (Kliewer et al., 1994). The RARs and RXRs act as heterodimers and recognize consensus sequences known as retinoic acid response elements (RAREs) in the enhancer sequences of RA-responsive genes. tRA binds only to the RARs, whereas 9-*cis*-RA binds to both the RARs and the RXRs. Thus, an active heterodimer is, for example, RARα/RXRβ, but there is a further complication because the RXRs can also heterodimerize with a variety of other related receptors such as the thyroid hormone receptors, the vitamin D receptors, the peroxisome proliferator-activated receptors (whose ligands are fatty acids), the LXRs (whose ligands are oxysterols), FXRs (whose ligands are farnesoids), or other orphan receptors such as NGF1B (which is a nerve growth factor-induced receptor). Thus, retinoids can elicit a diversity of biological responses with the RXRs at the center of a hormone cascade (Mangelsdorf and Evans, 1995). The retinoid signalling pathway is summarized in Figure 5.1.

III. HOW TO DISRUPT THE PATHWAY

There are, therefore, multiple ways in which to disrupt the retinoid signalling pathway in the embryo, and these are identified with red arrows in Figure 5.1.

1. Disrupting ligand supply (arrow 1). This can be performed by removal of the ligand and, conversely, by the administration of excess ligand. Because the ligand is derived from vitamin A, then removing vitamin A from the maternal diet will, once the liver stores have been used up, result in an embryo that develops in the absence of tRA. Administering excess tRA will overstimulate the signalling pathway, and because these retinoids are lipophilic and readily diffusible, they can be administered either to the

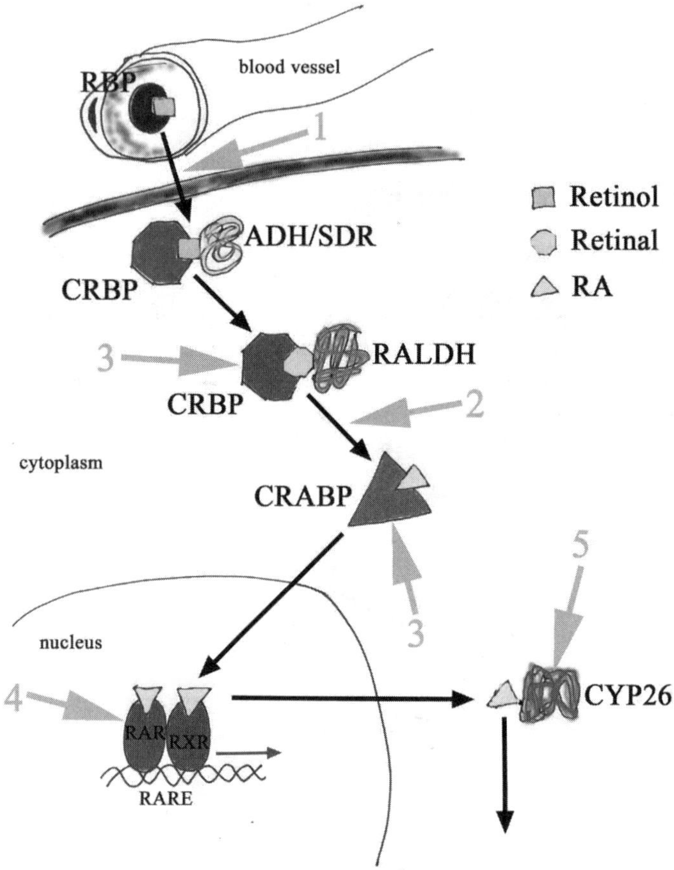

FIGURE 5.1 (See color insert following page 144.) The RA signalling pathway and the points at which it has been disrupted. Retinol (green cube) is transported in the bloodstream bound to retinol binding protein (RBP: blue circle). Retinol is transferred through the cell membrane (grey line) and enters the cell cytoplasm, where it binds to cellular retinol binding protein (CRBP: purple heptagon). The retinol metabolizing enzymes ADHs and SDRs bind to this retinol-CRBP complex and metabolize retinol into retinaldehyde (small green heptagon), which remains bound to CRBP. The retinaldehyde enzymes then bind to the retinaldehyde-CRBP complex and metabolize retinaldehyde into RA (small yellow triangle), which becomes bound to cellular retinoic acid binding protein (CRABP: pink triangle). This complex enters the nucleus, and RA is transferred to the heterodimer composed of a retinoic acid receptor (RAR) and a retinoid X receptor (RXR), which induces a conformational change resulting in the transcription of the gene (red arrow). The RA is then passed back to the cytoplasm, where it is metabolized by the Cyp26 enzymes into more polar compounds. The golden arrows and numbers mark the positions where the RA signalling pathway has been disrupted, with the results described in the text. 1 = preventing retinol entering the cell or supplying too much ligand. 2 = Inhibiting the action of or overexpressing the synthetic enzymes. 3 = Knocking out or overexpressing the binding proteins. 4 = Inhibiting the action of or overexpressing the RARs. 5 = Inhibiting the action of or overexpressing the catabolic enzymes.

mother (in the case of mammals) or directly to the embryo (birds, amphibians, and fish).

2. Disrupting ligand synthesis (arrow 2). This can be performed by the use of pharmacological inhibitors of the synthesizing enzymes. Although enzyme inhibition is not necessarily specific, inhibitors have provided valuable data, and as they are small molecules, they can easily be administered to the embryo. The enzymes can also be eliminated by homologous recombination to generate null mutant embryos. This is a precise method for eliminating the enzyme function one by one but can only be performed in the mouse. One enzyme has also been overexpressed.
3. Disrupting the retinoid binding proteins by homologous recombination to generate null mutant embryos or by overexpression (arrow 3).
4. Disrupting receptor signalling (arrow 4). This can be performed by homologous recombination to generate null mutant receptor embryos, use of antisense morpholinos to knock down receptors, application of synthetic retinoids that do not transactivate the receptor heterodimers and thus prevent receptor signalling, or down-regulation of receptor signalling by overexpressing mutant receptors that are inactive. Up-regulation of receptor signalling can be performed by overexpression of mutant receptors that are constitutively active or stimulating individual receptor activity with receptor selective ligands.
5. Disrupting the catabolic enzymes by knocking out the Cyps, using pharmacological inhibitors of the Cyps, or overexpressing them (arrow 5).

The great advantage of studying signalling pathways in a whole range of embryological systems is that one can pick and choose the experimental system according to which method is to be utilized, because many of the strategies described above are more suitable to some types of embryos than others. For example, gene knockouts were designed for mouse embryos and not other types of embryo, whereas antisense morpholinos are injected into *Xenopus* or zebrafish embryos with the same intent. Overexpression of genes is most readily performed in *Xenopus* or zebrafish embryos simply by injecting mRNA at the one- or two-cell stage. Conveniently, natural and synthetic retinoids are lipophilic and can be directly administered to all types of embryo whether they are laid in water, inside eggs, or in a uterus. Thus, there is a vast amount of literature on the effects of excess retinoids on embryos, some of which is described below as each of these methods of disrupting retinoid signalling is summarized.

IV. DISRUPTING LIGAND SUPPLY

A. Removal of the Ligand

The first retinoid signalling pathway disruptions on embryos through removal of vitamin A from the maternal diet were conducted on farm animals in the 1930s with the report that a litter of pigs was born with no eyes (Hale, 1933). Subsequently, it was shown that a wide range of embryonic defects appeared in vitamin A-deficient

(VAD) embryos of sheep, cattle, rabbits, rats, and humans (reviewed in Kalter and Warkany, 1959). Defects appeared in the following systems of the embryo: the central nervous system (CNS; hydrocephalus, spina bifida); the eyes (anophthalmia, microphthalmia); the face (harelip, cleft palate); the dentition; the ears (accessory ears, otosclerosis); the limbs; the urinogenital system (cryptorchidism, ectopic ovaries, pseudohermaphroditism, renal defects); the skin (subcutaneous cysts); the lungs (hypoplasia); the heart (incomplete ventricular spetation, spongy myocardium, aortic arch defects, aorticopulmonary spetal defects, valvulus communis); and the vasculature (failure of the vitelline veins to form, seen primarily in chicks and quails).

In recent years, this method of disrupting signalling has been returned to with modern molecular tools for assessing the precise anatomical details of the altered structures. Chick embryos were the first developmental system to be analyzed in this fashion (Thompson et al., 1969), followed by quails (Dersch and Zile, 1993; Heine et al., 1985) and latterly rats (Dickman et al., 1997; White et al., 1998; White et al., 2000). When the adult birds were fed a diet deficient in vitamin A, but with enough all-*trans*-RA to permit sperm and egg production, then the eggs produced had white yolks and no retinoids could be detected either by HPLC or a reporter system (Chen et al., 1995; Dong and Zile, 1995). Upon incubation, such embryos lived for only 3 to 4 days because the heart inflow tract and vitelline veins failed to develop, so there was no circulation established between the embryo and the developing area vasculosa, and no blood cells circulated into the embryo (Heine et al., 1985; Dersch and Zile, 1993). In rats, VAD females were generated and maintained on sufficient all-*trans*-RA to allow fertility. After fertilization, RA may then be removed at various times to generate a sudden acute RA deficiency during a selected gestational window (Dickman et al., 1997) or allowed to develop under conditions of varying RA levels (White et al., 1998; White et al., 2000).

In both the bird and the rat embryo, there were multiple organ defects, many of which have been carefully analyzed to discover the gene pathways that are affected. For example, the hindbrain, which normally consists of seven segments or rhombomeres, failed to develop the posterior four rhombomeres, leaving only the first three rhombomeres, which expanded to fill in the gap left (Gale et al., 1999; Maden et al., 1996; White et al., 1998; White et al., 2000) (Figure 5.2b). The neural crest cells died (Maden et al., 1996; Dickman et al., 1997), and neurite outgrowth from the neural tube failed (Maden et al., 1998a). Neural crest apoptosis was also responsible for craniofacial defects that appeared, such as frontonasal hypoplasia and mandibular arch hypoplasia (Dickman et al., 1997). The posterior pharyngeal arches did not form due to a patterning defect in the endoderm (Quinlan et al., 2002; White et al., 1998) that involves genes such as *Tbx1* (Roberts et al., 2005). Pituitary development was abnormal (Dickman et al., 1997), and in the deficient quail embryo, Rathke's pouch, the forerunner of the pituitary, was missing (Maden, unpublished data). Microphthalmia was seen, caused by apoptosis in the lens, neural retina, and pigmented retina (Dickman et al., 1997), and in the deficient quail embryo, the ventral third of the eye including the choroid fissure failed to develop (Maden, unpublished data). The lungs failed to develop, and there was a persistent laryngeal-tracheal groove (Dickman et al., 1997; Antipatis et al., 1998). The somites were about half the size of normal (Maden et al., 2000), and the down-regulation of *Eph4A* in the lateral plate may be responsible for this phenomenon (Schmidt et al., 2001). Dorsoventral patterning in the neural tube was highly abnormal.

The domains of ventrally expressed genes such as *shh* and *Nkx6.1* were expanded; the domains of dorsally expressed genes such as *Bmps*, several *pax* genes, and *Wnt* genes were reduced; and several populations of interneurons were completely missing (Wilson et al., 2004). At the posterior end of the embryo where the neural tube continues to develop by the addition of new cells and the somites continue to form, there is normally a negative interaction between RA and *Fgf8* (Diez and Storey, 2004). In the absence of RA, the domain of expression of *Fgf8* in the caudal stem cell zone was expanded, and neural differentiation in the newly formed neural tube was abnormal (Diez et al., 2003). Instead of looping to the right and subdividing into left and right chambers, the VAD heart, in the majority of cases, reversed its left-right asymmetry (Zile et al., 2000) and formed a single, distended tube (Heine et al., 1985; Dersch and Zile, 1993), confirming a role for RA in this early embryonic patterning event. Three genes in the left-right asymmetry pathway were down-regulated in RA deficiency: *nodal*, *SnR*, and *Pitx2* (Zile et al., 2000). The vitelline veins did not form so the blood that formed out in the periphery of the embryo never circulated around the body of the embryo, and the heart pumped a colorless fluid.

At least one gene, GATA-4, has been identified as being strongly down-regulated in the RA-deficient developing heart (Kostetskii et al., 1999), and the early patterning event that was responsible for the heart defect was in the anterior endoderm (Ghatpande et al., 2000). The limb buds failed to develop because of the down-regulation of genes such as *shh*, *Bmp2*, *Fgf4*, and *Hoxd* (Power et al., 1999; Stratford et al., 1999), and they showed double dorsal polarity (Stratford et al., 1999). Finally, in contrast to all the other abnormalities that involve a failure of development, there were multiple otic vesicles, the forerunner of the ear, instead of the normal one on each side, and extra ribs were also seen (White et al., 1998).

Virtually all of these defects can be recapitulated either by knocking out the major RA synthesizing enzyme, *Raldh2*, or by knockouts of the retinoic acid receptors (Sections V.B and VII.A).

B. Excess Ligand

It was not until 1953 that the first experiments involving the provision of excess RA to the rat embryo were conducted (Cohlan, 1953), and since then, a whole range of abnormalities in mammalian species (mouse, rat, hamster, monkey, human) have been described. Strikingly, the abnormalities seen are remarkably similar to, and in the same organ systems as, those generated by a lack of vitamin A, although when they have been analyzed in more detail, some differences became apparent. The abnormalities included the central nervous system (hydrocephalus, anencephaly, exencephaly, spina bifida); the eyes (anophthalmia, microphthalmia, defects of the retina); the face (harelip, cleft palate, brachygnathia, hypoplastic maxilla); the dentition; the ears (absent or deformed); the limbs (phocomelia); the urinogenital system (hypoplastic kidney, polycystic kidney, absent/hypoplastic genitalia); the heart (incomplete ventricular septation, transposition of the great vessels, double aortic arch, hypoplastic aortic valves); the thyroid gland (hypoplasia); the axial skeleton

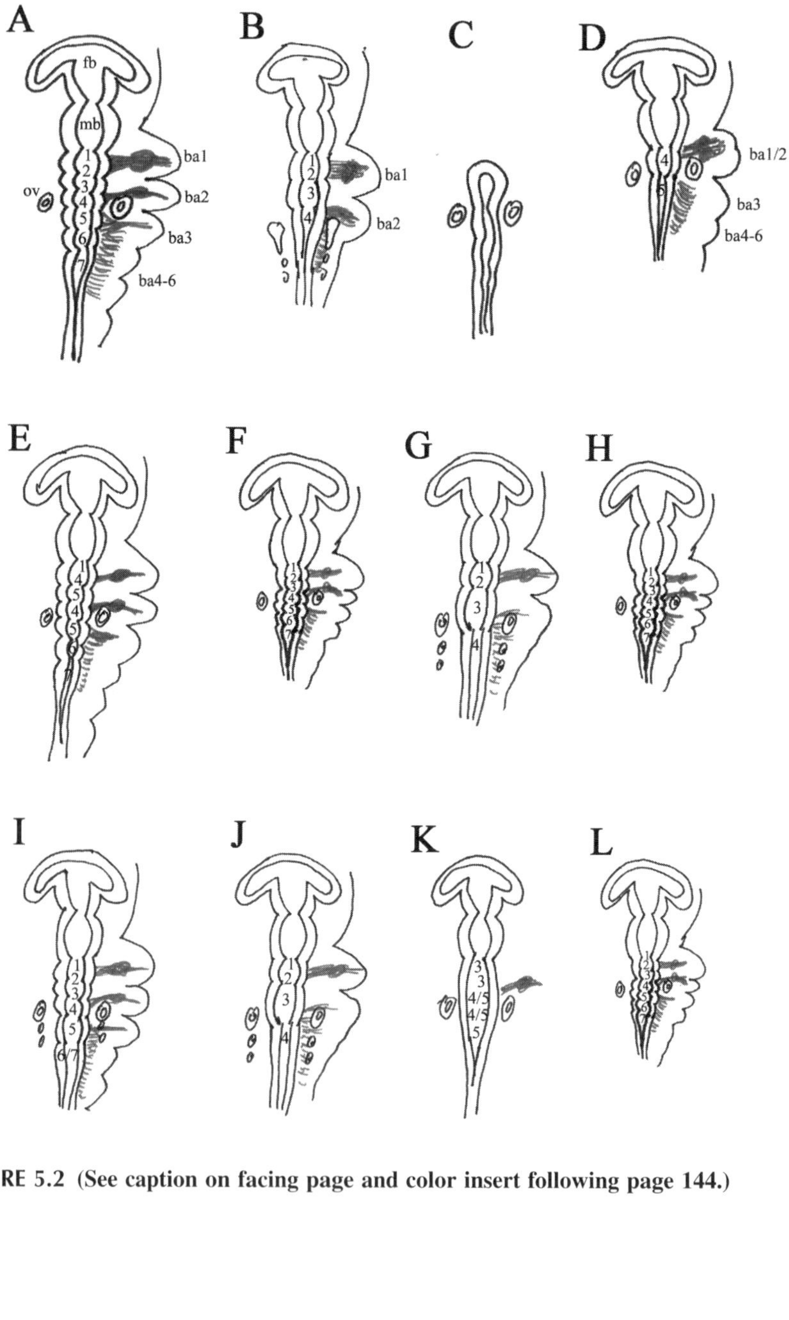

FIGURE 5.2 (See caption on facing page and color insert following page 144.)

(vertebral and rib fusions, extra vertebrae and ribs, hypoplastic tail) (Fantel et al., 1977; Knudsen, 1966; Kochhar, 1973; Rosa et al., 1986; Shenfelt, 1972). In humans whose mothers have taken 13-*cis*-RA, a similar range of abnormalities were seen (Lammer et al., 1985; Lammer and Armstrong, 1992), and because these babies have been followed up as children, it was revealed they had marked motor and sensory developmental delays, and severe mental retardation, and even the supposedly unaffected children had cognitive impairments (Adams and Lammer, 1995; Adams et al., 2001).

FIGURE 5.2 **(See facing page.)** Diagrams summarizing the effects of increasing or decreasing RA signalling on the anterior CNS, particularly the hindbrain, branchial arches, and associated neural crest. The drawings are of an idealized embryo even though the experiments concern mouse, quail, and zebrafish embryos. **A**, a normal embryo showing the forebrain and eyes (fb), the midbrain (mb), the hindbrain rhombomeres marked 1–7, the otic vesicles beside rhombomere 4, the anterior neural crest stream (red) emanating from rhombomere1/2, the middle neural crest stream (red) emanating from rhombomere 4, the post-otic neural crest (red) emanating from rhombomere 5 posteriorly, and the branchial arches (ba), which are numbered. **B**, the structure of the vitamin A-deficient quail where all RA signalling has been removed. There are only four rhombomeres and two branchial arches, and the otic vesicles are malformed and duplicated. From Gale et al. (1999). **C**, the effects of high levels of excess RA at early embryonic stages. The eyes, forebrain, and midbrain are lost. Data from several experiments, especially Durston et al. (1989). **D**, the effects of medium levels of excess RA or high levels at later stages than C. There is a loss of anterior rhombomeres (1–3), the posterior rhombomeres lose their boundaries, the otic vesicle shifts anteriorly, branchial arches 1 and 2 fuse together, and the two anterior neural crest streams are fused into one. Data from several experiments, especially Wood et al. (1994) on mouse embryos. **E**, the effect of a low dose of excess RA. The anterior rhombomeres 2 and 3 are transformed into more posterior ones, numbers 4 and 5. The anterior neural crest stream changes its character to resemble that from rhombomere 4 and so produces an ectopic facial ganglion. Data from several experiments, especially Marshall et al. (1992) on mouse embryos. **F**, the effect of the RA synthesis inhibitor DEAB on zebrafish embryos. The anterior rhombomeres are reduced in size, and there are no rhombomeric boundaries posterior to r5. Data from Begemann et al. (2004). **G**, the phenotype of *Raldh2-/-* null mutant mouse embryo. Posterior rhombomeres are missing, the anterior ones are expanded, and there are ectopic otic vesicles. This is virtually identical to the VAD quail in B. From Niederreither et al. (2000). **H**, the phenotype of the *neckless* or *no-fin* zebrafish mutant. The rhombomeres are all reduced in size especially in the posterior region, resulting in the anterior shift of the somites. The posterior branchial arches are lost. Data from Begemann et al. (2001) and Grandel et al (2002). **I**, the phenotype of the RARα/RARβ null mutant mouse. Posterior rhombomeres 6 and 7 are fused and r5 is enlarged, whereas the anterior rhombomeres are normal. Data from Dupe et al. (1999). **J**, the phenotype of the RARα/RARγ null mutant mouse. The posterior rhombomeres are missing and the anterior rhombomeres are enlarged. This is the same phenotype as the VAD quail (B) and the *Raldh2-/-* mouse (G). Data from Wendling et al. (2001). **K**, the effect of a dominant negative RARβ in *Xenopus*. The rhombomere boundaries are lost, and the anterior hindbrain takes on the characteristics of posterior rhombomeres. Data from van der Wees et al. (1998). **L**, the phenotype of the *giraffe* zebrafish mutant or the effect of injecting Cyp26A1 antisense morpholinos in zebrafish. The rhombomeres are all reduced in size. Data from Emoto et al. (2005).

More recent analyses of the effects of excess ligand began with the demonstration that the developing chick limb bud can be induced to form two limbs instead of one after a bead soaked in all-*trans*-RA was grafted into the bud (Tickle et al., 1982). The normal chick wing has only three digits, referred to as digits 2, 3, and 4, and when the RA bead was grafted into the anterior margin of the bud, then a 6-digit, double posterior limb was generated with the digit structure of 4,3,2,2,3,4. This effect was time dependent, dose dependent, position dependent, and stage dependent (Eichele et al., 1985; Summerbell, 1983; Tickle et al., 1985). Systemic administration of RA rather than local administration also induced duplication of the zebrafish fin (Vandersea et al., 1998), and in the mouse embryo, RA administered at stages before implantation generated multiple hindlimbs including some that emerged from the abdominal body wall (Niederreither et al., 1996; Rutledge et al., 1994).

This stimulatory effect of RA was not seen in other species such as amphibians or humans, and in these cases it acts as a teratogen by causing defective limb development (Die-Smulders et al., 1995; Scadding and Maden, 1986). Interestingly, if the tip of the amphibian limb bud was cut off before RA was administered, thereby turning the developing limb bud into a damaged (regenerating ?) developing limb bud, then this duplication phenomenon appeared, and more than one limb bud was produced (Scadding and Maden, 1986). Duplication of the limb is the characteristic response of the regenerating amphibian limb to systemic administration of RA (Maden, 1982), and even the regenerating tadpole tail can be induced to sprout limbs after RA treatment (Maden, 1993; Mohanty-Hejmadi et al., 1992). The duplication of the chick limb bud by RA involves the induction of a host of genes including *shh* (Riddle et al., 1993), the *Hoxd* genes (Izpisua-Belmonte et al., 1991; Nohno et al., 1991), *Fgf4*, *Bmp2*, and many others.

The effects of excess ligand on the development of the repeating mesodermal somites of the embryo, which subsequently become the individually patterned vertebrae and ribs, has been analyzed in the mouse embryo with a view to understanding more about the function of the *Hox* genes. Treatment of mouse embryos via the pregnant female during gastrulation stages induced anterior or posterior transformations of the vertebrae depending on the precise time of treatment (Kessel and Gruss, 1991; Kessel, 1992). The vertebrae can be characterized in terms of their number within each region of the body (7 cervical, 13 thoracic, 6 lumbar, 4 sacral in the mouse), and some can be individually identified (the atlas and axis, the first and last of each segment). Early treatments with all-*trans*-RA affected the more anterior regions of the vertebral column, and the alterations were interpreted as posterior transformations. For example, the atlas often had a structure resembling more posterior cervical vertebrae, the 5th cervical vertebra resembled the 6th, the 7th cervical vertebra had ribs on it and resembled the 1st thoracic.

When RA was administered at slightly later stages of gastrulation, anterior transformations of vertebrae were generally observed. For example, the 8th thoracic vertebra, whose rib does not normally connect to the sternum, did produce such a rib and thus came to resemble the 7th thoracic vertebra. The embryo has 14 or 15 ribs instead of the normal 13, so the 1st and 2nd lumbar vertebrae came to resemble thoracic vertebrae. These anatomical changes were preceded by changes in the expression patterns of various *Hox* genes in the mesoderm, and the gene expressions were anteriorized by early RA treatment and posteriorized by later RA treatment. There was a good correlation between the spread of particular genes and the level at which the anatomical abnormalities appeared. These studies led to the concept of the *Hox* code for each segment of the body, with changes in the code induced by RA administration leading to anatomical changes. This type of molecular/anatomical study on the somites has only been performed on the mouse embryo, but in two other systems, the chick and *Xenopus*, excess RA causes the somites to be abnormally large (Moreno and Kintner, 2004; Reijntjes et al., 2005), which, interestingly, is exactly the opposite effect to that of too little RA (Maden et al., 2000; Vermot and Pourquie, 2005).

Analyses of the effects of excess ligand on the developing central nervous system has revealed three different phenotypes depending on the embryonic stage when the ligand was administered, the length of administration, and the dose of ligand used. Which one of them appeared depended on the stage and dose of treatment. The most severe was the loss of rostral tissue, which was also referred to as posteriorization of the CNS because anterior-specific genes were repressed and posterior-specific genes were up-regulated (Cho and De Robertis, 1990; Dekker et al., 1992; Leroy and De Robertis, 1992; Lopez and Carrasco, 1992; Lopez et al., 1995; Sive et al., 1990). This was first seen in *Xenopus* embryos after RA treatment at late blastula/early neurula stages (Durston et al., 1989) and resulted in the failure of development of the forebrain, midbrain, and eyes (Figure 5.2c).

Because it was estimated that the same volume of CNS tissue was present in treated embryos and controls, it was concluded that the missing anterior structures had been respecified to form more posterior structures, and the subsequent gene analyses referred to above confirmed this view. In rat and mouse embryos, this severe loss of rostral tissue was seen after RA treatment at the mid to late streak stage (Avantaggiato et al., 1996; Cunningham et al., 1994; Simeone et al., 1995). The expression of forebrain genes such as *Emx1*, *Emx2*, and *Dlx1* were lost, and genes such as *Wnt-1*, *En-1*, and *Pax2*, which are normally expressed in more posterior domains, were anteriorized in their expression. In zebrafish embryos, this posteriorization phenotype was also seen, but when the ligand is 9-*cis*-RA rather than all-*trans*-RA (Zhang et al., 1996).

The second CNS phenotype in response to excess ligand has long been recognized (Morriss, 1972) and is characterized externally by the shortening of the preotic hindbrain and the abnormally rostral position of the otic vesicle in rat and mouse embryos (Figure 5.2d). It has since been observed many times in mammalian embryos (Lee et al., 1995; Leonard et al., 1995; Morriss-Kay et al., 1991), chick embryos (Lopez et al., 1995; Sundin and Eichele, 1992), and *Xenopus* embryos (Lopez and Carrasco, 1992; Papalopulu et al., 1991). It is caused by a loss of anterior hindbrain tissue due to the abnormal expression of *Hox* genes. The typical response of embryos to RA treatment was to induce an anterior expansion of *Hox* gene expression into the midbrain and forebrain, where they are not normally expressed, and then they retract, leaving behind an abnormal expression pattern (Conlon and Rossant, 1992). It was suggested that the hindbrain tissue anterior to the otocyst took on the *Hox* gene characteristics of a single large rhombomere 4, thereby losing rhombomeres 1–3, and there is often a loss of rhombomere boundaries in the posterior hindbrain, as well (Figure 5.2d) (Wood et al., 1994).

It is very likely that ectopic *Hox* gene expression is indeed the cause of these abnormalities, because in zebrafish embryos the overexpression of one *Hox* gene mimicked the effect of RA in respecifying the anterior rhombomeres (see page 99). In the zebrafish embryo treated with RA at gastrulation stages, there was a similar deletion of tissue at the midbrain/hindbrain border including rhombomeres 1–3, resulting in the absence of the cerebellum (Holder and Hill, 1991). Cerebellar abnormalities are the hallmark of the teratogenic effect of 13-*cis*-RA in humans (Lammer and Armstrong, 1992).

Associated with this anterior shift of the otic vesicle was the fusion of the trigeminal and facial ganglia. These ganglia are derived from the neural crest, which migrates from the dorsal neural tube of rhombomeres 1 and 2 (trigeminal ganglion) and rhombomeres 4 (facial ganglion) and out into the periphery, filling up the first and second branchial arches. As well as ganglia fusions, there was a fusion of the first and second branchial arches after RA treatment of embryos (Kessel, 1993; Pratt et al., 1987; Seegmiller et al., 1991). The ganglia fusion could be brought about by a loss of rhombomere 3, from which no neural crest emerges, as this rhombomere acts as a crest-free zone, thereby separating crest from rhombomeres 1/2 and 4. Indeed, the anterior stripe of *Krox-20*, which is normally expressed in rhombomere 3, was absent in RA-treated embryos (Morriss-Kay et al., 1991; Papalopulu et al., 1991; Wood et al., 1994).

In addition, the neural crest is known to migrate abnormally in the presence of excess ligand and either did not leave the neural epithelium or remained near to it as if migration was inhibited or slowed down (Moro Balbas et al., 1993; Webster et al., 1986; Morriss and Thorogood, 1978), or the migratory pathway displaced (Lee et al., 1995). In the chick embryo, RA- induced mismigration of the neural crest results in ganglia fusion without any abnormality in the hindbrain rhombomeres (Gale et al., 1996), so it is likely that that excess RA, when applied systemically, is causing both effects: abnormal patterning of the hindbrain caused by ectopic *Hox* gene expression and abnormal migration of the neural crest. This is likely to be the explanation of the craniofacial teratological effects that are seen, which include hypoplastic maxilla or mandible, microtia, thyroid or thymus malformations, and facial nerve palsies.

The third CNS phenotype of excess ligand affected the anterior hindbrain, but did not cause a loss of tissue. Instead, one or two anterior rhombomeres were respecified and assumed the characteristics of more posterior rhombomeres, a phenomenon that has been characterized thanks to the availability of molecular markers for individual rhombomeres (Figure 5.2e). This effect has also shown how exquisitely sensitive embryos are to altered ligand levels. Thus, the effect described above of loss of anterior hindbrain tissue in the zebrafish embryo was seen when a dose of RA of 1.5×10^{-7}M was administered. When the dose was slightly lowered to 1×10^{-7}M tRA, then a different phenotype was generated. Rather than a deletion of tissue in the hindbrain, a respecification of one rhombomere occurred so that the normal rhombomere sequence of 1,2,3,4,5,6,7 was transformed into 1,4,3,4,5,6,7 (Hill et al., 1995).

This conclusion was reached because lower vertebrates have a unique neuron present in rhombomere 4 called the Mauthner neuron, and after RA treatment, the embryos contained extra Mauthner neurons in the position of rhombomere 2. In the mouse embryo, this anterior respecification effect was observed when the embryos were treated at slightly later stages than for the other two effects (Lee et al., 1995; Simeone et al., 1995; Wood et al., 1994). This phenomenon was observed in the mouse using a lacZ reporter mouse that was generated using the *Hoxb1* promoter (Marshall et al., 1992). This gene is expressed in just one rhombomere, rhombomere 4, in the day 9.5 mouse embryo and in the facial nerve that derives from rhombomere 4. After RA treatment, there was the typical anteriorization of *Hox* gene expression (Conlon and Rossant, 1992), but subsequently two stripes of *Hoxb1* expression were revealed rather than the normal one (Marshall et al., 1992). This second stripe was anterior to the normal one and in rhombomere 2. The trigeminal nerve that normally exits from rhombomere 2 was also changed and came to resemble the facial nerve. Thus, rhombomere 2 was transformed into rhombomere 4, and in addition, the characteristics of the motoneurons in rhombomere 3 changed (Kessel, 1993), suggesting that the sequence of rhombomeres has become 1,4,5,4,5,6,7 (Figure 5.2e). This phenomenon may well explain the appearance of facial palsies in Accutane-exposed children.

As described above, the typical response of embryos to RA exposure is to undergo an anterior spread of *Hox* genes, and the response is precisely concentration dependent. In *Xenopus*, each anterior neural plate gene, and particularly the *Hox* genes in the hindbrain, had a different concentration of RA at which they were maximally induced (Godsave et al., 1998). Anterior rhombomeric genes (e.g., *Hoxb1*) were induced at low concentrations, and more posterior rhombomeric genes (e.g., *Hoxb3* and *Hoxb4*) were induced at progressively higher concentrations. Spinal cord genes (e.g., *Hoxb9*) were virtually unresponsive. Experiments in zebrafish and mouse embryos have revealed that these alterations in *Hox* expression patterns are causal in inducing the phenotypic change. After RA treatment, *Hoxa1* (which behaves identically to *Hoxb1* described above) was up-regulated within 1 hour of exposure, and the normal single expression domain in rhombomere 4 was altered to two stripes in rhombomere 2 and rhombomere 4. Ectopic expression of *Hoxa1*, either by injecting the mRNA into zebrafish eggs (Alexandre et al., 1996) or by creating transgenic mouse embryos (Zhang et al., 1994), resulted in the same patterns of ectopic *Hoxb1* and *Hoxa2* expression that was caused by RA. Furthermore, the effect on the anatomy of the zebrafish embryos was exactly the same: duplicated Mauthner neurons, abnormal development of the jaw apparatus, fusion of the two streams of neural crest from rhombomeres 2 and 4 into one, and the fusion of the trigeminal and facial ganglia (Alexandre et al., 1996).

V. DISRUPTING LIGAND SYNTHESIS

This has been performed by the use of pharmacological inhibitors of the RALDH enzymes on a variety of vertebrate embryos, by knocking out the enzymes in mice or by overexpressing one of the enzymes.

A. Pharmacological Inhibitors

Pharmacological inhibitor studies have employed three compounds: citral and diethylaminobenzaldehyde (DEAB), which are reversible, competitive inhibitors of the retinaldehyde dehyrogenases, and disulphiram, which inhibits aldehyde dehydrogenases by inducing disulphide bonding. Because of the ease of application of these compounds to amphibian or fish embryos, whose development can be observed in the culture dish, most experiments have involved their administration to zebrafish or *Xenopus* embryos.

Citral treatment of zebrafish embryos at neurula stages resulted in the failure of the ventral half of the eye to develop due to decreased RA production (Marsh-Armstrong et al., 1994), and the pectoral fins did not develop (Vandersea et al., 1998). In *Xenopus* embryos, this compound also produces eye defects (microphthalmia) along with laterally expanded heads, heart defects, and blood pools (Schuh et al., 1993). They do not develop touch responses due to an inhibition of primary neurogenesis, which can be rescued with RA (Sharpe and Goldstone, 2000a). In chick embryos, citral has been used to generate microphthalmia (Abramovici et al., 1978), and it inhibited limb bud development by producing limbs with shorter cartilage elements after the buds had been cultured in citral and then grafted back onto normal embryos (Tanaka et al., 1996). Local treatment with citral-soaked beads placed at the anterior end of the stage 10 chick embryo resulted in the lack of frontonasal process and forebrain tissues and fused or absent eyes, whereas the maxilla and mandibular processes were unaffected (Schneider et al., 2001). This was caused by the down-regulation of *Fgf8* and *Shh* in the forebrain and ectoderm covering the frontonasal process, resulting in apoptosis of the local neural crest.

When citral is applied rather later, at stage 20, to the developing nasal pit, where the enzyme *Raldh3* is expressed, then the entire upper beak on the treated side was ablated (Song et al., 2004). This was caused by a down-regulation of *RAR*β*, Msx1, Msx2, Fgf8*, and *Bmp4* and resulted in apoptosis of the cells going to form these bones. The citral-induced defects can be rescued by RA, as might be expected, but also by the application of FGF8 protein, which is valuable information for elaborating the pathway of action of RA in facial development. In mouse embryos, citral inhibits the development of the organ of Corti in the ear (Raz and Kelley, 1999) and the olfactory system (Anchan et al., 1997).

DEAB treatment of zebrafish embryos during gastrulation resulted in the absence of fins, no touch response, no functioning heart tube due to cardiac edema (resulting in no circulation and enlarged blood islands), reduced frontonasal region anterior to the eye, reduced ventral eye due to a missing choroid fissure, and slightly rounded somites (Begemann et al., 2004; Perz-Edwards et al., 2001). The hindbrain was also abnormal, in that it was smaller with an associated anterior shift of the somites closer to the otic vesicle, implying defects in the posterior hindbrain just as in the absence of RA (Section IV.A). The posterior stripe of *Krox-20* in rhombomere 5 was missing, there were no rhombomere boundaries posterior to rhombomere 5, the brachiomotor neuron number was reduced in the vagus and facial nuclei and axons were missing from these two nerves, the spinal motor nerve pattern was abnormal in the first three spinal nerves, and the glossopharyngeal nerve was absent (Begemann et al., 2004) (Figure 5.2f). Most of these defects (but not cardiac edema) were rescued by RA, confirming the specificity of DEAB, and the effects were the same as treatment with RA receptor antagonists (Section VII.C).

Disulphiram treatment of the developing limb bud of the chick embryo inhibited RA synthesis and down-regulated *shh* and *Fgf4* expression, and limb outgrowth was completely abolished (Stratford et al., 1996). The somites that were generated in the presence of disulphiram were smaller than normal in the chick embryo, exactly as in the VAD quail (Section IV.A), and there was a disturbance in left/right asymmetry at one particular stage such that somite boundaries were shifted anteriorly on the left side (Vermot and Pourquie, 2005). This was caused by a disturbance in the cyclic gene *Lunatic fringe*. In the zebrafish embryo, disulphiram produced embryos with short tails and wavy notochords, shorter spinal cords, and rounded and abnormally segmented somites (Marsh-Armstrong et al., 1995; Costaridis et al., 1996). Disulphiram administration to mouse embryos generated heart abnormalities, and they lacked sinoatrial structures and had enlarged ventricles (Xavier-Neto et al., 1999). In the adult newt, disulphiram inhibited lens regeneration (Tsonis et al., 2000) and limb regeneration (Maden, 1998).

Although there has been some debate about the specificity of these inhibitors, they have, in several studies, been shown to specifically inhibit RA synthesis (McCaffery and Drager, 1994), and in many others it has been shown that RA rescues the defects. Furthermore, there is a striking consistency of effects between the three different compounds (inhibition of limb development, frontonasal development, posterior hindbrain development), and virtually all of the defects have been found in vitamin A-deficient (Section IV.A) or *Raldh2* knockout embryos (Section V.B). We can therefore conclude that they do indeed interrupt the retinoid signalling pathway.

B. Knockouts of the Enzymes

Because there are several enzymes in each of the families that are involved in the synthesis of RA from retinol (ADHs, SDRs, ALDHs), it is highly likely that when any one of them is mutated, the others will compensate for the loss and the embryos will be normal. This is certainly true for the ADHs, as there was little tissue specificity in their embryonic distribution, and *Adh1-/-*, *Adh3-/-* (Molotkov et al., 2002b; Molotkov et al., 2002a), and *Adh4-/-* (Deltour et al., 1999) mutant mouse embryos were normal. The *Raldhs*, however, show a very high degree of tissue specificity (e.g., Blentic et al., 2003), and so we might expect tissue-specific defects to appear in mutant embryos. However, despite the fact that *Raldh1* is expressed in the dorsal half of the developing eye and in the mesonephros, the *Raldh1-/-* embryo was normal (Fan et al., 2003).

Raldh3 is expressed in quite restricted areas of the embryo, including the epithelium ventral to the developing forebrain, isthmus, ventral retina, lateral ganglionic eminence, otic vesicle and nasal placodes, olfactory epithelium, dental epithelium, kidney, adrenal gland, and prostate (Grun et al., 2000; Li et al., 2000; Mic et al., 2000; Niederreither et al., 2002b; Suzuki et al., 2000). Surprisingly, the *Raldh3-/-* mutant mouse embryo only had defects in the nasal region including choanal atresia, absence of maxillary sinuses, absence of nasolacrimal ducts, and hypoplasia of the ethmoturbinates (Dupe et al., 2003).

Raldh2 is the major RA synthesizing enzyme in the embryo, as it begins expression in newly invaginated mesoderm just after gastrulation commences, with a sharp border of expression at the level of the future first somite posteriorly (Berggren et al., 1999; Niederreither et al., 1997; Swindell et al., 1999). The distribution of *Raldh2* correlates perfectly with the distribution of endogenous RA (Maden et al., 1998b) with, at least at early stages, no *Raldh2* expression and no RA in the head region, because the RA degrading enzyme *Cyp26A1* is expressed there (Swindell et al., 1999; Blentic et al., 2003). Subsequently, *Raldh2* is expressed in the heart (Moss et al., 1998), the developing optic cup (Mic et al., 2004), the maxillary region (Niederreither et al., 1997), the hindgut region, the developing lung (Malpel et al., 2000), the limb level motoneurons and meninges (Zhao et al., 1996), the inner ear (Romand et al., 2001), and the tooth buds.

In clear contrast to the other two enzymes, a mutation in the *Raldh2* gene generated embryos with defects in all of these embryonic systems. *Raldh2-/-* embryos died at mid-gestation without undergoing rotation, they had a shorter body length due to smaller somites, the frontonasal region was truncated, there was no network of blood vessels in the yolk sac, and the otocyst was hypoplastic (Niederreither et al., 1999). The heart was a single dilated tube showing no left/right asymmetry, the atrial and sinus venosus development was impaired, and the ventricular cardiomyocytes failed to differentiate (Niederreither et al., 2001). The optic cup failed to invaginate (Mic et al., 2004), and the limb buds did not form due to the absence of *shh*, *Fgf4*, *Meis2*, *dHand*, and *Tbx5* expression (Niederreither et al., 2002c). The posterior hindbrain was reduced in size or absent, as evidenced by the absence of posterior markers such as *Kreisler*, group 3/4 *Hox* genes, and *Krox20*, and the anterior hindbrain was expanded (Niederreither et al., 2000) (Figure 5.2g).

At the posterior end of the embryo, the spinal cord differentiated abnormally due to the absence of expression of *Olig2* and *Pax6*, the somites were smaller and more densely packed, and the domain of mesodermal *Fgf8* expression was expanded (Molotkova et al., 2005). These embryos could be largely, but not fully, rescued by the maternal administration of RA, and they could be administered with low levels of RA via the mother to keep them alive past day 10.5 so that later developmental defects could be studied. In this way, it was revealed that the posterior branchial arches and derived organs (thymus and parathyroid) were abnormal, the post-otic neural crest was misrouted, outgrowth of the posterior cranial nerves was abnormal (Niederreither et al., 2003), the pancreas gene *Pdx1* failed to be induced in the dorsal endoderm by the local splanchnic mesoderm, and as a result the pancreas failed to develop (Molotkov et al., 2005).

Interestingly, a natural mutation of the *Raldh2* gene has been discovered in zebrafish, which is called *neckless* (Begemann et al., 2001) or *no-fin* (Grandel et al., 2002). These mutants had a down-regulation of *Hox* gene expression in the spinal cord, the posterior hindbrain/spinal cord region was smaller than expected accompanied by a slight expansion of the anterior hindbrain rhombomeres, the posterior branchial arches and the posterior jaw cartilages were lost, and there were no fins (Figure 5.2h). As a result, the somites were moved anteriorly toward the otic vesicle. These defects were rescued by RA addition to the mutant embryo. Because the defects were not as severe as the *Raldh2-/-* mouse, it is likely that there is an additional source of RA in these mutants, perhaps from an additional *Raldh2* allele.

The *Raldh2-/-* phenotype is thus virtually identical to the VAD phenotype, demonstrating that *Raldh2* is the major enzyme involved in the generation of RA in the embryo.

C. Overexpressing the Enzymes

Two studies have overexpressed *Raldh2*. In *Xenopus*, the ideal model system for such studies, mRNA for *Raldh2* was injected into one cell of a two-cell embryo, which resulted in an anterior shift and a contraction of the anterior neural marker genes *Otx2, En2*, and *Krox20* with no alteration in the posterior gene *Hoxb9* (Chen et al., 2001) (Figure 5.2f). This *in situ* phenotype is identical to that produced by overexpression of RARα (Section VII.D), and the reduction in forebrain territory is the same as the effects of treatment with RA (Section IV.B).

In the second study, the involvement of *Raldh2* in the development of motor neurons was investigated. *Raldh2* is expressed in the motor neurons of the spinal cord at the limb levels in a particular class of neurons called lateral motor column (LMC) neurons. There are no LMCs and no *Raldh2* expression at the thoracic level of the spinal cord. When a viral construct containing *Raldh2* was electroporated into the chick spinal cord at the thoracic level, then LMCs were induced, mimicking the effect of RA itself (Sockanathan and Jessell, 1998). Interestingly, the LMCs that were generated did not originate from the cells that expressed *Raldh2*, but from adjacent cells that migrated through the *Raldh2*, expressing cells to a lateral location. This suggests that there is a paracine inductive event in which one cell generates RA from *Raldh2* and an adjacent cell is induced to become a specific type of motor neuron.

VI. DISRUPTING THE BINDING PROTEINS

A. Knockouts of the Binding Proteins

CRBP I is expressed in the developing embryo, particularly in the developing nervous system including the motoneurons of the spinal cord and quite widely in the later brain in regions such as the lamina terminalis, striatum, Purkinje cells, and choroid plexus (Maden et al., 1990; Ruberte et al., 1991; Ruberte et al., 1993). CRBPs are thought to play a role in the oxidation of retinol to RA (Napoli, 1996). CRBP I mutant mice were developmentally normal, however, and only showed abnormal retinyl palmitate levels in the embryonic and adult liver (Ghyselinck et al., 1999; Matt et al., 2005).

CRABP I and II are expressed in specific localizations in the developing embryo, again, particularly in the nervous system in regions such as the commissural neurons, specific rhombomeres, choroid plexus, and neural crest (Maden et al., 1992; Ruberte et al., 1992; Ruberte et al., 1993). Again, rather surprisingly, CRABP I mutant embryos were normal (Gorry et al., 1994), CRABP II mutant mice had only the most minor of defects (an occasional extra postaxial phalange on one limb (Fawcett et al., 1995), and CRABP I/II double mutants had the same extra phalange and showed an enhanced susceptibility to the effects of excess RA *in utero* (Lampron et al., 1995).

Thus, knockouts of the binding proteins do not provide a method of disrupting retinoid signalling in order to study embryogenesis, although it is possible that in addition to metabolic disturbances, more subtle defects (for example, in behavior,) may become apparent in future studies.

B. Overexpression of the Binding Proteins

There has only been one study where CRABP has been overexpressed by injecting mRNA in *Xenopus* embryos (Dekker et al., 1994). The result was that the embryos showed anterior defects with reduced or fused eyes, reduced cement gland, reduced forebrain and midbrain, and reduced or lacking heart. The hindbrain had lost its rhombomeric boundaries. All these are characteristics of excess RA treatment, and accordingly there was an up-regulation of *Hoxb4* and *Hoxb9* (Section IV.B).

VII. DISRUPTING RECEPTOR SIGNALLING

In the nucleus of RA responsive cells, RA binds to a heterodimer of a RAR and a RXR, which sit on the retinoic acid response element of RA responsive genes. As a result of RA binding, the previously inactive complex consisting of the RAR/RXR heterodimer, co-repressors, and histone deacetylation activity complexes becomes active due to ligand-induced conformational changes that cause the dissociation of the co-repressors and recruitment of co-activators (Bastien and Rochette-Egly, 2004). One can thus disrupt retinoid signalling by knocking out the RARs or the RXRs, putting in a dominant negative receptor, putting in a constitutively active receptor, and also using synthetic retinoids, which bind to the receptors but then do not allow transactivation of the receptor complex. The latter compounds are thus inhibitors of receptor signalling and have provided very interesting data because they can be applied at varying concentrations to obtain information about lowered signalling levels rather than a complete absence.

A. Knockouts of Receptors

Considering the very high levels of conservation between species in the sequences of the RARs and RXRs, one might have expected individual functions for not only the receptors themselves but the receptor isoforms, as well. However, knockouts of individual RAR isoforms gave no phenotype. Thus, RARα1-/- (Li et al., 1993; Lufkin et al., 1993), RARβ2-/- (Mendelsohn et al., 1994b), and RARγ2-/- (Lohnes et al., 1993) embryos were normal. More surprisingly, knockouts of all the isoforms of any one receptor also gave, with one exception, very minor embryonic phenotypes. Thus, RARα-/- embryos were normal, but later showed higher than normal levels of postnatal lethality and, as adults, showed male sterility, abnormalities of the vertebrae, and larger alveolar volumes in the lung (Lohnes et al., 1994; Massaro et al., 2003). RARβ-/- embryos were normal and as adults had eye defects, some vertebral abnormalities, increased numbers of lung alveoli, and impaired spatial learning, and memory defects caused by defective hippocampal functioning (Chiang et al., 1998; Ghyselinck et al., 1997; Luo et al., 1995; Massaro et al.,

2000). RARγ-/- embryos were normal and as adults were very similar to RARα knockouts with male sterility, abnormal vertebrae, and abnormal tracheal rings (Lohnes et al., 1993). RXRα mutant embryos died at day 15 due to defective ventricle wall development, known as spongy myocardium, and also had eye defects (Dyson et al., 1995; Kastner et al., 1994b; Sucov et al., 1994). Interestingly, the heart defects were caused by an up-regulation of TGFβ2 (Kubalak et al., 2002). RXRβ-/- embryos were normal, but showed higher than normal levels of postnatal lethality and as adults were sterile due to defective spermatogenesis (Kastner et al., 1996). RXRγ -/- embryos were normal and as adults had deficits in long-term depression and memory due to impaired hippocampal functioning (Chiang et al., 1998; Wietrzych et al., 2005).

It required double knockouts before embryonic RA signalling could be abolished because of the functional redundancy between the receptors, whereby they substitute for one another in the absence of any one of them. Thus, double null RAR mutants (RARα/RARβ, RARα/RARγ, and RARβ/RARγ) recapitulated all the defects characteristic of vitamin A deficiency in embryos, implying that the RARs mediate the developmental functions of RA *in vivo* (Kastner et al., 1995). These defects included those of the respiratory tract, heart, urogenital tract, eye, craniofacial region and teeth, thymus, thyroid and parathyroid, hindgut, CNS, vertebrae, and limbs (Grondona et al., 1996; Mark et al., 1995; Mendelsohn et al., 1994a; Mendelsohn et al., 1999). For example, in the hindbrain, double null RARα/RARβ mutants had fused rhombomeres 6 and 7 due to the loss of the r6/7 boundary and an expanded rhombomere 5 because the expression domain of *kreisler* (normally in rhombomeres 5 and 6) was twice the normal size, the caudal stripe of *Krox-20* (normally in rhombomere 5) was expanded, *Hoxb1* and *Hoxb3* were ectopically expressed, *Hoxb4* expression was abolished, and the postotic cranial nerves were disorganized (Dupe et al., 1999) (Figure 5.2i).

In contrast, double null RARα/RARγ mutants had the full-blown hindbrain phenotype as seen in RA-deficient embryos (Section IV.A), *Raldh2* knockout embryos (Section V.B), and receptor antagonist treated embryos (Section VII.C), with completely missing posterior rhombomeres and expanded anterior rhombomeres to compensate (Figure 5.2j). This was characterized by the lack of expression of *kreisler*, an expanded single domain of *Krox-20* (normally in rhombomeres 3 and 5), and ectopic and wider *Hoxb1* expression (Wendling et al., 2001). There are some defects observed in these experiments that have not been found in VAD embryos, including defects in the ocular glands and salivary glands, some limb defects, exencephaly, some ocular abnormalities, and abnormal cranial cartilages. This is attributed to the difficulty in obtaining completely vitamin A-deficient embryos that are compatible with pregnancy.

Double mutants between RARs and RXRs also recapitulated most of the abnormalities of the vitamin A deficiency syndrome (Kastner et al., 1994b; Kastner et al., 1995; Kastner et al., 1997), and so it seemed that the RAR/RXR heterodimer is the functional unit that transduces the RA signal *in vivo*. Furthermore, the RXRα receptor is functionally the most important RXR during development (Mascrez et al., 1998; Mascrez et al., 2001). Amazingly, the triple mutant RXRα+/-/RXRβ-/-/RXRγ-/- did not express any overt developmental abnormalities.

These results raise several interesting issues with regard to testing the function of the receptors. As mentioned above, it was always expected that because of the high degree of interspecies sequence conservation of the RAR isoforms, they would have individual functions and that the pleitropic effects of RA could be realized through the cell-specific expression of different subsets of RAR and RXR isoforms. So, the lack of effect of single knockouts (except for RXRα) was a surprise. Because the same malformations can occur in different combinations of double mutants, it means that the lack of an individual receptor can be compensated for by another receptor, and that they are therefore functionally redundant. This could mean that these receptors work, for example, in a quantitative manner, and the only requirement in any particular cell is for a certain threshold level of receptor to be reached, and that several combinations of RARs and RXRs could satisfy these requirements. Alternatively, it could be that knockouts are not the way to answer questions about the function of a gene, because the deletion of a protein may induce a compensatory/regenerative reaction in a cell that is not normal.

In support of this concept, there are some experiments that have suggested that individual isoforms have particular functions. Thus, RARβ2 has been shown to be responsible for inducing neurite outgrowth in embryonic neurons rather than the other receptors that are present (RARα1 and RARγ1) (Corcoran et al., 2000; Corcoran et al., 2002). In a fascinating series of studies in the regenerating newt limb where chimeric receptors were produced (e.g., a RARα1 isoform with a thyroid hormone ligand binding domain instead of a RA ligand binding domain), the RARα1 isoform has been shown to be responsible for the inhibition of proliferation, the RARδ1 isoform (equivalent to the mammalian γ receptor) responsible for an antigenic change in the wound epidermal cells, and the RARδ2 isoform responsible for positional change in the proximodistal axis (Gann et al., 1996; Pecorino et al., 1996; Pecorino et al., 1994).

B. Antisense Morpholinos

Instead of knocking out genes in the mouse embryo, a far quicker method to ablate gene function is to inject antisense morpholinos against the gene of interest into *Xenopus* or zebrafish embryos. This has been done using morpholinos to the RARα1 and RARα2 receptor using *Xenopus* embryos. There were two results. At the anterior end of the embryo, the effects of excess RA were mimicked in reducing anterior structures, namely cement gland, eyes, and forebrain (Koide et al., 2001). At the posterior end of the embryo, *Hoxb9* and *XCAD3* were down-regulated (Shiotsugu et al., 2004). This curious anterior/posterior difference in deleting the same gene was also revealed by the effects of the morpholinos on the *Fgfr4* gene; in the anterior region of the embryo, this gene was expanded, whereas in the posterior region, the same gene was strongly inhibited (Shiotsugu et al., 2004). This is because unliganded RARα seems to behave as a repressor that is required for anterior development (Koide et al., 2001), and in its absence (after antisense morpholino action), head development is inhibited, whereas at the posterior end of the embryo, liganded RARα acts to up-regulate genes such as *Hoxb9* and *Fgfr4*.

C. Receptor Antagonists

The administration of synthetic compounds that resemble the structure and shape of all-*trans*-RA and bind to the RARs, but which do not transactivate them, has not only allowed experiments concerned with the abolition of retinoid signalling, but has also provided extremely valuable data on what happens when the levels of retinoid signalling are altered. It has been confirmed, using the RARβ2-lacZ transgenic reporter mouse, that RAR antagonist treatment does indeed inhibit RA signalling, as the lacZ staining indicative of retinoid activity disappeared after mouse embryos were treated with these antagonists (Wendling et al., 2000).

When administered to pregnant mice, the RAR antagonist- treated embryos showed craniofacial defects that bore the hallmarks of RA deficiency, particularly the absence of eyes or microphthalmia (Section IV.A), median clefting, no secondary palate, and delayed maturation of the skin (Kochhar et al., 1998; Zhou and Kochhar, 2003). Further work on the branchial arches showed that when headfold stage mouse embryos were treated with the RAR antagonist, then all the components of the 3rd and 4th branchial arches, namely the mesenchyme, arteries, nerves, placodes, and the corresponding endodermal organs, failed to form (Wendling et al., 2000). The cause of this loss of branchial arch tissue in the absence of RA signalling was the down-regulation of *Hoxa1* and *Hoxb1* in the endoderm. An identical loss of posterior pharyngeal arches caused by the respecification of the endoderm was seen in *Amphioxus* (Escriva et al., 2002).

In lung development, RAR antagonist administration has revealed that RA is required for the initial induction of lung buds in day 8 mouse embryos, as in its presence the respiratory system failed to form (Mollard et al., 2000). Again, the site of action of RA in this induction is the endoderm, which signals to the surrounding mesenchyme to induce *Fgf10* expression (Desai et al., 2004). Subsequently, however, around day 12, in the phase of branching morphogenesis, RA is inhibitory to lung development, and the administration of the RAR antagonist stimulated branching morphogenesis and the formation of distal lung buds (Mollard et al., 2000). This occurred via the down-regulation of *Tgfb3*, *Foxa2*, and *Shh* and the up-regulation of *Bmp4* and *Fgf10* in either the mesenchyme or epithelium at the distal budding tips (Chazaud et al., 2003).

Different concentrations of the antagonist were administered to cultured chick embryos at different stages, and these experiments showed the power of the approach using antagonists when the hindbrain was examined (Dupe and Lumsden, 2001). A high concentration of antagonist or an early treatment time caused the loss of the posterior hindbrain rhombomeres, with rhombomeres 1, 2, and 3 remaining attached to a smooth spinal cord. Gradually decreasing concentrations of antagonist or treating at later times allowed more rhombomeres to appear: next rhombomere 4 appeared and the 4/5 border, next rhombomere 5 appeared and the 5/6 border, next rhombomeres 6 and 7 appeared (Figure 5.3). These results led to the idea that gradually increasing RA signalling allowed the gradual development of the posterior hindbrain, showing how valuable this type of analysis can be.

In the same fashion as in mouse embryos, there was a reduction or loss of eyes when *Xenopus* embryos were treated with these antagonists as well as a disorgani-

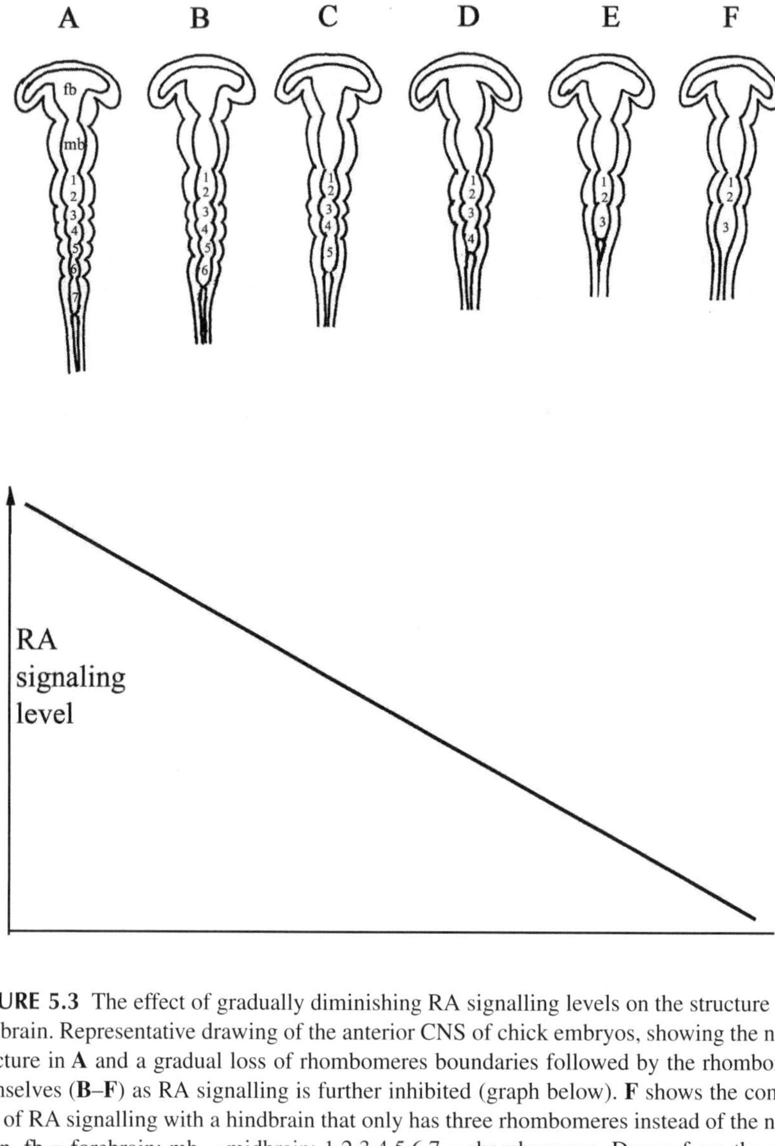

FIGURE 5.3 The effect of gradually diminishing RA signalling levels on the structure of the hindbrain. Representative drawing of the anterior CNS of chick embryos, showing the normal structure in **A** and a gradual loss of rhombomeres boundaries followed by the rhombomeres themselves (**B–F**) as RA signalling is further inhibited (graph below). **F** shows the complete lack of RA signalling with a hindbrain that only has three rhombomeres instead of the normal seven. fb = forebrain; mb = midbrain; 1,2,3,4,5,6,7 = rhombomeres. Drawn from the results of Dupe and Lumsden (2001).

zation of the branchial arches, the heart primordium was malformed or missing, and specific genes such as *XCad3* are suppressed (Lopez et al., 1995). In zebrafish embryos, RAR antagonist treatment induced a wide range of defects in the familiar RA-requiring embryonic systems, and the effects were identical to the inhibition of RA synthesis with DEAB (Section V.A). There were no fins, an expanded anterior hindbrain, and a reduced posterior hindbrain, which was manifest as a decrease in the distance between the first somite and the otic vesicle, a phenotype exactly like that of the mutant *neckless* (*Raldh2-/-* mutant) (Linville et al., 2004) (Figure 5.2h). In the hindbrain, there were severe defects in the branchiomotor neurons, including a 90% reduction in vagal BMNs; many cranial nerves showed aberrant projections; and there were ectopic Mauthner neurons, a loss of *Krox-20* and *valentino* expression in rhombomeres 5 and 6, a reduction in *Hoxb4* expression, and a corresponding expansion of the anterior rhombomeres.

Finally, a RARβ antagonist inhibited limb regeneration in axolotls (del Rincon and Scadding, 2002) and the outgrowth of nerves from the newt spinal cord, which occurs during limb regeneration (Dmetrichuk et al., 2005). A pan-RAR antagonist inhibited newt lens regeneration (Tsonis et al., 2000), and these regeneration experiments show that retinoid signalling is reused later in life when tissue regeneration is stimulated.

D. Alteration of Receptor Levels

Instead of knocking out receptors and perhaps generating aberrant results due to redundancy or compensation (see Section VII.A), an alternative strategy that has been used to investigate receptor function is to alter their levels. This has been accomplished by injecting high levels of mRNA for the receptor itself or injecting a constitutively active receptor (engineered to not require ligand activation or with a viral transactivation domain), which results in increased retinoid signalling. Conversely, injecting mRNA for a mutant receptor construct (engineered to have a nonfunctional ligand activated transactivation domain) to generate a dominant negative situation results in the lowering of retinoid signalling. Such experiments are ideally suited to embryos whose eggs can be readily injected with mRNA, namely amphibians and fish, and most of these experiments have been performed on *Xenopus* embryos.

In *Xenopus*, an even more elegant situation is exploited where one cell of the two-cell embryo is injected with the relevant mRNA along with β-galactosidase and the other cell is the control. These two cells become the left and right sides of the embryo, and the treated side can subsequently be identified because it stains blue after staining for β-galactosidase, and the other side of the embryo is the control side. So, after an *in situ* hybridization, any embryo has its own internal control (Figure 5.4). Less frequently, constructs of this type have been electroporated into chick embryos, and very occasionally transgenic mouse embryos have been generated.

Most of these receptor studies have been concerned with patterning in the early nervous system. In the neuroepithelium of the *Xenopus* embryo, there are three stripes of primary neurons, being the first ones to differentiate. They are arranged parallel to the midline from the posterior hindbrain backwards (Figure 5.4a). The row nearest the midline forms ventral motor neurons, the intermediate row forms interneurons, and the outside row forms dorsal sensory neurons. They are visualized with N-tubulin or islet1 antibody staining, and there are about 120 in total on each side of the embryo. The injection of mRNA encoding RARα2 and RXRβ into one cell of the two-cell *Xenopus* embryo increased the numbers of these primary neurons (Figure 5.4b). Conversely, decreasing receptor signalling by injecting dominant negative RARα2 reduced the numbers of primary neurons (Figure 5.4c) (Sharpe and

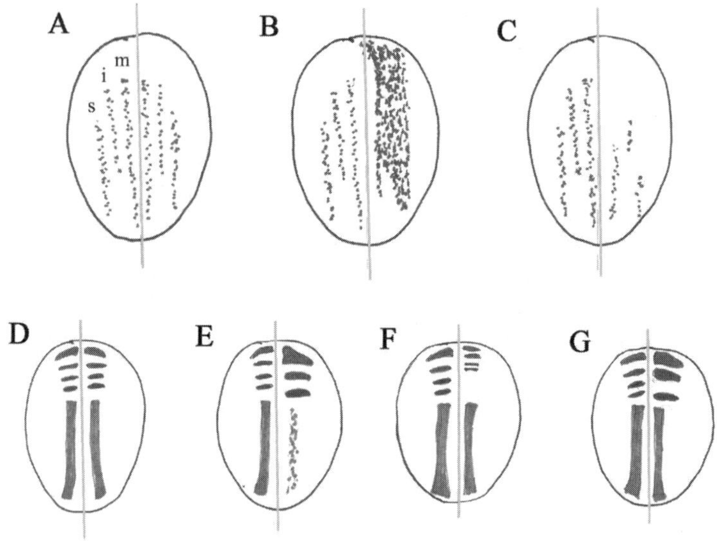

FIGURE 5.4 (See color insert following page 144.) The effects of increasing and decreasing RA signalling on neural patterning in *Xenopus* embryos. Drawings of *Xenopus* embryos at the neural plate (**A–C**) or late neurula stage (**D–G**). On the left of each figure is the left side of the embryo, which serves as an internal control. On the right side of each embryo is the experimental side, which was injected with various constructs at the two-cell stage. **A**, normal embryo stained with an anti-tubulin antibody, which picks out the first neurons to differentiate, the primary neurons. These are arranged in three parallel rows on each side of the embryo. The row nearest to the midline will produce motoneurons (m); the row in the middle (i) will produce interneurons; and the lateral row (s) will produce sensory neurons. **B**, the effect of injecting mRNA encoding for RARα2 and RXRβ into the right side of the embryo is to increase RA signalling and overproduce primary neurons. The two rows have fused into one large block due to their excessive production. **C**, the effect of injecting a dominant negative RARα2 construct into the right side of the embryo is to decrease RA signalling and underproduce primary neurons. There are three rows just visible, and reduced numbers in each row. **A–C** from the results of Sharpe and Goldstone (1997; 2000b). **D**, a normal embryo with *in situ* hybridization domains to *Otx-2* (green), a forebrain marker; *En-2* (purple), a midbrain/hindbrain border marker; *Krox20* (2 blue stripes), a marker of rhombomeres 3 and 5; and *Hoxb9* (brown), a spinal cord marker. **E**, the result of injecting a dominant negative RARα1 on the right side of the two-cell stage embryo, thereby decreasing RA signalling. At the neural plate stages, the left side of the embryo is normal, but on the right side the anterior brain markers have expanded and shifted posteriorly whereas the spinal cord marker has diminished. There is also a loss of the posterior *Krox20* stripe. **F**, the result of injecting a constitutively active RARα1, thereby increasing RA signalling. The left side is normal, but on the right side the anterior brain markers have contracted and shifted anteriorly whereas the spinal cord marker remains the same. The same result is produced when mRNA encoding *Raldh2* is injected. Results drawn from Blumberg (1997), Blumberg et al. (1997), and Chen et al. (2001). **G**, the result of injecting mRNA encoding *Cyp26A1* produces a result almost the same as decreasing RA signalling with a dn receptor (as in **E**) in expanding the size of the anterior neural plate and shifting the borders of gene expression posteriorly with a loss of the posterior *Krox20* stripe and no change in the spinal cord marker. Results drawn from Hollemann et al. (1998).

Goldstone, 2000b; Sharpe and Goldstone, 1997), showing that retinoid signalling is crucial for the generation of the appropriate pattern and cell numbers in this early neuronal event.

This expansion/contraction phenomenon according to increasing/decreasing receptor signalling also occurs in anteroposterior patterning of the early nervous system. Patterning can be assessed with various region-specific markers such as *Otx-2* (forebrain), *En-2* (midbrain/hindbrain junction), *Krox-20* (rhombomeres 3 and 5 in the hindbrain), and *Hoxb9* (spinal cord) (Figure 5.4d). Injection of dominant negative RARα1 resulted in the expansion of anterior gene domains, enlargement of anterior structures, and shortening of the tail (Figure 5.4e). Conversely, injection of a constitutively active RARα1 reduced the anterior structures, and the anterior gene expression domains became compressed (Figure 5.4f) (Blumberg, 1997; Blumberg et al., 1997). Within the hindbrain itself, a dominant negative RARβ injected into *Xenopus* embryos resulted in a shorter and thicker hindbrain with no rhombomere boundaries, an increased number of Mauthner neurons (normally in rhombomere 4), and expanded *Krox-20* domains (Figure 5.2k) (van der Wees et al., 1998). This was interpreted as an anteriorization of the posterior hindbrain.

On the other hand, a dominant negative RARα2 receptor construct down-regulated *Hoxd1*, and the posterior stripe of *Krox-20* was lost, suggesting a defective posterior hindbrain (Kolm et al., 1997). The RARα1 dominant negative receptor also inhibited neural crest cell production in *Xenopus*, and the neural crest markers *Cpl-1* and *Xslug* were down-regulated while the injection of excess RARα1 up-regulated the expression of these markers (Villaneuva et al., 2002). These results on primary neurons and anteroposterior axis patterning are precisely the same as those obtained with receptor agonists/antagonists and with excess RA (Sections VII.C and IV.B).

Retinoid signalling is also involved in determining the neuronal subsets that differentiate in the dorsoventral axis of the chick spinal cord (Wilson and Maden, 2005). RA acts by enhancing the expression of so-called class I genes expressed dorsally in the neural tube such as *pax6*, *Irx3*, and *Dbx*. Electroporating a human dominant repressor RARα into the chick embryonic spinal cord decreased levels of these class I genes as well as the genes required for motor neuron development, *Olig2*, *Mnx*, *Lim3*, and *Islet1/2* (Novitch et al., 2003). Conversely, electroporation of a RAR that activated genes in a ligand-independent manner induced ectopic *Olig2* and *Mnx* expressing cells. Within the motor neuron pathways itself, retinoid signalling is involved in the generation of specific subsets of motor neurons called the lateral motor column neurons. Electroporation of a dominant negative RARα blocked LMC specification in the chick spinal cord (Sockanathan et al., 2003), and as a result, the electroporated cells transformed into other types of motor neurons instead. This phenomenon only occurred at the regions of the spinal cord where the limbs develop because LMCs specifically innervate the limbs.

In the mouse, dominant negative receptor constructs have been linked to cell-specific promoters to make transgenic embryos. A dominant negative RARα controlled by the type II collagen promoter was used to drive the construct in cartilage cells (Yamaguchi et al., 1998). The homozygotes showed shortened skeletons, as bone development was inhibited by a lack of retinoid signalling, and in the prevertebrae, there was a reduction in *Hoxb4* expression followed by anterior and posterior transformations of vertebral identity. When a constitutively active RARα was linked to the αA-crystallin promoter, then the embryos showed the effects of excess RA on eyes, namely microphthalmia and cataracts (Balkan et al., 1992). A general overexpression of the dominant negative RARα produced effects on the skin (Imakado et al., 1995). Such embryos had a red shiny skin at birth, with a thinner stratum corneum with a very low ability to prevent water loss and an absence of a lipid barrier. Thus, in addition to patterning in the nervous system and other embryonic systems, retinoid signalling is involved in cartilage and skin development.

E. Stimulating Specific Receptors

Synthetic retinoids have been developed that, at particular concentrations, show increased affinity for particular RARs or RXRs and are known as agonists. These have been produced by several different companies and allocated code names (e.g. BMS189453; CD437; Am80; AGN191347). Pan-RAR agonists, individual RAR agonists (RARα, RARβ, RARγ agonists), and pan-RXR agonists have been produced. These compounds have been extremely valuable in identifying which receptor or whether specific RAR/RXR heterodimers are responsible for particular gene inductions or biological effects.

For example, tRA is well known to stimulate the outgrowth of neurites from cultured embryonic neurons (review, Maden, 2001). When individual receptor agonists were applied instead of tRA, a RARα agonist had no effect, a RARβ agonist precisely mimicked the effects of tRA, a RARγ agonist had no effect, and a pan-RXR agonist had no effect (Corcoran et al., 2000). From this, it was concluded that the RARβ receptor is involved in transducing the effect of RA in stimulating neurite outgrowth. In another neural situation, TTNPB, a pan-RAR agonist, doubled the number of striatal neurons differentiating in a culture system derived from embryonic lateral ganglionic eminence cells. A RXR agonist also doubled the number of neurons, but when the two were added together, there was no further increase (Toresson et al., 1999).

On the other hand, many cell culture studies have shown synergy between the RAR and RXR agonists. In P19 and F9 cells, low concentrations of RARα, RARβ, or RARγ agonists only inefficiently induced the expression of several RA target genes or induced differentiation. A RXR agonist was similarly inefficient, but at the same concentration, the combination of the two resulted in strong synergy of gene induction (Roy et al., 1995). The combination of RARα/RXR compounds had the greatest synergistic effect, and RARβ/RXR compounds had the lowest synergistic effect, although there were differences according to the gene of interest. Similar data has been obtained studying apoptosis and neuronal differentiation of P19 cells, namely a synergy between a RAR selective and a RXR selective compound (Horn et al., 1996).

In the embryo, these compounds vary in their ability to induce developmental defects when administered to mice, hamsters, or chick embryos or to regenerating amphibian limbs. The RAR agonists are much more potent that tRA itself. For example, TTNPB is 1000X more potent than tRA at inducing limb and craniofacial abnormalities in the mouse and hamster embryo (Kochhar et al., 1996; Willhite et al., 1996) and at inducing limb duplications in the chick embryo and regenerating limb (Maden et al., 1991). RARβ/γ selective compounds were as potent as TTNPB, and the RXR agonists were inactive (Jiang et al., 1995). Similarly, in *Xenopus* embryos, TTNPB characteristically enhanced posterior gene expression in the embryonic nervous system (Shiotsugu et al., 2004), whereas a RXR agonist was inactive (van der Wees et al., 1998). Studying a whole range of teratogenic effects in the mouse embryo including CNS, craniofacial, visceral, and vertebral defects, it was shown that the most potent compounds were RARα agonists, followed by RARβ agonists, followed by RARγ agonists (Elmazar et al., 1996). Furthermore, individually they produced a different spectrum of effects, with α compounds generating the most varied effects, β compounds inducing a greater frequency of urogenital and liver defects, and γ compounds generating a greater frequency of vertebral defects. As in the case of the studies on cell cultures, although a RXR compound was inactive as a teratogen for mouse embryos, it could nevertheless potentiate the effects of a RARα and a RARγ selective compound (Elmazar et al., 1997; Elmazar et al., 2001). More detailed studies in mouse embryos have revealed that the pan-RAR agonist TTNPB induced a fusion of the first and second branchial arches and ectopic expression of *Hoxb1* and *Hoxa1* in the anterior pharyngeal endoderm and ectopic expression of *Hoxb1* in rhombomere 3 of the anterior hindbrain (Matt et al., 2003). This phenotype also occurred with a RARβ agonist, but not a RARα agonist, RARγ agonist, or RXR agonist. The RARβ agonists and the RXR agonist synergized in generating branchial arch fusions, from which it was concluded that this teratological effect of RA was mediated through RARβ/RXR heterodimers in the anterior endoderm.

VIII. DISRUPTION OF RA CATABOLISM

A. Knockouts of the Cyps

There are three Cyp enzymes in higher vertebrates — Cyp26A1, Cyp26B1, and Cyp26C1 — and they have a non-overlapping distribution in the embryo (e.g., Reijntjes et al., 2004). The expression of *Cyp26A1* and *B1* are induced by their substrate, tRA, whereas *Cyp26C1* is repressed (Reijntjes et al., 2005), so they clearly have a complex functional relationship in retinoid signalling. Their suggested function is to catabolize tRA into more polar, inactive compounds such as 4-*oxo*-RA and 4-OH-RA in preparation for excretion, but these polar compounds are themselves biologically active in transactivating the RARs, inducing abnormalities in the embryo, and rescuing the VAD quail (Pijnappel et al., 1993).

The *Cyp26A1* null mouse embryo had several abnormalities (Abu-Abed et al., 2001; Sakai et al., 2001). One strong expression domain in the embryo was the tail bud, and accordingly, the knockout embryo had a defective tail bud that manifested itself as sirenomelia, namely fused hindlimbs, and other severe defects of the posterior gut and urogenital system. There were also transformations of cervical vertebrae, and in the hindbrain, rhombomere 4 was wider than normal whereas rhombomeres 2 and 3 were smaller than normal. Outgrowth of the trigeminal nerve seemed reduced, and the trigeminal and facial ganglia were sometimes fused. These abnormalities are characteristic of excess RA effects (Section IV.B), which implies that the role of

Cyp26A1 is to keep tRA levels low in specific domains. Indeed, the *Cyp26A1-/-* caudal phenotype can be rescued by crossing this line with the *Raldh2* null mutant heterozygote to produce *Cyp26A1-/- / Raldh2+/-* embryos; that is, reducing the levels of endogenous tRA prevents these defects (Niederreither et al., 2002a).

Coincidentally, there is a natural mutation of the *Cyp26A1* gene in zebrafish that generates a truncated, nonfunctional protein. The mutant is called *giraffe* (Emoto et al., 2005). Embryos had additional defects to the mouse, such as lack of blood cells circulating in the embryo because of the defective development of the common cardinal vein, no fins, and malformed Meckel's cartilage in the jaw. In the hindbrain, the distance between rhombomeres 5 or 6 and the first somite was expanded even though the size of individual rhombomeres themselves was reduced. The up-regulation and expansion of genes such as *Hoxb4a*, *Hoxb5a*, and *Hoxb6a*, which are expressed in the spinal cord and have their rostral boundaries at the rhombomeres 6/7 border, suggested that the spinal cord extended further rostrally than normal, hence the name giraffe (Figure 5.21). The *gir* mutant can be phenocopied by injecting *Cyp26A1* antisense morpholinos into the embryo, and it can be rescued by injecting antisense morpholinos to the *Raldh2* gene and thus reducing the amount of RA present, exactly as the *Raldh2* heterozygote rescues the *Cyp26A1-/-* mutant in the mouse, as described above.

The *Cyp26B1-/-* mutant mouse has also been generated, and this embryo had severe limb deformities (meromelia), as the limb bud is a major site of *Cyp26B1* expression and craniofacial abnormalities including micrognathia (Yashiro et al., 2004). In the mutant limb bud, the levels of distal *Hox* genes involved in digit development — *Hoxd12*, *Hoxd13*, and *Hoxd13* — were reduced, and on day 12 there was a great increase in apoptosis in the chondrogenic precursor cells.

B. Antisense Morpholinos

As described in Section VIII.A, antisense morpholinos to *Cyp26A1* have been injected into zebrafish embryos, with the result that they phenocopied the *gir* mutant in decreasing the size of the rhombomeres and expanding the anterior spinal cord (Figure 5.21) (Emoto et al., 2005). Prior to this morphological change, the forebrain marker *Otx2* was down-regulated and the rostral border of *Hoxb1b* was expanded anteriorly as were *Meis3* and *Iro1* (Kudoh et al., 2002). This suggested an anterior shift in the anterior neural plate had occurred.

C. Overexpression of Cyps

In very neat complementation to antisense morpholinos, overexpression of *Cyp26A1* in *Xenopus* seemed to expand the size of the anterior neural plate and shifted the borders of gene expression posteriorly (Hollemann et al., 1998) (Figure 5.4g). The forebrain marker *Otx2* was expanded in domain size, and the two stripes of *Krox20* in rhombomeres 3 and 5 were shifted posteriorly while *Hoxb9*, the spinal cord marker, was not altered. There was a duplication of the trigeminal ganglion, and *Cyp* overexpression rescued the RA induced developmental defects, suggesting it is an inactivating enzyme. Indeed, this phenotype (Figure 5.4g) is identical, at least within the hindbrain, to the effects of reducing RA signalling (Figure 5.4e). Overexpression of *Cyp26A1* in zebrafish produced the opposite effect of excess RA in suppressing the expression of *Hoxb1b* and *Meis3* (Kudoh et al., 2002).

IX. CONCLUSIONS

I have described here a whole range of different ways in which retinoid signalling has been altered in embryos: providing excess or generating a deficiency of the ligand, tRA; disrupting ligand synthesis by interfering with the RA synthetic enzymes; disrupting receptor signalling by interfering with the nuclear receptors; and altering the catabolic enzymes. The striking finding is that the same embryonic systems are affected by these diverse alterations: anteroposterior patterning in the CNS, the hindbrain, the heart, the limb buds, the upper jaw and frontonasal region, the posterior pharyngeal arches, the vertebrae, the eyes, and the lungs. Some of the effects on the hindbrain across this spectrum of alterations in signalling are brought together in Figure 5.2.

The second striking finding is that within each of these organs, the level of retinoid signalling is crucial for determining the correct anatomical structure because either too little or too much retinoid signalling prevents the organ from developing properly. Only the correct level allows development to proceed appropriately because too little retinoid signalling does not permit the full realization of the organ, and interestingly, excess signalling often causes abnormal development because there is "too much" of the organ and it duplicates itself. This is particularly apparent in the limb bud and the hindbrain. In the limb bud, excess retinoid signalling duplicates the limb and two appear instead of one, whereas preventing signalling prevents the limb bud from developing at all. In the hindbrain, inhibiting retinoid signalling prevents the posterior hindbrain from developing at all, and excess retinoid signalling causes an overgrowth of posterior characteristics in the anterior hindbrain. Thus, these analyses have told us a great deal about the role that retinoids play in embryogenesis and show how valuable these analytical tools are in understanding the complexities of pattern formation in the embryo.

REFERENCES

Abramovici A, Kam J, Liban E, Barishak RY (1978) Incipient histopathological lesions in citral-induced microphthalmos in chick embryos. Dev Neurosci 1:177–185.

Abu-Abed S, Dolle P, Metzger D, Beckett B, Chambon P, Petkovich M (2001) The retinoic acid-metabolizing enzyme, CYP26A1, is essential for normal hindbrain patterning, vertebral identity, and development of posterior structures. Genes Dev 15:226–240.

Adams J, Baer S, Gavin JAL, Janulewicz PA, Meyer E, Lammer EJ (2001) Neuropsychological characteristics of children embryonically exposed to isotretinoin (Accutane): outcome at age 10. Neurotoxicol Teratol 23:296.

Adams J, Lammer EJ (1995) Human isotretinoin exposure: the teratogenesis of a syndrome of cognitive defects. Neurotoxicol Teratol 17:386.

Alexandre D, Clarke JDW, Oxtoby E, Yan Y-L, Jowett T, Holder N (1996) Ectopic expression of Hoxa-1 in the zebrafish alters the fate of the mandibular arch neural crest and phenocopies a retinoic acid-induced phenotype. Development 122:735–746.

Anchan RM, Drake DP, Haines CF, Gerwe EA, LaMantia A-S (1997) Disruption of local retinoid-mediated gene expression accompanies abnormal development in the mammalian olfactory pathway. J Comp Neurol 379:171–184.

Antipatis C, Ashworth CJ, Grant G, Lea RG, Hay SM, Rees WD (1998) Effects of maternal vitamin A status on fetal heart and lung: changes in expression of key developmental genes. Am J Physiol 275:L1184–L1191.

Avantaggiato V, ACampora D, Tuorto F, Simeone A (1996) Retinoic acid induces stage-specific repatterning of the rostral central nervous system. Dev Biol 175:347–357.

Balkan W, Klintworth GK, Bock CB, Linney E (1992) Transgenic mice expressing a constitutively active retinoic acid receptor in the lens exhibit ocular defects. Dev Biol 151:622–625.

Bastien J, Rochette-Egly C (2004) Nuclear retinoid receptors and the transcription of retinoid-target genes. Gene 328:1–16.

Begemann G, Marx M, Mebus K, Meyer A, Bastmeyer M (2004) Beyond the neckless phenotype: influence of reduced retinoic acid signaling on motor neuron development in the zebrafish hindbrain. Dev Biol 271:119–129.

Begemann G, Schilling TF, Rauch G-J, Geisler R, Ingham PW (2001) The zebrafish neckless mutation reveals a requirement for raldh2 in mesodermal signals that pattern the hindbrain. Development 128:3081–3094.

Berggren K, McCaffery P, Drager U, Forehand CJ (1999) Differential distribution of retinoic acid synthesis in the chicken embryo as determined by immunolocalization of the retinoic acid synthetic enzyme, RALDH-2. Dev Biol 210:288–304.

Blentic A, Gale E, Maden M (2003) Retinoic acid signalling centres in the avian embryo identified by sites of expression of synthesising and catabolising enzymes. Dev Dynam 227:114–127.

Blumberg B (1997) An essential role for retinoid signalling in anteroposterior neural specification and neuronal differentiation. Sem in Cell & Dev Biol 8:417–428.

Blumberg B, Bolado J, Moreno TA, Kintner C, Evans RM, Papalopulu N (1997) An essential role for retinoid signalling in anteroposterior neural patterning. Development 124:373–379.

Chazaud C, Dolle P, Rossant J, Mollard R (2003) Retinoic acid signaling regulates murine bronchial tubule formation. Mech Dev 120:691–700.

Chen Y, Kostetskii I, Zile MH, Solursh M (1995) Comparative study of Msx-1 expression in early normal and vitamin A-deficient avian embryos. J Exp Zool 272:299–310.

Chen Y, Pollet N, Niehrs C, Pieler T (2001) Increased XRALDH2 activity has a posteriorizing effect on the central nervous system of Xenopus embryos. Mech of Dev 101:91–103.

Chiang M-Y, Misner D, Kempermann G, Schikorski T, Giguere V, Sucov HM, Gage FH, Stevens CF, Evans RM (1998) An essential role for retinoid receptors RARb and RXRg in long-term potentiation and depression. Neuron 21:1353–1361.

Cho KWY, De Robertis EM (1990) Differential activation of Xenopus homeobox genes by mesoderm-inducing growth factors and retinoic acid. Genes & Dev 4:1910–1916.

Cohlan SQ (1953) Excessive intake of vitamin A as a cause of congenital abnormalities in the rat. Science 117:535–536.

Conlon RA, Rossant J (1992) Exogenous retinoic acid rapidly induces anterior ectopic expression of murine Hox-2 genes in vivo. Development 116:357–368.

Corcoran J, Shroot B, Pizzey J, Maden M (2000) The role of retinoic acid receptors in neurite outgrowth from different populations of embryonic mouse dorsal root ganglia. J Cell Sci 113:2567–2574.
Corcoran J, So P-L, Barber RD, Vincent KJ, Maxarakis ND, Mitrophanous KA, Kingsman SM, Maden M (2002) Retinoic acid receptor b2 and neurite outgrowth in the adult mouse spinal cord in vitro. J Cell Sci 115:3779–3786.
Costaridis P, Horton C, Zeitlinger J, Holder N, Maden M (1996) Endogenous retinoids in the zebrafish embryo and adult. Dev Dynam 205:41–51.
Cunningham ML, MacAuley A, Mirkes PE (1994) From gastrulation to neurulation: transition in retinoic acid sensitivity identifies distinct stages of neural patterning in the rat. Dev Dynam 200:227–241.
Dekker E-J, Pannese M, Houtzager E, Timmermans A, Boncinelli E, Durston A (1992) Xenopus Hox-2 genes are expressed sequentially after the onset of gastrulation and are differentially inducible by retinoic acid. Development Supplement:195–202.
Dekker E-J, Vaessen M-J, van den Berg C, Timmermans A, Godsave S, Holling T, Niewkoop P, van Kessel AG, Durston A (1994) Overexpression of a cellular retinoic acid binding protein (xCRABP) causes anteroposterior defects in developing Xenopus embryos. Development 120:973–985.
del Rincon SV, Scadding SR (2002) Retinoid antagonists inhibit normal patterning during limb regeneration in the axolotl, Ambystoma mexicanum. J Exp Zool 292:435–443.
Deltour L, Foglio MH, Duester G (1999) Impaired retinol utilisation in Adh4 alcohol dehydrogenase mutant mice. Dev Genet 25:1–10.
Delva L, Bastie J-N, Rochette-Egly C, Kraiba R, Balitrand N, Despouy G, Chambon P, Chomienne C (1999) Physical and functional interactions between cellular retinoic acid binding protein II and the retinoic acid-dependent nuclear complex. Mol Cell Biol 19:7158–7167.
Dersch H, Zile MH (1993) Induction of normal cardiovascular development in the vitamin A-deprived quail embryo by natural retinoids. Dev Biol 160:424–433.
Desai TJ, Malpel S, Flentke GR, Smith SM, Cardoso WV (2004) Retinoic acid selectively regulates Fgf10 expression and maintains cell identity in the prospective lung field of the developing foregut. Dev Biol 273:402-415.
Dickman ED, Thaller C, Smith SM (1997) Temporally-regulated retinoic acid depletion produces specific neural crest, ocular and nervous system defects. Development 124:3111–3121.
Die-Smulders CEM, Sturkenboom MCJM, Veraart J, van Katwijk C, Sastrowijoto P, van der Linden E (1995) Severe limb defects and craniofacial anomalies in a fetus conceived during acitretin therapy. Teratology 52:215–219.
Diez del Corral, Olivera-Martinez I, Goriely A, Gale E, Maden M, Storey K (2003) Opposing FGF and retinoid pathways control ventral neural pattern, neuronal differentiation, and segmentation during body axis extension. Neuron 40:65–79.
Diez del Corral, Storey KG (2004) Opposing FGF and retinoid pathways: a signalling switch that controls differentiation and patterning onset in the extending vertebrate body axis. Bioessays 26:857–869.
Dmetrichuk JM, Spencer GE, Carlone RL (2005) Retinoic acid-dependent attraction of adult spinal cord axons towards regenerating newt limb blastemas in vitro. Dev Biol 281:112–120.
Dong D, Ruuska SE, Levinthal DJ, Noy N (1999) Distinct roles for cellular retinoic acid-binding proteins I and II in regulating signaling by retinoic acid. J Biol Chem 274:23695–23698.

Dong D, Zile MH (1995) Endogenous retinoids in the early avian embryo. Biochem Biophys Res Comm 217:1026–1031.

Duester G (2000) Families of retinoid dehydrogenases regulating vitamin A function. Production of visual pigment and retinoic acid. Eur J Biochem 267 4315–4324.

Duester G (2001) Genetic dissection of retinoid dehydrogenases. Chemico-Biological Interactions 130–132:469–480.

Dupe V, Ghyselinck N, Wendling O, Chambon P, Mark M (1999) Key roles of retinoic acid receptors alpha and beta in the patterning of the caudal hindbrain, pharyngeal arches and otocyst in the mouse. Development 126:5051–5059.

Dupe V, Lumsden A (2001) Hindbrain patterning involves graded responses to retinoic acid signalling. Development 128:2199–2208.

Dupe V, Matt N, Garnier J-M, Chambon P, Mark M, Ghyselinck NB (2003) A newborn lethal defect due to inactivation of retinaldehyde dehydrogenase type 3 is prevented by maternal retinoic acid treatment. Proc Natl Acad Sci USA 100:14036–14041.

Durston AJ, Timmermans JPM, Hage WJ, Hendricks HFJ, de Vries NJ, Heideveld M, Nieuwkoop PD (1989) Retinoic acid causes an anteroposterior transformation in the developing nervous system. Nature 340:140–144.

Dyson E, Sucov HM, Kubalak SW, Schmid-Schonbein GW, LeLano FA, Evans RM, Ross J, Chein KR (1995) Atrial-like phenotype is associated with embryonic ventricular failure in retinoid X receptor a -/- mice. Proc Natl Acad Sci USA 92:7386–7390.

Eichele G, Tickle C, Alberts BM (1985) Studies on the mechanism of retinoid-induced pattern duplications in the early chick limb bud: temporal and spatial aspects. J Cell Biol 101:1913–1920.

Elmazar MM, Ruhl R, Nau H (2001) Synergistic teratogenic effects induced by retinoids in mice by coadministration of a RARalpha- or RARgamma-selective agonist with a RXR-selective agonist. Toxicol Appl Pharmacol 170:2–9.

Elmazar MM, Ruhl R, Reichert U, Shroot B, Nau H (1997) RARalpha-mediated teratogenicity in mice is potentiated by an RXR agonist and reduced by an RAR antagonist: dissection of retinoid receptor-induced pathways. Toxicol Appl Pharmacol 146:21–28.

Elmazar MMA, Reichert U, Shroot B, Nau H (1996) Pattern of retinoid-induced teratogenic effects: possible relationship with relative selectivity for nuclear retinoid receptors RARa, RARb, and RARg. Teratology 53:158–167.

Emoto Y, Wada H, Okamoto H, Kudo A, Imai Y (2005) Retinoic acid-metabolizing enzyme Cyp26a1 is essential for determining territories of hindbrain and spinal cord in zebrafish. Dev Biol 278:415–427.

Escriva H, Holland ND, Gronemeyer H, Laudet V, Holland LZ (2002) The retinoic acid signaling pathway regulates anterior/posterior patterning in the nerve cord and pharynx of amphioxus, a chordate lacking neural crest. Development 129: 2905–2916.

Fan X, Molotkov A, Manabe S, Donmoyer CM, Deltour L, Foglio MH, Cuenca AE, Blaner WS, Lipton SA, Duester G (2003) Targeted disruption of Aldh1a1 (Raldh1) provides evidence for a complex mechanism of retinoic acid synthesis in the developing retina. Mol Cell Biol 23:4637–4648.

Fantel AG, Shepard TH, Newell-Morris LL, Moffett BC (1977) Teratogenic effects of retinoic acid in pigtail monkeys (Macaca nemestrina). 15:65–72.

Fawcett D, Pasceri P, Fraser R, Colbert M, Rossant J, Giguere V (1995) Postaxial polydactyly in forelimbs of CRABP-II mutant mice. Development 121:671–679.

Fujii H, Sato T, Kaneko S, Gotoh O, Fujii-Kuriyama Y, Osawa K, Kato S, Hamada H (1997) Metabolic inactivation of retinoic acid by a novel P450 differentially expressed in developing mouse embryos. EMBO J 16:4163–4173.

Gale E, Prince V, Lumsden A, Clarke J, Holder N, Maden M (1996) Late effects of retinoic acid on neural crest and aspects of rhombomere identity. Development 122:783–793.

Gale E, Zile M, Maden M (1999) Hindbrain respecification in the retinoid-deficient quail. Mech Dev 89:43–54.

Gann AA, Gates PB, Stark D, Brockes JP (1996) Receptor isoform specificity in a cellular response to retinoic acid. Proc Biol Sci 263:729–734.

Ghatpande S, Ghatpande A, Zile M, Evans T (2000) Anterior endoderm is sufficient to rescue foregut apoptosis and heart tube morphogenesis in an embryo lacking retinoic acid. Dev Biol 219:59–70.

Ghyselinck NB, Bavik C, Sapin V, Mark M, Bonnier D, Hindelang C, Dierich A, Nilsson CB, Hakansson H, Sauvant P, Azias-Braesco V, Frasson M, Picaud S, Chambon P (1999) Cellular retinol-binding protein I is essential for vitamin A homeostasis. EMBO J 18:4903–4914.

Ghyselinck NB, Dupe V, Dierich A, Massaddeq N, Garnier J-M, Rochette-Egly C, Chambon P, Mark M (1997) Role of the retinoic acid receptor beta (RARb) during mouse development. Int J Dev Biol 41:425–447.

Godsave SF, Koster CH, Getahun A, Mathu M, Hooiveld M, van der Wees J, Hendriks J, Durston AJ (1998) Graded retinoid responses in the developing hindbrain. Dev Dyn 213:39–49.

Gorry P, Lufkin T, Dierich A, Rochette-Egly C, Decimo D, Dolle P, Mark M, Durand B, Chambon P (1994) The cellular retinoic acid binding protein I is dispensable. Proc Natl Acad Sci USA 91:9032–9036.

Grandel H, Lun K, Rauch G-J, Rhinn M, Piotrowski T, Houart C, Sordino P, Kuchler AM, Schulte-Merker S, Geisler R, Holder N, Wilson SW, Brand M (2002) Retinoic acid signalling in the zebrafish embryo is necessary during pre-segmentation stages to pattern the anterior-posterior axis of the CNS and to induce a pectoral fin bud. Development 129:2851–2865.

Grondona JM, Kastner P, Gansmuller A, Decimo D, Chambon P, Mark M (1996) Retinal dysplasia and degeneration in RARb2/RARg2 compound mutant mice. Development 122:2173–2188.

Grun F, Hirose Y, Kawauchi S, Ogura T, Umesono K (2000) Aldehyde dehydrogenase 6, a cytosolic retinaldehyde dehydrogenase prominently expressed in sensory neuroepithelia during development. J Biol Chem 275:41210–41218.

Hale F (1933) Pigs born without eyeballs. J Hered 24:105–106.

Heine UI, Roberts AB, Munoz EF, Roche NS, Sporn MB (1985) Effects of retinoid deficiency on the development of the heart and vascular system of the quail embryo. Virchow's Arch B [Cell Pathol] 50:135–152.

Heyman RA, Mangelsdorf DJ, Dyck JA, Stein RB, Eichele G, Evans RM, Thaller C (1992) 9-cis retinoic acid is a high affinity ligand for the retinoid X receptor. Cell 68:397–406.

Hill J, Clarke JDW, Vargesson N, Jowett T, Holder N (1995) Exogenous retinoic acid causes specific alterations in the development of the midbrain and hindbrain of the zebrafish embryo including positional respecification of the Mauthner neuron. Mech Dev 50:3–16.

Holder N, Hill J (1991) Retinoic acid modifies development of the midbrain-hindbrain border and affects cranial ganglion formation in zebrafish embryos. Development 113:1159–1170.

Hollemann T, Chen Y, Grunz H, Pieler T (1998) Regionalized metabolic activity establishes boundaries of retinoic acid signalling. EMBO J 17:7361–7372.

Horn V, Minucci S, Ogryzko VV, Adamson ED, Howard BH, Levin AA, Ozato K (1996) RAR and RXR selective ligands cooperatively induce apoptosis and neuronal differentiation in P19 embryonal carcinoma cells. FASEB J 10:1071–1077.

Imakado S, Bickenbach JR, Bundman DS, Rothnagel JA, Attar PS, Wang X-J, Walczak VR, Wisniewski S, Pote J, Gordon JS, Heyman RA, Evans RM, Roop DR (1995) Targeting expression of a dominant-negative retinoic acid receptor mutant in the epidermis of transgenic mice results in loss of barrier function. Genes & Dev 9:317–329.

Izpisua-Belmonte J-C, Tickle C, Dolle P, Wolpert L, Duboule D (1991) Expression of the homeobox Hox-4 genes and the specification of position in chick wing development. Nature 350:585–589.

Jiang H, Penner JD, Beard RL, Chandraratna RAS, Kochhar DM (1995) Diminished teratogenicity of retinoid X receptor-selective synthetic retinoids. Biochem Pharmacol 50:669–676.

Kalter H, Warkany J (1959) Experimental production of congenital malformations in mammals by metabolic procedure. Physiol Rev 39:69–115.

Kastner P, Chambon P, Leid M (1994a) Role of nuclear retinoic acid receptors in the regulation of gene expression. In Vitamin A in Health and Disease (Blomhoff R, Ed.), pp. 189–238. New York: Marcel Dekker Inc.

Kastner P, Grondona JM, Mark M, Gansmuller A, LeMeur M, Decimo D, Vonesch J-L, Dolle P, Chambon P (1994b) Genetic analysis of RXRa developmental function: convergence of RXR and RAR signalling pathways in heart and eye morphogenesis. Cell 78:987–1003.

Kastner P, Mark M, Chambon P (1995) Nonsteroid nuclear receptors: what are genetic studies telling us about their role in real life? Cell 83:859–869.

Kastner P, Mark M, Ghyselinck N, Kretzel W, Dupe V, Grondona JM (1997) Genetic evidence that the retinoid signal is transduced by heterodimeric RXR/RAR functional units during mouse development. Development 124:313–326.

Kastner P, Mark M, Leid M, Gansmuller A, Chin W, Grondona JM, Decimo D, Krezel W, Dierich A, Chambon P (1996) Abnormal spermatogenesis in RXRb mutant mice. Genes & Dev 10:80–92.

Kessel M (1992) Respecification of vertebral identities by retinoic acid. Development 115:487–501.

Kessel M (1993) Reversal of axonal pathways from rhombomere 3 correlates with extra Hox expression domains. Neuron 10:379–393.

Kessel M, Gruss P (1991) Homeotic transformations of murine vertebrae and concomitant alteration of Hox codes induced by retinoic acid. Cell 67:89–104.

Kliewer SA, Umesono K, Evans RM, Mangelsdorf DJ (1994) The retinoid X receptors: modulators of multiple hormonal signaling pathways. In Vitamin A in Health and Disease (Blomhoff R, Ed.), pp. 239–255. New York: Marcel Dekker Inc.

Knudsen PA (1966) Congenital malformations of lower incisors and molars in exencephalic mouse embryos, induced by hypervitaminosis A. Acat Odont Scand 24:55–71.

Kochhar DM (1973) Limb development in mouse embryos. I. Analysis of teratogenic effects of retinoic acid. Teratology 7:289–298.

Kochhar DM, Jiang H, Penner JD, Beard RL, Chandraratna RAS (1996) Differential teratogenic response of mouse embryos to receptor selective analogs of retinoic acid. Chemico-Biological Interactions 100:1–12.

Kochhar DM, Jiang H, Penner JD, Johnson AT, Chandraratna RAS (1998) The use of a retinoid receptor antagonist in a new model to study vitamin A-dependent developmental events. Int J Dev Biol 42:601–608.

Koide T, Downes M, Chandraratna RAS, Blumberg B, Umesono K (2001) Active repression of RAR signalling is required for head formation. Genes Dev 15:2111–2121.

Kolm PJ, Apekin V, Sive H (1997) Xenopus hindbrain patterning requires retinoid signalling. Dev Biol 192:1–16.

Kostetskii I, Jiang Y, Kostetskaia E, Yuan S, Evans T, Zile M (1999) Retinoid signalling required for normal heart development regulates GATA-4 in a pathway distinct from cardiomyocyte differentiation. Dev Biol 206:206–218.

Kostetskii I, Yuan S-Y, Kostetskaia E, Linask KK, Blanchet S, Seleiro E, Michaille J-J, Brickell P, Zile M (1998) Initial retinoid requirement for early avian development coincides with retinoid receptor coexpression in the precardiac fields and induction of normal cardiovascular development. Dev Dynam 213:188–198.

Kubalak SW, Hutson DR, Scott KK, Shannon RA (2002) Elevated transforming growth factor b2 enhances apoptosis and contributes to abnormal outflow tract and aortic sac development in retinoic X receptor a knockout embryos. Development 129:733–746.

Kudoh T, Wilson SW, Dawid IB (2002) Distinct roles for Fgf, Wnt and retinoic acid in posteriorizing the neural ectoderm. Development 129:4335–4346.

Lammer EJ, Armstrong DL (1992) Malformations of hindbrain structures among humans exposed to isotretinoin (13-cis-retinoic acid) during early embryogenesis. In Retinoids in Normal and Development and Teratogenesis (Morriss-Kay G, Ed.), pp. 281–295. Oxford: Oxford University Press.

Lammer EJ, Chen DT, Hoar RM, Agnish AD, Benke PJ, Braun JT, Curry CJ, Fernhoff PM, Grix AW, Lott IT, Richard JM, Sun SC (1985) Retinoic acid embryopathy. New Eng J Med 313:837–841.

Lampron C, Rochette-Egly C, Gorry P, Dolle P, Mark M, Lufkin T, LeMeur M, Chambon P (1995) Mice deficient in cellular retinoic acid binding protein II (CRABP II) or in both CRABP I and CRABP II are essentially normal. Development 121:539–548.

Lee YM, Osumi-Yamashita N, Ninomiya Y, Moon CK, Eriksson U, Eto K (1995) Retinoic acid stage-dependently alters the migration pattern and identity of hindbrain neural crest cells. Development 121:825–837.

Leonard L, Horton C, Maden M, Pizzey JA (1995) Anteriorization of CRABP-I expression by retinoic acid in the developing mouse central nervous system and its relationship to teratogenesis. Dev Biol 168:514–528.

Leroy P, De Robertis EM (1992) Effects of lithium chloride and retinoic acid on the expression of genes from the Xenopus Hox 2 complex. Dev Dynam 194:21–32.

Li E, Sucov HM, Lee K-F, Evans RM, Jaenisch R (1993) Normal development and growth of mice carrying a targeted disruption of the a1 retinoic acid receptor gene. Proc Natl Acad Sci USA 90:1590–1594.

Li H, Wagner E, McCaffery P, Smith D, Andreadis A, Drager UC (2000) A retinoic acid synthesising enzyme in ventral retina and telencephalon of the embryonic mouse. Mech Dev 95:283–289.

Linville A, Gumusaneli E, Chandraratna RA, Schilling TF (2004) Independent roles for retinoic acid in segmentation and neuronal differentiation in the zebrafish hindbrain. Dev Biol 270:186–199.

Lohnes D, Kastner P, Dierich A, Mark M, LeMeur M, Chambon P (1993) Function of retinoic acid g in the mouse. Cell 73:643–658.

Lohnes D, Mark M, Mendelsohn C, Dolle P, Dierich A, Gorry P, Gansmuller A, Chambon P (1994) Function of the retinoic acid receptors (RARs) during development (I) Craniofacial and skeletal abnormalities in RAR double mutants. Development 120:2723–2748.

Lopez SL, Carrasco AE (1992) Retinoic acid induces changes in the localization of homeobox proteins in the antero-posterior axis of Xenopus laevis embryos. Mech of Dev 36:153–164.

Lopez SL, Dono R, Zeller R, Carrasco AE (1995) Differential effects of retinoic acid and a retinoid antagonist on the spatial distribution of the homeoprotein Hoxb-7 in vertebrate embryos. Dev Dynam 204:457–471.

Lufkin T, Lohnes D, Mark M, Dierich A, Gorry P, Gaub M-P, LeMeur M, Chambon P (1993) High postnatal lethality and testis degeneration in retinoic acid receptor a mutant mice. Proc Natl Acad Sci USA 90:7225–7229.

Luo J, Pasceri P, Conlon RA, Rossant J, Giguere V (1995) Mice lacking all isoforms of retinoic acid receptor b develop normally and are susceptible to the teratogenic effects of retinoic acid. Mech of Dev 53:61–71.

Maden M (1982) Vitamin A and pattern formation in the regenerating limb. Nature 295:672–675.

Maden M (1993) The homeotic transformation of tails into limbs in *Rana temporaria* by retinoids. Developmental Biology 159:379–391.

Maden M (1998) Retinoids as endogenous components of the regenerating limb and tail. Wound Rep Reg 6:358–365.

Maden M (2001) Role and distribution of retinoic acid during CNS development. Int Rev Cytol 209:1–77.

Maden M, Gale E, Kostetskii I, Zile M (1996) Vitamin A-deficient quail embryos have half a hindbrain and other neural defects. Current Biol 6:417–426.

Maden M, Gale E, Zile M (1998a) The role of vitamin A in the development of the central nervous system. J Nutr 128:471S–475S.

Maden M, Graham A, Zile M, Gale E (2000) Abnormalities of somite development in the absence of retinoic acid. Int J Dev Biol 44:151–159.

Maden M, Horton C, Graham A, Leonard L, Pizzey J, Siegenthaler G, Lumsden A, Eriksson U (1992) Domains of cellular retinoic acid-binding protein I (CRABP I) expression in the hindbrain and neural crest of the mouse embryo. Mech of Dev 37:13–23.

Maden M, Ong DE, Chytil F (1990) Retinoid-binding protein distribution in the developing mammalian nervous system. Development 109:75–80.

Maden M, Sonneveld E, van der Saag PT, Gale E (1998b) The distribution of endogenous retinoic acid in the chick embryo: implications for developmental mechanisms. Development 125:4133–4144.

Maden M, Summerbell D, Maignan J, Darmon M, Shroot B (1991) The respecification of limb pattern by new synthetic retinoids and their interaction with cellular retinoic acid-binding protein. Differentiation 47:49–55.

Malpel S, Mendelsohn C, Cardoso WV (2000) Regulation of retinoic acid signaling during lung morphogenesis. Development 127:3057–3067.

Mangelsdorf DJ, Evans RM (1995) The RXR heterodimers and orphan receptors. Cell 83:841–850.

Mark M, Lohnes D, Mendelsohn C, Dupe V, Vonesch J-L, Kastner P, Rijli F, Bloch-Zupan A, Chambon P (1995) Roles of retinoic acid receptors and of Hox genes in the patterning of the teeth and of the jaw skeleton. Int J Dev Biol 39:111–121.

Marsh-Armstrong N, McCaffery P, Gilbert W, Dowling JE, Drager UC (1994) Retinoic acid is necessary for development of the ventral retina in zebrafish. Proc Natl Acad Sci USA 91:7286–7290.

Marsh-Armstrong N, McCaffery P, Hyatt G, Alonso L, Dowling JE, Gilbert W, Drager UC (1995) Retinoic acid in the anteroposterior patterning of the zebrafish trunk. Roux's Arch Dev Biol 205:103–113.

Marshall H, Nonchev S, Sham MH, Muchamore I, Lumsden A, Krumlauf R (1992) Retinoic acid alters hindbrain Hox code and induces transformation of rhombomeres 2/3 into a 4/5 identity. Nature 360:737–741.

Mascrez B, Mark M, Dierich A, Ghyselinck NB, Kastner P, Chambon P (1998) The RXRa ligand-dependent activation function 2 (AF02) is important for mouse development. Development 125:4691–4707.

Mascrez B, Mark M, Krezel W, Dupe V, LeMeur M, Ghyselinck N, Chambon P (2001) Differential contributions of AF-1 and AF-2 activities to the developmental functions of RXRa. Development 128:2049–2062.

Massaro GD, Massaro D, Chambon P (2003) Retinoic acid receptor-a regulates pulmonary alveolus formation in mice after, but not during, perinatal period. Am J Physiol Lung Cell Mol Physiol 284:L431–L433.

Massaro GDC, Massaro D, Chan W-Y, Clerch LB, Ghyselinck N, Chambon P, Chandraratna RA (2000) Retinoic acid receptor-b: an endogenous inhibitor of the perinatal formation of pulmonary alveoli. Physiol Genomics 4:51–57.

Matt N, Ghyselinck N, Wendling O, Chambon P, Mark M (2003) Retinoic acid-induced developmental defects are mediated by RARb/RXR heterodimers in the pharyngeal endoderm. Development 130:2083–2093.

Matt N, Schmidt CK, Dupe V, Dennefeld C, Nau H, Chambon P, Mark M, Ghyselinck NB (2005) Contribution of cellular retinol-binding protein type 1 to retinol metabolism during mouse development. Dev Dyn 233:167–176.

McCaffery P, Drager UC (1994) Hot spots of retinoic acid synthesis in the developing spinal cord. Proc Natl Acad Sci USA 91:7194–7197.

Mendelsohn C, Batourina E, Fung S, Gilbert T, Dodd J (1999) Stromal cells mediate retinoid-dependent functions essential for renal development. Development 126:1139–1148.

Mendelsohn C, Lohnes D, Decimo D, Lufkin T, LeMeur M, Chambon P, Mark M (1994a) Function of the retinoic acid receptors (RARs) during development (II) Multiple abnormalities at various stages of organogenesis in RAR double mutants. Development 120:2749–2771.

Mendelsohn C, Mark M, Dolle P, Dierich A, Gaub MP, Krust A, Lampron C, Chambon P (1994b) Retinoic acid receptor b2 (RARb2) null mutant mice appear normal. Dev Biol 166:246–258.

Mic FA, Molotkov A, Benbrook DM, Duester G (2003) Retinoid activation of retinoic acid receptor but not retinoid X receptor is sufficient to rescue lethal defect in retinoic acid synthesis. Proc Natl Acad Sci USA 100:7135–7140.

Mic FA, Molotkov A, Fan X, Cuenca AE, Duester G (2000) RALDH3, a retinaldehyde dehydrogenase that generates retinoic acid, is expressed in the ventral retina, otic vesicle and olfactory pit during mouse development. Mech Dev 97:227–230.

Mic FA, Molotkov A, Molotkova N, Duester G (2004) Raldh2 expression in optic vesicle generates a retinoic acid signal needed for invagination of retina during optic cup formation. Dev Dyn 231:270–277.

Mohanty-Hejmadi P, Dutta SK, Mahapatra P (1992) Limbs generated at site of tail amputation in marbled balloon frog after vitamin A treatment. Nature 355:352–353.

Mollard R, Ghyselinck N, Wendling O, Chambon P, Mark M (2000) Stage-dependent responses of the developing lung to retinoic acid signaling. Int J Dev Biol 44:457–462.

Molotkov A, Fan X, Deltour L, Foglio MH, Martras S, Farres J, Pares X, Duester G (2002a) Stimulation of retinoic acid production and growth by ubiquitously expressed alcohol dehydrogenase Adh3. Proc Natl Acad Sci USA 99:5337–5342.

Molotkov A, Fan X, Duester G (2002b) Excessive vitamin A toxicity in mice genetically deficient in either alcohol dehydrogenase Adh1 or Adh3. Eur J Biochem 269:2607–2612.

Molotkov A, Molotkova N, Duester G (2005) Retinoic acid generated by Raldh2 in mesoderm is required for mouse dorsal endodermal pancreas development. Dev Dyn 232:950–957.

Molotkova N, Molotkov A, Sirbu IO, Duester G (2005) Requirement of mesodermal retinoic acid generated by Raldh2 for posterior neural transformation. Mech Dev 122:145–155.

Moreno TA, Kintner C (2004) Regulation of segmental patterning by retinoic acid signaling during Xenopus somitogenesis. Dev Cell 6:205–218.

Moro Balbas JA, Gato A, Alonso Revuelta MI, Pastor JF, Represa JJ, Barbosa E (1993) Retinoic acid induces changes in the rhombencephalic neural crest cells migration and extracellular matrix composition in chick embryos. Teratology 48:197–206.

Morriss GM (1972) Morphogenesis of the malformations induced in rat embryos by maternal hypervitaminosis A. J Anat 113:241–250.

Morriss GM, Thorogood PV (1978) An approach to cranial neural crest migration and differentiation in mammalian embryos. In Development in Mammals (Johnson M, Ed.), pp. 363–412.

Morriss-Kay GM, Murphy P, Hill RE, Davidson DR (1991) Effects of retinoic acid excess on expression of Hox-2.9 and Krox-20 and on morphological segmentation in the hindbrain of mouse embryos. EMBO J 10:2985–2995.

Moss JB, Xavier-Neto J, Shapiro MD, Nayeem SM, McCaffery P, Drager UC, Rosenthal N (1998) Dynamic patterns of retinoic acid synthesis and response in the developing mammalian heart. Dev Biol 199:55–71.

Napoli JL (1994) Retinoic acid homeostasis. In Vitamin A in Health and Disease (Blomhoff R, Ed.), pp. 135–188. New York: Marcel Dekker Inc.

Napoli JL (1996) Retinoic acid biosynthesis and metabolism. FASEB J 10:993–1001.

Niederreither K, Abu-Abed S, Schuhbaur B, Petkovich M, Chambon P, Dolle P (2002a) Genetic evidence that oxidative derivatives of retinoic acid are not involved in retinoid signaling during mouse development. Nat Genet 31:84–88.

Niederreither K, Fraulob V, Garnier J-M, Chambon P, Dolle P (2002b) Differential expression of retinoic acid-synthesising (RALDH) enzymes during fetal development and organ differentiation. Mech of Dev 110:165–171.

Niederreither K, McCaffery P, Drager UC, Chambon P, Dolle P (1997) Restricted expression and retinoic acid-induced downregulation of the retinaldehyde dehydrogenase type 2 (RALDH-2) gene during mouse development. Mech of Dev 62:67–78.

Niederreither K, Subbarayan V, Dolle P, Chambon P (1999) Embryonic retinoic acid synthesis is essential for early mouse post-implantation development. Nature Genetics 21:444–448.

Niederreither K, Vermot J, Le Roux I, Schuhbaur B, Chambon P, Dolle P (2003) The regional pattern of retinoic acid synthesis by RALDH2 is essential for the development of posterior pharyngeal arches and the enteric nervous system. Development 130:2525–2534.

Niederreither K, Vermot J, Messaddeq N, Schubaur B, Chambon P, Dolle P (2001) Embryonic retinoic acid synthesis is essential for heart morphogenesis in the mouse. Development 128:1031.

Niederreither K, Vermot J, Schubaur B, Chambon P, Dolle P (2000) Retinoic acid synthesis and hindbrain patterning in the mouse embryo. Development 127:75–85.

Niederreither K, Vermot J, Schuhbaur B, Chambon P, Dolle P (2002c) Embryonic retinoic acid synthesis is required for forelimb growth and anteroposterior patterning in the mouse. Development 139:3563–3574.

Niederreither K, Ward SJ, Dolle P, Chambon P (1996) Morphological and molecular characterization of retinoic acid-induced limb duplications in mice. Dev Biol 176:185–198.

Nohno T, Noji S, Koyama E, Ohyama K, Myokai F, Kuriowa A, Saito T, Taniguchi S (1991) Involvement of the Chox-4 chicken homeobox genes in determination of anteroposterior axial polarity during limb development. Cell 64:1197–1205.

Novitch BG, Wichterle H, Jessell TM, Sockanathan S (2003) A requirement for retinoic acid-mediated transcriptional activation in ventral neural patterning and motor neuron specification. Neuron 40:81–95.

Papalopulu N, Clarke JDW, Bradley L, Wilkinson D, Krumlauf R, Holder N (1991) Retinoic acid causes abnormal development and segmental patterning of the anterior hindbrain in Xenopus embryos. Development 113:1145–1158.

Pecorino L, Entwistle A, Brockes JP (1996) Activation of a single retinoic acid receptor isoform mediates proximodistal respecification. Curr Biol 6:563–569.

Pecorino LT, Lo DC, Brockes JP (1994) Isoform-specific induction of a retinoid-responsive antigen after biolistic transfection of chimaeric retinoic acid/thyroid hormone receptors into a regenerating limb. Development 120:325–333.

Perz-Edwards A, Hardison NL, Linney E (2001) Retinoic acid-mediated gene expression in transgenic reporter zebrafish. Dev Biol 229:89–101.

Pijnappel WWM, Hendricks HFJ, Folkers GE, van den Brink CE, Dekker EJ, Edelenbosch C, van der Saag PT, Durston AJ (1993) The retinoid ligand 4-oxo-retinoic acid is a highly active modulator of positional specification. Nature 366:340–344.

Power SC, Lancman J, Smith SM (1999) Retinoic acid is essential for Shh/Hoxd signalling during rat limb outgrowth but not for limb initiation. Dev Dynam 216:469–480.

Pratt RM, Goulding EH, Abbott BD (1987) Retinoic acid inhibits migration of cranial neural crest cells in the cultured mouse embryo. J Craniofac Gen Dev Biol 7:205–217.

Quinlan R, Gale E, Maden M, Graham A (2002) Deficits in the posterior pharyngeal endoderm in the absence of retinoids. Dev Dyn 225:54–60.

Raz Y, Kelley MW (1999) Retinoic acid signalling is necessary for the development of the organ of Corti. Dev Biol 213:180–193.

Reijntjes S, Blentic A, Gale E, Maden M (2005) The control of morphogen signalling: Regulation of the synthesis and catabolism of retinoic acid in the developing embryo. Dev Biol 285:224–237.

Reijntjes S, Gale E, Maden M (2004) Generating gradients of retinoic acid in the chick embryo: Cyp26C1 expression and a comparative analysis of the Cyp26 enzymes. Dev Dyn 230:509–517.

Riddle RD, Johnson RL, Laufer E, Tabin C (1993) Sonic hedgehog mediates the polarizing activity of the ZPA. Cell 75:1401–1416.

Roberts C, Ivins SM, James CT, Scambler PJ (2005) Retinoic acid down-regulates Tbx1 expression in vivo and in vitro. Dev Dyn 232:928–938.

Romand R, Albuisson E, Niederreither K, Fraulob V, Chambon P, Dolle P (2001) Specific expression of the retinoic acid-synthesising enzyme RALDH2 during mouse inner ear development. Mech of Dev 106:185–189.

Rosa FW, Wilk AL, Kelsey FO (1986) Teratogen update: vitamin A congeners. Teratology 33:355–364.

Roy B, Taneja R, Chambon P (1995) Synergistic activation of retinoic acid (RA)-responsive genes and induction of embryonal carcinoma cell differentiation by an RA receptor a (RARa)-, RARb-, or RARg-selective ligand in combination with a retinoid X receptor-specific ligand. Mol Cell Biol 15:6481–6487.

Ruberte E, Dolle P, Chambon P, Morriss-Kay G (1991) Retinoic acid receptors and cellular retinoid binding proteins. II Their differential pattern of transcription during early morphogenesis in mouse embryos. Development 111:45–60.

Ruberte E, Friederich V, Chambon P, Morriss-Kay G (1993) Retinoic acid receptors and cellular retinoid binding proteins. III Their differential transcript distribution during mouse nervous system development. Development 118:267–282.

Ruberte E, Friederich V, Morris-Kay G, Chambon P (1992) Differential distribution patterns of CRABP I and CRABP II transcripts during mouse embryogenesis. Development 115:973–987.

Rutledge JC, Shourbaji AG, Hughes LA, Polifka JE, Cruz YP, Bishop JB, W.M. G (1994) Limb and lower-body duplications induced by retinoic acid in mice. Proc Natl Acad Sci USA 91:5436–5440.

Sakai Y, Meno C, Fujii H, Nishino J, Shiratori H, Saijoh Y, Rossant J, Hamada H (2001) The retinoic acid-inactivating enzyme CYP26 is essential for establishing an uneven distribution of retinoic acid along the anterio-posterior axis within the mouse embryo. Genes Dev 15:213–225.

Scadding SR, Maden M (1986) Comparison of the effects of vitamin A on limb development and regeneration in the axolotl, Ambystoma mexicanum. J Embryol Exp Morph 91:19–34.

Schmidt C, Christ B, Maden M, Brand-Saberi B, Patel K (2001) Regulation of EphA4 expression in paraxial and lateral plate mesoderm by ectoderm-derived signals. Dev Dyn 220:377–386.

Schneider RA, Hu D, Rubenstein JLR, Maden M, Helms JA (2001) Local retinoid signalling coordinates forebrain and facial morphogenesis by maintaining FGF8 and SHH. Development 128:2755–2767.

Schuh TJ, Hall BL, Kraft JC, Privalsky ML, Kimelman D (1993) v-erbA and citral reduce the teratogenic effects of all-trans-retinoic acid and retinol, respectively, in Xenopus embryogenesis. Development 119:785–798.

Seegmiller RE, Harris C, Luchtel DL, Juchau MR (1991) Morphological differences elicited by two weak acids, retinoic acid and valproic acid, in rat embryos grown in vitro. Teratology 43:133–150.

Sharpe C, Goldstone K (1997) Retinoid receptors promote primary neurogenesis in Xenopus. Development 124:515–523.

Sharpe C, Goldstone K (2000a) Retinoid signalling acts during the gastrula stages to promote primary neurogenesis. Int J Dev Biol 44:463–470.

Sharpe C, Goldstone K (2000b) The control of Xenopus embryonic primary neurogenesis is mediated by retinoid signalling in the neurectoderm. Mech Dev 91:69–80.

Shenfelt RE (1972) Morphogenesis of malformations in hamsters caused by retinoic acid: relation to dose and stage at treatment. Teratology 5:103–118.

Shiotsugu J, Katsuyama Y, Arima K, Baxter A, Koide T, Song J, Chandraratna RA, Blumberg B (2004) Multiple points of interaction between retinoic acid and FGF signaling during embryonic axis formation. Development 131:2653–2667.

Simeone A, Avantaggiato V, Moroni MC, Mavilio F, Arra C, Cotelli F, Nigro V, ACampora D (1995) Retinoic acid induces stage-specific antero-posterior transformation of rostral central nervous system. Mech of Dev 51:83–98.

Sive HL, Draper BW, Harland RM, Weintrub H (1990) Identification of a retinoic acid-sensitive period during primary axis formation in Xenopus laevis. Genes & Dev 4:932–942.

Sockanathan S, Jessell TM (1998) Motor neuron-derived retinoid signalling specifies the subtype identity of spinal motor neurons. Cell 94:503–514.

Sockanathan S, Perlmann T, Jessell TM (2003) Retinoid receptor signaling in postmitotic motor neurons regulates rostrocaudal positional identity and axonal projection pattern. Neuron 40:97–111.

Song Y, Hui JN, Fu KK, Richman JM (2004) Control of retinoic acid synthesis and FGF expression in the nasal pit is required to pattern the craniofacial skeleton. Dev Biol 276:313–329.

Sonneveld E, van den Brink CE, Tertoolen LGJ, van der Burg B, van der Saag PT (1999) Retinoic acid hydroxylase (CYP26) is a key enzyme in neuronal differentiation of embryonal carcinoma cells. Dev Biol 213:390–404.

Stratford T, Horton C, Maden M (1996) Retinoic acid is required for the initiation of outgrowth in the chick limb bud. Current Biology 6:1124–1133.
Stratford T, Logan C, Zile M, Maden M (1999) Abnormal anteroposterior and dorsoventral patterning of the limb bud in the absence of retinoids. Mech of Dev 81:115–125.
Sucov HM, Dyson E, Gumeringer CL, Price J, Chien KR, Evans RM (1994) RXRa mutant mice establish a genetic basis for vitamin A signaling in heart morphogenesis. Genes & Dev 8:1007–1018.
Summerbell D (1983) The effects of local application of retinoic acid to the anterior margin of the developing chick limb. J Embryol Exp Morph 78:269–289.
Sundin O, Eichele G (1992) An early marker of axial pattern in the chick embryo and its respecification by retinoic acid. Development 114:841–852.
Suzuki S, Shintani T, Sakuta H, Kato A, Ohkawara T, Osumi N, Noda M (2000) Identification of RALDH-3, a novel retinaldehyde dehydrogenase, expressed in the ventral region of the retina. Mech Dev 98:37–50.
Swindell EC, Thaller C, Sockanathan S, Petkovich M, Jessell TM, Eichele G (1999) Complementary domains of retinoic acid production and degradation in the early chick embryo. Dev Biol 216:282–296.
Tanaka M, Tamura K, Ide H (1996) Citral, an inhibitor of retinoic acid synthesis, modifies chick limb development. Dev Biol 175:239–247.
Thompson JN, Howell JM, Pitt GAJ, McLaughlin CI (1969) The biological activity of retinoic acid in the domestic fowl and the effects of vitamin A deficiency on the chick embryo. Br J Nutr 23:471–490.
Tickle C, Alberts B, Wolpert L, Lee J (1982) Local application of retinoic acid to the limb bond mimics the action of the polarizing region. Nature 296:564–566.
Tickle C, Lee J, Eichele G (1985) A quantitative analysis of the effect of all-trans-retinoic acid on the pattern of chick wing development. Dev Biol 109:82–95.
Toresson H, Mata de Urquiza A, Fagerstrom C, Perlmann T, Campbell K (1999) Retinoids are produced by glia in the lateral ganglionic eminence and regulate striatal neuron differentiation. Development 126:1317–1326.
Tsonis PA, Trombley MT, Rowland T, Chandraratna RAS, Del Rio-Tsonis K (2000) Role of retinoic acid in lens regeneration. Dev Dyn 219:588–593.
van der Wees J, Schilthuis JG, Koster CH, Diesveld-Schipper H, Folkers GE, van der Saag PT, Dawson MI, Shudo K, van der Burg B, Durston AJ (1998) Inhibition of retinoic acid receptor-mediated signalling alters positional identity in the developing hindbrain. Development 125:545–556.
Vandersea MW, Fleming P, McCarthy RA, Smith DG (1998) Fin duplications and deletions induced by disruption of retinoic acid signaling. Dev Genes Evol 208:61–68.
Vermot J, Pourquie O (2005) Retinoic acid coordinates somitogenesis and left-right patterning in vertebrate embryos. Nature 435:215–220.
Villaneuva S, Glavic A, Ruiz P, Mayor R (2002) Posteriorization by FGF, Wnt, and retinoic acid is required for neural crest induction. Dev Biol 241:289–301.
Webster WS, Johnston MC, Lammer EJ, Sulik KK (1986) Isotretinoin embryopathy and the cranial neural crest: an in vivo and in vitro study. J Craniofac Gen Dev Biol 6:211–222.
Wendling O, Dennefeld C, Chambon P, Mark M (2000) Retinoid signaling is essential for patterning the endoderm of the third and fourth pharyngeal arches. Development 127:1553–1562.
Wendling O, Ghyselinck N, Chambon P, Mark M (2001) Roles of retinoic acid receptors in early embryonic morphogenesis and hindbrain patterning. Development 128:2031–2038.

White JA, Guo Y-D, Baetz K, Beckett-Jones B, Bonasoro J, Hsu KE, Dilworth FJ, Jones G, Petkovich M (1996) Identification of the retinoic acid-inducible all-trans-retinoic acid 4-hydroxylase. J Biol Chem 271:29922–29927.

White JA, Ramshaw H, Taimi M, Stangle W, Zhang A, Everingham S, Creighton S, Tam S-P, Jones G, Petkovich M (2000) Identification of the human cytochrome P450, P450RAI-2, which is predominantly expressed in the adult cerebellum and is responsible for all-trans-retinoic acid metabolism. Proc Natl Acad Sci USA 97:6403–6408.

White JC, Highland M, Kaiser M, Clagett-Dame M (2000) Vitamin A deficiency results in the dose-dependent acquisition of anterior character and shortening of the caudal hindbrain of the rat embryo. Dev Biol 220:263–284.

White JC, Shankar VN, Highland M, Epstein ML, DeLuca HF, Clagett-Dame M (1998) Defects in embryonic hindbrain development and fetal resorption resulting from vitamin A deficiency in the rat are prevented by feeding pharmacological levels of all-trans-retinoic acid. Proc Natl Acad Sci USA 95:13459–13464.

Wietrzych M, Meziane H, Sutter A, Ghyselinck N, Chapman PF, Chambon P, Krezel W (2005) Working memory deficits in retinoid X receptor gamma-deficient mice. Learn Mem 12:318–326.

Willhite CC, Dawson MI, Reichert U (1996) Receptor-selective retinoid agonists and teratogenic activity. Drug Metab Rev 28:105–119.

Wilson L, Gale E, Chambers D, Maden M (2004) Retinoic acid and the control of dorsoventral patterning in the avian spinal cord. Dev Biol 269:433–446.

Wilson L, Maden M (2005) The mechanisms of dorsoventral patterning in the vertebrate neural tube. Dev Biol 282:1–13.

Wood H, Pall G, Morriss-Kay G (1994) Exposure to retinoic acid before or after the onset of somitogenesis reveals separate effects on rhombomeric segmentation and 3′ HoxB gene expression domains. Development 120:2279–2285.

Xavier-Neto J, Neville CM, Shapiro MD, Houghton L, Wang GF, Nikovits W, Stockdale FE, Rosenthal N (1999) A retinoic acid-inducible transgenic marker of sino-atrial development in the mouse heart. Development 126:2677–2687.

Yamaguchi M, Nakamoto M, Honda H, Nakagawa T, Fujita H, Nakamura T, Hirai H, Narumiya S, Kakizuka A (1998) Retardation of skeletal development and cervical abnormalities in transgenic mice expressing a dominant-negative retinoic acid receptor in chondrogenic cells. Proc Natl Acad Sci USA 95:7491–7496.

Yashiro K, Zhao X, Uehara M, Yamashita K, Nishijima M, Nishino J, Saijoh Y, Sakai Y, Hamada H (2004) Regulation of retinoic acid distribution is required for proximodistal patterning and outgrowth of the developing mouse limb. Dev Cell 6:411–422.

Zhang M, Kim H-J, Marshall H, Gendron-Maguire M, Lucas DA, Baron A, Gudas LJ, Gridley T, Krumlauf R, Grippo JF (1994) Ectopic Hoxa-1 induces rhombomere transformation in mouse hindbrain. Development 120:2431–2442.

Zhang Z, Balmer JE, Lovlie A, Fromm SH, Blomhoff R (1996) Specific teratogenic effects of different retinoic acid isomers and analogs in the developing anterior central nervous system of zebrafish. Dev Dynam 206:73–86.

Zhao D, McCaffery P, Ivins KJ, Neve RL, Hogan P, Chin WW, Drager UC (1996) Molecular identification of a major retinoic acid-synthesising enzyme, a retinaldehyde-specific dehydrogenase. Eur J Biochem 240:15–22.

Zhou J, Kochhar DM (2003) Regulation of AP-2 and apoptosis in developing eye in a vitamin A-deficiency model. Birth Defects Res A Clin Mol Teratol 67:41–53.

Zile MH, Kostetskii I, Yuan S, Kostetskaia E, St. Amand TR, Chen Y-P, Jiang W (2000) Retinoid signaling is required to complete the vertebrate cardiac left/right asymmetry pathway. Dev Biol 223:323–338.

6 Wnt Signaling via the Rho Family of GTPases during Embryonic Development

Raymond Habas and Xi He

CONTENTS

I. INTRODUCTION

The establishment of the vertebrate body plan during embryogenesis requires a complex series of cell movements. These movements involve cell polarization and migration, and are often based on alterations of the cellular cytoskeletal architecture and adhesive properties. One example is the so-called convergent extension (CE) movements during vertebrate gastrulation. The molecular machinery and the signaling pathways that control CE movements remain poorly deciphered, but recent studies have uncovered a role for the non-canonical (β-catenin independent) Wnt signaling pathway. In particular, Wnt/Frizzled (Fz) activation of the small GTPases Rho and Rac has emerged as a key mechanism that regulates cell polarity and movements during gastrulation. In this chapter, we briefly discuss advances in our understanding of Wnt/Fz regulation of Rho GTPases and describe in detail biochemical assays for the detection of activation of the GTPases in *Xenopus* embryo explants.

II. BASICS OF CONVERGENT EXTENSION MOVEMENTS

Several distinct forms of cell movements occur during *Xenopus* gastrulation and neurulation. Here we limit our discussion on CE movements, which have been shown to be regulated by non-canonical Wnt signaling and are responsible for generating the extended anterior-posterior axis of the vertebrate embryo. During gastrulation, dorsal mesodermal cells polarize and elongate along the mediolateral axis, and align and intercalate towards the midline, resulting in mediolateral narrowing (convergence) and anterioposterior lengthening (extension) of the embryonic axis (Keller, 2002; Wallingford et al., 2002). The net directionality of these cell movements involves the stabilization of protrusions (lamellopodia) on the mediolateral surfaces of the gastrulating mesodermal cells (Wallingford et al., 2000). CE movements also occur during primary neurulation, which involves folding of the neural plate to form the neural tube (Keller, 2002; Wallingford et al., 2002). For closure of the neural tube, the bilateral folds of the neural plate elevate, appose in the midline, and fuse to form a sealed tube that is subsequently covered by the future epidermis. CE movements provide force for elevation of the neural folds and subsequent closure of the neural tube (Wallingford and Harland, 2002).

III. WNT/PCP SIGNALING

The Wnt family of secreted ligands play crucial roles during embryonic development, and deregulation of Wnt signaling has profound consequences resulting in numerous human pathologies (Logan and Nusse, 2004). One major Wnt signaling route operates via the so-called canonical β-catenin signaling pathway, which participates in cell fate determination, cell proliferation, and various human cancers (Giles et al., 2003; Logan and Nusse, 2004; He et al., 2004) and is thoroughly discussed by Clements and Kimelman in Chapter 1. Other Wnt signaling pathways, which operate independent of the β-catenin protein and thus are distinct from the canonical pathway, have emerged and are sometimes referred to as non-canonical Wnt signaling. One such non-canonical pathway is the Wnt/planar cell polarity (PCP) pathway, which involves the activation of the Rho family of GTPases and is essential for cell polarity and movements during vertebrate gastrulation. This Wnt/PCP pathway

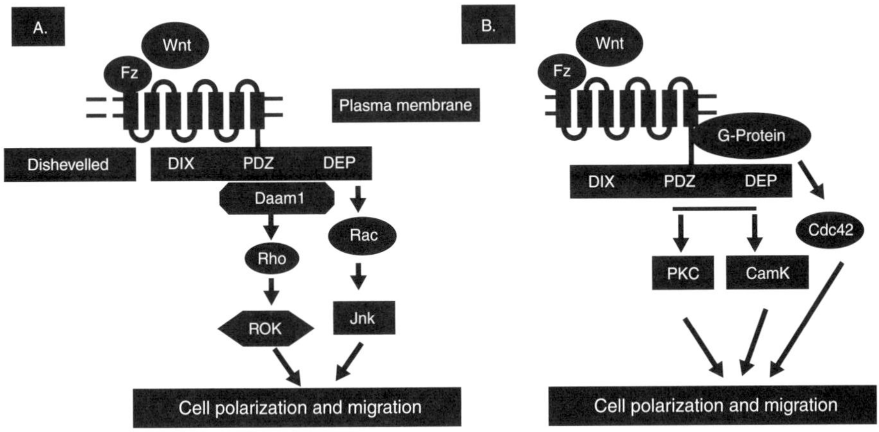

FIGURE 6.1 A schematic representation of the non-canonical Wnt pathways, the Wnt/planar cell polarity (PCP) pathway (a) and the Wnt/calcium pathway (b). (a) Wnt signaling is transduced through Fz and Dsh/Dvl to mediate cytoskeletal changes for cell polarization and movements. This involves the activation of small GTPases Rho and Rac, which in turn activate Rho kinase (ROK) and Jun N-terminal Kinase (JNK), respectively. (b) For the Wnt/calcium pathway, Wnt signaling via Fz mediates the activation of G-proteins to activate CaMK2 and PKC. Cdc42 is also activated downstream of Fz and may work in concert with CaMK2 and PKC for cell adhesion and cytoskeletal changes. The involvement of Dsh in this pathway is debated.

derives its name from the fact that it shares common components similar to those involved in the establishment of epithelial planar cell polarity in *Drosophila* (Mlodzik, 2002). Besides the PCP pathway, non-canonical Wnt signaling pathways include those that activate calcium calmodulin dependent kinase 2 (CaMK2) and protein kinase C (PKC) (Sheldahl et al., 2003), and protein kinase A (PKA) (Chen et al., 2005). Whether these pathways overlap with one another or represent distinct signaling routes remains an open question (Veeman et al., 2003).

For Wnt/PCP signaling (Figure 6.1), Wnt and Fz mediated activation of RhoA and Rac requires the cytoplasmic phosphoprotein Dishevelled (Dsh or Dvl), which acts downstream of Fz (Wharton, 2003). There are three Dsh proteins in mammals (Dvl-1, -2, and -3), and all Dsh family members are comprised of three highly conserved domains: an N-terminal DIX domain (for Dishevelled and Axin), a central PDZ domain (for Post synaptic density-95, Discs-large, and Zonula occludens-1), and a carboxyl DEP domain (for Dishevelled, Egl-10 and Pleckstrin) (Wallingford and Habas, 2005; Wharton, 2003). The DIX domain is dispensable for PCP and for RhoA and Rac activation, but is critical for canonical β-catenin signaling; whereas the PDZ and DEP domains are, generally speaking, involved in non-canonical signaling (Veeman et al., 2003; Wharton, 2003). It should be pointed out that we lack a clear understanding of how Dsh channels Wnt/Fz signaling to different pathways. It is conceivable that particular Wnt-Fz combinations, Fz coreceptors, and Dsh cellular localization together determine its ability to signal to distinct branches (Habas and Dawid, 2005; Wallingford and Habas, 2005).

For PCP signaling downstream of Dsh, two independent and parallel pathways lead to the activation of RhoA and Rac. The first pathway involves a molecule called Daam1 that binds to the PDZ domain of Dsh and can activate RhoA (Habas et al., 2001), which in turn leads to the activation of the Rho associated kinase Rock (Habas et al., 2003; Marlow et al., 2002). The second pathway requires the DEP domain of Dsh and activates Rac, which in turn stimulates Jun N-terminal Kinase (JNK) activity (Boutros et al., 1998; Habas et al., 2003). A subsequent integration of the signals triggered by these two GTPases and their effectors results in modification of the cellular cytoskeletal architecture for polarity and motility control, although the nature of this integration remains poorly understood. Additionally for PCP signaling, other components such as Prickle and Strabismus are required, but how they integrate with the small GTPases remains unknown (Veeman et al., 2003; Wallingford and Habas, 2005).

IV. THE RHO FAMILY OF GTPASES

The Rho GTPases comprises a large family of proteins that play important roles in regulating cytoskeletal architectures associated with cell polarity, adhesion, and motility. To date, 20 members of the Rho GTPase family have been identified in mammals. The prototypes of this family include Rho, Rac, and Cdc42, but the family can be further subdivided into eight distinct groups: (1) the Rho group, RhoA, RhoB, and RhoC; (2) the Rac group, Rac1, Rac2, Rac3, and RhoG; (3) the Cdc42 group, Cdc42, TC10, TCL, Chp, and Wrch-1; (4) the RhoD group, RhoD and Rif; (5) RhoH/TTF, (6) the RhoBTB group, RhoBTB1 and RhoBTB2; (7) the Miro group, Miro-1 and Miro-2; and (8) the Rnd group, which includes Rnd1, Rnd2, and Rnd3/RhoE (Sorokina and Chernoff, 2005). The Rho GTPases act as bimolecular switches cycling between an active or GTP-bound form and an inactive or GDP-bound form (Raftopoulou and Hall, 2004). This cycling is catalyzed by three classes of proteins that include guanine nucleotide exchange factors (GEFs), which promote the formation of GTP-bound Rho GTPases; GTPase activating proteins (GAPs), which inactivate Rho family proteins; and guanine nucleotide disassociation factors (GDIs) (Schwartz, 2004).

In the GTP-bound form, the Rho GTPases perform their biological functions by interacting with effector molecules that subsequently relay changes to the cellular cytoskeleton. Such components include Rho kinase (Matsui et al., 1996) and Formin proteins for Rho (Watanabe et al., 1997), and JNK, p38, Pak, and Wasp for Rac and Cdc42 (Bagrodia et al., 1995; Coso et al., 1995; Minden et al., 1995; Symons et al., 1996). Although the classic biological effects of the Rho GTPases has been linked to cytoskeletal regulation, it is noteworthy that they can function in a plethora of biological responses including proliferation, cytokinesis, phagocytosis, endocytosis, cell cycle progression, and gene transcription (Sorokina and Chernoff, 2005).

V. EXPRESSION PATTERN OF THE RHO GTPASES DURING EMBRYOGENESIS

With such a large family of Rho GTPase proteins, one logical question would be whether these proteins are expressed in any temporal or spatial manner that may be indicative of their functions during embryogenesis. We, however, lack significant details of the embryonic expression patterns of these GTPases, and to date only those for RhoA (*Xenopus*) (Wunnenberg-Stapleton et al., 1999), RhoB (mouse), Rac1 (mouse and *Xenopus*) (Lucas et al., 2002; Sugihara et al., 1998), Rnd1 (*Xenopus*) (Wunnenberg-Stapleton et al., 1999), and Cdc42 (*Xenopus*) (Choi and Han, 2002) have been described. For these members described, RhoA, Rac1, Cdc42, and Rnd1 show ubiquitous expression in *Xenopus* and are slightly upregulated in dorsal mesodermal tissues undergoing gastrulation. The expression patterns are consistent with potential roles for these factors in cytoskeletal reorganization and cell movements during gastrulation. In the mouse, Rac1 is ubiquitously expressed and Rac2 is expressed in cells of the hematopoietic lineage (Williams et al., 2000). Much work remains to examine the patterns of expression of these factors during embryogenesis, but no doubt these studies are under way.

VI. FUNCTIONS OF THE RHO GTPASES DURING EMBRYONIC DEVELOPMENT

Detailed analysis of the effects of the Rho family of GTPases on the cytoskeleton have been uncovered over the last 2 decades using mammalian culture cells as primary model systems. For simplicity, we will focus on the effects of Rho, Rac, and Cdc42, as they represent the best understood cases. Overexpression of Rho in mammalian cells induces the formation of actin stress fibers, whereas overexpression of Rac induces the formation of lamellipodia and membrane ruffles, and overexpression of Cdc42 induces the formation of filopodia. Thus, whereas each of the Rho GTPases appears to have distinct effects on the actin cytoskeleton and cell morphology, it remains unclear how these effects are coordinated in cell migrations during embryogenesis. It is likely that the interplay between the activated Rho, Rac, and Cdc42 and their effectors dictates cell polarity and movements.

The biological roles of some of the Rho family members have been elucidated in mutant studies in the mouse. Knockout of Rac1 demonstrates a requirement of this protein in germ layer formation and gastrulation (Sugihara et al., 1998). Knockout of Cdc42 results in early embryonic lethality and defective cytoskeletal organization in embryonic stem cells (Chen et al., 2000). Mice mutant for Rac2 exhibit no obvious abnormality in early embryonic development but have severe myeloid cell dysfunction exemplified by leukocytosis and defective neutrophil chemotaxis (Gu and Williams, 2002). A knockout of Rac3 revealed no role for this protein in early development but uncovered a role of Rac3 in Bcr/Abl-causing lymphoblastic leukemia (Cho et al., 2005). Surprisingly, to date no knockout mutants of RhoA-C have been reported.

Studies in *Xenopus* and zebrafish have further advanced our understanding of the roles of the Rho family of GTPases for gastrulation cell movements during embryogenesis. The first hint of the roles of Rho and Cdc42 were uncovered during studies in which activated or dominant negative forms of these factors were microinjected into the developing *Xenopus* embryo. Expression of these mutant proteins was found to potently inhibit cytokinesis, implicating a role for these factors in cytoskeletal rearrangements associated with the embryonic cleavage (Drechsel et al., 1997). A role for RhoA and Rnd1 in cell adhesion and migration was subsequently demonstrated (Wunnenberg-Stapleton et al., 1999). This study also revealed an antagonistic relationship between RhoA and Rnd1 on the cytoskeleton and demonstrated a potential role for RhoA in head formation, possibly via functioning as an antagonist of canonical Wnt signaling (Wunnenberg-Stapleton et al., 1999).

A further understanding of the pathway that controls gastrulation came when molecules of the PCP pathway including the Rho GTPases were shown to regulate gastrulation. The Dishevelled protein was genetically implicated upstream of the PCP pathway involving the Rho GTPases Rho, Rac, and CDC42 in mediating cell polarization during *Drosophila* embryogenesis (Mlodzik, 2002). In *Xenopus*, a dominant negative Dvl protein lacking a portion of the PDZ domain impairs gastrulation (Sokol, 1996). Subsequent studies further revealed a role of Dvl in mediating stability of cellular protrusions required for CE movements during gastrulation (Wallingford et al., 2000). In addition, a role for Dvl in neural fold closure and CE movements during neurulation was also identified (Wallingford and Harland, 2002). A direct role for the Wnt involvement, specifically Wnt11, in regulating CE movements during gastrulation was uncovered in genetic studies in zebrafish (Heisenberg et al., 2000) and using dominant negative mutant Wnt11 in *Xenopus* (Tada and Smith, 2000).

These studies paved the way for biochemical demonstration that Wnt-11 and Fz are able to activate both RhoA and Rac to regulate CE movements in *Xenopus* embryos (Habas et al., 2001; Habas et al., 2003). Wnt-11 activation of RhoA and Rac can be demonstrated in dorsal embryo explants, in which interfering with Wnt-11 or Fz function prevents RhoA and Rac activation (Habas et al., 2003; Habas et al., 2001). Conversely, overexpression of Wnt-11 or *Xenopus* Fz7 (Xfz7) in the embryo ventral region, which neither exhibits CE movements nor expresses Wnt-11 and Xfz7, is sufficient to activate RhoA and Rac (Habas et al., 2003; Habas et al., 2001). Cell biological studies of explants of *Xenopus* also revealed a crucial interplay of Rho, Rac, and Cdc42 in mediating specific behaviors during gastrulation (Tahinci and Symes, 2003). Indeed, Cdc42 was also shown to play a role in regulating gastrulation cell movements and cell adhesion in *Xenopus* embryos but functions in a non-canonical Wnt/calcium signaling pathway that may be independent of the Wnt/PCP pathway (Choi and Han, 2002).

Although Rho, Rac, and Cdc42 are regulated the Wnt/PCP and Wnt/calcium pathways (Figure 6.1), other molecules not yet linked to these Wnt pathways have been shown to be capable of modulating CE movements during gastrulation. These include transmembrane receptors/proteins such as NRH1 (Chung et al., 2005; Sasai et al., 2004), PTK7 (Lu et al., 2004), ROR2 (Hikasa et al., 2002), and paraxial protocadherin (PAPC) (Hukriede et al., 2003; Medina et al., 2004; Unterseher et al., 2004). The relationship between these molecules and the Wnt/PCP or Wnt/calcium pathway remains enigmatic.

Another key puzzle is how Wnt signaling activates the Rho GTPases. Although a Formin protein, Daam1, binds Dvl and may be involved in Wnt activation of RhoA (Figure 6.1) (Habas et al., 2001), factors that activate the Rho GTPases (GEFs) or inactivate the GTPases (GAPs) have remained poorly deciphered. To date, two *Xenopus* GEFs, XLfc and XNet1, have been reported that modulate CE movements during gastrulation (Kwan and Kirschner, 2005; Miyakoshi et al., 2004). Whether they act in the non-canonical Wnt pathways remains unclear.

VII. ASSAYS TO INVESTIGATE THE ACTIVATION STATES OF RHO GTPASES

Careful analyses of cell behavior during gastrulation have been performed over the years and have offered a great deal of information regarding cellular mechanisms of CE movements during gastrulation and neural tube closure (Keller et al., 2003; Wallingford et al., 2002). However, our understanding of the molecular mechanisms that control and execute these cell movements remains poor. Although new molecules involved in gastrulation are being uncovered, novel imaging and biochemical assays allow us to begin to dissect the molecular mechanisms of cell polarization and movements. In this portion of the chapter, we describe two assays for monitoring the activation state of the Rho GTPases. We will briefly describe the concept for fluorescence resonance energy transfer (FRET) assay before we go into details of biochemical assays.

The FRET assay relies on the knowledge that Rho, Rac, or Cdc42 bind to specific effectors in their active or GTP-bound form (Kaibuchi et al., 1999). Thus, for FRET imaging, one can use a Rho effector such as PKN or Rhotekin fused to a fluorescent probe such as yellow fluorescence protein (YFP) and utilize a fluorescently tagged version of Rho such as Rho-GFP (green fluorescence protein). If GFP-Rho is activated, it will bind to the YFP-PKN, bringing the fluorescent moieties within close proximity and an energy transfer detected by a change of fluorescence can be measured. This approach that relies on microscopic imaging provides two advantages; first, one can detect the levels of activated Rho, and second, one can visualize where in the cell the activated Rho is localized (Benink and Bement, 2005; Miyagi et al., 2004). A similar strategy can be utilized using specific effectors for Rac such as the p21 kinase PAK and for Cdc42 such as WASP (Benard et al., 1999). This FRET strategy has been shown to be effective in the Zebrafish embryo and *Xenopus* oocyte (Benink and Bement, 2005; Miyagi et al., 2004).

A biochemical assay to investigate activation of Rho, Rac, and Cdc42 in the *Xenopus* embryo provides a useful means to detect the levels of activated forms of Rho, Rac, and Cdc42 temporally and, to some degree, spatially within the embryo, although this assay lacks the cellular resolution FRET offers. Here, we provide a detailed protocol for these biochemical assays, which are similar or slightly modified to those recently published. These assays employ a glutathione S-transferase (GST)-pulldown strategy using fusion proteins that specifically bind to the activated- or GTP-bound forms of Rho, Rac, and Cdc42. For the Rho assay, a Rho binding fragment of the Rho-effector Rhotekin is fused to GST and termed GST-RBD (Ren et al., 1999) (Figure 6.2). For the Rac/Cdc42 assay, the Rac/Cdc42 binding fragment of p21 (PAK) is fused to GST and termed GST-PBD (Akasaki et al., 1999; Benard et al., 1999) (Figure 6.2). The GST-RBD and GST-PBD fusion proteins are produced in bacterial cells, purified, and incubated with cell lysates derived from *Xenopus* embryo explants (Figure 6.2). GST-RBD and GST-PBD bind specifically to the GTP-bound forms of Rho and Rac/Cdc42, respectively, and are precipitated using GST-

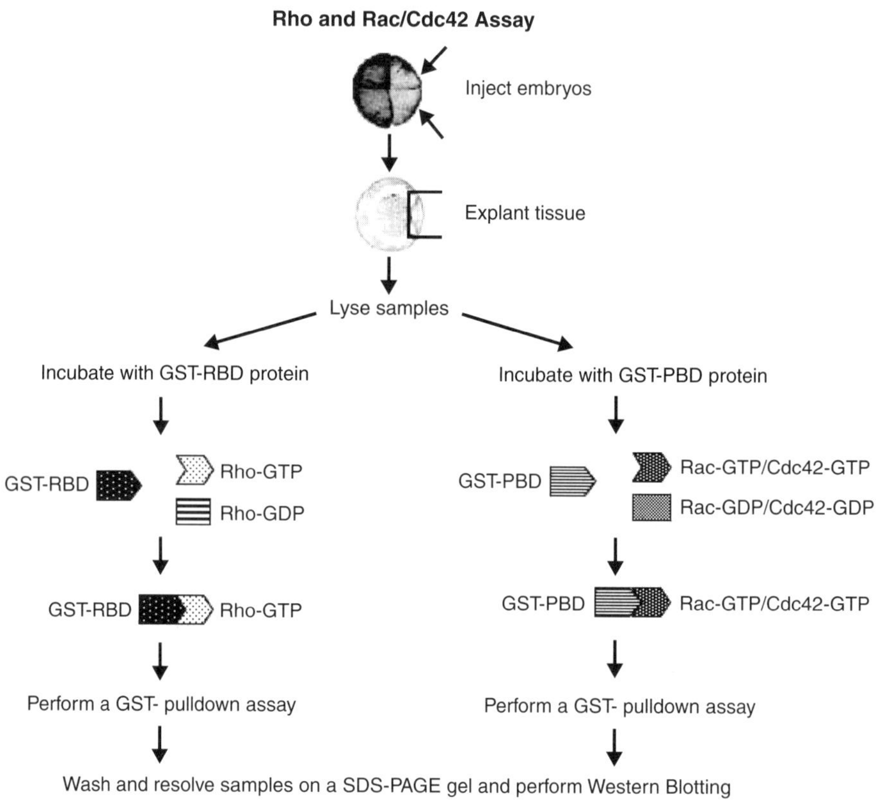

FIGURE 6.2 A schematic overview of the Rho and Rac/Cdc42 assays. Embryos can be injected with RNAs and tissues can then be explanted from the dorsal marginal zone (DMZ) or ventral marginal zone (VMZ).

agarose beads. Thus, co-precipitated and activated forms of the GTPases can be detected via conventional immunoblotting (Figures 6.2 and 6.3). This assay provides an efficient means to access levels of activated Rho, Rac, and Cdc42.

VIII. METHODS

A. Rho and Rac/Cdc42 Assays

1. Recombinant GST-RBD Protein Preparation

1. An overnight culture of a single colony of BL21 bacterial cell containing the GST-RBD plasmid is grown in 20 ml LB-amp (100 μg/ml) at 30°C.
2. This culture is then diluted into 1 l of LB-amp (100 μg /ml) and grown at 30°C until the optical density of the culture at 600 nm is 1.0. This growth will take approximately 5–7 hours depending on starting optical density.

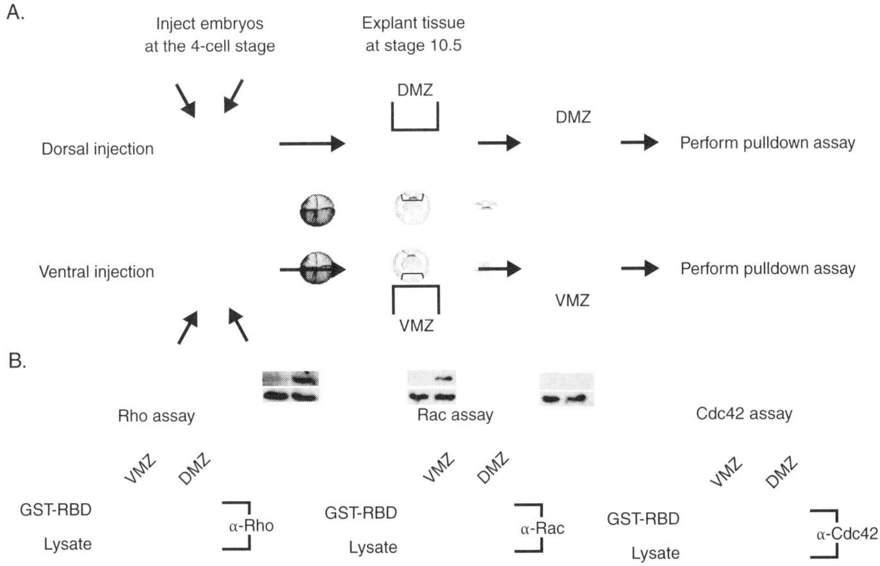

FIGURE 6.3 Rho and Rac/Cdc42 assays in *Xenopus*. (a.) Embryos are injected with RNAs at the four-cell stage into the two dorsal or ventral cells and allowed to develop to stage 10.5. At this stage, the dorsal marginal zone (DMZ) or ventral marginal zone (VMZ) are explanted and used for Rho or Rac/Cdc42 pulldown assays. (b.) Examples of Rho, Rac, and Cdc42 assays performed with DMZ and VMZ samples. Rho, Rac, and Cdc42 activation detected from GST-RBD or GST-PBD pulldown samples show Rho and Rac but not Cdc42 is activated preferentially in the DMZ. Equal endogenous levels of Rho, Rac, and Cdc42 are shown in the lysate samples. (With permission from Elsevier Ltd. and Cold Spring Harbor Press.)

3. The bacterial culture is then induced to produce the fusion protein using 1 ml of 1 M IPTG for 34 hours at 30°C.
4. The bacterial culture is then aliquotted into 50 ml fractions, placed into Falcon tubes, and spun at 4000 rpm for 10 minutes to pellet the bacteria. The supernatant is discarded and the pellets are flash frozen in liquid nitrogen.
5. The pellets are immediately stored at –80°C and are stable for up to 1 year.

2. Recombinant GST-PBD Fusion Protein Preparation

1. An overnight culture of a single colony of BL21 bacterial cell containing the GST-PBD plasmid is grown in 20 ml LB-amp (100 μg/ml) at 30°C.
2. This culture is then diluted into 1 l of LB-amp (100 μg /ml) and grown at 30°C until the optical density of the culture at 600 nm is 1.0. This growth will take approximately 5–7 hours depending on starting optical density.
3. The bacterial culture is then induced to produce the fusion protein using 1 ml of 1 M IPTG for 3–4 hours at 30°C.

4. The bacterial cells are then lysed in 1X PBS and protease inhibitors using either sonication or a French press.
5. After lysis, the cellular debris is removed by spinning the lysate at 15,000 rpm for 15 minutes.
6. The supernatant is isolated, aliquotted into 1.5 ml ependorf tubes, and flash frozen in liquid nitrogen.
7. The aliquotted supernatant is stored at –80°C and is stable for up to 1 year.

B. Isolation of GST-RBD and GST-PBD Fusion Proteins

1. GST-RBD

1. Glutathione sepharose beads are prepared by swelling approximately 100 µl beads with 1X PBS/10 mM DTT/1% Triton-X 100 for at least 30 minutes on ice. The beads are then washed three times with 500 µl of 1X PBS/10 mM DTT/1% Triton-X 100. After a final wash, the beads can be stored on ice as a 1X slurry. Take care not to spin beads higher than 3000 rpm during the pelleting and washing stages, for this will damage the beads. Also prepare two tubes of beads, for you will have 2 ml of lysate.
2. Thaw one aliquot of frozen GST-RBD pellet on ice and resuspend the pellet in 2 ml 1X PBS.
3. To the resuspended pellet, add 20 µl 1M DTT, 20 µl protease inhibitor cocktail (Boehringer-Mannheim), and 40 µl of 50 mg/ml lysozyme.
4. Mix well by vortexing briefly and incubate on ice for 30 minutes.
5. Add 225 µl 10% Triton-X 100, 22.5 µl 1M $MgCl_2$, and 22.5 µl of 10 mg/ml DNase1 to the resuspension.
6. Mix well by vortexing briefly and incubate on ice for 30 minutes.
7. The suspension is then spun at 14,000 rpm at 4°C for 2 minutes, and 1 ml of supernatant is added to each of the two tubes of the preswollen beads.
8. The beads and suspension are then incubated on a Nutator at 4°C for 45 minutes (do not exceed 1 hour).
9. The beads are then spun and washed three times with 500 µl of 1X PBS/10 mM DTT/1% Triton-X 100. Following the final wash, store the samples on ice in a 1X slurry with the final volume approximately 500 µl.

2. GST-PBD

1. Glutathione sepharose beads are prepared by swelling approximately 100 µl beads with 1X PBS/10 mM DTT/1% Triton-X 100 for at least 30 minutes on ice. The beads are then washed three times with 500 µl of 1X PBS/10 mM DTT/1% Triton-X 100. After a final wash, the beads can be stored on ice as a 1X slurry. Take care not to spin beads higher than 3000 rpm during the pelleting and washing stages, for this will damage the beads. Also prepare two tubes of beads, for you will have approximately 1.5 ml of lysate.
2. Thaw one aliquot of frozen bacterial supernatant containing the GST-PBD on ice.

3. Add 750 μl of supernatant to each of the two tubes of the preswollen beads.
4. The beads and suspension are then incubated on a Nutator at 4°C for 45 minutes (do not exceed 1 hour).
5. The beads are then spun and washed three times with 500 μl of 1X PBS/10 mM DTT/1% Triton-X 100. Following the final wash, store the samples on ice in a 1X slurry with the final volume approximately 500 μl.

C. Preparation of Samples for the Pulldown Assays

1. *Xenopus* Embryos and Explants

1. *Xenopus* embryos are injected with RNAs at the four-cell stage into the two dorsal cells (for dorsal marginal zone [DMZ] explants) or into the two ventral cells (for ventral marginal zone [VMZ] explants) in 3% Ficoll/0.5X MMR.
2. Two hours following the injections, the embryos are changed into 0.1X MMR solution and cultured to stage 10.5.
3. Embryonic vitelline membranes are removed and DMZ or VMZ are explanted using forceps. The explants are pooled and stored on ice until they are lysed. All embryos are dissected on agarose-coated culture dishes in a solution of 0.5X MMR/1% BSA. Additional details of handling *Xenopus* embryos and explants are as described elsewhere (Sive, 2000).

2. GST-RBD and GST-PBD Binding Assay

1. The explants are lysed in 500 μl of Rho or Rac lysis buffer, and 10 μl of 10 mg/ml DNase1 solution is added to each sample. The samples are then incubated on ice for 10 minutes and spun at 14,000 rpm at 4°C. For *Xenopus* explants assays, we typically use 50 explants (DMZ or VMZ) for each sample.
2. 25 μl of the lysate is removed and 25 μl of 2X sample buffer is added, heat at 90°C for 5 minutes, and store at –80°C. This is your whole cell lysate for control immunoblotting.
3. The remaining 475 μl of supernatant is removed and added to tubes containing approximately 50 μl of GST beads coupled to the RBD or to PBD.
4. The samples are incubated on a Nutator at 4°C for 1 hour and washed three times with Rho or Rac wash buffer. After the final wash, the samples are resuspended in 50 μl of 2X sample buffer and heated at 90°C for 5 minutes. The samples can then be stored at –80°C.
5. Western Blot analysis is then performed.

D. Western Blot Analysis

1. The samples are resolved on a 12% SDS-PAGE gel and the gels are run until the Bromophenol dye is approximately 1 inch from the bottom of the gel.

2. The proteins are then transferred to nitrocellulose membrane, which is incubated for 1 hour with 5% non-fat dry milk.
3. The blots are washed twice for 5 minutes with 1X PBST.
4. The blots are incubated with the primary antibody (Rho monoclonal, Santa Cruz Laboratories, for mammalian cell extracts or Rho polyclonal, Santa Cruz Laboratories, for *Xenopus* explant extracts; Rac or Cdc42 monoclonal, Transduction Labs, for both mammalian cell and *Xenopus* explant extracts) at a 1/500 dilution for 1 hour at room temperature or overnight at 4°C.
5. Wash the blots once for 15 minutes and then four times for 5 minutes with 1X PBST.
6. The blots are then incubated with secondary antibody at 1/5000 solution for 1 hour at room temperature.
7. Wash the blots once for 15 minutes and then four times for 5 minutes with 1X PBST.
8. Perform the ECL reaction. We typically use SuperSignal PicoWest (Pierce).
9. The endogenous Rho/Rac/Cdc42 can be detected within 3–5 minutes in total lysates or 10 minutes in the GST-RBD/PBD pulldown samples for *Xenopus* explant.

IX. CAUTIONARY NOTES

1. For each preparation of GST-RBD or GST-PBD bacteria, use a single aliquot and follow the protocol to extract the fusion protein and resolve a fraction of the sample on a 12% SDS-PAGE gel and perform a Coomassie Blue stain. The GST-RBD protein migrates at approximately 36–37 kD and GST-PBD protein at approximately 34–35 kD. If there is extensive degradation, redo the protein preparation.
2. Each 50 ml bacterial aliquot of GST-RBD will yield approximately 200–300 μg of fusion protein, and each aliquot is enough for 10 samples for the GST-RBD assay. Each 1 ml aliquot of GST-PBD will yield approximately 150–250 μg of protein, and each aliquot is enough for 10 samples for the GST-PBD assay.
3. Always keep samples on ice whenever possible and do not exceed the incubation times.

X. BUFFERS

Rho lysis buffer: 50 mM Tris-HCl pH 7.2, 500 mM NaCl, 1% Triton-X 100, 0.5% sodium deoxycholic acid, 0.1% SDS, 10 mM $MgCl_2$, and 1X protease inhibitors (added fresh each time for this and other buffers).

Rho wash buffer: 50 mM Tris-HCl pH 7.2, 1% Triton-X 100, 150 mM NaCl, 10 mM $MgCl_2$, and 1X protease inhibitors.

Rac/CDC42 lysis buffer: 50 mM Tris, pH 7.5, 200 mM NaCl, 2% NP40, 10% glycerol, 10 mM $MgCl_2$, and 1X protease inhibitors.
Rac wash buffer: 25 mM Tris, pH 7.5, 40 mM NaCl, 1% NP40, 30 mM $MgCl_2$, and 1X protease inhibitors.
10X PBST: 995 ml of 10X PBS and 5 ml Tween-20 in 1 l of water.
10X MMR: 1M NaCl, 20 mM KCl, 20 mM CaCl2, 10 mM MgCl2, 50 mM Hepes, pH to 7.6.

ACKNOWLEDGMENTS

The authors wish to thank Steve Shamah and Michael Lin for their guidance in the GTPase assays. This work was supported by grants from the American Heart Association, March of Dimes and the National Science Foundation to R. H. and S. H. is supported by grants for NIH, and is a W. M. Keck Foundation Distinguished Young Scholar and a Leukemia and Lymphoma Society Scholar.

REFERENCES

Akasaki, T., Koga, H., and Sumimoto, H. (1999). Phosphoinositide 3-kinase-dependent and -independent activation of the small GTPase Rac2 in human neutrophils. J Biol Chem *274*, 18055–18059.

Bagrodia, S., Taylor, S. J., Creasy, C. L., Chernoff, J., and Cerione, R. A. (1995). Identification of a mouse p21Cdc42/Rac activated kinase. J Biol Chem *270*, 22731–22737.

Benard, V., Bohl, B. P., and Bokoch, G. M. (1999). Characterization of rac and cdc42 activation in chemoattractant-stimulated human neutrophils using a novel assay for active GTPases. J Biol Chem *274*, 13198–13204.

Benink, H. A., and Bement, W. M. (2005). Concentric zones of active RhoA and Cdc42 around single cell wounds. J Cell Biol *168*, 429–439.

Boutros, M., Paricio, N., Strutt, D. I., and Mlodzik, M. (1998). Dishevelled activates JNK and discriminates between JNK pathways in planar polarity and wingless signaling. Cell *94*, 109–118.

Chen, A. E., Ginty, D. D., and Fan, C. M. (2005). Protein kinase A signalling via CREB controls myogenesis induced by Wnt proteins. Nature *433*, 317–322.

Chen, F., Ma, L., Parrini, M. C., Mao, X., Lopez, M., Wu, C., Marks, P. W., Davidson, L., Kwiatkowski, D. J., Kirchhausen, T., et al. (2000). Cdc42 is required for PIP(2)-induced actin polymerization and early development but not for cell viability. Curr Biol *10*, 758–765.

Cho, Y. J., Zhang, B., Kaartinen, V., Haataja, L., de Curtis, I., Groffen, J., and Heisterkamp, N. (2005). Generation of rac3 null mutant mice: role of Rac3 in Bcr/Abl-caused lymphoblastic leukemia. Mol Cell Biol *25*, 5777–5785.

Choi, S. C., and Han, J. K. (2002). *Xenopus* Cdc42 regulates convergent extension movements during gastrulation through Wnt/Ca2+ signaling pathway. Dev Biol *244*, 342–357.

Chung, H. A., Hyodo-Miura, J., Nagamune, T., and Ueno, N. (2005). FGF signal regulates gastrulation cell movements and morphology through its target NRH. Dev Biol *282*, 95–110.

Coso, O. A., Chiariello, M., Yu, J. C., Teramoto, H., Crespo, P., Xu, N., Miki, T., and Gutkind, J. S. (1995). The small GTP-binding proteins Rac1 and Cdc42 regulate the activity of the JNK/SAPK signaling pathway. Cell *81*, 1137–1146.

Drechsel, D. N., Hyman, A. A., Hall, A., and Glotzer, M. (1997). A requirement for Rho and Cdc42 during cytokinesis in *Xenopus* embryos. Curr Biol *7*, 12–23.

Giles, R. H., van Es, J. H., and Clevers, H. (2003). Caught up in a Wnt storm: Wnt signaling in cancer. Biochim Biophys Acta *1653*, 1–24.

Gu, Y., and Williams, D. A. (2002). RAC2 GTPase deficiency and myeloid cell dysfunction in human and mouse. J Pediatr Hematol Oncol *24*, 791–794.

Habas, R., and Dawid, I. B. (2005). Dishevelled and Wnt signaling: is the nucleus the final frontier? J Biol *4*, 2.

Habas, R., Dawid, I. B., and He, X. (2003). Coactivation of Rac and Rho by Wnt/Frizzled signaling is required for vertebrate gastrulation. Genes Dev *17*, 295–309.

Habas, R., Kato, Y., and He, X. (2001). Wnt/Frizzled activation of Rho regulates vertebrate gastrulation and requires a novel Formin homology protein Daam1. Cell *107*, 843–854.

Heisenberg, C. P., Tada, M., Rauch, G. J., Saude, L., Concha, M. L., Geisler, R., Stemple, D. L., Smith, J. C., and Wilson, S. W. (2000). Silberblick/Wnt11 mediates convergent extension movements during zebrafish gastrulation. Nature *405*, 76–81.

He, X., Semenov, M., Tamai, K., and Zeng, X. (2004). LDL receptor related proteins 5 and 6 in Wnt signaling: ARROWs point the way. Development *131*, 1663–1677.

Hikasa, H., Shibata, M., Hiratani, I., and Taira, M. (2002). The *Xenopus* receptor tyrosine kinase Xror2 modulates morphogenetic movements of the axial mesoderm and neuroectoderm via Wnt signaling. Development *129*, 5227–5239.

Hukriede, N. A., Tsang, T. E., Habas, R., Khoo, P. L., Steiner, K., Weeks, D. L., Tam, P. P., and Dawid, I. B. (2003). Conserved requirement of Lim1 function for cell movements during gastrulation. Dev Cell *4*, 83–94.

Kaibuchi, K., Kuroda, S., and Amano, M. (1999). Regulation of the cytoskeleton and cell adhesion by the Rho family GTPases in mammalian cells. Annu Rev Biochem *68*, 459–486.

Keller, R. (2002). Shaping the vertebrate body plan by polarized embryonic cell movements. Science *298*, 1950–1954.

Keller, R., Davidson, L. A., and Shook, D. R. (2003). How we are shaped: the biomechanics of gastrulation. Differentiation *71*, 171–205.

Kwan, K. M., and Kirschner, M. W. (2005). A microtubule-binding Rho-GEF controls cell morphology during convergent extension of *Xenopus* laevis. Development *132*, 4599–4610.

Logan, C. Y., and Nusse, R. (2004). The Wnt Signaling Pathway in Development and Disease. Annu Rev Cell Dev Biol *20,* 781–810.

Lu, X., Borchers, A. G., Jolicoeur, C., Rayburn, H., Baker, J. C., and Tessier-Lavigne, M. (2004). PTK7/CCK-4 is a novel regulator of planar cell polarity in vertebrates. Nature *430*, 93–98.

Lucas, J. M., Nikolic, I., and Hens, M. D. (2002). cDNA cloning, sequence comparison, and developmental expression of *Xenopus* rac1. Mech Dev *115*, 113–116.

Marlow, F., Topczewski, J., Sepich, D., and Solnica-Krezel, L. (2002). Zebrafish Rho kinase 2 acts downstream of Wnt11 to mediate cell polarity and effective convergence and extension movements. Curr Biol *12*, 876–884.

Matsui, T., Amano, M., Yamamoto, T., Chihara, K., Nakafuku, M., Ito, M., Nakano, T., Okawa, K., Iwamatsu, A., and Kaibuchi, K. (1996). Rho-associated kinase, a novel serine/threonine kinase, as a putative target for small GTP binding protein Rho. Embo J *15*, 2208–2216.

Medina, A., Swain, R. K., Kuerner, K. M., and Steinbeisser, H. (2004). *Xenopus* paraxial protocadherin has signaling functions and is involved in tissue separation. Embo J *23*, 3249–3258.

Minden, A., Lin, A., Claret, F. X., Abo, A., and Karin, M. (1995). Selective activation of the JNK signaling cascade and c-Jun transcriptional activity by the small GTPases Rac and Cdc42Hs. Cell *81*, 1147–1157.

Miyagi, C., Yamashita, S., Ohba, Y., Yoshizaki, H., Matsuda, M., and Hirano, T. (2004). STAT3 noncell-autonomously controls planar cell polarity during zebrafish convergence and extension. J Cell Biol *166*, 975–981.

Miyakoshi, A., Ueno, N., and Kinoshita, N. (2004). Rho guanine nucleotide exchange factor xNET1 implicated in gastrulation movements during *Xenopus* development. Differentiation *72*, 48–55.

Mlodzik, M. (2002). Planar cell polarization: do the same mechanisms regulate *Drosophila* tissue polarity and vertebrate gastrulation? Trends Genet *18*, 564–571.

Raftopoulou, M., and Hall, A. (2004). Cell migration: Rho GTPases lead the way. Dev Biol *265*, 23–32.

Ren, X. D., Kiosses, W. B., and Schwartz, M. A. (1999). Regulation of the small GTP-binding protein Rho by cell adhesion and the cytoskeleton. Embo J *18*, 578–585.

Sasai, N., Nakazawa, Y., Haraguchi, T., and Sasai, Y. (2004). The neurotrophin-receptor-related protein NRH1 is essential for convergent extension movements. Nat Cell Biol *6*, 741–748.

Schwartz, M. (2004). Rho signalling at a glance. J Cell Sci *117*, 5457–5458.

Sheldahl, L. C., Slusarski, D. C., Pandur, P., Miller, J. R., Kuhl, M., and Moon, R. T. (2003). Dishevelled activates Ca2+ flux, PKC, and CamKII in vertebrate embryos. J Cell Biol *161*, 769–777.

Sive, H. L., Grainger, R.M. and Harland, R.M. (2000). Early development of *Xenopus* laevis. A laboratory manual (Cold Spring Harbor, NY: Cold Spring Harbor Press).

Sokol, S. Y. (1996). Analysis of Dishevelled signalling pathways during *Xenopus* development. Curr Biol *6*, 1456–1467.

Sorokina, E. M., and Chernoff, J. (2005). Rho-GTPases: new members, new pathways. J Cell Biochem *94*, 225–231.

Sugihara, K., Nakatsuji, N., Nakamura, K., Nakao, K., Hashimoto, R., Otani, H., Sakagami, H., Kondo, H., Nozawa, S., Aiba, A., and Katsuki, M. (1998). Rac1 is required for the formation of three germ layers during gastrulation. Oncogene *17*, 3427–3433.

Symons, M., Derry, J. M., Karlak, B., Jiang, S., Lemahieu, V., McCormick, F., Francke, U., and Abo, A. (1996). Wiskott-Aldrich syndrome protein, a novel effector for the GTPase CDC42Hs, is implicated in actin polymerization. Cell *84*, 723–734.

Tahinci, E., and Symes, K. (2003). Distinct functions of Rho and Rac are required for convergent extension during *Xenopus* gastrulation. Dev Biol *259*, 318–335.

Tada, M., and Smith, J. C. (2000). Xwnt11 is a target of *Xenopus* Brachyury: regulation of gastrulation movements via Dishevelled, but not through the canonical Wnt pathway. Development *127*, 2227–2238.

Unterseher, F., Hefele, J. A., Giehl, K., De Robertis, E. M., Wedlich, D., and Schambony, A. (2004). Paraxial protocadherin coordinates cell polarity during convergent extension via Rho A and JNK. Embo J *23*, 3259–3269.

Veeman, M. T., Axelrod, J. D., and Moon, R. T. (2003). A second canon. Functions and mechanisms of beta-catenin-independent Wnt signaling. Dev Cell *5*, 367–377.

Wallingford, J. B., Fraser, S. E., and Harland, R. M. (2002). Convergent extension: the molecular control of polarized cell movement during embryonic development. Dev Cell *2*, 695–706.

Wallingford, J. B., and Habas, R. (2005). The developmental biology of Dishevelled: an enigmatic protein governing cell fate and cell polarity. Development *132*, 4421–4436.

Wallingford, J. B., and Harland, R. M. (2002). Neural tube closure requires Dishevelled-dependent convergent extension of the midline. Development *129*, 5815–5825.

Wallingford, J. B., Rowning, B. A., Vogeli, K. M., Rothbacher, U., Fraser, S. E., and Harland, R. M. (2000). Dishevelled controls cell polarity during *Xenopus* gastrulation. Nature *405*, 81–85.

Watanabe, N., Madaule, P., Reid, T., Ishizaki, T., Watanabe, G., Kakizuka, A., Saito, Y., Nakao, K., Jockusch, B. M., and Narumiya, S. (1997). p140mDia, a mammalian homolog of *Drosophila* diaphanous, is a target protein for Rho small GTPase and is a ligand for profilin. Embo J *16*, 3044–3056.

Wharton, K. A., Jr. (2003). Runnin' with the Dvl: proteins that associate with Dsh/Dvl and their significance to Wnt signal transduction. Dev Biol *253*, 1–17.

Williams, D. A., Tao, W., Yang, F., Kim, C., Gu, Y., Mansfield, P., Levine, J. E., Petryniak, B., Derrow, C. W., Harris, C., *et al.* (2000). Dominant negative mutation of the hematopoietic-specific Rho GTPase, Rac2, is associated with a human phagocyte immunodeficiency. Blood *96*, 1646–1654.

Wunnenberg-Stapleton, K., Blitz, I. L., Hashimoto, C., and Cho, K. W. (1999). Involvement of the small GTPases XRhoA and XRnd1 in cell adhesion and head formation in early *Xenopus* development. Development *126*, 5339–5351.

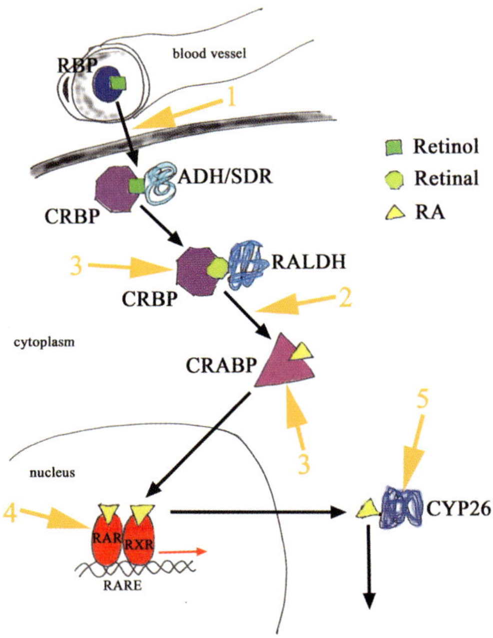

FIGURE 5.1 The RA signalling pathway and the points at which it has been disrupted.

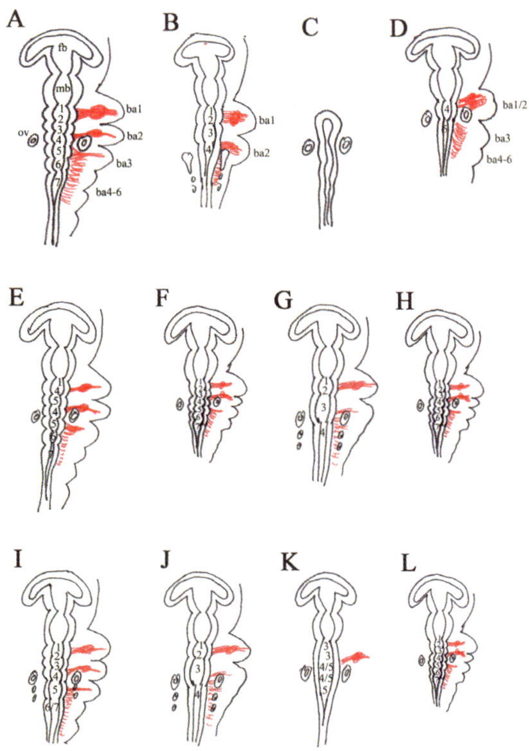

FIGURE 5.2 Diagrams summarizing the effects of increasing or decreasing RA signalling on the anterior CNS, particularly the hindbrain, branchial arches, and associated neural crest.

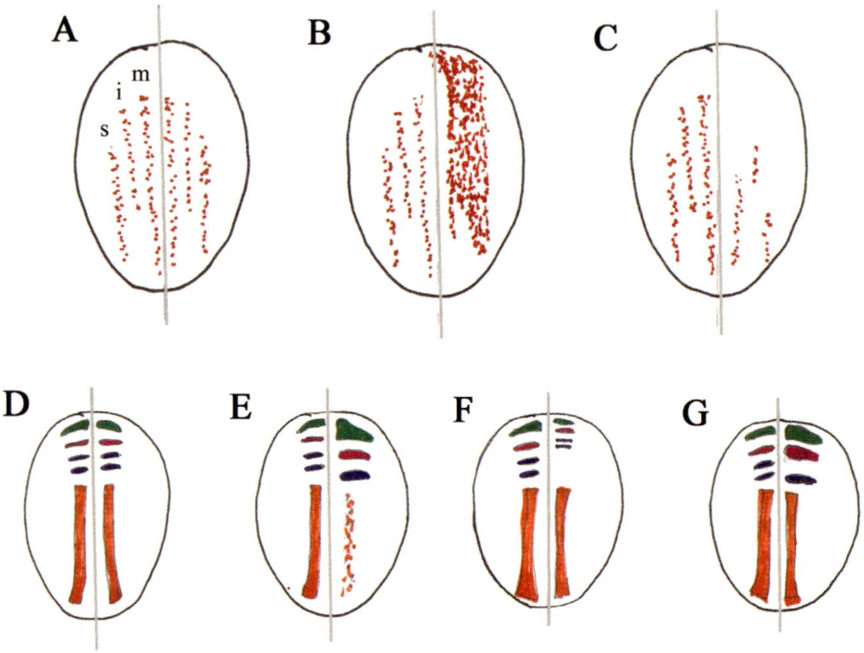

FIGURE 5.4 The effects of increasing and decreasing RA signalling on neural patterning in *Xenopus* embryos.

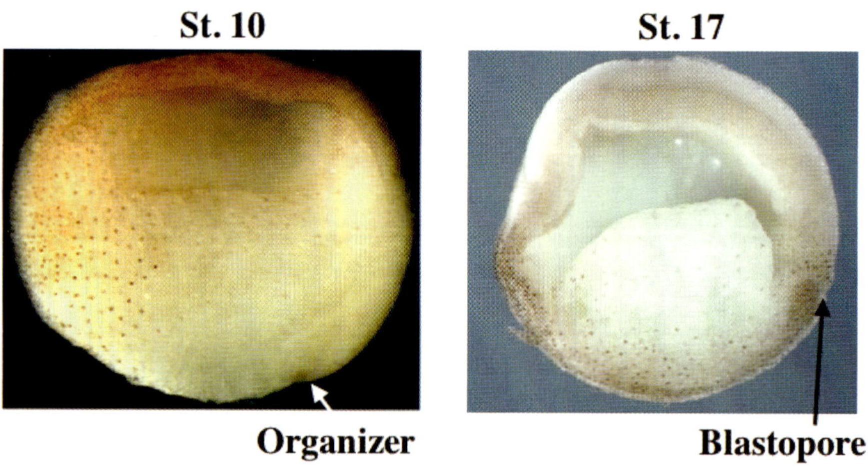

FIGURE 7.1 Anti-phospho Smad1/5/8 semi-whole mount staining of *Xenopus* embryos.

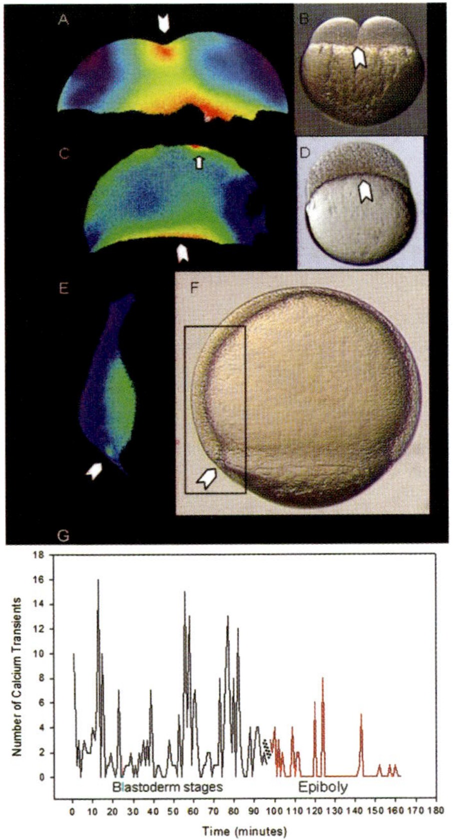

FIGURE 8.2 Endogenous Ca^{2+} release events in zebrafish development.

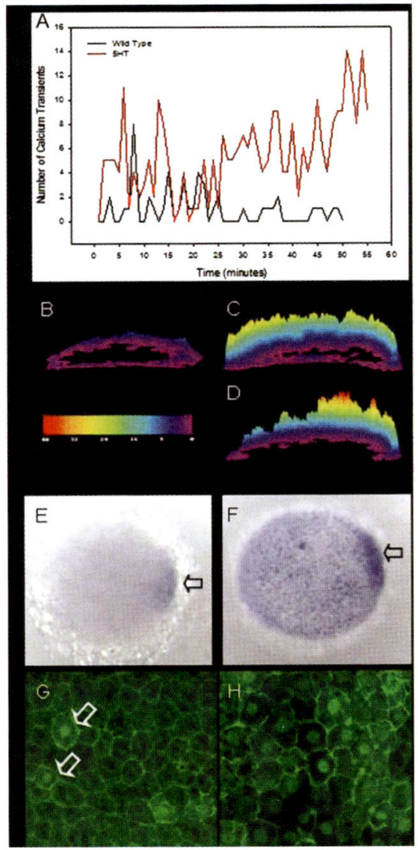

FIGURE 8.3 Manipulation of Ca^{2+} activity and downstream responders.

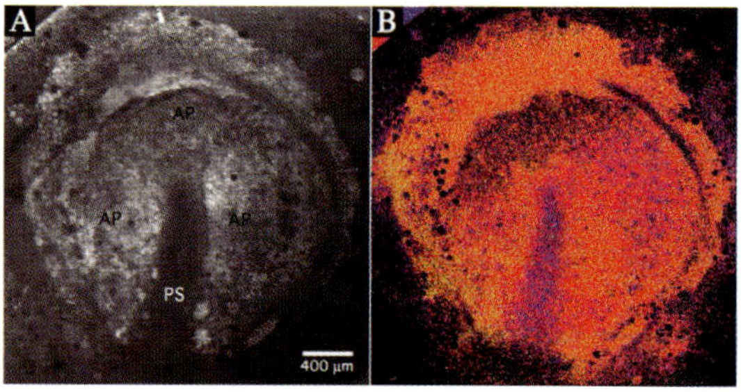

FIGURE 9.2 Chick Stage 3 blastoderm stained with cSNARF-1-AM and illuminated with 546 nm light.

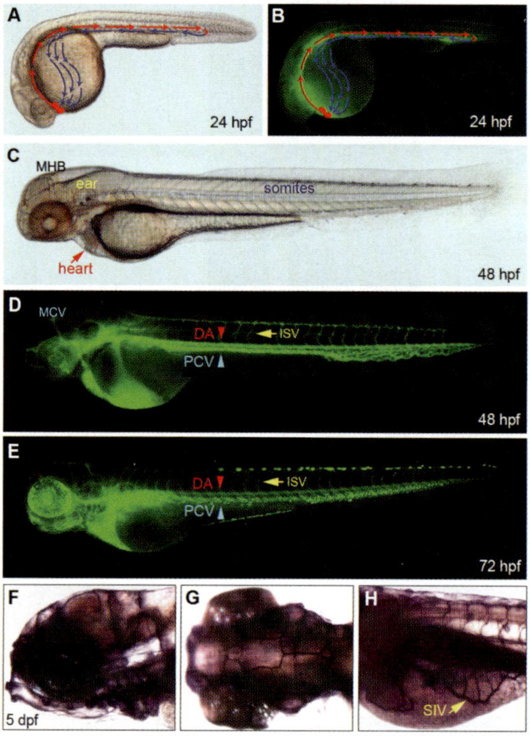

FIGURE 12.1

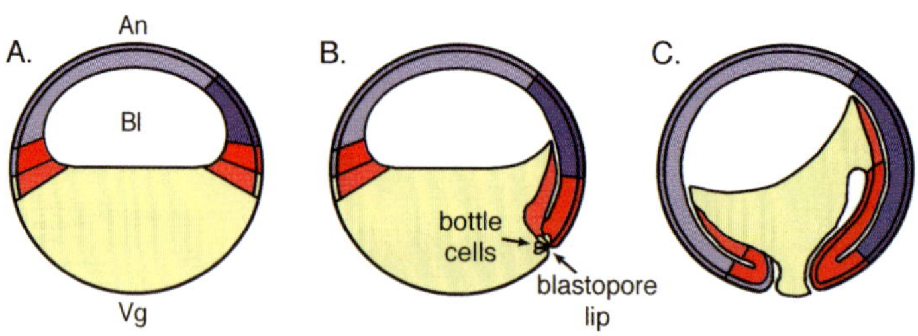

FIGURE 13.1 ***Xenopus*** **gastrulation.**

7 Phosphospecific Antibodies as Tools for the Study of Signal Transduction during Development

Malcolm Whitman

CONTENTS

I. INTRODUCTION

In the study of mechanisms of signal transduction, whether in embryos or elsewhere, protein phosphorylation stands as the predominant mechanism of post-translational protein modification, mediating responses to extracellular signals. Historically, the study of protein phosphorylation depended heavily on radioisotopic labeling, which was poorly suited for the study of embryogenesis, both because of the unwieldiness of metabolic labeling of embryos and because of the very poor spatial resolution provided by dissection of labeled embryos. Although *in situ* hybridization has long been established as a tool for spatial and temporal localization of complex patterns of gene expression during development, comparable tools for the study of the state of signal transduction pathways have not been available. The advent of site-specific, phosphorylation state-specific antibodies has revolutionized the study of protein phosphorylation during embryogenesis and made possible a direct means to visualize states of signaling pathway activation. Visualization of states of activity of signaling pathways helps to both identify developmental events in which specific pathways participate and test specific models for how patterns of signaling participate in morphogenesis.

II. GENERAL TYPES, USES, AND LIMITATIONS OF PHOSPHOSPECIFIC ANTIBODIES

The essential feature of phosphospecific antibodies is that they recognize phosphorylated sites in proteins but not the corresponding de-phospho site. There are, however, a number of important variations on this basic feature. The first widely used phosphospecific antibodies were directed against phosphotyrosine (Draetta et al., 1988; White et al., 1985) and specifically designed to be as context-independent as possible; that is, they recognize phosphorylated tyrosines independent of surrounding residues. The advantage of antibodies of this type is their general utility: They can be applied to the study of any tyrosine phosphoprotein. The disadvantage, however, is that they yield information about specific proteins in combination with other techniques, such as immunoprecipitation of a protein of interest.

For immunohistochemical studies of endogenous signaling, however, anti-phosphotyrosine antibodies are most useful to visualize activation of tyrosine kinases generally, but do not in general provide much information about signaling through specific pathways. Although general phosphoserine and phosphothreonine specific antibodies have been available commercially for many years (from Sigma, Zymed, and others), they are not completely context independent and generally less sensitive than antibodies against phosphotyrosine or against specific phosphoepitopes. Also, because most proteins in the cell have multiple sites of serine phosphorylation, the inability of general phosphoserine antibodies to distinguish among individual sites limits the useful information that can be obtained with them. Although anti-phosphoserine and anti-phosphothreonine antibodies have been used effectively in conjunction with immunoprecipitation of specific proteins in a limited number of cases, they have not been generally useful for the study of endogenous phosphorylation.

Antibodies directed against specific phosphorylation sites in proteins, although narrower in their range of application, have been far more powerful than general phosphoantibodies for examining signaling during embryonic development. In this case, antibodies are raised against the phosphorylated forms of peptides derived from individual phosphorylation sites in proteins of interest. The activation loop TEYVATR site in ERK kinases, and the C-terminal SSXS phosphorylation sites in the receptor regulated Smads, have provided particularly good phosphoantibodies that have been widely useful for the study of endogenous regulation of these signal transducers (Gabay et al., 1997; Christen and Slack, 1999; Faure et al., 2000; Schohl and Fagotto, 2002; Corson et al., 2003; Dorfman and Shilo, 2001; Faure et al., 2002; Teleman and Cohen, 2000). In both these examples, the target peptides contain two phosphorylation sites, which may help to account for the particular effectiveness of these antibodies.

A third type of phosphorylation-specific antibody that has recently come into wider use are antibodies directed against preferred recognition motifs for specific protein kinases. For example, the kinase AKT recognizes the preferred motif RxRxxS/T (CST), and an antibody directed against the phosphorylated form of such a generic motif (Cell Signaling Technologies) allows the selective identification of proteins phosphorylated at this preferred motif. Like the fully site-specific antibodies, the generic motif directed antibodies allow the distinction of one type of phosphorylation site among many potential sites in a protein. Like the general p-Tyr antibodies, however, they require an additional identification procedure, such as immunoprecipitation of the protein of interest, to distinguish phosphorylation of one particular protein from all the others that may contain the same kinase recognition motif.

An additional limitation of generic motif directed phosphoantibodies is that different classes of protein kinases can target motifs that are similar. For example, cyclic-AMP dependent protein kinase, like AKT, phosphorylates motifs with basic residues N-terminal to the phosphorylation site (www.scansite.mit.edu). Although the optimal motifs for AKT- and cyclic AMP-dependent protein kinase are different, some naturally occurring phosphorylation sites are similar enough to both motifs that antibodies directed against the optimal motif for one kinase may react with sites phosphorylated by another. Additional techniques (e.g., pharmacological blockade or activation of specific kinase pathways) are therefore necessary to confirm that phosphorylation sites recognized by generic motif directed antibodies are in fact regulated by the kinase of interest.

III. USES OF PHOSPHOSPECIFIC ANTIBODIES

A. Western Blots

Phosphospecific antibodies are useful for a range of immunochemical assays, with distinct advantages and disadvantages for the developmental biologist. Because most phosphospecific antibodies are initially characterized by Western blot, they are generally useful for this purpose. Western blot analysis provides information about the size of the proteins reacting with the antibody, which is useful to confirm that signal observed is due to the phosphoprotein of interest and not a cross-reacting protein. Western blots can also, if used in conjunction with a relatively quantitative detection technique (e.g., fluorescent secondary antibodies rather than chemiluminescence), provide fairly reliable quantitation of changes in signaling in an isolated tissue. The spatial resolution of Western blots, however, is limited to the resolution with which individual regions or tissues can be dissected for analysis. Similarly, the sensitivity with which changes in phosphorylation can be detected by Western blot is limited by the ability to isolate sufficient quantities of the tissue of interest. For small or early-stage embryos (e.g., zebrafish, pre-gastrula mouse embryos), examination of region-specific phosphorylation by Western blot can be prohibitively difficult, but this is feasible in some larger embryos (e.g., *Xenopus*) (Lee et al., 2001).

B. Immunohistochemistry

The immunohistochemical application of phosphospecific antibodies allows for visualization of complex or highly localized patterns of signal activation in embryos. With a high-affinity antibody, the data are of comparable quality to *in situ* hybridization detection of mRNAs (an example of staining directed against pSmad1 in *Xenopus* semi-whole mounts is shown in Figure 7.1). Phosphospecific antibodies have been used in a wide range of histochemical applications, including detection of pERK and pMAD in fly embryos; pERK, pSmad1, and pSmad2 in *Xenopus* embryos; pSmad1 and pStat3 in zebrafish embryos; pSmad1 and pSmad2 in chick embryos; and pERK, pSmad1, and pSmad2 in early mouse embryos (Gabay et al., 1997; Christen and Slack, 1999; Faure et al., 2000; Corson et al., 2003; Dorfman and Shilo, 2001; Faure et al., 2002; Mintzer et al., 2001; Raftery and Sutherland, 1999; Teleman and Cohen, 2000; Yamashita et al., 2002). In early embryos, it is generally possible to do immunohistochemical analysis in whole mount. Frozen sections have also been successfully used (Schohl and Fagotto, 2002), and in later embryos, immunohistochemistry of fixed, thin sectioned tissues has been used as well (Goumans et al., 2003; Meno et al., 2001). Although data from immunohistochemistry are typically less quantitatively reliable than Western or ELISA analysis, the use of fluorescent secondary antibody detection and relatively planar tissues (e.g., *Drosophila* wing disc) has made possible quite elegant quantitation of the graded distribution of a specific signaling during pattern formation (Dorfman and Shilo, 2001; Teleman and Cohen, 2000).

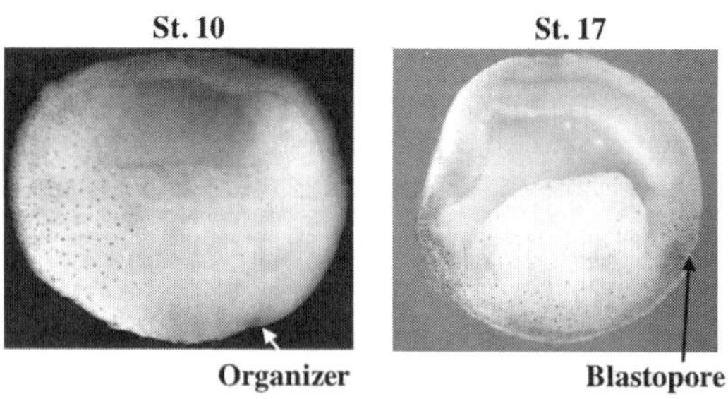

FIGURE 7.1 (See color insert following page 144.) Anti-phospho Smad1/5/8 semi-whole mount staining of *Xenopus* embryos. Embryos were fixed, bisected in sagittal section, processed for semi-whole mount as described in the accompanying protocol. Staining was visualized using diaminobenzidine. Brown punctate stain reflects nuclearly localized phospho Smad1. At gastrulation (left panel), staining is distributed on the contra-Organizer side of the embryo and excluded from the Organizer side (the prospective dorso-anterior of the embryo). In neural embryos (right panel), pSmad1 staining is absent from neural tissue and axial/paraxial mesoderm, but is strong at the prospective anterior and posterior tips of the embryo, as well as ventrally.

C. Other Applications: Immunoprecipitation/Purification/ELISA

Although immunoprecipitation is well established as an application of anti-phosphotyrosine antibodies and has been used for immunopurification of tyrosine kinase targets, it has been less widely used for antibodies directed against specific phosphorylation sites. In our hands, anti-phospho-Smad antibodies that are effective for both Western analysis and immunohistochemistry are useless for immunoprecipitation. Because sites of phosphorylation are recognition sites for other proteins, it may be common for these sites to be masked in native cell extracts, making immunoprecipitation difficult. Commercial ELISA kits are available for a number of phosphospecific antibodies, but these are intended for analysis of cultured cells and have not, to our knowledge, been used for the study of embryos.

IV. IDENTIFICATION OR PREPARATION OF SUITABLE PHOSPHOSPECIFIC ANTIBODIES

A. Commercial Antibodies

A wide variety of phosphospecific antibodies are now available from commercial vendors (including, but not limited to, Sigma, Cell Signaling Technologies, Upstate Biotechnologies Inc., Santa Cruz Laboratories, Zymed, Promega, BD Biosciences, Calbiochem, Transduction Laboratories). These antibodies are in general prepared against phosphorylation sites in mouse or human proteins, but because phosphorylation sites are generally well conserved, these antibodies usually recognize sites in orthologous proteins across vertebrate species and often recognize appropriate sites in invertebrates, as well. Nevertheless, for any particular application in a nonmammalian model organism, it is a good idea to check on how well the epitope of interest is conserved before purchase.

Commercial phosphospecific antibodies are generally tested against cultured cells that have been given a stimulus that maximally stimulates the relevant kinase. They can generally be counted on to detect strong signal activation, but whether they are adequate to detect phosphorylation at levels that occur endogenously is variable and difficult to predict. The care with which commercial phosphospecific antibodies are checked for specificity with regard to other phosphoepitopes is variable as well, and claims regarding specificity should be checked by the investigator rather than taken at face value. In addition, most phosphospecific antibodies are polyclonal, which means that different serum lots can vary widely in their sensitivity and specificity. Both rabbit and mouse monoclonal phosphospecific antibodies are coming into increasing use, which should help to alleviate variability, and sometimes permanent changes, in the quality of available antibodies over time.

B. Making Your Own Phosphospecific Antibodies

The purchase of commercial phosphospecific antibodies has the advantage that a wide range of them are available immediately, but for many regulated phosphorylation sites of interest, commercial antibodies are either unavailable or inadequate for detection of endogenous signals. As an alternative, phosphospecific antibodies can be prepared by the investigator. Antigen preparation and immunization protocols are typically the same as for preparation of other antipeptide antibodies, with the exception that a phosphopeptide corresponding to the site of interest is synthesized (Harlow and Lame, 1988). Immunizing peptides for phosphospecific antibodies are often shorter than their nonphospho counterparts, to minimize the possibility of antibody recognition outside the phosphorylation site region. Use of too short a peptide may result in an antibody that is not specific for the phosphorylation site of interest, although one of the most widely used of the anti-phospho-Smad2 antibodies utilized only four amino acids at the phosphorylation site (Persson et al., 1998). This was probably only possible, however, because the location of the epitope at the C-terminus of the protein provided an additional degree of specificity.

One of the major practical issues distinguishing the use of phosphopeptides as antigens compared to the use of standard peptides is the potential for dephosphorylation of peptides *in vivo* following immunization. Although the author is not aware of any systematic examination of this issue, and it has not prevented the successful development of many phosphospecific antibodies, there have been numerous anecdotal cases (Harlow and Lane, 1988), including in the author's own laboratory, in which efficient dephosphorylation of immunizing phosphopeptides appeared to preclude phosphoantibody production. The use of non-hydrolyzable phosphonates or phosphonate derivatives (e.g., l-2-amino-4-phosphono-4, 4-difluorobutanoic acid) instead of phosphoserine for peptide synthesis provides one solution to this problem (Higashimoto et al., 2000; Sakaguchi et al., 1996). The author is not aware, however, of a commercial source for such phosphonate derivatives in a form suitable for use on a peptide synthesizer, and therefore this approach is not widely used. In the absence of direct measurement of antigen dephosphorylation in the immunized rabbit, it is also difficult to assess whether the failure to successfully raise a phosphoantibody results from antigen dephosphorylation or another problem.

Once an anti-serum directed against a phosphoepitope is successfully generated, the utility of the serum is usually dramatically improved by affinity purification. This can include negative selection (against the de-phospho form of the peptide epitope) as well as positive selection (using the immunizing phosphopeptide immobilized on solid support), using standard antibody affinity purification protocols (Harlow and Lane, 1988). The success of purification is most easily monitored by Western blot with appropriate controls (e.g., recognition of phospho versus de-phospho target protein).

V. USING PHOSPHOSPECIFIC ANTIBODIES: SAMPLE PREPARATION/FIXATION

Use of phosphospecific antibodies for Western blot analysis is identical to other Western applications, with the exception that an appropriate cocktail of phosphatase inhibitors should be included in the cell lysis buffer (Faure et al., 2000). For immunohistochemistry, the effectiveness of phosphospecific antibodies may be very sensitive to fixation conditions, but optimal conditions need to be determined empirically. In our hands, formaldehyde-based fixatives work quite well for several different phosphospecific antibodies, whereas methanol- or acid-based fixatives work poorly. A substantial loss of signal generally can result from embedding in wax for thin sectioning, but this can be alleviated by both standard antigen reactivation procedures and signal amplification kits (Harlow and Lame, 1988). Cryo-sectioning has also been effectively used for immunohistochemistry with phosphospecific antibodies, but this may be more labor intensive and in some cases less satisfactory for preservation of morphology.

To what extent phosphatase activity during fixation or postfixation handling of whole mount embryos is significant is not known. Sulfhydryl alkylating agents such as N-ethyl maleimide or iodoacetamide irreversibly inhibit tyrosine phosphatases and can be added during or after fixation, but these are not effective against most serine-threonine phosphatases. If a loss of signal during embryo fixation and processing is suspected, a cocktail of serine-threonine phosphatase inhibitors can be added to fixation and wash buffers, but unless there is direct evidence that this improves signal, it is probably a waste of reagents. For whole mount *Xenopus* embryos, we find that a standard procedure for bleaching embryos substantially improves immunohistochemical signals, but this appears to be due to an improvement in permeability of embryos, and the effect is not limited to phosphospecific antibodies.

A. Validation of Phosphospecific Antibodies

Establishing criteria for confirmation that a phosphospecific antibody is in fact recognizing the target of interest is a critical and often neglected aspect of studying endogenous signaling. Although phosphopeptides corresponding to the target site are commonly used as blocking reagents to establish specificity, blocking with a specific phosphoepitope antibody does not adequately establish that the antiserum is not recognizing a different phosphoepitope. Pharmacological inhibition or genetic tools to either inactivate the upstream signaling pathway or eliminate the endogenous target protein provide the best means to establish specificity of phosphospecific antibody recognition. Confirmation of immunohistochemical signals by Western blotting (and confirming that perturbations that alter the Western blot signal cause corresponding changes in immunohistochemical signal) is also useful. Unfortunately, in a developmental context, it is not always practical to apply these criteria rigorously. In mammalian embryos, application of pharmacological inhibitors is problematic, and although there are many genetic tools for altering specific signaling pathways, the study of events in late development can be complicated by disrupting signaling at earlier stages. When studying highly localized signaling in specific tissues, obtaining material for Western blotting may also not be feasible.

Although practical issues may interfere with full implementation of criteria for validation of phosphospecific antibody recognition of endogenous signals, it is nevertheless important to recognize that without some attempt to confirm signal dependence of antibody recognition, phosphospecific antibodies can yield seriously misleading information. In our hands, affinity purified phospho-Smad2 antibody provides strong, nucleus-specific staining of whole mounted zebrafish embryos, but this staining is unaffected by genetic manipulations (e.g., cyc/sqt or MZoep mutants) that completely eliminate phospho-Smad2 detection by Western blot and that eliminate expression of transcriptional targets of Smad2 activation (M. Whitman and A. Schier, unpublished). This staining is almost certainly artifactual, even though it is completely blocked by free phospho-Smad2 peptide. Although phosphospecific antibodies are clearly powerful tools for studying endogenous signals, without validating techniques, they can yield dramatically inaccurate information.

VI. CONCLUSION

Phosphospecific antibodies provide the opportunity to identify new potential roles for signaling pathways in developmental events, to confirm existing models of how signals act endogenously, to test the function of specific genes in signal regulation in a developmental context, and to directly visualize gradients or discontinuities in signaling in embryos as a clue to mechanisms of pattern formation. Although the number and range of commercial phosphospecific antibodies continue to grow, only a subset of these are likely to be adequate for the study of endogenous signaling, and therefore custom preparation of desired phosphospecific sera is likely to be necessary, as well. The development of monoclonal phosphospecific antibodies is of particular importance for the generation of consistent, reproducible data over time and among different investigators.

VII. PROTOCOL

Semi-whole mount immunohistochemistry (8/22/00) (M. Whitman, unpublished, but adapted from procedure described in Faure et al., 2000):

1. Fix embryos in MEMFA+EDTA (4% formaldehyde, made fresh) for 2–3 hours at room temperature (RT).[1]
2. Wash 2X in PBST, 1X in 50% MeOH (5 minutes), then 1X in 100% MeOH (5 minutes), then store at –20°C in MeOH. Rehydrate in PBS through five washes: 30% PBS/MeOH, 50% PBS/MeOH, 70% PBS/MeOH, 90% PBS/10% MeOH, then 100% PBS.
3. Soak embryos 15 minutes in 0.2X PBS, 0.3 M Sucrose.[2]
4. Embed embryos in a drop of 2% low melt agarose in PBS/0.3 M sucrose/0.05% Triton and section with Micro Feather scalpel (Electron Microscopy Sciences) on a silanized microscope slide. Recover embryos out of agarose with fine forceps in a petri dish under PBT.
5. Bleach in 1% H_2O_2 (fresh), 5% formamide, 0.5X SSC for 60 minutes on nutator under bright light.[3] All steps at room temperature unless specified otherwise.
6. Wash 3X in PBT (PBS + 2 mg/ml BSA + 0.1% Triton) on rocker. Leave in PBT at final wash for 15 minutes.
7. Incubate with PBT + 10% goat (or donkey for D-anti-M secondary) serum 1 hour on rocker.

8. Incubate 1–4 hours (or 12–16 hours in cold) in primary antibody + PBT + 10% goat (or donkey) serum on nutator (0.5 ml final volume).
9. Wash with five changes PBT/rocking over 2 hours.
10. Incubate 40 minutes to 2 hours/nutation with secondary (HRP)[4] antibody (1:250) in PBT + 10% goat (donkey) serum (0.5 ml final volume).
11. Wash with five changes PBT over 2 hours.
12. Stain with metal enhanced DAB (Pierce 34065) per instructions, 5–20 minutes. For more even staining, use the Vector kit and preincubate embryos with substrate and no H_2O_2. Staining is also more even if you cool the substrate/H_20_2 mixture before adding to embryos, although with the metal enhanced kit, I find that at 4°C the substrate comes out of solution.

An example of semi-whole mount anti-pSmad1 stained embryos can be found in Figure 7.1.

VIII. NOTES

1. We use MEMFA with EDTA added to inhibit divalent cation-dependent phosphatases during fixation. The addition of EDTA has no obvious effect on morphology of fixed embryos. In our hands, formaldehyde and paraformaldehyde yield indistinguishable results; for some applications, freshly prepared paraformaldehyde may yield more consistent fixation. Fixed embryos can be counterstained with Nile Blue while in MeOH (1 μl 1% Nile Blue in 500 μl MeOH). I have not found a way to counterstain in aqueous solution that works well.
2. Steps 2 and 3 are specifically used for the preparation of semi-whole mount embryos (embryos that have been bisected across an axis of choice before immunostaining). For intact whole mounts, these steps are unnecessary. Incubation of fixed embryos in sucrose softens them for cutting with a scalpel.
3. Bleaching considerably increases permeability of whole mounted frog embryos to antibody, as well as reducing background pigmentation. Although it is not necessary for the latter purpose in albino embryos, it is worth doing even in albinos for the improvement in permeability.
4. DAB or metal-enhanced DAB provides good contrast against the yolky background of frog embryos, but suffers from the disadvantages of carcinogenicity and nonlinearity of signal. Signal from fluorescent secondary antibodies can be more effectively quantitated, and the antibodies are not toxic, but in pre-tadpole embryos, the fluorescent background from yolk can substantially reduce sensitivity of detection. In tadpoles state embryos, fluorescent secondary antibodies are preferable to HRP/DAB detection.

REFERENCES

Christen, B. and Slack, J. M. (1999). Spatial response to fibroblast growth factor signalling in *Xenopus* embryos. *Development* 126, 119–25.

Corson, L. B., Yamanaka, Y., Lai, K. M. and Rossant, J. (2003). Spatial and temporal patterns of ERK signaling during mouse embryogenesis. *Development* 130, 4527–37.

Dorfman, R. and Shilo, B. Z. (2001). Biphasic activation of the BMP pathway patterns in the *Drosophila* embryonic dorsal region. *Development* 128, 965–72.

Draetta, G., Piwnica-Worms, H., Morrison, D., Druker, B., Roberts, T. and Beach, D. (1988). Human cdc2 protein kinase is a major cell-cycle regulated tyrosine kinase substrate. *Nature* 336, 738–44.

Faure, S., de Santa Barbara, P., Roberts, D. J. and Whitman, M. (2002). Endogenous patterns of BMP signaling during early chick development. *Dev Biol* 244, 44–65.

Faure, S., Lee, M. A., Keller, T., ten Dijke, P. and Whitman, M. (2000). Endogenous patterns of TGFβ superfamily signaling during early *Xenopus* development. *Development* 127, 2917–31.

Gabay, L., Seger, R. and Shilo, B. Z. (1997). MAP kinase *in situ* activation atlas during *Drosophila* embryogenesis. *Development* 124, 3535–41.

Goumans, M. J., Valdimarsdottir, G., Itoh, S., Lebrin, F., Larsson, J., Mummery, C., Karlsson, S. and ten Dijke, P. (2003). Activin receptor-like kinase (ALK)1 is an antagonistic mediator of lateral TGFbeta/ALK5 signaling. *Mol Cell* 12, 817–28.

Harlow, E. and Lane, D. (1988). *Antibodies: A Laboratory Manual.* Cold Spring Harbor, NY: Cold Spring Harbor Laboratory.

Higashimoto, Y., Saito, S., Tong, X. H., Hong, A., Sakaguchi, K., Appella, E. and Anderson, C. W. (2000). Human p53 is phosphorylated on serines 6 and 9 in response to DNA damage-inducing agents. *J Biol Chem* 275, 23199–203.

Lee, M. A., Heasman, J. and Whitman, M. (2001). Timing of endogenous activin-like signals and regional specification of the *Xenopus* embryo. *Development* 128, 2939–52.

Meno, C., Takeuchi, J., Sakuma, R., Koshiba-Takeuchi, K., Ohishi, S., Saijoh, Y., Miyazaki, J., ten Dijke, P., Ogura, T. and Hamada, H. (2001). Diffusion of nodal signaling activity in the absence of the feedback inhibitor Lefty2. *Dev Cell* 1, 127–38.

Mintzer, K. A., Lee, M. A., Runke, G., Trout, J., Whitman, M. and Mullins, M. C. (2001). *lost-a-fin* encodes a type I BMP receptor, Alk8, acting maternally and zygotically in dorsoventral pattern formation. *Development* 128, 859–69.

Persson, U., Izumi, H., Souchelnytskyi, S., Itoh, S., Grimsby, S., Engström, U., Heldin, C.-H., Funa, K. and ten Dijke, P. (1998). The L45 loop in type I receptors for TGF-β family members is a critical determinant in specifying Smad isoform activation. *FEBS Letters* 434, 83–87.

Raftery, L. A. and Sutherland, D. J. (1999). TGF-beta family signal transduction in *Drosophila* development: from Mad to Smads. *Dev Biol* 210, 251–68.

Sakaguchi, K., Roller, P. P. and Appella, E. (1996). Chemical synthesis and applications of phosphopeptides. *Genet Eng (NY)* 18, 249–78.

Schohl, A. and Fagotto, F. (2002). Beta-catenin, MAPK and Smad signaling during early *Xenopus* development. *Development* 129, 37–52.

Teleman, A. A. and Cohen, S. M. (2000). Dpp gradient formation in the *Drosophila* wing imaginal disc. *Cell* 103, 971–80.

White, M. F., Maron, R. and Kahn, C. R. (1985). Insulin rapidly stimulates tyrosine phosphorylation of a Mr-185,000 protein in intact cells. *Nature* 318, 183–86.

Yamashita, S., Miyagi, C., Carmany-Rampey, A., Shimizu, T., Fujii, R., Schier, A. F. and Hirano, T. (2002). Stat3 controls cell movements during zebrafish gastrulation. *Dev Cell* 2, 363–75.

Section II

Ionic Signals

8 Dynamic Analysis of Calcium Signaling in Animal Development

Diane C. Slusarski

CONTENTS

I. SUMMARY

Calcium (Ca^{2+}) release is a key signal for many cellular processes including neuronal synapse, muscle contraction, cell division, and fertilization. As an essential second messenger molecule, the dynamics of Ca^{2+} release inside a cell are tightly regulated. The goal of this review is to introduce the phenomenon of Ca^{2+} release in the embryo, predominantly vertebrate. Topics to be covered include sources of intracellular Ca^{2+}

and downstream responders, methods used to define endogenous Ca^{2+} release in whole embryos, tools to manipulate Ca^{2+} release, and finally, Ca^{2+} release dynamics placed in the context of signaling pathways known to be of developmental importance.

II. INTRODUCTION TO CALCIUM MODULATION

A. PATHWAYS THAT LEAD TO CALCIUM RELEASE

Although Ca^{2+} is an almost universal intracellular messenger that regulates an enormous range of cellular processes, the cell does not metabolize this ion. Instead, Ca^{2+} levels are predominantly controlled by a gradient of Ca^{2+} concentration either across the plasma membrane or across the membrane of intracellular Ca^{2+} stores. The opening of specialized ion channels and the release from intracellular organelles generates bursts of Ca^{2+} into the cytosol. The location, extent, and duration of the ion channel opening can result in a local or global signaling event.

Figure 8.1 is a schematic of a few components that contribute to Ca^{2+} dynamics in a cell. A majority of intracellular Ca^{2+} release in nonexcitable cells (i.e., nonneuronal) is from the endoplasmic reticulum (ER), whereas release from the mitochondria (not shown) and influx from external stores through the plasma membrane also contribute.

A major signal transduction pathway that leads to Ca^{2+} release from intracellular organelles is the phosphatidylinositol (PI) cycle. The PI cycle is activated in response to many hormones and growth factors that bind to cell surface receptors. Two predominant receptor classes are the G protein-coupled receptor (GPCR) class and the receptor tyrosine kinase (RTK) class. Extracellular ligand stimulation of these receptors

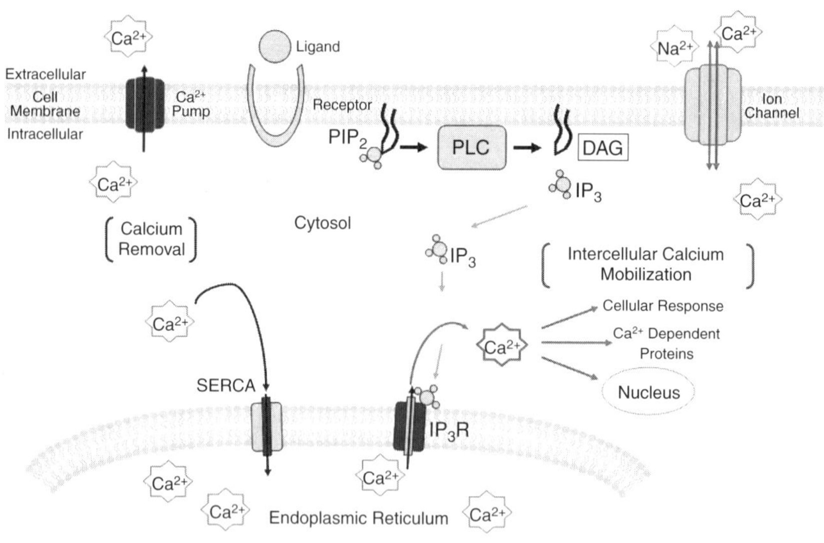

FIGURE 8.1 (See color insert following page 144.) Schematic of calcium mobilization and uptake/removal in a cell.

activates a phosphatidylinositol-specific phospholipase C (PLC) (Figure 8.1). GPCRs generally activate PLC, whereas RTKs generally stimulate PLC. Activated PLC converts membrane bound phosphatidylinositol (4,5) bisphosphate (PIP_2) into soluble inositol 1,4,5-triphosphate (IP_3) and lipophilic diacylglycerol (DAG).[1] IP_3 subsequently binds to receptors located principally on the endoplasmic reticulum and activates a ligand-gated Ca^{2+} channel, the IP_3 Receptor (IP_3R). Activation of this channel then triggers the rapid release of Ca^{2+} into the cytosol of the cell. Another class of intracellular Ca^{2+} release channel is the ryanodine receptor (RyR), which plays a major role in striated muscle function. Expression studies demonstrate broad expression of IP_3R subtypes from early developmental stages on, whereas RyR expression initiates as organogenesis proceeds and is enhanced in skeletal and cardiac muscle.[2–4]

Ca^{2+} released from the ER can bind back to receptors (IP_3R and RyR) and stimulate Ca^{2+}-induced Ca^{2+} release, influencing neighboring receptors and potentially triggering a regenerative Ca^{2+} wave.[5–7] Continued stimulation or depletion of ER stores activates a store operated Ca^{2+} (SOC) entry influx pathway located at the plasma membrane.[8]

B. Pathways to Clear Intracellular Calcium

Ca^{2+} release is tightly coupled with Ca^{2+} uptake. This is a critical function, as prolonged elevation of Ca^{2+} levels can lead to cell death.[9] To facilitate Ca^{2+} clearance from the cytosol, pumps are utilized for reuptake into intracellular organelles including the efficient P-type Ca^{2+}-ATPase, the Sarco-Endoplasmic Reticulum Ca^{2+}-ATPase (SERCA) pump. Additionally, pumps on the plasma membrane operate to actively extrude/transport Ca^{2+} extracellularly (Figure 8.1).[10] Thus, activation-dependent signals coupled with clearance mechanisms can lead to Ca^{2+} release of varying amplitude and duration. The complex release activity of this simple ion is then decoded downstream by Ca^{2+}-dependent pathways.

C. Calcium Release and its Many Effectors

Upon cellular stimulation, the cytosolic Ca^{2+} concentration undergoes either a transient increase or a set of Ca^{2+} oscillations. Much of the newly released cytosolic Ca^{2+} is quickly bound by Ca^{2+} binding proteins.[11] Some act to buffer the amount of Ca^{2+} in the cytosol, whereas others become activated components of signal transduction pathways. In many cases, the impact of Ca^{2+} is mediated by calmodulin (CaM), a member of the EF-hand protein family that represents the most abundant family of eukaryotic Ca^{2+} binding proteins.[12] Calmodulin is activated by cooperative binding of Ca^{2+} ions and subsequently activates protein kinases, phosphatases, ion transporters, and cytoskeletal proteins with one particularly notable class, the Ca^{2+}/calmodulin-dependent kinase (CaMK) family.[13] A few of the many other molecular targets of cytosolic Ca^{2+} release include the protein kinase C (PKC) isozymes, calcineurin, and calcineurin-dependent nuclear factor of activated T cells (NF-AT).[14] In addition, proteins involved with protein synthesis and folding are modulated by Ca^{2+} and ER luminal Ca^{2+}.[15] Specific gene expression in culture has been shown to be sensitive to Ca^{2+} oscillations.[16–18]

In development, growth factors and their associated receptors can modulate Ca^{2+} release, leading to activation of Ca^{2+} binding proteins. Feedback from these activated Ca^{2+} binding proteins adds another layer of complexity to the dynamics of Ca^{2+} release and removal. For example, protein kinase C (PKC) and Ca^{2+}/calmodulin-dependent protein kinase II (CaMK-II) modulate IP_3 Receptor activity, and PKA and casein kinase II modulate SERCA pump activity.[19, 20]

III. DETECTION OF ENDOGENOUS CALCIUM RELEASE IN ANIMAL SYSTEMS

Signaling in cells can result in fast and efficient spatio-temporal changes in Ca^{2+}, and these changes can be monitored in embryos using a variety of tools. I will focus on Ca^{2+}-sensitive dyes coupled with microscope-based applications to highlight Ca^{2+} activity in intact embryos. Then, I will outline some of the endogenous Ca^{2+} dynamics observed in embryos.

The basic technical approach to detect endogenous Ca^{2+} release activity uses a light source to excite a Ca^{2+} indicator in a sample, coupled with a detector that monitors indicator emission. Variants of Fura, aequorin, Fluo, Oregon Green, and Calcium Green are among the most common forms of Ca^{2+} indicators to date, and in general neither the dye nor the fluorescence imaging has any detrimental effect on the developing embryo.[21, 22] Changes in Ca^{2+} can be detected by single wavelength and ratiometric approaches. Single wavelength excitation allows imaging in the visible range and the use of confocal microscopy. Thus, precise spatial resolution can be obtained.

However, one disadvantage of single wavelength measurements is the potential to interpret signal artifacts as Ca^{2+}-dependent changes. Artificial changes in fluorescence can be due to alterations in cell thickness and distribution of the indicator (which can vary in the rapidly dividing and often motile cells of an embryo). Luminescence studies use a Ca^{2+}-stimulated photoprotein, such as aequorin, and have the advantage of rapid measurement and detection of Ca^{2+} gradients. In addition, they can be genetically encoded. However, aequorin is photon limited, is prone to photobleaching (the irreversible destruction of fluorophores), and requires the incorporation of the cofactor coelenterazine.[23] Ratiometric dyes such as Fura-2, a fluorescent derivative of the Ca^{2+} chelator EGTA developed by Tsien and collegues,[24] have reduced sensitivity to signal artifacts, and they enable quantitative measurement of Ca^{2+} concentrations. However, the Fura-2 excitation wavelengths are not common on confocal microscopes, and Fura-2 cannot be genetically encoded.

Another emerging technology is the use of genetically encoded chimera proteins engineered to produce a fluorescent signal, using green fluorescent protein (GFP) or its mutant forms (YFP, CFP), upon activation of a physiological sensor. The most common sensors are forms of calmodulin that alter GFP fluorescence upon Ca^{2+} binding (camgaroos, G-Cam, and pericams) or promote the reversible association of two GFP forms using fluorescence resonance energy transfer (cameleons).[25] Transgenic animals with genetically encoded Ca^{2+} sensors have been reported for worm, fly, fish, and recently mouse (reviewed in Reference 25). Although a majority of these studies focus on the excitable cells of the nervous system, this technology holds great promise for the future. However, in order to be accurate physiological sensors, improvements to binding kinetics are necessary to increase their dynamic range and off-rates.

IV. TOOLS TO MANIPULATE CALCIUM RELEASE

Defining endogenous Ca^{2+} release activity is the first step in evaluating its role in developmental processes. To move from correlative studies to functional studies, Ca^{2+} release mechanisms can be delineated with the use of pharmacological reagents, transgenes, antisense oligos, and genetics. This section outlines pharmacological and molecular tools used to manipulate Ca^{2+} release. The advantage of coupling pharmacological and molecular manipulations with *in vivo* imaging is that the investigator can evaluate if a particular manipulation alters endogenous Ca^{2+} activity and the extent/duration of the alteration. In addition, by use of whole animal imaging, the impacted tissue can also be identified in the context of its natural environment, not as an isolated tissue or cell line.

A. Calcium Release Activation

Ca^{2+} release can be activated to determine if increased activity is sufficient to influence a specific process or test candidate genes to determine if they can modulate Ca^{2+} release. Caged compounds can be used to release exogenous IP_3 or Ca^{2+} in a tissue-specific manner or at a specific developmental time frame. Caged compounds have the advantage of defined regional release, but are often of short duration, especially because some developmental processes may span several minutes to hours. Inducible receptors can be ubiquitously expressed and activated at a specific developmental time, for as long as the agonist is present. The serotonin type 1c receptor (5HT) has been used for this purpose in studies with sea urchin, *Xenopus*, and zebrafish.[26–28]

Because Ca^{2+} image analysis is not a common technique, I will briefly describe the processing relevant to understanding the following discussion. A more detailed explanation of the software and equipment is in Slusarski and Corces.[29] Image pairs collected from a time course are used to generate a sequence of ratio images, which represent a pixel-by-pixel match of both excitation wavelengths calculated by computer software. Panels of Figure 8.2 represent Fura-2 ratio images, where warmer colors indicate high Ca^{2+} levels and cooler colors indicate lower Ca^{2+} levels, and live photos of selected zebrafish developmental stages. Excitation spectra are different between the Ca^{2+}-bound Fura-2 (340 nm) and Ca^{2+}-free (380 nm) forms, and taking the ratio of the fluorescence intensity at these two wavelengths allows an estimate of intracellular free Ca^{2+}. For analysis of Ca^{2+} release dynamics, sequential ratio images are subtracted from each other using a subtractive analog (patterned after References 30 and 31), and Ca^{2+} fluxes (transients) are determined. A new transient (white arrow, Figures 8.2c and 8.2e) is defined as a feature approximately the size of a cell with an increase in fluorescence intensity that was not present in the previous frame. Ca^{2+} fluxes can be represented as the number of new transients

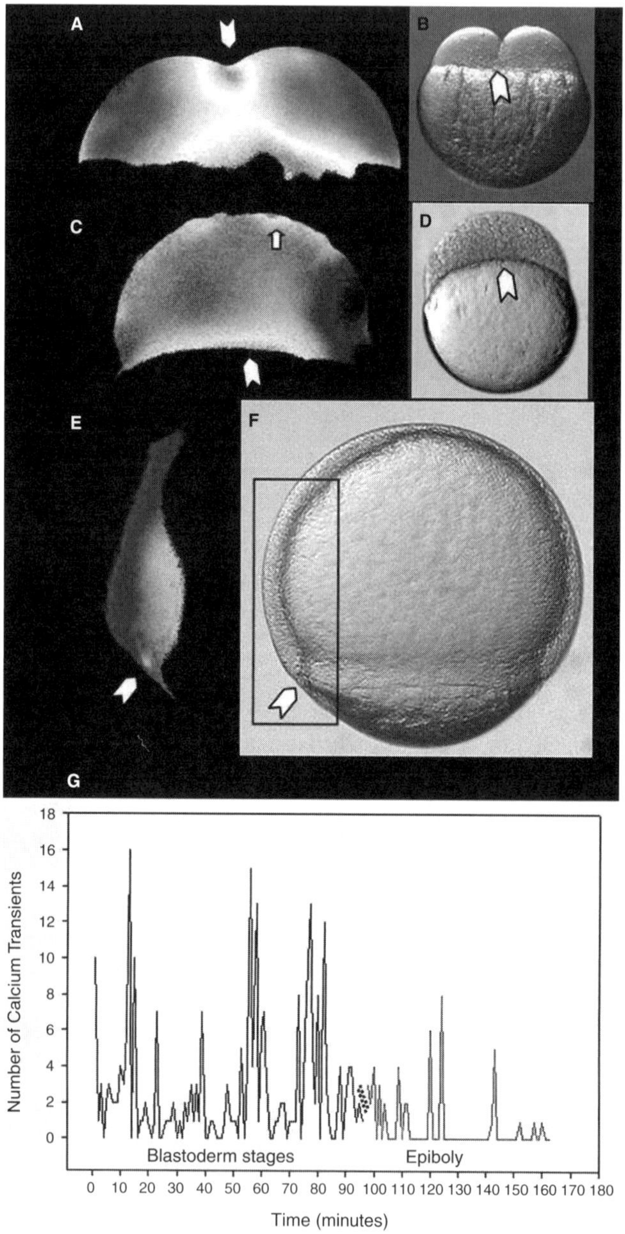

FIGURE 8.2 (See color insert following page 144.) Endogenous Ca^{2+} release events in zebrafish development. A. Ratiometric image at two-cell stage. B. Live image of two-cell stage embryo (arrowheads note cleavage furrow). C. Ratiometric image at the cellular blastoderm stage. D. Live image at cellular blastoderm stage (arrow highlights a Ca^{2+} transient in the EVL, and arrowheads note the YSL). E. Ratiometric image at late epiboly. F. Live image of late epiboly (arrow notes dorsal forerunner cells). G. Graphical output of the number of transients as a function of time during blastula and epiboly stages.

per image as a function of time (Figure 8.2g). Demonstrated in Figure 8.3 is a time course of an untreated embryo (black line) and a serotonin receptor injected embryo (red line). Upon addition of agonist to the chamber, there is a rapid and sustained increase in Ca^{2+} release in the injected embryo (Figure 8.3a, after t = 25 min).

Monitoring Ca^{2+} release *in vivo* in the whole embryo can also determine if a specific growth factor modulates Ca^{2+} release and, if so, in what tissue type. The Wnt gene family of secreted growth factors has noted roles in development and disease and potentially activates several intracellular effectors described in more detail in Chapter 4. In fact, mis-expressions of a subset of Wnts were found to stimulate an increase in intracellular Ca^{2+} signaling, whereas mis-expression of Wnts that stimulate Wnt/-catenin activity did not stimulate Ca^{2+} release in the zebrafish.[28, 32, 33] Analysis of different classes of zebrafish Frizzleds (Wnt interacting receptors) likewise demonstrated differential activity in stimulating Ca^{2+} release (Figure 8.3b compared to Figure 8.3c). Panels 8.3b and 8.3c represent changes in Ca^{2+} as a surface plot. In this fashion, the spatial distribution of the total number of Ca^{2+} transients during the time course is mapped relative to the location in the embryo, and the resultant topographical image is rendered in pseudocolor where a low peak and purple color represents a small number and a high peak and red color represents a high number of Ca^{2+} transients (Figure 8.3: color bar inset). Reagents can be injected unilaterally into an embryo with lineage marker, and activity can be localized to one half of the embryo (the right side of Figure 8.3d), using the other side of the same embryo as a control. One caveat to mis-expression studies is that they determine if a reagent is sufficient to activate Ca^{2+} release but should be coupled with loss-of-function analyses to determine necessity in a specific process.

One can utilize Ca^{2+} activation as a functional assay to determine which intracellular components are necessary to stimulate Ca^{2+} release. Indeed, although Wnts interact with seven transmembrane-spanning Fz receptors, the involvement of G-proteins was first shown by our demonstration that Wnt/Fz-induced Ca^{2+} release was blocked by both pertussis toxin (PTX), which inactivates a subset of G proteins, and over-expression of α-transducin, presumably inhibiting by interaction with $\beta\gamma$ subunits.[32] This work was further substantiated with experiments identifying the intracellular effectors of G-protein signaling and the demonstration that inhibition of these effectors generated cell movement defects.[34] An inducible chimeric receptor, containing the ligand binding and transmembrane spanning domains of the hamster β_2-adrenergic receptor covalently joined to the cytoplasmic domains of the *Rfz-2* receptor (β_2*AR-Rfz2*), demonstrated that the cytoplasmic sequences of *Rfz-2* are sufficient to stimulate Ca^{2+} release in the presence of agonist.[34, 35]

B. Inhibition of Calcium Release

A majority of pathways that lead to Ca^{2+} release can be targeted with pharmacological reagents or dominant-negative constructs. For general Ca^{2+} inhibition, Ca^{2+} chelaters (such as BAPTA and EGTA) can shut down all signaling. SERCA pump inhibitors (such as thapsigargin and cyclopiazonic acid) will ultimately lead to depletion of internal stores, and heparin, although it has other targets, inhibits IP_3R function. There is a wide spectrum of G-protein signaling inhibitors, or reagents

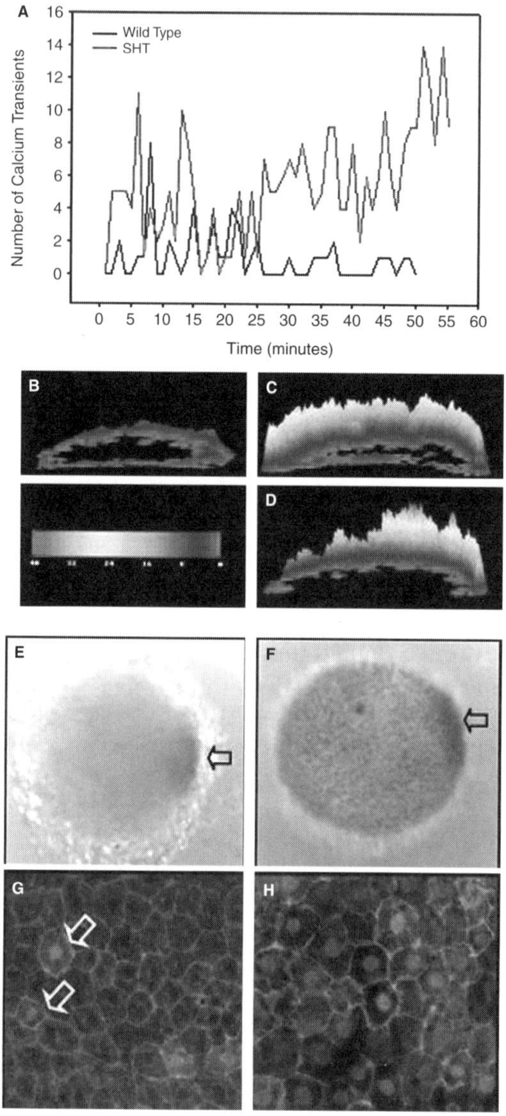

FIGURE 8.3 (See color insert following page 144.) Manipulation of Ca^{2+} activity and downstream responders. A. Graphical output of the number of Ca^{2+} transients as a function of time with control-injected embryo (black) and 5HT-injected and agonist activated embryo (red). B–D are topographical representations of the total number and location of Ca^{2+} transients in Frizzled receptor-injected embryos imaged during the cellular blastoderm stages. Color bar inset notes low numbers (1) as purple, and high numbers of transients as yellow (30) and red (40). Ca^{2+} output of a B. canonical Frizzled-injected embryo, C. a non-canonical Frizzled, and D. unilateral (right-sided) injection of a non-canonical Frizzled. *Bozozok* expression analysis by whole mount of sphere-stage embryos, animal pole view. E. Control-treated and F. Ca^{2+}-inhibited (arrowhead notes the endogenous dorsal domain). β-catenin protein immunolocalization in G. control-treated and H. Ca^{2+}-inhibited embryos.

that target specific intracellular pathways such as the PI cycle. Use of these inhibitors in embryos can identify Ca^{2+}-dependent processes and necessary intracellular components. When working with pharmacological reagents, several aspects need to be considered; one is to determine by a functional assay (such as *in vivo* imaging) the extent and duration of the inhibition. It is particularly important when studying processes that span long time frames, as some reagents are more potent than others, similar to comparing phenotypes between hypomorphic and null genetic alleles. Another caution is to use different reagents that target the same pathway, perhaps at differing steps or by differing mechanisms, and confirm overlapping phenotypes, because some reagents have multiple cellular targets, as we will discuss with lithium. A strong indicator of reagent specificity is rescue of the inhibitor-induced phenotype by supplying an exogenous component of the targeted pathway downstream of the block.

V. CALCIUM DYNAMICS IN DEVELOPMENT AND IDENTIFICATION OF BIOLOGICAL RESPONDERS

It is clear from the body of published studies in several animal models that there exists an amazing variety of Ca^{2+} release dynamics associated with specific developmental events. For example, much has been learned about Ca^{2+} release dynamics during egg activation, fertilization, and neuronal signaling, and these will not be discussed in detail here. Rather, I will outline Ca^{2+} signaling dynamics in the context of whole embryos in the following embryonic stages: early cleavage, cellular blastoderm, epiboly/gastrulation, and organogenesis. This discussion will relay some of the descriptive analyses, but focus more on downstream targets of Ca^{2+} signaling that potentially influence or respond to growth factor signaling.

A. Cleavage Stage

After fertilization, there are dramatic Ca^{2+} increases associated with the forming cleavage furrow during the first few cell cycles.[29, 30, 36, 37] Smaller prepatterning Ca^{2+} signals can be observed flashing across the region preceding furrow formation, followed by an intense sustained increase in Ca^{2+} as furrow formation and cell division progresses (Figure 8.2a, arrow). Early cleavage is dependent upon PI cycle function, as application of heparin inhibits cytokinesis in *Xenopus* and zebrafish embryos.[30, 38] Consistent with PI cycle requirement is the observation of ER accumulation localized near the deepening furrow in zebrafish.[39] Three distinct steps in early cleavage seem to involve (a) prepatterning/positioning of the furrow, (b) lateral extension of the furrow, followed by (c) deepening of the furrow, and finally cleavage. Several mutations in zebrafish have been identified with early defects influencing egg activation, nuclear division, or cell division.[40–44] The question of whether the reported endogenous Ca^{2+} fluxes correlate with specific steps of egg maturation and patterning, mitosis, or furrow formation may be answered with in-depth analysis of these genetic mutations.

B. Cellular Blastoderm

1. PI Cycle Activity and Dorsal-Ventral Patterning

After the 32-cell stage in zebrafish, the furrow-associated Ca^{2+} increases become less prominent, and a group of cells initiate rapid aperiodic Ca^{2+} release that persists until the midblastula transition stage.[28, 32, 45] This Ca^{2+} release activity is localized to the enveloping layer and yolk syncytial layer of the embryo (Figure 8.2c, arrows). Manipulation of Ca^{2+} release, via manipulation of phosphatidylinositol cycle, during this stage of development has suggested an involvement in the establishment of the body plan of the vertebrate embryo. Classical studies linking PI cycle activity to body plan determination predominantly used lithium as a teratogen. Lithium treatment of cleavage stage *Xenopus* embryos induced a secondary axis and rescued UV-ventralized embryos.[46–48]

In addition, zebrafish show a localized sensitivity to lithium injection.[49] Lithium is an inhibitor of inositol turnover, preventing the generation of free inositol and the subsequent recycling back into the plasma membrane.[50] Lithium-induced expansion of dorsal structures in the embryo can be rescued by supplying an intermediate of the PI cycle (*myo*-inositol).[51] These data were taken as evidence that the PI cycle is required for defining the dorsal-ventral (D-V) axis; however, lithium also blocks the function of many other cellular factors including glycogen synthase kinase-3 (Gsk-3).[52, 53] The effect of lithium on Gsk-3 function is clearly biologically relevant. Interestingly, inhibition of Gsk-3 function by another means (via a dominant negative construct, dnGsk) can generate an ectopic dorsal axis, whereas introduction of *myo*-inositol along with dnGsk rescues the dorsalizing effects.[54] Thus, given that lithium treatment targets both the PI cycle and Gsk, investigators should be careful not to suggest it as either a specific Wnt/β-catenin inhibitor or a specific PI-cycle inhibitor.

To clarify the role of the PI cycle in D-V patterning, additional studies with inhibitors other than lithium were necessary. Several pieces of evidence in zebrafish and *Xenopus* further confirmed a requirement for PI cycle activity in proper D-V patterning. *Xenopus* embryos injected with antibodies that disrupt IP_3 Receptor function displayed expanded dorsal structures with the loss of ventral structures.[55] This sensitivity to IP_3 activity during axis formation is consistent with biochemical data noting a spontaneous increase in IP_3 levels in the *Xenopus* embryo at this stage.[51, 56] In zebrafish, the use of IP_3R blocking antibodies as well as several other PI-cycle inhibitors dorsalized embryos.[57]

Inhibition of G-protein signaling suppresses Ca^{2+} release in zebrafish[32, 34] and dorsalizes *Xenopus* embryos.[58] Given sensitivity of dorsal-ventral patterning to modulation of PI cycle and G-protein signaling activity, it could be predicted that some of the genetic mutations with dorsalized or ventralized phenotypes would also display altered Ca^{2+} dynamics.[59] In fact, in zebrafish, the ventralized maternal effect mutation *hecate* demonstrates increased Ca^{2+} release dynamics. In addition, the *hecate* phenotype can be reversed upon inhibition of Ca^{2+} release and G-protein signaling.[59]

2. Link between Calcium and Wnt Signaling

PI cycle inhibition leads to reduced Ca^{2+} release. Evidence for the biological target of Ca^{2+} release comes from similarities between PI cycle manipulation and Wnt manipulation. Although this review focuses on Ca^{2+} in development, a brief description of Wnt signaling is necessary to place the following phenotypes into perspective. In vertebrate embryos, over-expression of a subset of Wnts induces hyperdorsalization and ectopic axes.[60–63] These have been termed the canonical Wnt class, consisting of the *Drosophila wingless* (*wg*) and vertebrate *Wnt-1*, *-3a*, and *-8*.[64] This class is also referred to as Wnt/β-catenin, as regulation of β-catenin protein abundance and localization is central to canonical Wnt activity. The common element of so-called non-canonical Wnt signaling is that this class (including *Wnt-5A*, *-4*, and *-11*) appears to be β-catenin-independent.[65] Non-canonical Wnt activity can be viewed as a complex network with cellular outputs identified by calcium modulation and polarized cell movement (Wnt/Ca^{2+} and planar cell polarity [PCP][66]).

Non-canonical Wnts have dual functions: They are antagonistic to canonical Wnt signaling and modulate cell movement/polarity. In *Xenopus* and zebrafish, co-injection of RNA encoding *Xwnt-8* with *Xwnt-5A* inhibits the dorsalizing effects of *Xwnt-8*.[28, 60] Stimulating Ca^{2+} release, via activated serotonin receptor, also antagonizes *Xwnt-8* induced expansion of dorsal domains,[28] suggesting that *Wnt-5* antagonism of Wnt/β-catenin is mediated by Ca^{2+} release. In fact, in zebrafish embryos, pharmacological or genetic reduction of Ca^{2+} release generates ectopic accumulation of nuclear β-catenin and activation of β-catenin transcriptional targets.[33, 57] At sphere stage, zebrafish embryos express *bozozok* (*boz*, also called *neukoid/dharma*) in the endogenous dorsal domain (Figure 8.3e, arrow). Inhibition of Ca^{2+} release results in ectopic activation of boz (Figure 8.3f, arrow notes endogenous dorsal domain). Ca^{2+} release inhibition at later developmental stages also results in ectopic nuclear β-catenin accumulation; compare control (Figure 8.3g) to Ca^{2+} inhibited (Figure 8.3h).

Use of molecular, genetic, and pharmacological tools will aid in the identification of the Ca^{2+}-sensitive mediators of the observed Wnt/-catenin antagonism. Obvious candidates would include Ca^{2+}-binding proteins that can decipher cytosolic Ca^{2+} concentration changes. The Ca^{2+}-sensitive protein CaMKII promotes ventral fates[67] and is antagonistic to Wnt/β-catenin by activating a mitogen-activated protein kinase cascade.[68] Introduction of constitutively active CaMKII into zebrafish is sufficient to ventralize embryos and rescue the Wnt-5 (*pipetail*) mutant phenotype.[33] Calcineurin, a Ca^{2+}/calmodulin-dependent protein phosphatase, modulates nuclear factor of activated T cells, leading to NF-AT accumulation in the nucleus. Over-expression of constitutively active NF-AT in *Xenopus* perturbs cell movements, ventralizes embryos, and antagonizes Wnt/β-catenin activity.[14]

C. Epiboly/Gastrulation

During epiboly/gastrula stages, cells are both advancing to spread over the yolk and initiating polarized cell movements toward the future dorsal side, a critical step in the patterning of the embryo. Pan-embryonic intercellular Ca^{2+} waves in the marginal zone of zebrafish embryos as well as waves of Ca^{2+} release in *Xenopus* explants have been described during these stages.[69, 70] There is the exciting possibility that these Ca^{2+} waves coordinate convergent extension (C-E), the lengthening and narrowing of groups of cells as well as their directed migration to the future dorsal side.[70, 71] Genetic data supports this idea, in that zebrafish homozygous zygotic Wnt-

5 genetic mutants (*pipetail*, *ppt*) display C-E defects[72] and reduced Ca^{2+} release frequency in the embryo at this stage of development.[33] Interestingly, the embryonic region displaying the greatest reduction of Ca^{2+} activity in *ppt* homozygotes encompasses the yolk syncytial layer (YSL), which plays important roles in epiboly and patterning in zebrafish (reviewed in Reference 73).

Convergent extension is often considered the vertebrate equivalent to planar cell polarity in the *Drosophila* cuticle.[74] Mis-expression of non-canonical Wnt genes is sufficient to activate Ca^{2+} release[28, 33] and alter morphogenetic movements.[28, 75] The genetic mutants of these same Wnts demonstrate cell movement defects in zebrafish.[76, 77] PCP-specific genes including *Strabismus* (*stbm*), a putative transmembrane protein (identified as the zebrafish *trilobite* mutant),[78, 79] and *Prickle* (*pk*), an intracellular protein, influence morphogenetic movements and stimulate Ca^{2+} release in zebrafish[80] (DCS, unpublished results). Likewise, the PCP-specific form of *Dishevelled* (*Dsh*) activates the Wnt/Ca^{2+} cascade in *Xenopus* and zebrafish,[81] raising the intriguing possibility that Wnt/Ca^{2+} and PCP either substantially overlap or are part of the same signaling network. Recently, a requirement for G-protein signaling in gastrulation movements was demonstrated by antisense morpholino oligonucleotide knockdown of G_{12} and G_{13} and use of dominant negative constructs.[82] Investigation of the Ca^{2+} release dynamics in zebrafish epiboly and convergence extension mutants would further correlate intracellular Ca^{2+} dynamics with coordinated or polarized cell movements.

D. Left-Right Asymmetry

Although vertebrates appear bilaterally symmetric from the outside, our heart, lungs, liver, and gut are carefully positioned across the left-right (L-R) axis, and the development of this asymmetric pattern is highly conserved across species. Deviations from the normal L-R asymmetric arrangement of internal organs can cause lethality and human diseases, including complex congenital heart defects.[83] Molecular indications of L-R axis determination are found in a group of asymmetrically expressed genes.[84] Of note is the early left-sided expression of Transforming Growth Factor-β (TGF-β) related signal tranducers (*nodal*, *lefty*, and *southpaw*).[85] The following will summarize data to suggest that Ca^{2+} modulation influences the left-sided expression of these genes, thus resulting in asymmetric organ placement.

Several vertebrate models implicate a role for Ca^{2+} signaling in the establishment of L-R asymmetry.[86] In mice, the symmetry-breaking event is thought to arise from a directional flow generated by the rotation of monocilia in the embryonic node.[87, 88] It has been proposed that the leftward nodal flow stimulates mechanosensory cilia to trigger elevation in intracellular Ca^{2+} levels at the left edge of the mouse node.[89] Elevated Ca^{2+} is thought to act as a second messenger, via an unknown mechanism, to ultimately induce left-sided gene expression. The asymmetry in node Ca^{2+} levels is lost in mouse embryos that are homozygous for mutations in the *Pkd-2* gene (polycystic kidney disease), a Ca^{2+}-permeable ion channel, and these mutants exhibit laterality defects.[89]

In chick embryos, it is not known if there is a similar asymmetry of intracellular Ca^{2+} as observed in the mouse node, but extracellular Ca^{2+} levels appear to be higher transiently on the left side. This asymmetry was abolished after treatment with ompremazole, an inhibitor of H^+/K^+ ATPase that also caused L-R defects (reversal of heart looping). This led the authors to propose that differential H^+/K^+ ATPase activity sets up a spatial gradient of extracellular Ca^{2+}, which is subsequently transduced to activate asymmetric gene expression on the left side.[90]

In the zebrafish, a ciliated structure called the Kupffer's Vesicle (KV) is proposed to serve a similar function as the mouse node, relative to L-R asymmetry, and the loss of KV cilia abolishes L-R patterning.[91] Cilia in the KV beat in the same direction, possibly to establish an asymmetric Ca^{2+} flux, as in chick and mouse.[92] In fact, an intracellular Ca^{2+} flux with a left-sided bias near the zebrafish KV has recently been detected and is proposed to be required for normal L-R patterning.[93] Thus, evidence of a role for Ca^{2+} in L-R patterning is very tantalizing, but many questions and issues remain to be addressed, such as the Ca^{2+} sources, the Ca^{2+}-dependent responders, and whether there are multiple layers/stages of Ca^{2+}-dependent events involving the initiation of asymmetric expression and the subsequent maintenance of laterality signals.

E. Somite and Neural Patterning

Other aspects of organogenesis impacted by Ca^{2+} release involve neural induction, which will give rise to the peripheral and central nervous systems, and somite formation, which will give rise to muscle and cartilage elements. Periodic Ca^{2+} fluxes are observed in anterior dorsal ectoderm during stages of presumptive neural patterning in *Xenopus* and zebrafish (reviewed in Reference 94). In *Xenopus* embryos, Ca^{2+} release from L-type Ca^{2+} channels is required to induce neural-specific genes.[95] Inhibition of Ca^{2+} release suppresses neural induction; however, these manipulations also alter gastrulation movements.[70, 95, 96] In these studies, Ca^{2+} release inhibition may target multiple Ca^{2+}-sensitive events occurring in parallel, or perhaps gastrulation and neural patterning are too tightly coupled to separate with pharmacological reagents. Neural induction involves interaction between Bone Morphogenetic Proteins (BMPs) and their antagonists such as chordin and noggin.[97] In *Pleurodeles* explants, noggin application triggers an increase in Ca^{2+} release.[98] Whether the Ca^{2+} flux is due to noggin-mediated suppression of BMP signaling or activation of noggin has yet to be determined, as well as whether these events occur in the context of the whole animal.

Somites are derived from paraxial mesoderm, and Ca^{2+} release activity during the segmentation period has been reported.[37, 99] In mature somites, Ca^{2+} release activity has been reported in isolated *Xenopus* myocytes[100] and in whole zebrafish embryos.[101] Ca^{2+} release inhibition alters myotome patterning.[100] In addition, elimination of calcineurin activity in *Xenopus* embryos abolished somite formation as well as causing additional later organogenesis defects in the heart, kidney, and gut looping.[102] Recent work has linked bilateral somite formation to L-R asymmetry signals.[103–105] It has yet to be determined whether this coupling of L-R and somite formation processes are directly linked to Ca^{2+} fluxes.

VI. CONCLUSIONS AND FUTURE DIRECTIONS

In vivo imaging studies are a critical step in the comprehensive analysis of Ca^{2+} signaling in development. Coupling *in vivo* imaging with molecular, genetic, or pharmacological tools will determine the mechanism by which Ca^{2+} signaling is modulated and interpreted in the embryo. Future studies will reconstruct the spatial and temporal dynamics of Ca^{2+} release and incorporate this activity into known signaling pathways, thus providing the capacity to discern the true nature of the cellular basis of pattern formation.

ACKNOWLEDGMENTS

I am grateful to Ms. T. Westfall for her contributions to the drawings and graphs shown in the manuscript, to Ms. H. Griesbach and Mr. I. Schneider for contributing images, and to the laboratories of the University of Iowa for their collaboration and continuous support. The Slusarski Laboratory is supported by the March of Dimes, American Cancer Society, and the National Institutes of Health.

REFERENCES

1. Berridge, M. J., Inositol trisphosphate and calcium signalling, *Nature* 361, 315–25, 1993.
2. Kume, S., Muto, A., Aruga, J., Nakagawa, T., Michikawa, T., Furuichi, T., Nakade, S., Okano, H., and Mikoshiba, K., The *Xenopus* IP3 receptor: structure, function, and localization in oocytes and eggs, *Cell* 73 (3), 555–70, 1993.
3. Kume, S., Muto, A., Okano, H., and Mikoshiba, K., Developmental expression of the inositol 1,4,5-trisphosphate receptor and localization of inositol 1,4,5-trisphosphate during early embryogenesis in *Xenopus* laevis, *Mech Dev* 66 (1–2), 157–68, 1997.
4. Rosemblit, N., Moschella, M. C., Ondriasa, E., Gutstein, D. E., Ondrias, K., and Marks, A. R., Intracellular calcium release channel expression during embryogenesis, *Dev Biol* 206 (2), 163–77, 1999.
5. Berridge, M. J., The AM and FM of calcium signalling, *Nature* 386 (6627), 759–60, 1997.
6. Roderick, H. L., Berridge, M. J., and Bootman, M. D., Calcium-induced calcium release, *Curr Biol* 13 (11), R425, 2003.
7. Berridge, M. J., Bootman, M. D., and Roderick, H. L., Calcium signalling: dynamics, homeostasis and remodelling, *Nat Rev Mol Cell Biol* 4 (7), 517–29, 2003.
8. Parekh, A. B. and Putney, J. W., Jr., Store-operated calcium channels, *Physiol Rev* 85 (2), 757–810, 2005.
9. Berridge, M. J., Bootman, M. D., and Lipp, P., Calcium—a life and death signal, *Nature* 395, 645–8, 1998.
10. Strehler, E. E. and Treiman, M., Calcium pumps of plasma membrane and cell interior, *Curr Mol Med* 4 (3), 323–35, 2004.
11. Falcke, M., Building a wave—models of the puff-to-wave transition, in *Understanding Calcium Dynamics Experiments and Theory*, Falcke, M. and Malchow, D. (Eds.), Springer-Verlag, Berlin, 2003, pp. 253–90.
12. Haiech, J., Moulhaye, S. B., and Kilhoffer, M. C., The EF-Handome: combining comparative genomic study using FamDBtool, a new bioinformatics tool, and the network of expertise of the European Calcium Society, *Biochim Biophys Acta* 1742 (1–3), 179–83, 2004.

13. Hoeflich, K. P. and Ikura, M., Calmodulin in action: diversity in target recognition and activation mechanisms, *Cell* 108 (6), 739–42, 2002.
14. Saneyoshi, T., Kume, S., Amasaki, Y., and Mikoshiba, K., The Wnt/calcium pathway activates NF-AT and promotes ventral cell fate in *Xenopus* embryos, *Nature* 417 (6886), 295–9, 2002.
15. Roderick, H. L., Berridge, M. J., and Bootman, M. D., The endoplasmic reticulum: a central player in cell signalling and protein synthesis, in *Understanding Calcium Dynamics Experiments and Theory*, Falcke, M. and Malchow, D. (Eds.), Springer-Verlag, Berlin, 2003, pp. 17–36.
16. Li, W., Llopis, J., Whitney, M., Zlokarnik, G., and Tsien, R. Y., Cell-permeant caged InsP3 ester shows that Ca2+ spike frequency can optimize gene expression, *Nature* 392 (6679), 936–41, 1998.
17. Hu, S. C., Chrivia, J., and Ghosh, A., Regulation of CBP-mediated transcription by neuronal calcium signaling, *Neuron* 22 (4), 799–808, 1999.
18. Dolmetsch, R. E., Xu, K., and Lewis, R. S., Calcium oscillations increase the efficiency and specificity of gene expression, *Nature* 392 (6679), 933–6, 1998.
19. Colyer, J., Phosphorylation states of phospholamban, *Ann N Y Acad Sci* 853, 79–91, 1998.
20. Patterson, R. L., Boehning, D., and Snyder, S. H., Inositol 1,4,5-trisphosphate receptors as signal integrators, *Annu Rev Biochem* 73, 437–65, 2004.
21. Tsien, R. Y., Pozzan, T., and Rink, T. J., Calcium homeostasis in intact lymphocytes: cytoplasmic free calcium monitored with a new, intracellularly trapped fluorescent indicator, *J Cell Biol* 94 (2), 325–34, 1982.
22. Thomas, D., Tovey, S. C., Collins, T. J., Bootman, M. D., Berridge, M. J., and Lipp, P., A comparison of fluorescent Ca2+ indicator properties and their use in measuring elementary and global Ca2+ signals, *Cell Calcium* 28 (4), 213–23, 2000.
23. Shimomura, O., The discovery of aequorin and green fluorescent protein, *J Microsc* 217 (Pt 1), 1–15, 2005.
24. Grynkiewicz, G., Poenie, M., and Tsien, R. Y., A new generation of Ca2+ indicators with greatly improved fluorescence properties, *J Biol Chem* 260 (6), 3440–50, 1985.
25. Griesbeck, O., Fluorescent proteins as sensors for cellular functions, *Curr Opin Neurobiol* 14 (5), 636–41, 2004.
26. Shilling, F., Mandel, G., and Jaffe, L. A., Activation by serotonin of starfish eggs expressing the rat serotonin 1c receptor, *Cell Regul* 1 (6), 465–9, 1990.
27. Ault, K. T., Durmowicz, G., Harger, P. L., Galione, A., and Busa, W. B., Modulation of *Xenopus* embryo mesoderm-specific gene expression and dorso-anterior patterning by receptors that activate the phosphatidylinositol cycle signal transduction pathway, *Development* 122, 2033–41, 1996.
28. Slusarski, D. C., Yang-Snyder, J., Busa, W. B., and Moon, R. T., Modulation of embryonic intracellular Ca2+ signaling by Wnt-5A, *Dev Biol* 185, 114–20, 1997.
29. Slusarski, D. C. and Corces, V. G., Calcium imaging in cell-cell signaling, in *Developmental Biology Protocols*, Tuan, R. S. and Lo, C. W. (Eds.), Humana Press, Totowa, NJ, 2000, pp. 253–61.
30. Chang, D. C. and Meng, C., A localized elevation of cytosolic free calcium is associated with cytokinesis in the zebrafish embryo, *J Cell Biol* 131, 1539–45, 1995.
31. Lechleiter, J., Girard, S., Peralta, E., and Clapham, D., Spiral calcium wave propagation and annihilation in *Xenopus laevis* oocytes, *Science* 252, 123–6, 1991.
32. Slusarski, D. C., Corces, V. G., and Moon, R. T., Interaction of Wnt and a Frizzled homologue triggers G-protein-linked phosphatidylinositol signalling, *Nature* 390, 410–3, 1997.

33. Westfall, T. A., Brimeyer, R., Twedt, J., Gladon, J., Olberding, A., Furutani-Seiki, M., and Slusarski, D. C., Wnt-5/pipetail functions in vertebrate axis formation as a negative regulator of Wnt/β-catenin activity, *J Cell Biol* 162 (5), 889–98, 2003.
34. Ahumada, A., Slusarski, D. C., Liu, X. X., Moon, R. T., Malbon, C. C., and Wang, H. Y., Signaling of rat Frizzled-2 through phosphodiesterase and cyclic GMP, *Science* 298 (5600), 2006–10, 2002.
35. Liu, X. X., Liu, T., Slusarski, D. C., Yang-Snyder, J., Malbon, C. C., Moon, R. T., and Wang, H. Y., Activation of a Frizzled-2/beta-adrenergic receptor chimera promotes Wnt signaling and differentiation of mouse F9 teratocarcinoma cells via G alpha o and G alpha t, *PNAS USA* 96 (25), 14383–8, 1999.
36. Webb, S. E., Lee, K. W., Karplus, E., and Miller, A. L., Localized calcium transients accompany furrow positioning, propagation, and deepening during the early cleavage period of zebrafish embryos, *Dev Biol* 192 (1), 78–92, 1997.
37. Créton, R., Speksnijder, J. E., and Jaffe, L. F., Patterns of free calcium in zebrafish embryos, *J Cell Sci* 111, 1613–22, 1998.
38. Muto, A., Kume, S., Inoue, T., Okano, H., and Mikoshiba, K., Calcium waves along the cleavage furrows in cleavage-stage *Xenopus* embryos and its inhibition by heparin, *J Cell Biol* 135 (1), 181–90, 1996.
39. Lee, K. W., Webb, S. E., and Miller, A. L., Ca2+ released via IP3 receptors is required for furrow deepening during cytokinesis in zebrafish embryos, *Int J Dev Biol* 47 (6), 411–21, 2003.
40. Dosch, R., Wagner, D. S., Mintzer, K. A., Runke, G., Wiemelt, A. P., and Mullins, M. C., Maternal control of vertebrate development before the midblastula transition: mutants from the zebrafish I, *Dev Cell* 6 (6), 771–80, 2004.
41. Pelegri, F., Dekens, M. P., Schulte-Merker, S., Maischein, H. M., Weiler, C., and Nusslein-Volhard, C., Identification of recessive maternal-effect mutations in the zebrafish using a gynogenesis-based method, *Dev Dyn* 231 (2), 324–35, 2004.
42. Kishimoto, Y., Koshida, S., Furutani-Seiki, M., and Kondoh, H., Zebrafish maternal-effect mutations causing cytokinesis defect without affecting mitosis or equatorial vasa deposition, *Mech Dev* 121 (1), 79–89, 2004.
43. Dekens, M. P., Pelegri, F. J., Maischein, H. M., and Nusslein-Volhard, C., The maternal-effect gene futile cycle is essential for pronuclear congression and mitotic spindle assembly in the zebrafish zygote, *Development* 130 (17), 3907–16, 2003.
44. Pelegri, F., Knaut, H., Maischein, H. M., Schulte-Merker, S., and Nusslein-Volhard, C., A mutation in the zebrafish maternal-effect gene nebel affects furrow formation and vasa RNA localization, *Curr Biol* 9 (24), 1431–40, 1999.
45. Reinhard, E., Yokoe, H., Niebling, K. R., Allbritton, N. L., Kuhn, M. A., and Meyer, T., Localized calcium signals in early zebrafish development, *Dev Biol* 170 (1), 50–61, 1995.
46. Kao, K., Masui, Y., and Elinson, R., Lithium-induced respecification of pattern in *Xenopus laevis* embryos, *Nature* 322, 371–3, 1986.
47. Kao, K. R. and Elinson, R. P., Dorsalization of mesoderm induction by lithium, *Dev Biol* 132 (1), 81–90, 1989.
48. Kao, K. R. and Elinson, R. P., The legacy of lithium effects on development, *Biology of the Cell* 90, 585–90, 1998.
49. Aanstad, P. and Whitaker, M., Predictability of dorso-ventral asymmetry in the cleavage stage zebrafish embryo: an analysis using lithium sensitivity as a dorso-ventral marker, *Mech Dev* 88 (1), 33–41, 1999.
50. Berridge, M., Downes, C., and Hanley, M., Neural and developmental actions of lithium: a unifying hypothesis, *Cell* 59, 411–9, 1989.

51. Busa, W. and Gimlich, R., Lithium-induced teratogenesis in frog embryos prevented by a polyphosphoinositide cycle intermediate or a diacylglycerol analog, *Dev. Biol.* 132, 315–24, 1989.
52. Klein, P. S. and Melton, D. A., A molecular mechanism for the effect of lithium on development, *PNAS USA* 93, 8455–9, 1996.
53. Stambolic, V., Ruel, L., and Woodgett, J. R., Lithium inhibits glycogen synthase kinase-3 activity and mimics wingless signalling in intact cells [published erratum appears in *Curr Biol* 1997 Mar 1;7(3):196], *Curr Biol* 6 (12), 1664–8, 1996.
54. Hedgepeth, C. M., Conrad, L. J., Zhang, J., Huang, H.-C., Lee, V. M., and Klein, P. S., Activation of the Wnt signaling pathway: A molecular mechanism for lithium action, *Dev. Biol* 185, 82–91, 1997.
55. Kume, S., Muto, A., Inoue, T., Suga, K., Okano, H., and Mikoshiba, K., Role of inositol 1,4,5-trisphosphate receptor in ventral signaling in *Xenopus* embryos, *Science* 278 (5345), 1940–3, 1997.
56. Maslanski, J., Leshko, L., and Busa, W., Lithium-sensitive production of inositol phosphates during amphibian embryonic mesoderm induction, *Science* 256, 243–5, 1992.
57. Westfall, T. A., Hjertos, B., and Slusarski, D. C., Requirement for intracellular calcium modulation in zebrafish dorsal-ventral patterning, *Dev Biol* 259 (2), 380–91, 2003.
58. Kume, S., Inoue, T., and Mikoshiba, K., Galphas family G proteins activate IP(3)-Ca(2+) signaling via gbetagamma and transduce ventralizing signals in *Xenopus*, *Dev Biol* 226 (1), 88–103, 2000.
59. Lyman Gingerich, J., Westfall, T. A., Slusarski, D. C., and Pelegri, F., Hecate, a zebrafish maternal effect gene, affects dorsal organizer induction and intracellular calcium transient frequency, *Dev Biol* 286 (2), 427, 2005.
60. Moon, R. T., Christian, J. L., Campbell, R. M., McGrew, L. L., DeMarais, A. A., Torres, M., Lai, C. J., Olson, D. J., and Kelly, G. M., Dissecting Wnt signaling pathways and Wnt-sensitive developmental processes through transient misexpression analyses in embryos of *Xenopus laevis*, *Dev. Suppl.*, 85–94, 1993.
61. Du, S., Purcell, S., Christian, J., McGrew, L., and Moon, R., Identification of distinct classes and functional domains of Wnts through expression of wild-type and chimeric proteins in *Xenopus* embryos, *Mol. Cell. Biol.* 15, 2625–34, 1995.
62. Kelly, G. M., Greenstein, P., Erezyilmaz, D. F., and Moon, R. T., Zebrafish *wnt-8* and *wnt-8b* share a common activity but are involved in distinct developmental pathways, *Development* 121, 1787–99, 1995.
63. Moon, R. T. and Kimelman, D., From cortical rotation to organizer gene expression: toward a molecular explanation of axis specification in *Xenopus*, *Bioessays* 20 (7), 536–45, 1998.
64. Dale, T., Signal transduction by the Wnt family of ligands, *Biochem. J.* 329, 209–23, 1998.
65. Kuhl, M., Sheldahl, L. C., Park, M., Miller, J. R., and Moon, R. T., The Wnt/Ca2+ pathway—a new vertebrate Wnt signaling pathway takes shape, *Trends in Genet* 16 (7), 279–83, 2000.
66. Mlodzik, M., Planar cell polarization: do the same mechanisms regulate Drosophila tissue polarity and vertebrate gastrulation? *Trends in Genet* 18 (11), 564–71, 2002.
67. Kuhl, M., Sheldahl, L. C., Malbon, C. C., and Moon, R. T., Ca^{2+}/calmodulin-dependent protein kinase II is stimulated by Wnt and Frizzled homologs and promotes ventral cell fates in *Xenopus*, *J Biol Chem* 275 (17), 12701–11, 2000.
68. Ishitani, T., Kishida, S., Hyodo-Miura, J., Ueno, N., Yasuda, J., Waterman, M., Shibuya, H., Moon, R. T., Ninomiya-Tsuji, J., and Matsumoto, K., The TAK1-NLK mitogen-activated protein kinase cascade functions in the Wnt-5A/Ca2+ pathway to antagonize Wnt/β-catenin signaling, *Molecular and Cellular Biology* 23 (1), 131–9, 2003.

69. Gilland, E., Miller, A. L., Karplus, E., Baker, R., and Webb, S. E., Imaging of multicellular large-scale rhythmic calcium waves during zebrafish gastrulation, *PNAS USA* 96 (1), 157–61, 1999.
70. Wallingford, J. B., Ewald, A. J., Harland, R. M., and Fraser, S. E., Calcium signaling during convergent extension in *Xenopus*, *Curr Biol* 11 (9), 652–61, 2001.
71. Wallingford, J. B., Fraser, S. E., and Harland, R. M., Convergent extension. The molecular control of polarized cell movement during embryonic development, *Dev Cell* 2 (6), 695–706, 2002.
72. Hammerschmidt, M., Pelegri, F., Mullins, M. C., Kane, D. A., Brand, M., van Eeden, F. J., Furutani-Seiki, M., Granato, M., Haffter, P., Heisenberg, C. P., Jiang, Y. J., Kelsh, R. N., Odenthal, J., Warga, R. M., and Nusslein-Volhard, C., Mutations affecting morphogenesis during gastrulation and tail formation in the zebrafish, Danio rerio, *Development* 123, 143–51, 1996.
73. Sakaguchi, T., Mizuno, T., and Takeda, H., Formation and patterning roles of the yolk syncytial layer, *Results Probl Cell Differ* 40, 1–14, 2002.
74. Solnica-Krezel, L., Conserved patterns of cell movements during vertebrate gastrulation, *Curr Biol* 15 (6), R213–28, 2005.
75. Ungar, A. R. and Moon, R. T., *Wnt4* affects morphogenesis when misexpressed in the zebrafish embryo, *Mech Dev 52*, 153–64, 1995.
76. Heisenberg, C. P., Tada, M., Rauch, G. J., Saude, L., Concha, M. L., Geisler, R., Stemple, D. L., Smith, J. C., and Wilson, S. W., Silberblick/Wnt11 mediates convergent extension movements during zebrafish gastrulation, *Nature* 405 (6782), 76–81, 2000.
77. Kilian, B., Mansukoski, H., Barbosa, F. C., Ulrich, F., Tada, M., and Heisenberg, C. P., The role of Ppt/Wnt5 in regulating cell shape and movement during zebrafish gastrulation, *Mech Dev* 120 (4), 467–76, 2003.
78. Park, M. and Moon, R. T., The planar cell-polarity gene stbm regulates cell behavior and cell fate in vertebrate embryos, *Nat Cell Biol* 4, 20–5, 2002.
79. Jessen, J. R., Topczewski, J., Bingham, S., Sepich, D. S., Marlow, F., Chandrasekhar, A., and Solnica-Krezel, L., Zebrafish trilobite identifies new roles for Strabismus in gastrulation and neuronal movements, *Nat Cell Biol* 4 (8), 610–5, 2002.
80. Veeman, M. T., Slusarski, D. C., Kaykas, A., Louie, S. H., and Moon, R. T., Zebrafish prickle, a modulator of noncanonical wnt/fz signaling, regulates gastrulation movements, *Curr Biol* 13 (8), 680–5, 2003.
81. Sheldahl, L. C., Slusarski, D. C., Pandur, P., Miller, J. R., Kuhl, M., and Moon, R. T., Dishevelled activates Ca2+ flux, PKC, and CamKII in vertebrate embryos, *J Cell Biol* 161 (4), 769–77, 2003.
82. Lin, F., Sepich, D. S., Chen, S., Topczewski, J., Yin, C., Solnica-Krezel, L., and Hamm, H., Essential roles of Gα12/13 signaling in distinct cell behaviors driving zebrafish convergence and extension gastrulation movements, *J Cell Biol*, 169, 777–87, 2005.
83. Bisgrove, B. W., Morelli, S. H., and Yost, H. J., Genetics of human laterality disorders: insights from vertebrate model systems, *Annu Rev Genomics Hum Genet* 4, 1–32, 2003.
84. Levin, M., Left-right asymmetry in embryonic development: a comprehensive review, *Mech Dev* 122 (1), 3–25, 2005.
85. Ahmad, N., Long, S., and Rebagliati, M., A southpaw joins the roster: the role of the zebrafish nodal-related gene southpaw in cardiac LR asymmetry, *Trends Cardiovasc Med* 14 (2), 43–9, 2004.
86. Shimeld, S. M., Calcium turns sinister in left-right asymmetry, *Trends Genet* 20 (7), 277–80, 2004.

87. Nonaka, S., Tanaka, Y., Okada, Y., Takeda, S., Harada, A., Kanai, Y., Kido, M., and Hirokawa, N., Randomization of left-right asymmetry due to loss of nodal cilia generating leftward flow of extraembryonic fluid in mice lacking KIF3B motor protein, *Cell* 95 (6), 829–37, 1998.
88. Okada, Y., Nonaka, S., Tanaka, Y., Saijoh, Y., Hamada, H., and Hirokawa, N., Abnormal nodal flow precedes situs inversus in iv and inv mice, *Mol Cell* 4 (4), 459–68, 1999.
89. McGrath, J., Somlo, S., Makova, S., Tian, X., and Brueckner, M., Two populations of node monocilia initiate left-right asymmetry in the mouse, *Cell* 114 (1), 61–73, 2003.
90. Raya, A., Kawakami, Y., Rodriguez-Esteban, C., Ibanes, M., Rasskin-Gutman, D., Rodriguez-Leon, J., Buscher, D., Feijo, J. A., and Izpisua Belmonte, J. C., Notch activity acts as a sensor for extracellular calcium during vertebrate left-right determination, *Nature* 427 (6970), 121–8, 2004.
91. Amack, J. D. and Yost, H. J., The T box transcription factor no tail in ciliated cells controls zebrafish left-right asymmetry, *Curr Biol* 14 (8), 685–90, 2004.
92. Kramer-Zucker, A. G., Olale, F., Haycraft, C. J., Yoder, B. K., Schier, A. F., and Drummond, I. A., Cilia-driven fluid flow in the zebrafish pronephros, brain and Kupffer's vesicle is required for normal organogenesis, *Development* 132 (8), 1907–21, 2005.
93. Sarmah, B., Latimer, A. J., Appel, B., and Wente, S. R., Inositol polyphosphates regulate zebrafish left-right asymmetry, *Dev Cell* 9 (1), 133–45, 2005.
94. Webb, S. E., Moreau, M., Leclerc, C., and Miller, A. L., Calcium transients and neural induction in vertebrates, *Cell Calcium* 37 (5), 375–85, 2005.
95. Leclerc, C., Webb, S. E., Daguzan, C., Moreau, M., and Miller, A. L., Imaging patterns of calcium transients during neural induction in *Xenopus* laevis embryos, *J Cell Sci* 113 (19), 3519–29, 2000.
96. Palma, V., Kukuljan, M., and Mayor, R., Calcium mediates dorsoventral patterning of mesoderm in *Xenopus*, *Curr Biol* 11 (20), 1606–10, 2001.
97. De Robertis, E. M. and Kuroda, H., Dorsal-ventral patterning and neural induction in *Xenopus* embryos, *Annu Rev Cell and Dev Biol* 20 (1), 285–308, 2004.
98. Leclerc, C., Duprat, A. M., and Moreau, M., Noggin upregulates Fos expression by a calcium-mediated pathway in amphibian embryos, *Dev Growth Differ* 41 (2), 227–38, 1999.
99. Webb, S. E. and Miller, A. L., Calcium signalling during zebrafish embryonic development [Review], *Bioessays* 22 (2), 113–23, 2000.
100. Ferrari, M. B. and Spitzer, N. C., Calcium signaling in the developing *Xenopus* myotome, *Dev Biol* 213 (2), 269–82, 1999.
101. Ashworth, R., Approaches to measuring calcium in zebrafish: focus on neuronal development, *Cell Calcium* 35 (5), 393–402, 2004.
102. Yoshida, Y., Kim, S., Chiba, K., Kawai, S., Tachikawa, H., and Takahashi, N., Calcineurin inhibitors block dorsal-side signaling that affect late-stage development of the heart, kidney, liver, gut and somitic tissue during *Xenopus* embryogenesis, *Dev Growth Differ* 46 (2), 139–52, 2004.
103. Kawakami, Y., Raya, A., Raya, R. M., Rodriguez-Esteban, C., and Belmonte, J. C., Retinoic acid signalling links left-right asymmetric patterning and bilaterally symmetric somitogenesis in the zebrafish embryo, *Nature* 435 (7039), 165–71, 2005.
104. Vermot, J. and Pourquie, O., Retinoic acid coordinates somitogenesis and left-right patterning in vertebrate embryos, *Nature* 435 (7039), 215–20, 2005.
105. Vermot, J., Gallego Llamas, J., Fraulob, V., Niederreither, K., Chambon, P., and Dolle, P., Retinoic acid controls the bilateral symmetry of somite formation in the mouse embryo, *Science* 308 (5721), 563–6, 2005.

9 Strategies and Techniques for Investigation of Biophysical Signals in Patterning

Dany S. Adams and Michael Levin

CONTENTS

I. INTRODUCTION

Great strides have been made in understanding the mechanisms of morphogenesis operating during embryonic development and regeneration. The strategies of forward or reverse genetics have allowed a significant degree of insight into gene transcription networks and secreted protein signaling factors. However, an important component has not yet been integrated with the body of work arising from molecular genetics: the physiological and biophysical events resulting from the function of ion transporters. Ion flows set up by channels and pumps produce pH and voltage gradients within cells and across cell fields, and provide an important but still poorly understood signaling system. Endogenous bioelectrical control mechanisms are now known to not simply be physiological house-keeping but rather to play crucial, instructive roles in embryonic patterning, regeneration, and neoplasm throughout the animal and plant kingdoms [1–5]. Ion flows have been shown to control cell proliferation, differentiation, and migration [6–9]. Here, we provide information on a number of powerful techniques designed to enable investigators to determine whether ion flux plays a role in their favorite morphogenetic system and to characterize the mechanisms upstream and downstream of the relevant bioelectrical events. The strategies covered here include (1) reverse drug screens to determine whether ion flux is involved and, if so, to implicate specific transporter genes for further analysis; (2) techniques for the characterization of endogenous gradients and ion flows *in vivo*; and (3) tools for loss- and gain-of-function experiments designed to probe the function of specific bioelectrical events.

When combined with well-established methods and strategies in molecular and cell biology, this paradigm allows the unification of physiology and biophysics with genetic information at the molecular level and adds a crucial physiomic component to the understanding of any process. The integration of these techniques is required to foster interdisciplinary and systems approaches to the understanding of biological phenomena. Because potential gradients and current flows add a rich set of epigenetic control mechanisms operating on time scales faster than changes in gene expression, but ultimately feed into known transcription cascades, the investigation of the roles of ion flow provides a unique opportunity for enriching both the basic understanding of any process and the functional control over cell behavior of clinical relevance.

A. Vocabulary

We assume that the reader is familiar with the standard molecular- and cell-biological techniques utilized in their model system. We will use the phrase "your phenomenon" to refer to the pattern, process, tissue, or morphological feature that is the focus of investigation. "Phenotype" will refer to any induced effect on your phenomenon. "Drug" will refer to any chemical used to generate a phenotype, including inorganics, such as GdCl, and both natural and synthetic pharmacological agents, such as tetrodotoxin and Omeprazole. "Ion flux" is the movement of ions through an area per unit of time. It does not imply anything about the mechanism of that movement and includes diffusion as well as active transport. A "gradient" is a difference in concentration of ions across a distance (including step changes, such as a difference in ion concentration on either side of a membrane). A "channelopathy" is a phenotype arising from a mutation in a transporter gene [10]. The following section is a short review of ion flux in cells and the translocators that regulate ion movement.

B. The Mechanics of Ion Flux in Biological Systems

To generate an electric field, cell plasma membranes or epithelia provide a power source and generate a voltage potential by segregating ions across the barrier (resulting in membrane voltage or transepithelial potential, respectively). Because ions are charged, they cannot cross membranes by themselves; by establishing membrane bound compartments, cells can establish ion concentration differences or gradients. Cells use stored energy in the form of these gradients (by carefully regulating the flow of ions across membranes) to accomplish everything, either by coupling local ion flux directly to local physiology (e.g., Ca^{2+} regulation of secretion) or by siphoning energy from the gradient and storing it as ATP. This process is so fundamental, and so adaptable, that a staggering number of mechanisms for creating and controlling transmembrane ion flux have evolved.

These mechanisms, involving proteinaceous ion translocators, are described based on four characters: (1) the ion or ions that move; (2) the number of ions moved, either one or more, and, if more than one, the direction the ions move relative to each other; (3) whether or not the translocator undergoes a conformational change during transport; and (4) whether at least one of the ions is moving up its concentration gradient. This last character is what determines whether the move will require energy, either from ATP hydrolysis or from coupling the movement of the ion that is moving up its gradient (an energy-requiring process) to the movement of an ion moving down its gradient (an energy-releasing process).

1. The ions that move through biological membranes are H^+, Cl^-, Na^+, K^+, and Ca^{2+}, and to a lesser degree, Fe^{2+}, Cu^{2+}, and Zn^{2+}. Most ion movement takes place via mechanisms that are highly ion specific; however, nonspecific ion flux can also take place through gap junctions [11, 12]. Other charged species that are relevant to biological function include HCO_3^- and OH^-. Finally, ion movement can be coupled to the movement of nonionic molecules, such as peptides and carbohydrates.
2. If one ion is translocated, it is called a uniport. If two or more ions are moved, and they move in the same direction, it is called a symport or a cotransporter. If multiple ions are moved, but they move in opposite directions, it is called an antiport or an exchanger.
3. If, during transport, the translocator is essentially a transmembrane tunnel, open at both ends,* it is called a channel. If, during transport, the translocator is only open to one side of the membrane at a time and, like a subway turnstile, must undergo a conformational change in order to release the ion to the destination compartment, it is called a transporter. If, during transport, the translocator physically travels from one side of the membrane to the other, it is called a transporter, a carrier, or an ionophore.
4. Movement down gradients requires no input of energy. When this free movement of ions takes place across a membrane, it is called passive

* Channels can be opened and closed ("gated") when there is no active translocation.

transport or facilitated diffusion. If the ion is moving up its concentration gradient, energy is required, and it is called active transport. The active transporters are often called pumps. When the energy for pumping comes from the enzymatic hydrolysis of ATP,* the translocation is called primary active transport, and the pump is often called an ATPase. If the energy source is the gradient of another ion, the translocation is called secondary active transport.

Note that, depending on the research history, different names may be used to refer to the same translocator, each emphasizing a different one of the four characters. Electrophysiologists and geneticists often have different names for the same transporter, and efforts at unification of nomenclature have been made [13–15]. A further complication arises with respect to invertebrate transporters. Although evolutionary relationships among ion transporter families cannot be discussed here for reasons of space, several excellent reviews exist [16] and can be very useful. For example, the vertebrate P-ATPase family contains Na^+/K^+-ATPases, H^+/K^+-ATPases, and Ca^+-ATPases, which can be recognized on the basis of sequence, but invertebrate pumps tend to be basal forms that do not cleanly fall into the Na^+/K^+-ATPase versus H^+/K^+-ATPase distinction and must be characterized electrophysiologically to determine which ions they actually translocate.

Another consequence of establishing and maintaining ion gradients across membranes is that there can be a net difference in the number of positive and negative charges on the two sides. That is, in addition to the many ion-specific concentration differences (the chemical gradients), there is also a single, all-inclusive charge difference (the electrical gradient) across the membrane. An electrical gradient is called a voltage; hence, cellular membranes are described as having a membrane voltage (V_m). The V_m of a cell when nothing special is happening is called the resting potential (a voltage is a potential). Many cells have resting potentials on the order of –60 mV, although there is a wide range, and indeed interesting relationships exist between the differentiation states of cells and their membrane voltage [17–19]. A number of basic texts give detailed expositions of this fundamental property, but they are almost always restricted to the Na^+/K^+-ATPase and a small subset of channels [20, 21]. In fact, cells can utilize a large number of diverse transporters to regulate their ion profiles and, thus, their V_m.

V_m is determined by the gradients, and therefore the concentrations, of all the ions, sometimes necessitating use of the Hodgkin-Katz-Goldman equation (the expanded version of the Nernst equation). It will not be necessary to use that equation to perform the analyses described below, but the form of the equation does illustrate how the one electrical gradient is related to all the many chemical gradients. The electrical gradient affects ion flux just as the chemical gradients affect it. Hence, ions are often spoken of as being in electrochemical (EC) gradients.

Another layer of complication is the following: Because the electrical gradient is determined by the combined affects of all the ion gradients, the direction of the electrical gradient may be opposite that of the chemical gradient for a given ion,

* This is different from certain channels for which ATP is a ligand, but not a source of energy.

putting that ion in continuous conflict. That is the case for K^+ ions in most animal cells, where there are more K^+ ions in the cytoplasm than outside, but there are also more negative charges. So, entropy favors K^+s leaving the cytoplasm, whereas the attraction of opposite charges favors K^+s staying put. Its distribution ends up such that the two forces are balanced.* The resting potential of a membrane is the combined result of all the ions in the cell finding their own balance,† which depends on their distributions, which in turn depend on the activities of all the ion translocators.

Probably the most familiar example of the elegant ways cells exploit ion gradients is the action potential (AP), beginning with the opening of Na^+ channels, followed by the opening of K^+ channels and Ca^{2+} channels, ultimately leading to the re-establishing of the resting potential accomplished by pumps. Other well-known examples include sperm capacitation (K^+, Ca^{2+}, and HCO_3^- gradients); both the fast and slow blocks to polyspermy (Na^+ and Ca^{2+}); uptake of glucose into intestinal epithelial cells (Na^+); and the synthesis of ATP (H^+). It is important to note that in most patterning systems examined to date, the key parameter is the standing, long-term membrane voltage across cells and steady pH/voltage gradients across cell fields. Thus, most of the discussion below focuses not on rapid action potential-like depolarizations but steady-state electrical properties of cells and tissues.

Voltages across cell membranes are often important autocrine control parameters. At the next level of complexity, trans-epithelial potentials (TEPs) exert their effects non-cell-autonomously. Whereas plasma membranes are usually inside negative, the TEP is often positive on the basal side of the epithelial monolayer. Tight and adherens junctions create high-resistance regions and ensure that the epithelium is able to generate a difference across itself, and thus control the paths that the ion currents must follow to complete the circuit [22].

In addition to the wide variety of channels, pumps, and carriers, there are gap junctions (GJs). These are plasma membrane protein complexes that directly connect the cytoplasm of neighboring cells, allowing for cell-cell exchange known as gap junctional communication (GJC). GJs can be opened or closed post-translationally (in response to changes in local Ca^{2+} concentration, pH, or membrane voltage) and provide for fairly complex gating of the flow of ions and small molecules (< 1 kD) directly between cells [23–27]. If the GJs connecting two cells are opened, any ion gradient that previously existed will quickly, although not instantaneously, dissipate. Thus, ion flux through GJs can be an important determinant of ion distribution in a tissue, and expression patterns of GJs can establish isopotential cell fields.

C. Mechanisms of Ion Action: How Might Ion Flows Feed into Canonical Pathways?

1. Direct regulation of protein function. When an ion interacts with an oppositely charged point on a peptide, that ion is changing the charge

---

* Balanced does not mean "not moving across membranes" but only that the flux causes no net change in the K^+ gradient.

† In many cell types, the gradient of K^+ has by far the greatest influence on the resting potential, but the contribution of the other ions is not zero.

distribution and thus can change the structure of the protein, thereby affecting its function. Familiar examples are the binding of Ca^{2+} to calmodulin and receptors with ions as ligands, such as the Ca^{2+}-activated channels that mediate the spread of cortical granule exocytosis at fertilization. The proteins regulated can also affect gene expression; for example, Ca^{2+}-calmodulin-dependent protein kinase IV (CaMK IV) upregulates c-fos expression [28]. pH gradients can exert their effects through regulation of G-protein-coupled receptors [29], ion channel conductivity [30–34], and many enzyme families.

2. Membrane voltage regulation. The activity of membrane-associated proteins can be affected by changes in V_m. For example, the dramatic change in V_m caused by increased Na^+ flux during an action potential increases the activity of the voltage-sensitive K^+ channels. The transduction of repeated depolarizations into gene expression are now also beginning to be understood and involve cAMP and CREB pathways [35–37], although it is still largely unknown how long-term changes in membrane voltage are transduced into differentiation pathways. The control of activity of voltage-gated transporters of small molecules (such as the serotonin transporter) provides a mechanism to transduce V_m into the flow of biochemical signals into the cell.
3. Electrophoresis. A third mechanism by which ion flux can influence downstream events is by turning a cell or tissue into an electrophoresis apparatus. Ion separations across a zone of isolation can result in electromotive force being applied within conductive cytoplasm or extracellular fluids, within GJ-connected cell fields, or even within a lipid membrane. This can result in the electrophoretic movement of small molecules, the redistribution of which can induce downstream events (if the small molecules possess signaling activity). This mechanism is likely to be involved in the redistribution of small molecule determinants in left-right (LR) patterning [38–42].

Thus, it is clear that extremely complex and recursive feedback loops can exist whereby ion flux modifies the action and expression of other ion transporters, affording the possibility of rapid amplification or inhibition of signals. It is important to keep in mind that in many cases, the important bioelectrical signals are the result of the combination of a number of transporters acting in tandem. For example, the activity of the H^+/K^+-ATPase can require a co-expressed Kir4.1 or KCNQ1 potassium channel to allow for the recycling of the symported K^+ ion during H^+ efflux [40, 43, 44].

D. Examples of Bioelectrical Events as Control Factors in Morphogenesis

Electrical activity due to ion channels and pumps has been extensively studied in the function and structure of the nervous system. However, a variety of classical and more recent observations have suggested that bioelectric phenomena have multiple instructive roles in patterning [1, 22, 45–48]. The discovery of strong endogenous

DC electric fields within living systems has been augmented by functional experiments, suggesting that these fields have a causal role in physiology and development [46]. A wide variety of embryonic and other patterning systems drive specific ion currents and generate voltage gradients. For example, most epithelia exhibit extracellular current densities on the order of 10–100 μA/cm^2. Indeed, a comparative analysis of membrane voltage properties of various kinds of cells reveals a relationship between depolarization and control of differentiation and proliferation: Quiescent, highly differentiated cells tend to be strongly polarized, whereas proliferating, embryonic, and tumor cells tend to be depolarized.

Functional data (reviewed in references cited above) suggest not that this relationship is epiphenomenal but rather that such ion flux events are a causal factor in determining crucial aspects of cell behavior [49, 50]. Developing systems generally drive steady ion currents and produce substantial fields within themselves; examples include currents that enter both the prospective and active growth point of several tip-growing plant cells; voltage across the cytoplasmic bridge between an insect oocyte and its nurse cell; current traversing a recently fertilized egg from animal to vegetal pole; and early potentials across embryonic epithelia. These currents can be anywhere from 1 to 1000 μA/cm (Jaffe, 1982).

Data from many contexts in embryonic development and regeneration have indicated that ion fluxes and endogenous voltage gradients correlate with, and are predictive of, specific morphogenetic events; most interestingly, in some cases, gain-of-function experiments have shown that altering the endogenous ion flux profile in developing embryos often has a direct and specific effect on embryonic morphology [3, 4, 19, 22, 51–59]. Some of the best studies include those of limb development [60–62] and neurulation. Voltage gradients within the neural tube are required for cranial development [63, 64]. Inhibition of the trans-neural tube potential [65] produces a remarkable disaggregation of internal morphology (otic primordia, brain, notochord) coupled with fairly normal external form in amphibian embryos [66]. Currents arising in the posterior intestinal portal are necessary for chick tail development [67].

Consistent with a wide conservation of bioelectrical control systems throughout the biosphere, many of the bioelectrical phenomena observed in embryonic morphogenesis also function in the guidance of regeneration of plant [68–72] and animal systems, from invertebrates [73–78] to humans [79], and to contexts from polarized organs (such as the limb) to entire segmented body plans [80]. One of the best examples of a gain-of-function control over morphogenesis in regeneration is the ability to control head/tail polarity in regenerating worms by experimentally orienting the endogenous anterior-posterior dipole electric field [74–78, 81].

Endogenous currents participate in patterning events through the control of differentiation by membrane voltage, [17, 18, 82–84] and by providing spatial guidance cues for cells [55–57, 85–100]. It has been suggested that three-dimensional systems of voltage gradients during amphibian neurulation may provide coordinates for cell migration and morphogenesis [287, 288]. In particular, neural crest cells are galvanotactic and are a good candidate for the target of endogenous electrical cues [289, 290]. Although cellular galvanotaxis was observed more than 100 years ago [101], more modern studies are now beginning to reveal underlying mechanisms [94, 102], including the roles of inositol phosphate second messengers, calcium entry, microfilament polymerization, [93, 102–105], and the integration of growth factor receptors and substrates with electrical signals in the growth cone [86, 87, 91, 106, 107]. These data have been incorporated into a predictive biophysical model, where a steerer determines the speed and an automatic controller determines the angle of migration [99].

A number of labs have made recent contributions to the molecular understanding of how membrane voltage controls cell differentiation [3, 6, 7, 57, 108]. Ion channel function controls the progression through the cell cycle [8] and the proliferation rate of a number of cell types [109–119]. Important advances in the merging of electrophysiology with molecular biology have been made in a few cases, implicating Ca^{2+} flux in amphibian neural induction and the dorsoventral identity of embryonic mesoderm [120–125], downstream of the neural inducer *noggin*, and in morphogenesis of hydra [126]. The Ca^{2+} ion has received the most attention in developmental studies [121, 127–132]. The activity of the Na^+/K^+-ATPase is involved in gastrulation and neuronal differentiation in amphibians [133–135]. In ascidians, analysis of developmental calcium currents [136] has led to the identification of a novel role for early expression of channel and pump mRNAs. Activity of the NHE1 exchanger controls the G_2/M transition by modulating intracellular pH fibroblasts [9].

Ion fluxes have also been implicated in the control of regeneration [22, 137–139] and in neoplasm [109–119, 140]. Regenerating appendages drive specific currents that are required for regeneration and are absent or reversed in nonregenerating species. Moreover, experimental imposition of these flows can induce or augment regeneration. Tumor cells differ from untransformed cells in terms of the type of ion channels and pumps they express and in the resulting membrane potential of the cells [108, 141–145]. Indeed, genetic subtypes of adenocarcinoma tumor cells can be distinguished on the basis of electrical characteristics [146], and introduction of ectopic H^+ pumping by misexpression of an H^+ pump induces tumorigenic properties in mammalian cells [147, 148]. Thus, the above-illustrated conservation of the roles of ion flux in the control of embryonic development, regeneration, and neoplasm suggests that such biophysical signals are part of a fundamental pattern control system.

The previous research implicating bioelectric events in patterning provides a fertile background for important advances in understanding of, and control over, tissue shape and cell behavior. However, much of that work took place before the advances of molecular biology, and the most important findings have neither been integrated with known molecular pathways nor received the benefit of powerful functional analysis tools. In contrast, the best modern work implicating specific channel/pump genes in cellular control have taken place in cell culture and have not explored the roles of ion fluxes in controlling three-dimensional patterns in morphogenesis.

To date, we have identified and characterized novel roles for gap junctions [38, 39] and endogenous electric fields in embryonic left-right asymmetry [39, 40, 149–151], tail regeneration in vertebrates [152], and anterior-posterior polarity in planaria [153, 154]. Certainly, there are more to be discovered. Below, we present information designed to allow a broad selection of workers, studying different aspects of morphogenesis, to bring this fascinating aspect of biology to their system.

E. Reverse Drug Screens to Implicate Specific Transporters in Your Phenomenon

Given the many fascinating roles in which ion flux has been implicated, what strategies can be used to investigate the possible involvement of bioelectrical signals in a given phenomenon? Neurobiologists have been using specific poisons and toxins to probe the function of individual ion channels and pumps for a long time. Fortunately, this immense body of pharmacological work can be utilized as a loss-of-function screen to rapidly and (usually) inexpensively determine whether ion flux is specifically involved in your favorite biological context [155, 156]. The key is to develop an assay that is amenable to application of pharmacological reagents, to define an optimal ordering of drug blockers into a hierarchy from least to most specific, and to use the results of each step to dictate the choice of the next step. Of course, pharmacological techniques cannot conclusively prove the involvement of any target; however, unlike expensive and time-consuming morpholino/RNAi approaches, this drug screen can be an extremely effective way to narrow down from many thousands of possible candidates to a manageable number of transporters that can then be characterized using molecular reagents.

This strategy can also be used to study other pathways, such as to discover novel roles of neurotransmitters [42]. In the case of bioelectricity, it has been successfully used to discover that ion flux is a novel component of left-right patterning in a number of species [40]. The assay chosen was the specific alteration of the normal *situs* (in the context of normal organogenesis) of the heart and asymmetric viscera. Broad inhibitors were first used to determine that Cl^-, Ca^{++}, and Na^+ ions were not crucial in the determination of vertebrate laterality, whereas K^+ and H^+ flux was involved. More specific compounds targeting different members of various K^+ and H^+ channel and pump families implicated the V-ATPase, the H^+,K^+-ATPase pumps, and the KvLQT-1 and K_{atp} channels. Molecular expression and loss-of-function analyses later confirmed the involvement of these targets and uncovered a number of new physiological and subcellular asymmetries, validating the screen as a powerful strategy. Similar screens have been successfully utilized to probe the roles of ion flux in regeneration [152].

The logical structure of the screen is as follows. You begin with drugs of low specificity, targeting each of several large families of transporter (Step 1) and apply the drugs to your preparation (making sure that the N is large enough for each drug to enable statistical comparison with controls). If toxicity or nonspecific effects result, the dose must be reduced or a different reagent chosen. If no effect is seen on the phenotype being assayed, the whole family can be crossed off the list and need not be considered further (although it is usually good to confirm the lack of involvement using a different drug of similar targeting, to reduce the number of false negatives due to problems specific to some peculiar compounds). On the other hand, if an interesting phenotype results from the treatment, you move on to treatment with more specific drugs that can distinguish among the members of the broad family implicated in the previous step (Step 2). This process continues as long as increasingly specific drugs exist (Steps 3, 4, etc.). This differs depending on which families happen to be involved in your phenomenon but usually allows at least two or three rounds of the screen. It is often possible to narrow down to a reasonably small family that can then be pursued using molecular techniques (see Section II). The specific information needed to carry out the screen, and examples of such screens, are given in Appendixes A–C.

For example, to determine if sodium flux plays a role in your process,* you might first treat with palytoxin, a reagent that converts the Na^+/K^+-ATPase pump into a Na^+/K^+ pore [157]. If it has an effect, you could move on to treatment with saxitoxin (Step 2), a drug that blocks Na^+ channels, but does not affect sodium pumps. If that also had an effect, you next try tricaine (Step 3), which blocks voltage-sensitive, but not Ca^{++}-activated, sodium channels. If that also causes a phenotype, you know that Na^+ flux through voltage-gated sodium channels is involved in your phenomenon. If, however, palytoxin does not cause a phenotype, you can cross sodium (and potassium) flux from your list of phenomena that might be involved in your process.†

If palytoxin (Step 1) causes a phenotype, but saxitoxin (Step 2) does not, you may be dealing with an active Na^+ transport process. In that case, the next step might be EIPA (Step 3) to implicate or rule out the Na^+/H^+ exchanger, KB-R7943 mesylate (Step 3) to implicate or rule out the Na/Ca exchanger, and bumetanide (Step 3) to implicate or rule out the $Na^+/K^+/Cl^-$ cotransporter. If any of those causes a phenotype, you have narrowed down your search to a very manageable number of gene products to explore further. Though the overall number of reagents is huge, this screen harnesses the power of tiered organization to proceed as a binary search, resulting in only a small number of the compounds needing to be tried. Thus, like using a dichotomous key, each iteration either reduces the number of alternatives that have to be considered or increases the efficiency and precision with which you can choose candidates to characterize using molecular techniques. Alongside blockers and inhibitors of channels and pumps, useful reagents also include chelators, ionophores, gap junctional communication inhibitors, and direct alterations of ion concentrations in the extracellular medium.

The structure of the strategy is fairly simple. The most important thing is knowing the drugs that are available for each category and having an effective ordering structure. Although no definitive list can be compiled (because new pharmacological reagents are constantly being developed, and new specificity information sometimes appears for known drugs), we have assembled a very detailed list that should greatly reduce the effort needed to carry out such a screen. This table (Appendix C) is generally applicable to most model systems, but of course, depending on the particular area of transporter biology to which the data will point, one must then delve deeper into the specific literature about the implicated family(ies) to understand the nuances specific to each group of transporters.

---

* Another example, for determining which calcium channels are involved in your phenomenon, is illustrated in Appendix B.

† See, however, information about false negatives, page 189.

F. The Screen Reagents

The known ion translocators are sorted into categories in the Step Screen Tables (Appendix A). On the right hand side of each, under Ion Translocators, the types and names of translocators are divided into four categories: channels (i.e., passive translocators), ATPases (the primary active transporters), and symports and antiports (secondary active transporters). Sorted into those four categories, which are present for each ion, are the known translocators of that ion. The format of these tables is meant to aid in experimental design by reflecting the steplike structure of the screen; the categorization is not meant to reflect evolutionary relationships among the proteins.

The left-hand sides of the Step Screen Tables categorize drugs that can be used to decrease or increase flux of an ion through specific translocators. The structure of the drug categories is identical to the structure of the translocator categories, with each entry containing drugs that target the corresponding translocators listed on the right. A dotted line connects the drugs to their targets, whereas arrows connecting the drugs indicate the temporal order in which drugs could be used to fully exploit the hierarchical paradigm. Appendix C is an alphabetical list of the same reagents, with details about action and other relevant information, along with useful references to the literature.

In our lab, Step Screens have been used successfully to study axial patterning in *Xenopus,* zebrafish, and chick [40, 158]. Others have extended them to *Ciona* and sea urchins [159, 160]. Each of these model organisms offers a different advantage. Below, we describe advantages and disadvantages of three model vertebrates to illustrate issues that may make Step Screens easier or more difficult in your organism of choice.

Xenopus laevis. Xenopus is excellent for a drug screen. Advantages: The cells of *Xenopus* embryos are relatively large, and thus easily manipulated, injected, and used for electrophysiological recording. They can also be collected in very large numbers, facilitating biochemical, pharmacological, and statistical analyses. There is a detailed fate map for *Xenopus* blastomeres, enabling some control over the distribution of injected reagents. Finally, and uniquely among model organisms, is the fact that *Xenopus* oocytes are frequently used as laboratories for the study of ion translocators (and other membrane proteins). This is a crucial advantage because the literature contains an incredible number of available plasmids encoding wildtype and mutant transporters that have already been physiologically characterized in *Xenopus* (although usually in oocytes, not embryos). Disadvantages: The cells of *Xenopus* embryos are relatively large and round, and this can make imaging difficult. In addition, the blastomeres are opaque, complicating light-microscopy, and the yolk is autofluorescent, which complicates the use of fluorescent markers to localize proteins or ions. It is also fairly difficult to de-vitellinize early embryos without affecting ion flux, and the vitelline membrane, which is both a physical and a chemical barrier, may affect the delivery of certain reagents to the cell surface (although it does not usually present much of a problem).

Chicks. Advantages: Chick embryos are flat and transparent, making them amenable to a variety of imaging techniques (most of which were designed for use on cultured cells), including the use of fluorescent indicators of ion flux. The cleavage pattern of chicks is similar to that of most mammals. Disadvantages: Chick embryos are not available for study until there are already several tens of thousands of cells, thus they cannot be used to study phenomena that occur during early cleavage stages. Delivery of reagents *in ovo* is possible, but the N will necessarily be smaller than for *Xenopus*.

Danio rerio (zebrafish). Advantages: Zebrafish embryos are available at all stages, from gamete onward, and cohorts can be synchronized if needed for subsequent biochemical analysis. The cells are relatively large and transparent, thus making them excellent for imaging with fluorescent indicators. Embryos can be collected in large numbers (for biochemistry and statistics), and it is possible to create transgenic lines of zebrafish. Disadvantages: Cell migration and mixing during gastrulation can make it difficult to relate early events to later effects.

It is often informative to use more than one species, both to overcome difficulties inherent in one type of embryo and to gain insight into the evolutionary biology of ion flux-related phenomena. Results achieved in one species (e.g., implicated targets from a screen) can often be directly applied to other species, but it must be remembered that divergence may result in different kinds of embryos utilizing different transporter genes to achieve similar physiological ends.

Once effective and specific drugs are found that perturb your process of interest, they will serve as important reagents in subsequent steps. For example, they can be used (1) to ensure that the ion flows you detect directly are in fact due to the target (by ensuring that the flows cease upon application of the relevant blocker) and (2) to determine the timing of your phenomenon. For example, if you discover that application of a particular blocker during early development results in craniofacial defects, you can apply the drug at different stages to determine which developmental process might be sensitive. One important issue is that because one can rarely demonstrate complete wash-out of the drug, it is most useful to compare exposures that begin early versus those that begin late, rather than exposures that end at various time periods.

G. Caveats and Troubleshooting

One of the difficult aspects of using pharmacological agents is determining the correct dose, that is, a dose that will clearly affect your phenomenon without affecting anything else, and without being toxic. Remarkably, however, this is often possible. Even drugs as potentially lethal as palytoxin (listed in the catalogs as the most toxic non-proteinaceous natural reagent now known) can be used at a low enough dosage to affect patterning without causing any other defects [158]. Thus, titrating the dose of each drug is a critical part of the preliminary work for a screen; a good starting point is the concentrations at which the drugs are used in the mammalian literature. Nonetheless, certain drugs will undoubtedly be toxic, and certain drugs will do nothing, in your system. Luckily, however, there is often another drug, with the same target but a slightly different mechanism of action, which can be used as an alternative.

For example, if your early steps (1 and 2) indicate that Ca^{2+} channels may have a role in your phenomenon, Step 3 might include treatment with Bay K 8644, an interesting drug because the R and S isomers have opposite effects on L-type Ca^{++} channels [161]. You might discover, though, that although there seems to be an effect, it is not clear enough to interpret. The next step might be to try FPL 64176, another L-type channel inhibitor that is known to be 40X more potent than Bay K 8644 [162]. Another approach to solving this problem is to use a less-potent analog of the drug of interest, or a slightly different member of the same family of drugs. Saxitoxin, a blocker of Na^+ channels, has many natural and synthetic analogs that might give results if saxitoxin itself proves toxic or inconclusive. Importantly, even if one drug gives a phenotype, it is still valuable to show that other drugs with the same target cause the same phenotype; these are positive controls showing that it is the desired effect of the drug that is causing the phenotype. Thus, you should plan on testing alternatives (if they exist) regardless of whether you choose well the first time. The goal is not to unequivocally rule targets in or out, but rather to accumulate evidence making a particular gene product worthwhile to investigate using more specific (and more time-consuming and expensive) molecular techniques.

Another difficulty is that there is often no positive control to prove that the drug being used has reached the target of interest. This may lead to false negatives that result from the failure of a drug to penetrate a chorion, vitelline membrane, cell layer, or subcellular compartment. Labeled drugs (for example, FITC-conjugated glibenclimide for the study of K-ATP channels) are increasingly available however, so before dismissing your favorite reagent, you can research whether a labeled version is available or possible to create. Nonetheless, false negatives can happen, especially because a few families of transporters have no good reagents that target them. Another source of false negatives is that some systems can adjust after a loss-of-function treatment by up-regulating a different ion transporter to do the same job [138]. Thus, this screen will not saturate; however, it is a very useful tool to focus later efforts on a manageable number of interesting candidates.

Also complicating the interpretation of some results is the lack of specificity of some of the drugs. For example, if Step 1 indicates that H^+ flux has a role in your phenomenon, Step 2 might include oligomycin, an inhibitor of the V-type H^+-ATPase. If that inhibitor causes an effect, however, you still need to control for the fact that oligomycin also inhibits Cl^- flux through the Cystic Fibrosis Transmembrane Conductance Receptor (CFTR); one solution would be to determine whether lonidamine, a CFTR-specific inhibitor, gives the same phenotype. Thus, negative results can be as important as positive results, and it is the intersection of your data-set that implicates specific targets. This is another reason it is important to know as much as possible about your reagents, and it is important to remember that each result must be interpreted in the context of the entire screen.

Finally, it is important to consider that much drug-mechanism data comes from study of mammalian neural cells. Thus, in extrapolating to nonneuronal contexts and other species, specificity may be somewhat wider (for example, as a result of sequence and structure divergence between mammalian and other versions of channels and pumps). In some cases (such as in *Drosophila*), data is available on the interaction of drug reagents with nonmammalian targets [163–166].

II. DETECTION OF IMPLICATED TARGETS

Once the step screen has narrowed down the list of proteins important for your phenomenon, it is critical that more specific techniques be employed to confirm the identification and to explore function. Once a gene product has been implicated, the resources and techniques of molecular biology can be brought to bear. If you were fortunate, the data may have implicated a family whose members have already been cloned in your model system. If not, you may have to clone them yourself (but the possible choices of what transporters to clone will be enormously narrowed by the screen results). For example, if the screen indicates that voltage-gated Na^+ channels are implicated (e.g., positive result for tricaine), a pan-NaV antibody may be used to confirm the result. If a positive signal is obtained, you will then want to clone the various NaV family members from your species to assay the expression of each. The clones will also prove useful for misexpression of w.t. and mutant constructs in functional experiments (see Section IV).

With clones and antibodies in hand, you will then want to validate the results of the screen by examining the expression pattern of the implicated targets in your system. *In situ* hybridization can be used to localize mRNA [167]. However, ion transporters are often highly regulated post-translationally and may be localized differently than the mRNA (this is especially true in early *Xenopus* embryos, but is usually not the case in chick or mouse). Therefore, it is highly advisable to use immunohistochemistry if antibodies are available (localization to cell membranes is an important feature of transporter function). For both techniques, picking a highly conserved sequence (for your DNA probe or for generating an antibody) may allow you to use the same reagent in more than one species. Antibodies can also be created to be useful in functional analysis as well as localization. Antibodies can often be obtained from labs working with mammalian transporters; a Western blot and other usual controls in your species can reveal whether the considerable conservation of these genes across phyla will enable you to use the existing antibodies or whether specific ones need to be made. It should be noted that it is often very useful to co-localize elements of the cytoskeleton with your transporter, because many polarized cell types (epithelia and early blastomeres) utilize cytoskeletal elements to control the localization of ion transport [168–170]. This information can provide invaluable clues to possible links between physiological and morphological polarity in your system [158, 171–173].

A fusion of your protein to a fluorescent protein such as GFP can often allow observation of your transporter *in vivo*, as well as tracking its movement subcellularly. An important consideration when designing a fusion protein, however, is any autofluoresence of your organism. *Xenopus* yolk, for example, is highly autofluorescent when viewed through the FITC/GFP cubes of many fluorescent microscopes. Luckily, there are now many choices of fluorescent protein, and it is not difficult to find a protein that will be visible in your organism. Also important to consider is that adding anything to your protein may interfere with its function. Many channels and pumps are part of a multiprotein complex; some fusion proteins can disrupt the binding to accessory subunits or oligomerization, and this is a potential source of problems when making GFP fusions for channels and pumps. Smaller labels, such as tetracysteine tags, may soon alleviate such problems [175].

III. DIRECT DETECTION OF ION FLUX

An important step in characterizing the involvement of bioelectrical events in your phenomenon is to directly examine the ion fluxes and voltage/ion gradients. These data will not only allow you to extend and refine the drug screen and expression results into more specific, predictive models of the role of bioelectricity in your patterning context but also may reveal existing physiological patterns that are not apparent in morphological or gene expression analyses. One common theme is that cells and tissues undergoing active morphogenesis, as well as organizing centers, tend to be depolarized relative to surrounding cells. These techniques are also crucial in allowing the monitoring of the effects of functional experiments (Section 6).

Ion fluxes can be detected in many ways. Electrophysiological techniques are the classic approach used to measure bioelectrical phenomena, often using KCl-filled microelectrodes connected via Ag-AgCl junctions to a very high input impedance preamplifier (to avoid draining current from the system) to measure voltage levels within cells or beneath epithelia. Intracellular recording and voltage clamping are used to measure membrane voltage and whole cell currents. Patch recording and patch clamping can be used to measure ion flux through a limited number of channels in a small area of membrane. These techniques are accepted and powerful; they are also invasive, and careful control of conditions is paramount. These techniques have been reviewed extensively [176, 177].

A newer tool, the self-referencing "vibrating" ion-selective probe (SERIS), is a noninvasive technique for detecting and measuring ion gradients at the surface of cells [178–182]. The technique utilizes ion-specific ionophore-filled microelectrodes. The tip of the electrode is vibrated at about 300 Hz between two points about 10 μm apart, one closer and one farther away from a cell's plasma membrane. A difference in concentration at the two points indicates a gradient, which is measured quantitatively. An ion gradient near the outer surface of a cell implies transport of that ion across that membrane (without transport, the gradient would quickly dissipate). This technique is powerful and allows the characterization of ion flux that can be used to infer the physiological state of the cell, as well as reveal the bioelectrical signals that the given cell or tissue is sending to its neighbors. However, this method does not directly measure cell membrane voltage, ion flux across internal membranes, or gap junction states. This technique has been used to study a number of important bioelectric controls of development [183–190].

Fluorescent ion-reporting dyes are being used more and more [191–201]. The ion-sensing abilities of these dyes are based on the principle that binding of an ion changes the conformation of the molecule sufficiently to alter its fluorescence spectrum, a phenomenon referred to as a "spectral shift" or a "spectral response." This is illustrated in Figure 9.1A: Each of the curves is a spectrum of the same (imaginary) dye, made by measuring the intensity of the light emitted by the dye at a range of different wavelengths. As the concentration of the sensed ion ("[Ion]") changes, the

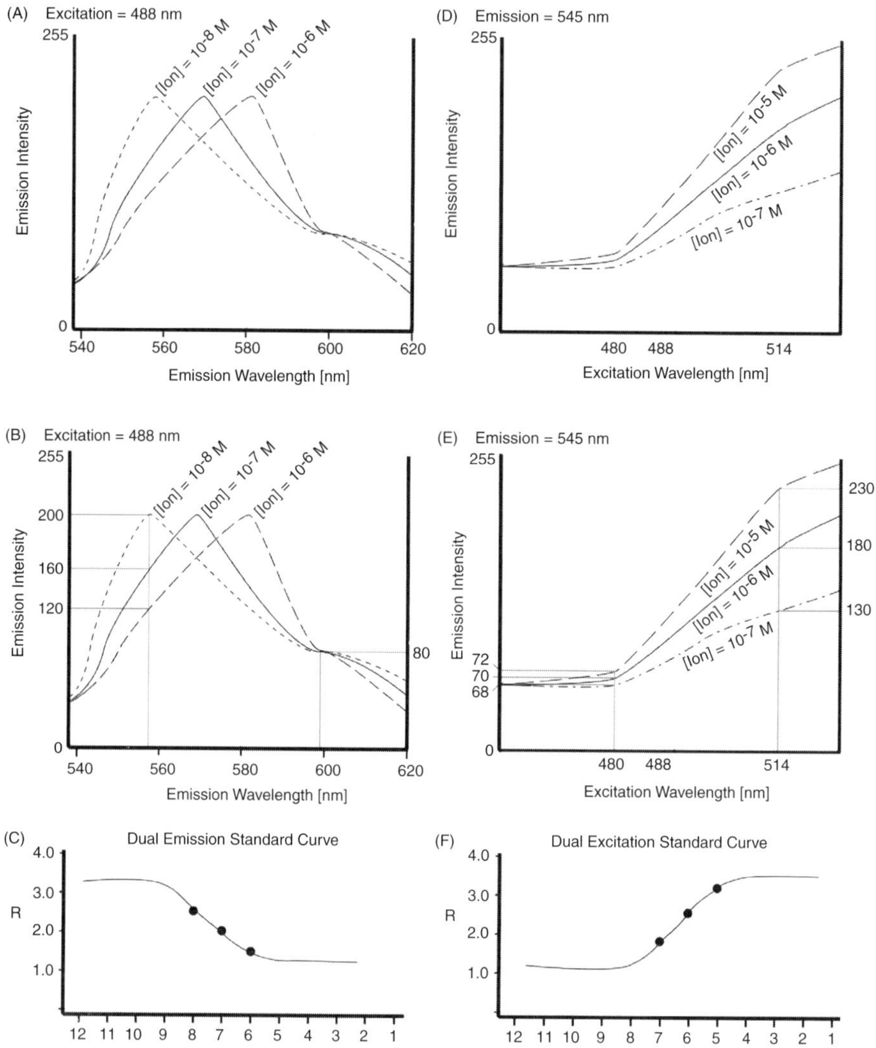

FIGURE 9.1 Schematic of properties of fluorescent reporter dyes. See main text for description.

curve moves, i.e., the spectrum changes; that change is the spectral shift. Once you know how the spectrum shifts, you can use a graph like the one in Figure 9.1A as a standard curve, measuring the spectral shift of an unknown and reading the ion concentration off the graph. These dyes can be loaded into cells by injection, by electroporation, or, most commonly, by soaking the cells in membrane permeant (acetoxymethyl, or AM) forms of the dye; once these AM dyes are internalized, esterases cleave off the AMs, trapping the active form of the dye.

A dye can be used for ion sensing if its bound and unbound states fluoresce differently. Referring again to Figure 9.1A, note that when excited by 488 nm light, this imaginary dye fluoresces most intensely at about 555 nm in the presence of 10^{-8}

M ion (when fewer dye molecules are bound), whereas it fluoresces most intensely at about 585 nm in the presence of 10^{-6} M ion (when more dye molecules are bound). The units of emission intensity are not important—in this figure they are the pixel values, on a scale of 0 to 255, given by a confocal microscope, but they might be quite different if measured using a fluorimeter or a cytometer—what is important is that there is a measurable difference between the curves. Once you have determined the family of curves that describes the behavior of your dye, you could measure an unknown, figure out which curve you have, then read the ion concentration from the graph. The manufacturers supply this family of curves with the dyes they sell, but it is neither this simple, nor this complicated, to use these dyes. It is simpler because you do not need to measure the entire spectrum, only two points; it is more complicated because you have many dyes to choose from, and calibration (that is, generating a standard curve for your system) is not nearly as straightforward as the theory makes it sound.

There are two ways that a fluorophore can fluoresce differently: its emission, when excited at a single wavelength, can be ion-concentration sensitive (Figure 9.1A), or its wavelength of maximal excitation, when its emission is monitored at a single wavelength, can be ion-concentration sensitive (Figure 9.1D). cSNARF-1, an H^+-reporting dye (see Figure 9.2), is a so-called "dual-emission" dye, whereas Fluo-3, the widely used Ca^{2+}-reporting dye, is a "dual-excitation" dye. Which type of dye you choose may depend on whether your imaging apparatus allows you to collect more than one wavelength or deliver more than one wavelength; your microscope salesperson can tell you about filter sets that are appropriate for particular

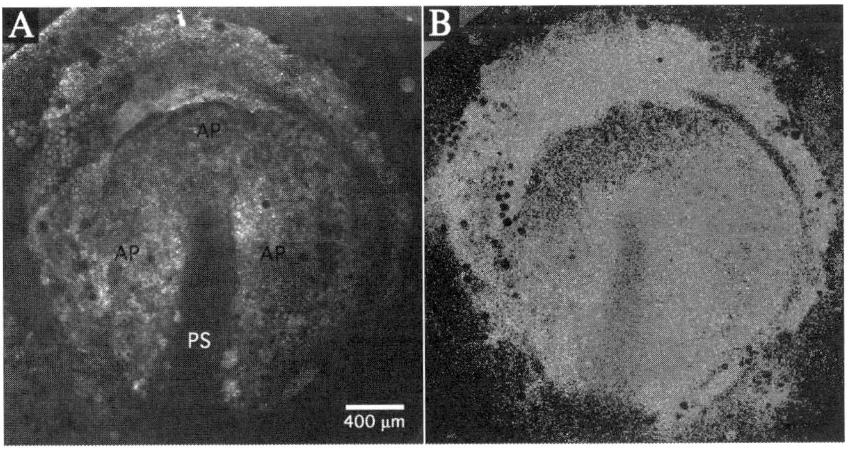

FIGURE 9.2 (See color insert following page 144.) Chick Stage 3 blastoderm stained with cSNARF-1-AM and illuminated with 546 nm light. (**A**) A transmitted light image showing the position of the primitive streak (PS) and the area pellucida (AP). (**B**) Different colors indicate different cellular pH in this projected image made by color-coding the pixel-by-pixel ratio of the fluorescence emission intensity at 620 nm and 590 nm. The primitive streak stands out clearly against the area pellucida, indicating a difference in pH of these two tissues.

dyes. If you have a laser scanning confocal microscope, you can likely do either, so you have a better chance of being able to use the dye that is best for your system rather than having to tailor your system to the dye.

The best ion-reporting dyes are ratiometric. The ratiometric dyes allow you to correct for the many artifacts that can affect the spectral shift, such as bleaching, differences in dye concentration or cell thickness, and spatial variation in instrument sensitivity. The effect of these artifacts can be substantial; if dye bleaching occurs, for example, the intensity of the emitted light will be artificially low, and you will get an incorrect value of concentration. However, these local conditions (which are different from background) can be accounted for and removed. The easiest way to correct for local conditions is to measure them and divide out their influence. That is, take a ratio: the intensity of the wavelength of interest (called λ_1) over the intensity of the local-condition light. To get the latter, you measure the same spot (the same dye molecules, in fact) at a second wavelength (called λ_2). If you are using a dual-emission dye, you collect emitted light of a different wavelength; if you are using a dual-excitation dye, you collect light emitted at the same wavelength, but excited by a different wavelength.

For the imaginary dual-emission dye of Figure 9.1A, you would illuminate with 488 nm light, then collect 555 nm light and 598 nm light (Figure 9.1B); for the imaginary dual-excitation dye of Figure 9.1D, you would excite with 480 nm light and collect 545 nm light, then excite with 514 nm light and collect 545 nm light (Figure 9.1E). Either way, you end up with two values of intensity, one for the wavelength that varies as a function of ion concentration (λ_1), and another that you will use to represent local conditions only (λ_2); again, the manufacturer of the dye will provide information about which two wavelengths to use. In Figure 9.1C, the intensity of the light emitted at 555 nm (λ_1) is very sensitive to ion concentration, whereas the light emitted at 590 nm (λ_2) is actually pH insensitive (this is called the isobestic point of the dye). In Figure 9.1E, the intensity of the 545 nm light emitted when the dye is excited at 514 nm is very sensitive to ion concentration (λ_1), whereas at 480 nm, it is not (λ_2). The great advantage of using the ratio λ_1 / λ_2 (often referred to as R) is that the two intensities will be similarly affected by local conditions, meaning their ratio will only be sensitive to ion concentration.

Figure 9.1C represents the calibration (or standard curve) for the imaginary dual-emission dye. The three points represent R from each of the three curves shown in Figure 9.1B: 120 / 80 = 1.5, 160 / 80 = 2.0, 200 / 80 = 2.5. They are graphed as a function of the negative log of the concentration (if the ion were H^+, this would be the pH scale). Figure 9.1F is the calibration resulting from taking the ratios for the dual-excitation dye, that is, the data from Figure 9.1E. The S-curves drawn through those points illustrate a critical point, which is that any given dye is only ion-concentration sensitive in a defined range of concentrations; above that concentration, the dye is all bound, so the spectrum cannot shift anymore, whereas below that concentration, the dye is all unbound, with the same result. The pK_d* of the dye is the value of ion concentration that gives a ratio exactly halfway between the completely bound and completely unbound states: For the imaginary dual-emission dye,

* If you are working with an H^+-reporting dye, this will be called the pK_a.

pK_d is approximately 7.2; for the dual-excitation dye, pK_d is approximately 6.2. This value will be available from the manufacturer, and it is a critical part of choosing a dye: To be useful, the concentrations you are measuring must be near the pK_d of the dye; that is the only range where the dye is ion-concentration sensitive.

If you can generate a standard curve based on *in situ* measurements in your system, you can use this curve to measure ion concentration. Another, more cumbersome but more reliable approach is to calibrate the dye in every specimen you measure. After measuring intensity at the two wavelengths, you change the bathing medium, replacing it with an ionophore-containing medium that will raise the internal concentration of ion to a known, high value (to bind up the dye). You then measure intensity at the two wavelengths to get the bound ratio, R_B. You then change the medium again, to a medium that will lower the internal concentration to the point where all of the dye is unbound, and you measure for R_U. From that information (the measured intensities, R, R_B, R_U, and the pK_d of the dye), you can calculate the ion concentration in the specimen.*

A final important consideration is background fluorescence, which can include autofluorescence. Background can be measured in any of the standard ways (defocusing, or measuring undyed autofluorescent specimens) and should be measured separately for each wavelength. The intensity of the background must be subtracted from the intensities measured at λ_1 and λ_2 before you calculate R.

Probably the most familiar of the ratiometric ion-sensing fluorescent probes are the calcium indicators, including the Fluo, Fura, and INDO dyes. pH-sensitive dyes, such as BCECF and the SNARF dyes, have also been used successfully in embryos. There are also probes for many other ions including Mg^{2+}, Na^+, K^+, and Cl^-. The Molecular Probes catalog is an invaluable resource for the latest data and available reagents.

In addition to ion-sensing dyes, there are also V_m-reporting dyes. These fluorophores undergo a spectral shift in response to V_m. There are two categories of V_m dyes, those that report fast changes, such as action potentials, and those that react more slowly, reporting V_m averages over longer periods of time. The fast-response dyes, such as Di-8-ANEPPS, localize to the membrane and undergo a spectral shift due to a redistribution of intramolecular charge caused by a change in V_m. Thus, their spectral shift occurs quickly and can be used as a measure of fast changes in V_m. The slow-response dyes, in contrast, are anionic or cationic molecules that accumulate inside the cell due, it is thought, to an electrophoretic mechanism driven by the voltage across the membrane. The spectrum of these dyes is affected by both environment (intra- versus extracellular) and concentration. For a cationic dye, such as TMRE, persistent hyperpolarization will cause the accumulation of more dye molecules, and the intensity of the shifted light will increase. For an anionic dye, such as $DiBAC_4(3)$, persistent depolarization will have the same effect. Therefore, one consideration when choosing a dye is the expected voltage and the expected direction of its change; using both an anionic and a cationic dye, and detecting opposite changes in their respective intensities, can be a good control. Because the spectral shift of the slow-response probes depends on the movement of molecules,

* This calculation, the Hill equation, is available from the supplier.

these dyes are appropriate for measuring longer term phenomena, such as changes in the resting potential.

IV. FUNCTIONAL APPROACHES TO TESTING SPECIFIC TRANSPORT

Once the characterization of the expression of implicated transporters and their resulting ion flows is complete, it is crucial to move to functional experiments to understand the instructive role of these biophysical parameters in your system. Ca^{++} is probably the most-studied ion signal, but because of the huge number of calcium-dependent mechanisms, it may be difficult to separate its biochemical signaling functions from its roles as an ion *per se*. In general, the questions to be addressed using functional experiments will include the locus of action (Is it plasma membrane transport that is important or flux elsewhere?), autonomy (Are the fluxes signaling to other cells or moving molecules across cell fields, or is the effect cell-autonomous?), contribution to membrane voltage (Is it the K^+ ion flow itself that is important, or its role in determining overall membrane voltage?), upstream steps (What determines the nature, magnitude, and direction of the currents?), and downstream steps (Which genetic and cell behavior programs are activated as a result of this ion flow?).

Traditional methods for loss-of-function have involved shunting endogenous electrical fields [202, 203] or depletion of ions from medium [204]. Gain-of-function is also crucial, as modifying fields in predictable ways, or introducing ion flows into regions where they are normally not present, can provide crucial information and uncover novel ways to control morphology. For example, in LR asymmetry, it is necessary to introduce normally right-sided pumps into the left-side cells to truly understand the contribution of ion flow to patterning. Classical methods have involved direct application of electric fields [89, 205–208] and magnetic fields [209–211]. However, although applied AC magnetic fields do indeed modify the behavior of ions in the cytoplasm, magnetic field effects can be extremely difficult to interpret because they are almost impossible to restrict spatially and because they interact with so many other poorly characterized biological mechanisms [212]. In contrast, direct application of electrical fields can be very informative if the considerable issues regarding electrode products are properly controlled.

Fortunately, the progress of neurobiology and molecular biology has now provided a number of highly specific and powerful techniques for altering ion flows *in vivo*. Loss-of-function using molecular reagents is a very good method to validate the targets implicated in the drug screen. The use of morpholinos and RNAi is now standard in many species, and a variety of knock-outs and mutants exist in genetic systems such as mouse and zebrafish. One important caveat is that in some species (notably, *Xenopus* embryos), significant maternal protein exists for many channels and pumps from before fertilization and throughout early development. In such systems, antisense techniques targeting mRNA will be ineffective, and loss-of-function requires dominant negative protein expression or introduction of blocking antibodies. Maternal protein is, however, not usually a problem with chick or mouse embryos and may or may not be an issue for zebrafish depending on the target involved.

A huge number of w.t. and mutant constructs are available for many channels and pumps through the efforts of the neurobiology community, who are using them to study neuronal functions and are usually happy to send them to people working on basic patterning. In many cases, they have already been characterized in *Xenopus* oocytes, making things much easier for people working with frog embryos. Useful variants include those with constitutively active (missing regulatory regions), inactive (pore mutants and truncations), and more subtle (such as altered pH or voltage gating) properties. For example, for the KvLQT-1 channel, gain-of-function [213], loss-of-function [214], and blocker-insensitive mutants [215] are available. The blocker-insensitive mutants are especially useful for rescue experiments. Because many channels, pumps, and GJs operate through oligomerization, dominant negative effects can often be achieved by over-expressing inactive mutants or mutants with ER retention signals (which titrate out native complexes). One important issue to keep in mind is that introducing ectopic ion conductances into the ER may have effects unrelated to the loss of ion transport at the cell membrane, and this has to be controlled for by testing combination ER-retention and pore mutant constructs. In all cases, targeting of your construct to the appropriate location may need to be confirmed (this may be especially relevant when misexpressing constructs into evolutionarily distant species).

In many cases, the desired ion flow can be recapitulated by misexpressing one or two subunits of the native channel/pump complex. However, H^+ flux illustrates a potential issue that may occur. Intracellular alkalinization and pH waves are important in development [216–226]. These fluxes function in the determination of anterior neural fate [227], gene expression [220, 221], and cell cycle [228]. In many cases, the pH of the cell (and sometimes the membrane voltage) is controlled by the action of the V-ATPase complex [229, 230]. However, this is a 13-subunit complex, and misexpressing all of the subunits together to recapitulate the necessary H^+ pumping is extremely difficult. In such cases, one strategy is to utilize a genetically distinct construct encoding a simpler protein with a similar physiological role. Where pH changes are required, a good choice is the P-type H^+ pump from yeast or *Neurospora*, which is a single-subunit pump and has been used in mammalian cells [231]. Because the V-ATPase is highly electrogenic, it may be necessary to distinguish between roles of its contribution to pH versus membrane voltage. In such cases, a construct like the Na^+/H^+ exchanger may be useful; for example, NHE3 [232] is an electroneutral proton pump [233].

V. DETECTION AND MANIPULATION OF GJC

Molecular techniques have been used to probe the role of GJC in a number of physiological processes [234, 235]. Several human syndromes have been identified as mutations in connexin genes [236], and one of the most tantalizing roles for GJC is in the processes that distinguish normal tissue from tumor cells [237–241]. Normal tissue possesses a much higher degree of GJC than tumor tissue, and a loss of GJC accompanies early steps in neoplastic transformation [242]. Moreover, a neoplastic phenotype can be induced in cell culture by ectopic closing of gap junctions using pharmacological agents or dominant negative constructs [243]. Most interestingly, neoplastic characteristics can be suppressed by ectopic induction of GJC in tumor tissue [244–246]. GJs also play an important role in several aspects of embryogenesis [247–249] in vertebrate and invertebrate systems.

Knowledge of available gap-junctional paths provides important information about distribution of ion flows and dissipation of potentials to neighboring cells (which can result in non-cell-autonomous effects of any bioelectrical event). Gap junctions can be formed by docking of a wide variety of connexins [11, 12, 250], innexins/pannexins [251, 252], and possibly ductin [253–255] proteins. Because of the many steps of regulation of GJC after transcription and translation [256, 257], evaluation of connexin expression in a tissue is often insufficient. It is usually best to evaluate GJC functionally, because even existing junctions in the membrane may be closed or of complex permeability. The ability of GJs to regulate the signals between adjacent cells, to provide unidirectional (chemically rectifying) channels [258, 259], and to establish voltage-gated paths [260, 261] makes them a highly versatile and important component to consider in the physiological model of your system.

Junctional paths can be traced by microinjection of fluorescent small molecule dyes in large cells [262] or scrape-loaded into smaller cells. Various fluorescent techniques such as FRAP and photo-bleaching/uncaging of permeable molecules are beginning to be used to study GJC *in vivo* [263–267]. It is important to utilize a system of two differentially fluorescent molecules, mixing a large molecule dye that does not pass through gap junctions (such as Molecular Probes' Rhodamine-linked 10 kDa Dextran), together with a small (< 1 kD) tracer, to test transfer of the small dye to the neighboring cell if GJC paths exist. The inclusion of the larger dye will allow you to rule out cytoplasmic bridges and incomplete cell division (only cell pairs that show transfer of the small dye but not the large dye represent a true instance of GJC). The permeability of the specific gap junctions involved in your system can be tested using a panel of fluorescent small molecule probes with different shape/charge/size characteristics. Below are some candidates [268]:

Substance	MW	Charge
Lucifer Yellow	443	–2
2′,7′-dichlorofluorescein	401	–1
Neurobiotin	287	+1
6-carboxyfluorescein	376	–2
DAPI	350	+1
Ethidium bromide	314	+1
Propidium iodide	414	+2
Biocytin	373	0
Biotin-X cadaverin	442	+1
Alexa 350 hydrazide	349	–1
Alexa 488 hydrazide	570	–1

If GJC is implicated in your process, there are a variety of issues to consider. A good molecular dominant negative construct targeting a number of connexin family members is available [269], as well as more specific ones such as for Cx31

[270–272] and Cx26 [273]. Innexins and pannexins (invertebrate connexins) can also be inhibited by reagents such as heptanol [274, 275] and carbenoxolone [276], but no molecular dominant negatives have been developed as yet. Gain-of-function can sometimes be achieved by inducing existing gap junctions to open, using specific peptides [277] or melatonin [278], or by misexpressing w.t. or constitutively active connexin constructs.

Aside from the transfer of ions, GJs are known to transfer other small molecules such as cAMP and metabolites [279, 280]. One example of how GJC interfaces to downstream signaling elements has been observed. Specifically, information on how GJC is transduced into gene expression has become available in mammalian cells [281]: There is a connexin response element (CxRE) in the promoter of the osteocalcin gene. It contains Sp1 and Sp3 (Sp1 activates transcription, Sp3 represses). Open GJC alters recruitment of Sp1/SP3 to the promoter. When GJC is active, Sp1 is phosphorylated and recruited to the CxRE. Disruption of Cx43-based GJC by over-expression of Cx45 leads to the preferential recruitment of the repressor Sp3. This process involves ERK/PI3K signaling [282].

VI. CONCLUSION

The techniques described above should allow the investigation of bioelectric fields at the molecular level in a wide variety of tractable model systems. Some work has been done in the fields of development and regeneration, and we anticipate important advances in tumor and adult stem cell biology from an increased consideration of the roles of ion fluxes. The merging of genomic and biochemical data with biophysical and electrophysiological techniques is sure to uncover new vistas in biological control, which will be fascinating in terms of both basic biology and clinical applications [22, 283, 284].

Understanding the how of genetic control of development has been the focus of developmental biology since the end of the 19th century [285]. Importantly, the screen strategy we have described in this chapter can easily be adapted to other phenomena for which drug discovery has been significant. Systematic use of hierarchical prescreens of this kind, to implicate specific, known protein functions in biological phenomena, could improve the efficiency with which probably important genes are chosen for more extensive molecular characterization. Obvious areas of application include the testing of neurotransmitter involvement in prenervous patterning and the development of predictive, quantitative computer models of bioelectric signals in morphogenesis [286]. By combining molecular genetics with real-time, *in vivo* techniques for manipulating and detecting epigenetic biophysical events, a truly integrative systems biology of complex patterning systems will be achieved.

ACKNOWLEDGMENTS

This chapter is dedicated to E.J. Lund, whose early work on bioelectric fields and growth provided much inspiration for this work. We are grateful for support from

NIH grants GM-0.6227 to M.L. and K22-DE016633 to D.S.A., and grant #6-FY04-65 from the March of Dimes to M.L. This work was written in a Forsyth Institute facility renovated with support from Research Facilities Improvement Grant Number CO6RR11244 from the National Center for Research Resources, National Institutes of Health.

REFERENCES

1. Levin, M., Bioelectromagnetic patterning fields: roles in embryonic development, regeneration, and neoplasm. Bioelectromagnetics, 2003. 24(5): pp. 295–315.
2. Robinson, K.R. and M.A. Messerli, Left/right, up/down: the role of endogenous electrical fields as directional signals in development, repair, and invasion. BioEssays, 2003. 25: pp. 1–8.
3. McCaig, C.D. and M. Zhao, Physiological electrical fields modify cell behaviour. Bioessays, 1997. 19(9): pp. 819–26.
4. Jaffe, L., Developmental currents, voltages, and gradients, in Developmental order: its origin and regulation, S. Subtelny (Ed.). 1982, New York: Alan R. Liss. pp. 183–215.
5. McCaig, C.D., et al., Controlling cell behavior electrically: current views and future potential. Physiol Rev, 2005. 85(3): pp. 943–78.
6. Arcangeli, A., et al., Soluble or bound laminin elicit in human neuroblastoma cells short- or long-term potentiation of a K^+ inwardly rectifying current: relevance to neuritogenesis. Cell Adhesion & Communication, 1996. 4(4–5): pp. 369–85.
7. Jones, S. and A. Ribera, Overexpression of a potassium channel gene perturbs neural differentiation. Journal of Neuroscience, 1994. 14(5): pp. 2789–2799.
8. Klimatcheva, E. and W. Wonderlin, An ATP-sensitive K(+) current that regulates progression through early G1 phase of the cell cycle in MCF-7 human breast cancer cells. Journal of Membrane Biology, 1999. 171(1): pp. 35–46.
9. Putney, L.K. and D.L. Barber, Na-H exchange-dependent increase in intracellular pH times G2/M entry and transition. J Biol Chem, 2003. 278(45): pp. 44645–9.
10. Takamori, M., An autoimmune channelopathy associated with cancer: Lambert-Eaton myasthenic syndrome. Internal Medicine, 1999. 38(2): pp. 86–96.
11. Nicholson, B.J., Gap junctions—from cell to molecule. J Cell Sci, 2003. 116(Pt 22): pp. 4479–81.
12. Bruzzone, R., T. White, and D. Goodenough, The cellular Internet: on-line with connexins. Bioessays, 1996. 18(9): pp. 709–718.
13. Conley, E.C. and W.J. Brammar, The ion channel factsbook. Factsbook series. 1996, London; San Diego: Academic Press Harcourt Brace & Co.
14. Bradley, J., et al., Nomenclature for ion channel subunits. Science, 2001. 294(5549): pp. 2095–6.
15. Pongs, O., Potassium channel nomenclature: a personal view. Trends Pharmacol Sci, 1993. 14(12): pp. 435.
16. Wheatly, M.G. and Y. Gao, Molecular biology of ion motive proteins in comparative models. J Exp Biol, 2004. 207(Pt 19): pp. 3253–63.
17. Cone, C.D., Unified theory on the basic mechanism of normal mitotic control and oncogenesis. Journal of Theoretical Biology, 1971. 30(1): pp. 151–81.
18. Cone, C.D., Variation of the transmembrane potential level as a basic mechanism of mitosis control. Oncology, 1970. 24(6): pp. 438–70.

19. Olivotto, M., et al., Electric fields at the plasma membrane level: a neglected element in the mechanisms of cell signalling. Bioessays, 1996. 18(6): pp. 495–504.
20. Nicholls, J.G., et al., From neuron to brain: a cellular and molecular approach to the function of the nervous system. 3rd ed. 1992, Sunderland, MA: Sinauer Associates.
21. Hille, B., Ion channels of excitable membranes. 3rd ed. 2001, Sunderland, MA: Sinauer.
22. Borgens, R., et al., Electric fields in vertebrate repair. 1989, New York: Alan R. Liss.
23. White, T.W., et al., Voltage gating of connexins. Nature, 1994. 371(6494): pp. 208–9.
24. Lampe, P.D. and A.F. Lau, Regulation of gap junctions by phosphorylation of connexins. Archives of Biochemistry & Biophysics, 2000. 384(2): pp. 205–15.
25. Bruzzone, R. and C. Giaume, Connexins and information transfer through glia. Advances in Experimental Medicine & Biology, 1999. 468: pp. 321–37.
26. Goldberg, G.S., P.D. Lampe, and B.J. Nicholson, Selective transfer of endogenous metabolites through gap junctions composed of different connexins. Nature Cell Biology, 1999. 1(7): pp. 457–9.
27. Jalife, J., G.E. Morley, and D. Vaidya, Connexins and impulse propagation in the mouse heart. Journal of Cardiovascular Electrophysiology, 1999. 10(12): pp. 1649–63.
28. Corcoran, E.E. and A.R. Means, Defining Ca2+/calmodulin-dependent protein kinase cascades in transcriptional regulation. J Biol Chem, 2001. 276(5): pp. 2975–8.
29. Ludwig, M.G., et al., Proton-sensing G-protein-coupled receptors. Nature, 2003. 425(6953): pp. 93–8.
30. Akopian, A.N., et al., A new member of the acid-sensing ion channel family. Neuroreport, 2000. 11(10): pp. 2217–22.
31. Berdiev, B.K., et al., Acid-sensing ion channels in malignant gliomas. J Biol Chem, 2003. 278(17): pp. 15023–34.
32. Chu, X.P., et al., Proton-gated channels in PC12 cells. J Neurophysiol, 2002. 87(5): pp. 2555–61.
33. Waldmann, R., Proton-gated cation channels—neuronal acid sensors in the central and peripheral nervous system. Adv Exp Med Biol, 2001. 502: pp. 293–304.
34. Aziz, Q., Acid sensors in the gut: a taste of things to come. Eur J Gastroenterol Hepatol, 2001. 13(8): pp. 885–8.
35. Wheeler, D.G. and E. Cooper, Depolarization strongly induces human cytomegalovirus major immediate-early promoter/enhancer activity in neurons. J Biol Chem, 2001. 276(34): pp. 31978–85.
36. Finkbeiner, S. and M.E. Greenberg, Ca2+ channel-regulated neuronal gene expression. J Neurobiol, 1998. 37(1): pp. 171–89.
37. Mermelstein, P.G., et al., Calmodulin priming: nuclear translocation of a calmodulin complex and the memory of prior neuronal activity. Proceedings of the National Academy of Sciences of the United States of America, 2001. 98(26): pp. 15342–7.
38. Levin, M. and M. Mercola, Gap junctions are involved in the early generation of left right asymmetry. Developmental Biology, 1998. 203(1): pp. 90–105.
39. Levin, M. and M. Mercola, Gap junction-mediated transfer of left-right patterning signals in the early chick blastoderm is upstream of Shh asymmetry in the node. Development, 1999. 126: pp. 4703–4714.
40. Levin, M., et al., Asymmetries in H+/K+-ATPase and cell membrane potentials comprise a very early step in left-right patterning. Cell, 2002. 111(1): pp. 77–89.
41. Fukumoto, T., R. Blakely, and M. Levin, Serotonin transporter function is an early step in left-right patterning in chick and frog embryos. Dev Neurosci, 2005. 27(6): pp. 349–63.

42. Fukumoto, T., I.P. Kema, and M. Levin, Serotonin signaling is a very early step in patterning of the left-right axis in chick and frog embryos. Curr Biol, 2005. 15(9): pp. 794–803.
43. Fujita, A., et al., Specific localization of an inwardly rectifying K+ channel, Kir4.1, at the apical membrane of rat gastric parietal cells; its possible involvement in K+ recycling for the H+-K+ pump. Journal of Physiology, 2002. 540(1): pp. 85–92.
44. Grahammer, F., et al., The cardiac K+ channel KCNQ1 is essential for gastric acid secretion. Gastroenterology, 2001. 120(6): pp. 1363–71.
45. Lund, E., Bioelectric fields and growth. 1947, Austin: Univ. of Texas Press.
46. Jaffe, L., The role of ionic currents in establishing developmental pattern. Philosophical Transactions of the Royal Society (Series B), 1981. 295: pp. 553–566.
47. Jaffe, L. and R. Nuccitelli, Electrical controls of development. Annual Review of Biophysics and Bioengineering, 1977. 6: pp. 445–476.
48. Reid, B., et al., Wound healing in rat cornea: the role of electric currents. Faseb J, 2005. 19(3): pp. 379–86.
49. Nuccitelli, R., Endogenous electric fields in embryos during development, regeneration and wound healing. Radiat Prot Dosimetry, 2003. 106(4): pp. 375–83.
50. Nuccitelli, R., A role for endogenous electric fields in wound healing. Curr Top Dev Biol, 2003. 58: pp. 1–26.
51. Robinson, K. and M. Messerli, Electric embryos: the embryonic epithelium as a generator of developmental information, in Nerve growth and guidance, C. McCaig (Ed.). 1996, Portland Press: London, UK. pp. 131–141.
52. Stern, C.D., Do ionic currents play a role in the control of development? Bioessays, 1986. 4(4): pp. 180–4.
53. Nuccitelli, R., Ionic currents in development. 1986, New York: Alan R. Liss.
54. Matzke, M.A. and A.J. Matzke, Electric fields and the nuclear membrane. Bioessays, 1996. 18(10): pp. 849–50.
55. McCaig, C.D. and A.M. Rajnicek, Electrical fields, nerve growth and nerve regeneration. Experimental Physiology, 1991. 76(4): pp. 473–94.
56. McCaig, C.D., Nerve guidance: a role for bio-electric fields? Progress in Neurobiology, 1988. 30(6): pp. 449–68.
57. McCaig, C.D., et al., Has electrical growth cone guidance found its potential? Trends in Neurosciences, 2002. 25(7): pp. 354–9.
58. Robinson, K.R. and C. McCaig, Electrical fields, calcium gradients, and cell growth. Annals of the New York Academy of Sciences, 1980. 339: pp. 132–8.
59. Nuccitelli, R., Ionic currents in morphogenesis. Experientia, 1988. 44(8): pp. 657–66.
60. Altizer, A.M., et al., Endogenous electric current is associated with normal development of the vertebrate limb. Developmental Dynamics, 2001. 221(4): pp. 391–401.
61. Borgens, R.B., Are limb development and limb regeneration both initiated by an integumentary wounding? A hypothesis. Differentiation, 1984. 28(2): pp. 87–93.
62. Robinson, K.R., Endogenous electrical current leaves the limb and prelimb region of the Xenopus embryo. Developmental Biology, 1983. 97(1): pp. 203–11.
63. Borgens, R., M. Metcalf, and R. Shi, Endogenous ionic currents and voltages in amphibian embryos. Journal of Experimental Zoology, 1994. 268: pp. 307–322.
64. Shi, R. and R. Borgens, Embryonic neuroepithelial sodium transport, the resulting physiological potential, and cranial development. Developmental Biology, 1994. 165(1): pp. 105–16.
65. Hotary, K.B. and K.R. Robinson, The neural-tube of the Xenopus embryo maintains a potential difference across itself. Developmental Brain Research, 1991. 59(1): pp. 65–73.

66. Borgens, R. and R. Shi, Uncoupling histogenesis from morphogenesis in the vertebrate embryo by collapse of the transneural tube potential. Developmental Dynamics, 1995. 203(4): pp. 456–67.
67. Hotary, K. and K. Robinson, Evidence of a role for endogenous electric fields in chick embryo development. Development, 1992. 114: pp. 985–996.
68. Rehm, W.S., Bud regeneration and electric polarities in Phaseolus multiflorus. Plant Physiology, 1938. 13: pp. 81–101.
69. Rehm, W.S., Electrical responses of Phaseolus multiflorus to electrical currents. Plant Physiology, 1939. 14: pp. 359–363.
70. Novak, B. and C. Sirnoval, Inhibition of regeneration of Acetabularia mediterranea enucleated posterior stalk segments by electrical isolation. Plant Science Letters. 5, 1975: pp. 183–188.
71. Novák, B. and F.W. Bentrup, An electrophysiological study of regeneration in Acetabularia Mediterranea. Planta, 1972. 108: pp. 227–244.
72. Nawata, T., T. Hishinuma, and S. Wada, Ionic currents during regeneration of thallus and rhizoid from cell segments isolated from the marine alga Bryopsis. Biological Bulletin, 1989. 176S: pp. 41–45.
73. Neufeld, D.A., S.K. Westley, and B. Clarke, The electrical potential gradient in regenerating Tubularia: spatial and temporal characteristics. Growth, 1978. 42: pp. 347–356.
74. Marsh, G. and H.W. Beams, Electrical control of axial polarity in a regenerating annelid. Anatomical Record, 1949. 105(3): pp. 513–514.
75. Marsh, G. and J. Dimmitt, Specific resistance of regenerating Dugesia in different media and the electrical work of polarity control. Federation Proceedings, 1949. 8(1): pp. 106–106.
76. Marsh, G. and H.W. Beams, Electrical control of growth axis in a regenerating annelid. Anatomical Record, 1950. 108(3): pp. 512–512.
77. Marsh, G. and H.W. Beams, Electrical control of morphogenesis in regenerating Dugesia tigrina. 1. Relation of axial polarity to field strength. Journal of Cellular and Comparative Physiology, 1952. 39(2): pp. 191–213.
78. Marsh, G. and H. Beams, Electrical control of morphogenesis in regenerating Dugesia tigrina. Journal of Cellular and Comparative Physiology, 1957. 39: pp. 191–211.
79. Illingworth, C.M. and A.T. Barker, Measurement of electrical currents emerging during the regeneration of amputated fingertips in children. Clinical Physics and Physiological Measurement, 1980. 1: pp. 87–89.
80. Kurtz, I. and A.R. Schrank, Bioelectrical properties of intact and regenerating earthworms Eisenia foetida. Physiological Zoology, 1955. 28: pp. 322–330.
81. Dimmitt, J. and G. Marsh, Electrical control of morphogenesis in regenerating Dugesia tigrina. 2. Potential gradient vs. current density as control factors. Journal of Cellular and Comparative Physiology, 1952. 40(1): pp. 11–23.
82. Cone, C.D. and M. Tongier, Control of somatic cell mitosis by simulated changes in the transmembrane potential level. Oncology, 1971. 25(2): pp. 168–82.
83. Cone, C.D. and M. Tongier, Contact inhibition of division: involvement of the electrical transmembrane potential. Journal of Cellular Physiology, 1973. 82(3): pp. 373–86.
84. Stillwell, E.F., C.M. Cone, and C.D. Cone, Stimulation of DNA synthesis in CNS neurones by sustained depolarisation. Nature—New Biology, 1973. 246(152): pp. 110–1.
85. Wang, E., et al., Re-orientation and faster, directed migration of lens epithelial cells in a physiological electric field. Experimental Eye Research, 2000. 71(1): pp. 91–8.

86. McCaig, C.D., L. Sangster, and R. Stewart, Neurotrophins enhance electric field-directed growth cone guidance and directed nerve branching. Developmental Dynamics, 2000. 217(3): pp. 299–308.
87. Zhao, M., et al., Electric field-directed cell motility involves up-regulated expression and asymmetric redistribution of the epidermal growth factor receptors and is enhanced by fibronectin and laminin. Molecular Biology of the Cell, 1999. 10(4): pp. 1259–76.
88. Rajnicek, A.M., K.R. Robinson, and C.D. McCaig, The direction of neurite growth in a DC electric field depends on the substratum. Molecular Biology of the Cell, 1998. 9: pp. 1308.
89. Zhao, M., et al., Human corneal epithelial cells reorient and migrate cathodally in a small applied electric field. Current Eye Research, 1997. 16(10): pp. 973–84.
90. Davenport, R.W. and C.D. McCaig, Hippocampal growth cone responses to focally applied electric fields. Journal of Neurobiology, 1993. 24(1): pp. 89–100.
91. McCaig, C.D. and R. Stewart, The effects of melanocortins and electrical fields on neuronal growth. Experimental Neurology, 1992. 116(2): pp. 172–9.
92. Rajnicek, A.M., N.A. Gow, and C.D. McCaig, Electric field-induced orientation of rat hippocampal neurones in vitro. Experimental Physiology, 1992. 77(1): pp. 229–32.
93. McCaig, C.D. and P.J. Dover, Factors influencing perpendicular elongation of embryonic frog muscle cells in a small applied electric field. Journal of Cell Science, 1991. 98(Pt 4): pp. 497–506.
94. McCaig, C.D., Nerve growth in a small applied electric field and the effects of pharmacological agents on rate and orientation. Journal of Cell Science, 1990. 95(Pt 4): pp. 617–22.
95. McCaig, C.D., Nerve branching is induced and oriented by a small applied electric field. Journal of Cell Science, 1990. 95(Pt 4): pp. 605–15.
96. McCaig, C.D., Studies on the mechanism of embryonic frog nerve orientation in a small applied electric field. Journal of Cell Science, 1989. 93(Pt 4): pp. 723–30.
97. McCaig, C.D., Nerve growth in the absence of growth cone filopodia and the effects of a small applied electric field. Journal of Cell Science, 1989. 93(Pt 4): pp. 715–21.
98. Djamgoz, M.B.A., et al., Directional movement of rat prostate cancer cells in direct-current electric field: involvement of voltage-gated Na+ channel activity. Journal of Cell Science, 2001. 114(Pt 14): pp. 2697–705.
99. Gruler, H. and R. Nuccitelli, The galvanotaxis response mechanism of keratinocytes can be modeled as a proportional controller. Cell Biochemistry and Biophysics, 2000. 33(1): pp. 33–51.
100. Poo, M.M. and K.R. Robinson, Electrophoresis of concanavalin-a receptors along embryonic muscle-cell membrane. Nature, 1977. 265(5595): pp. 602–605.
101. Verworn, M., Die polare Erregnung der Protisten durch den galvanischen Strom. Pflugers Archive, 1889. 45: pp. 1–36.
102. McCaig, C.D. and P.J. Dover, Raised cyclic-AMP and a small applied electric field influence differentiation, shape, and orientation of single myoblasts. Developmental Biology, 1993. 158(1): pp. 172–82.
103. Erskine, L. and C.D. McCaig, The effects of lyotropic anions on electric field-induced guidance of cultured frog nerves. Journal of Physiology, 1995. 486(Pt 1): pp. 229–36.
104. Stewart, R., L. Erskine, and C.D. McCaig, Calcium channel subtypes and intracellular calcium stores modulate electric field-stimulated and -oriented nerve growth. Developmental Biology, 1995. 171(2): pp. 340–51.

105. Erskine, L., R. Stewart, and C.D. McCaig, Electric field-directed growth and branching of cultured frog nerves: effects of aminoglycosides and polycations. Journal of Neurobiology, 1995. 26(4): pp. 523–36.
106. Rajnicek, A.M., K.R. Robinson, and C.D. McCaig, The direction of neurite growth in a weak DC electric field depends on the substratum: Contributions of adhesivity and net surface charge. Developmental Biology, 1998. 203(2): pp. 412–423.
107. Erskine, L. and C.D. McCaig, Growth cone neurotransmitter receptor activation modulates electric field-guided nerve growth. Developmental Biology, 1995. 171(2): pp. 330–9.
108. Arcangeli, A., et al., A novel inward-rectifying K+ current with a cell-cycle dependence governs the resting potential of mammalian neuroblastoma cells. Journal of Physiology, 1995. 489(Pt 2): pp. 455–71.
109. Schuldiner, S. and E. Rozengurt, Na+/H+ antiport in Swiss 3T3 cells: mitogenic stimulation leads to cytoplasmic alkalinization. Proceedings of the National Academy of Sciences of the United States of America, 1982. 79(24): pp. 7778–82.
110. Lopez-Rivas, A., E.A. Adelberg, and E. Rozengurt, Intracellular K+ and the mitogenic response of 3T3 cells to peptide factors in serum-free medium. Proceedings of the National Academy of Sciences of the United States of America, 1982. 79(20): pp. 6275–9.
111. Burns, C.P. and E. Rozengurt, Extracellular Na+ and initiation of DNA synthesis: role of intracellular pH and K+. Journal of Cell Biology, 1984. 98(3): pp. 1082–9.
112. Pillozzi, S., et al., HERG potassium channels are constitutively expressed in primary human acute myeloid leukemias and regulate cell proliferation of normal and leukemic hemopoietic progenitors. Leukemia, 2002. 16(9): pp. 1791–8.
113. Wohlrab, D., J. Wohlrab, and F. Markwardt, Electrophysiological characterization of human keratinocytes using the patch-clamp technique. Experimental Dermatology, 2000. 9(3): pp. 219–23.
114. Dalle-Lucca, S.L., et al., Abnormal proliferative response of the carotid artery of spontaneously hypertensive rats after angioplasty may be related to the depolarized state of its smooth muscle cells. Brazilian Journal of Medical & Biological Research, 2000. 33(8): pp. 919–27.
115. Wohlrab, D. and W. Hein, Der Einfluss von Ionenkanalmodulatoren auf das Membranpotential humaner Chondrozyten. Orthopade, 2000. 29(2): pp. 80–4.
116. MacFarlane, S.N. and H. Sontheimer, Changes in ion channel expression accompany cell cycle progression of spinal cord astrocytes. Glia, 2000. 30(1): pp. 39–48.
117. Kamleiter, M., et al., Voltage-dependent membrane currents of cultured human neurofibromatosis type 2 Schwann cells. Glia, 1998. 24(3): pp. 313–22.
118. Wang, S., et al., Evidence for an early G1 ionic event necessary for cell cycle progression and survival in the MCF-7 human breast carcinoma cell line. Journal of Cellular Physiology, 1998. 176(3): pp. 456–64.
119. Knutson, P., et al., K+ channel expression and cell proliferation are regulated by intracellular sodium and membrane depolarization in oligodendrocyte progenitor cells. Journal of Neuroscience, 1997. 17(8): pp. 2669–82.
120. Palma, V., M. Kukuljan, and R. Mayor, Calcium mediates dorsoventral patterning of mesoderm in Xenopus. Current Biology, 2001. 11(20): pp. 1606–10.
121. Leclerc, C., et al., Imaging patterns of calcium transients during neural induction in Xenopus laevis embryos. Journal of Cell Science, 2000. 113 Pt 19: pp. 3519–29.
122. Leclerc, C., et al., L-type calcium channel activation controls the in vivo transduction of the neuralizing signal in the amphibian embryos. Mechanisms of Development, 1997. 64(1–2): pp. 105–10.

123. Leclerc, C., A.M. Duprat, and M. Moreau, Noggin upregulates Fos expression by a calcium-mediated pathway in amphibian embryos. Development Growth & Differentiation, 1999. 41(2): pp. 227–38.
124. Moreau, M., et al., Increased internal Ca2+ mediates neural induction in the amphibian embryo. Proceedings of the National Academy of Sciences of the United States of America, 1994. 91(26): pp. 12639–43.
125. Drean, G., et al., Expression of L-type Ca2+ channel during early embryogenesis in Xenopus laevis. International Journal of Developmental Biology, 1995. 39(6): pp. 1027–32.
126. Zeretzke, S., et al., Ca2+-ions and pattern control in Hydra. International Journal of Developmental Biology, 2002. 46: pp. 705–710.
127. Webb, S.E., et al., Calcium transients and neural induction in vertebrates. Cell Calcium, 2005. 37(5): pp. 375–85.
128. Webb, S.E. and A.L. Miller, Calcium signalling during embryonic development. Nat Rev Mol Cell Biol, 2003. 4(7): pp. 539–51.
129. Webb, S.E. and A.L. Miller, Imaging intercellular calcium waves during late epiboly in intact zebrafish embryos. Zygote, 2003. 11(2): pp. 175–82.
130. Webb, S.E., et al., Calcium in development: from ion transients to gene expression. Bioessays, 2001. 23(4): pp. 372–4.
131. Leung, C.F., S.E. Webb, and A.L. Miller, On the mechanism of ooplasmic segregation in single-cell zebrafish embryos. Dev Growth Differ, 2000. 42(1): pp. 29–40.
132. Webb, S.E. and A.L. Miller, Calcium signalling during zebrafish embryonic development. Bioessays, 2000. 22(2): pp. 113–23.
133. Messenger, N.J. and A.E. Warner, Primary neuronal differentiation in Xenopus embryos is linked to the beta(3) subunit of the sodium pump. Developmental Biology, 2000. 220(2): pp. 168–82.
134. Uochi, T., et al., The Na+,K+-ATPase alpha subunit requires gastrulation in the Xenopus embryo. Development Growth & Differentiation, 1997. 39(5): pp. 571–80.
135. Burgener-Kairuz, P., et al., Polyadenylation of Na(+)-K(+)-ATPase beta 1-subunit during early development of Xenopus laevis. American Journal of Physiology, 1994. 266(1 Pt 1): pp. C157–64.
136. Simoncini, L., M.L. Block, and W.J. Moody, Lineage-specific development of calcium currents during embryogenesis. Science, 1988. 242(4885): pp. 1572–5.
137. Borgens, R.B., A.R. Blight, and M.E. McGinnis, Functional recovery after spinal cord hemisection in guinea pigs: the effects of applied electric fields. J Comp Neurol, 1990. 296(4): pp. 634–53.
138. Borgens, R.B., What is the role of naturally produced electric current in vertebrate regeneration and healing? International Review of Cytology, 1982. 76: pp. 245–98.
139. Borgens, R., The role of ionic current in the regeneration and development of the amphibian limb, in Limb development and regeneration. 1983, New York: Alan R. Liss. pp. 597–608.
140. Puro, D.G., F. Roberge, and C.C. Chan, Retinal glial cell proliferation and ion channels: a possible link. Investigative Ophthalmology & Visual Science, 1989. 30(3): pp. 521–529.
141. Killion, J.J., Electrical properties of normal and transformed mammalian cells. Biophysical Journal, 1984. 45(3): pp. 523–8.
142. Martinez-Zaguilan, R. and R.J. Gillies, A plasma membrane V-type H(+)-ATPase may contribute to elevated intracellular pH (pHin) in some human tumor cells. Annals of the New York Academy of Sciences, 1992. 671: pp. 478–80.

143. Martinez-Zaguilan, R., et al., Vacuolar-type H(+)-ATPases are functionally expressed in plasma membranes of human tumor cells. American Journal of Physiology, 1993. 265(4 Pt 1): pp. C1015–29.
144. Bianchi, L., et al., herg encodes a K+ current highly conserved in tumors of different histogenesis: a selective advantage for cancer cells? Cancer Research, 1998. 58(4): pp. 815–22.
145. McLean, L.A., et al., Malignant gliomas display altered pH regulation by NHE1 compared with nontransformed astrocytes. Am J Physiol Cell Physiol, 2000. 278(4): pp. C676–88.
146. Teichmann, E.M., et al., Untersuchung eines m&glichen mutagenen Potenzials von Magnetfeldern. Rofo. Fortschritte auf dem Gebiete der Rontgenstrahlen und der Neuen Bildgebenden Verfahren, 2000. 172(11): pp. 934–9.
147. Perona, R. and R. Serrano, Increased pH and tumorigenicity of fibroblasts expressing a yeast proton pump. Nature, 1988. 334(6181): pp. 438–40.
148. Gunn, J.M., et al., NIH3T3 cells transfected with the yeast H(+)-ATPase have altered rates of protein turnover. Arch Biochem Biophys, 1994. 314(2): pp. 268–75.
149. Chen, I. and M. Levin, The role of KATP channels in development of left-right asymmetry in Xenopus. Journal of Dental Research, 2004. 83: pp. A1340.
150. Adams, D.S. and M. Levin, Early patterning of the left/right axis, in Gastrulation: from cells to embryo, C.D. Stern (Ed.). 2004, Cold Spring Harbor, NY: Cold Spring Harbor Press. pp. 403–417.
151. Levin, M., Electric embryos: endogenous ion fluxes and voltage gradients in left-right asymmetry. Developmental Biology, 2003. 259: pp. 482.
152. Masi, A., D.S. Adams, and M. Levin, Molecular characterization of bioelectrical events required for regeneration. In preparation.
153. Nogi, T., K. Robinson, and M. Levin, The V-ATPase generates asymmetric H+ flux required for correct LR asymmetry in chick and frog embryos. In preparation.
154. Nogi, T., D.S. Adams, and M. Levin, Electric controls of regeneration in planaria. Developmental Biology, 2003. 259: pp. 586.
155. Tomlinson, M.L., R.A. Field, and G.N. Wheeler, Xenopus as a model organism in developmental chemical genetic screens. Molecular BioSystems, 2005. 1(3): pp. 223–228.
156. Mitchison, T.J., Towards a pharmacological genetics. Chem Biol, 1994. 1(1): pp. 3–6.
157. Tosteson, M.T., et al., Interactions of palytoxin with the Na,K-ATPase. Where are those sites? Annals of the New York Academy of Sciences, 1997. 834: pp. 424–5.
158. Adams, D.S., et al. Early H^+-V-ATPase-dependent proton flux is necessary for consistent left-right patterning of non-mammalian vertebrates. Development, 2006. 133(9): pp. 1657–1671.
159. Ishii, Y., et al. An inhibitor for H+/K+-ATPase disrupted the left-right asymmetry in the echinoid embryo. Annual Conference of Zoological Society of Japan, 2003. Japan.
160. Shimeld, S.M. Regulation of left right asymmetry in Ciona. International Urochordate Meeting, 2003. Carry le Rouet, France.
161. Bellemann, P. and G. Franckowiak, Different receptor affinities of the enantiomers of BAY K 8644, a dihydropyridine Ca channel activator. Eur J Pharmacol, 1985. 118(1–2): pp. 187–8.
162. Rampe, D., et al., Comparison of the in vitro and in vivo cardiovascular effects of two structurally distinct Ca++ channel activators, BAY K 8644 and FPL 64176. J Pharmacol Exp Ther, 1993. 265(3): pp. 1125–30.
163. Jiang, S., et al., Antileishmanial activity of the antiulcer agent omeprazole. Antimicrobial Agents & Chemotherapy, 2002. 46(8): pp. 2569–74.

164. Etter, A., et al., Picrotoxin blockade of invertebrate glutamate-gated chloride channels: subunit dependence and evidence for binding within the pore. J Neurochem, 1999. 72(1): pp. 318–26.
165. Alshuaib, W.B. and M.V. Mathew, Blocking effect of lanthanum on delayed-rectifier K+ current in Drosophila neurons. Int J Neurosci, 2004. 114(5): pp. 639–50.
166. Gasque, G., et al., Shal and shaker differential contribution to the K+ currents in the Drosophila mushroom body neurons. J Neurosci, 2005. 25(9): pp. 2348–58.
167. Rutenberg, J., S.M. Cheng, and M. Levin, Early embryonic expression of ion channels and pumps in chick and Xenopus development. Developmental Dynamics, 2002. 225(4): pp. 469–84.
168. Vitavska, O., H. Wieczorek, and H. Merzendorfer, A novel role for subunit C in mediating binding of the H(+)-V-ATPase to the actin cytoskeleton. J Biol Chem, 2003.
169. Lee, B.S., S.L. Gluck, and L.S. Holliday, Interaction between vacuolar H(+)-ATPase and microfilaments during osteoclast activation. J Biol Chem, 1999. 274(41): pp. 29164–71.
170. Holliday, L.S., et al., The amino-terminal domain of the B subunit of vacuolar H+-ATPase contains a filamentous actin binding site. J Biol Chem, 2000. 275(41): pp. 32331–7.
171. Al-Awqati, Q., et al., Phenotypic plasticity in the intercalated cell: the hensin pathway. American Journal of Physiology, 1998. 275(2 pt 2): pp. F183–90.
172. Alpern, R.J., Hensin: a matrix protein determinant of epithelial polarity. Journal of Clinical Investigation, 1996. 98(10): pp. 2189–90.
173. Vijayakumar, S., et al., Hensin remodels the apical cytoskeleton and induces columnarization of intercalated epithelial cells: processes that resemble terminal differentiation. Journal of Cell Biology, 1999. 144(5): pp. 1057–67.
174. Adams, D.S., et al., Early H^+-V-ATPase-dependent proton flux is necessary for consistent left-right patterning of non-mammalian vertebrates. Development, 2006. 133(9): pp. 1657–1671.
175. Andresen, M., R. Schmitz-Salue, and S. Jakobs, Short tetracysteine tags to beta-tubulin demonstrate the significance of small labels for live cell imaging. Mol Biol Cell, 2004. 15(12): pp. 5616–22.
176. Park, M.K., A.V. Tepikin, and O.H. Petersen, What can we learn about cell signalling by combining optical imaging and patch clamp techniques? Pflugers Arch, 2002. 444(3): pp. 305–16.
177. Jurkat-Rott, K. and F. Lehmann-Horn, The patch clamp technique in ion channel research. Curr Pharm Biotechnol, 2004. 5(4): pp. 387–95.
178. Nuccitelli, R., Vibrating probe—high spatial-resolution extracellular current measurement. Federation Proceedings, 1980. 39(6): pp. 2129–2129.
179. Nuccitelli, R., A two-dimensional extracellular vibrating probe for the detection of trans-cellular ionic currents. Biophysical Journal, 1987. 51(2): pp. A447–A447.
180. Smith, P.J. and J. Trimarchi, Noninvasive measurement of hydrogen and potassium ion flux from single cells and epithelial structures. American Journal of Physiology—Cell Physiology, 2001. 280(1): pp. C1–11.
181. Smith, P.J., et al., Self-referencing, non-invasive, ion selective electrode for single cell detection of trans-plasma membrane calcium flux. Microscopy Research & Technique, 1999. 46(6): pp. 398–417.
182. Hotary, K.B., R. Nuccitelli, and K.R. Robinson, A computerized 2-dimensional vibrating probe for mapping extracellular current patterns. Journal of Neuroscience Methods, 1992. 43(1): pp. 55–67.

183. Messerli, M.A. and K.R. Robinson, Cytoplasmic acidification and current influx follow growth pulses of Lilium longiflorum pollen tubes. Plant Journal, 1998. 16(1): pp. 87–91.
184. Brown, D., P.J. Smith, and S. Breton, Role of V-ATPase-rich cells in acidification of the male reproductive tract. Journal of Experimental Biology, 1997. 200(Pt 2): pp. 257–62.
185. Nuccitelli, R., Endogenous electric fields measured in developing embryos. Electromagnetic Fields: Biological Interactions and Mechanisms, Martin Black (Ed.), Oxford University Press and American Chemical Society, 1995. pp. 109–124.
186. Keefe, D., et al., Identification of calcium flux in single preimplantation mouse embryos with the calcium-sensitive vibrating probe. Biological Bulletin, 1995. 189(2): pp. 200.
187. Anderson, M., E. Bowdan, and J.G. Kunkel, Comparison of defolliculated oocytes and intact follicles of the cockroach using the vibrating probe to record steady currents. Developmental Biology, 1994. 162(1): pp. 111–22.
188. Nuccitelli, R., Endogenous ionic currents and DC electric-fields in multicellular animal tissues. Bioelectromagnetics, 1992: pp. 147–157.
189. Hotary, K.B. and K.R. Robinson, The neural tube of the Xenopus embryo maintains a potential difference across itself. Developmental Brain Research, 1991. 59(1): pp. 65–73.
190. Hotary, K.B. and K.R. Robinson, Endogenous electrical currents and the resultant voltage gradients in the chick embryo. Developmental Biology, 1990. 140(1): pp. 149–60.
191. Epps, D., M. Wolfe, and V. Groppi, Characterization of the steady-state and dynamic fluorescence properties of the potential-sensitive dye bis-(1,3-dibutylbarbituric acid)trimethine oxonol (Dibac4(3)) in model systems and cells. Chem. Phys. Lipids, 1994. 69(2): pp. 137–150.
192. Krotz, F., et al., Membrane-potential-dependent inhibition of platelet adhesion to endothelial cells by epoxyeicosatrienoic acids. Arterioscler Thromb Vasc Biol, 2004. 24(3): pp. 595–600.
193. Gasalla-Herraiz, J., S. Rhee, and C.M. Isales, Calcium-sensitive probes for the measurement of intracellular calcium: effects of buffer system and magnesium concentration. Biochem Biophys Res Commun, 1995. 214(2): pp. 373–88.
194. Franck, P., et al., Measurement of intracellular pH in cultured cells by flow cytometry with BCECF-AM. J Biotechnol, 1996. 46(3): pp. 187–95.
195. Amirand, C., et al., Intracellular pH in one-cell mouse embryo differs between subcellular compartments and between interphase and mitosis. Biol Cell, 2000. 92(6): pp. 409–19.
196. Edwards, L.J., D.A. Williams, and D.K. Gardner, Intracellular pH of the preimplantation mouse embryo: effects of extracellular pH and weak acids. Molecular Reproduction & Development, 1998. 50(4): pp. 434–42.
197. Bassnett, S., L. Reinisch, and D.C. Beebe, Intracellular pH measurement using single excitation-dual emission fluorescence ratios. Am J Physiol, 1990. 258(1 Pt 1): pp. C171–8.
198. Bassnett, S., Intracellular pH regulation in the embryonic chicken lens epithelium. J Physiol, 1990. 431: pp. 445–64.
199. Dascalu, A., Z. Nevo, and R. Korenstein, The control of intracellular pH in cultured avian chondrocytes. J Physiol, 1993. 461: pp. 583–99.
200. De Clerck, L.S., et al., Use of fluorescent dyes in the determination of adherence of human leucocytes to endothelial cells and the effect of fluorochromes on cellular function. J Immunol Methods, 1994. 172(1): pp. 115–24.
201. Sater, A.K., J.M. Alderton, and R.A. Steinhardt, An increase in intracellular pH during neural induction in Xenopus. Development, 1994. 120(2): pp. 433–42.

202. Hotary, K.B. and K.R. Robinson, Evidence of a role for endogenous electrical fields in chick embryo development. Development, 1992. 114(4): pp. 985–996.
203. Borgens, R.B., J.W. Vanable, Jr., and L.F. Jaffe, Role of subdermal current shunts in the failure of frogs to regenerate. Journal of Experimental Zoology, 1979. 209(1): pp. 49–56.
204. Borgens, R.B., J.W. Vanable, Jr., and L.F. Jaffe, Reduction of sodium dependent stump currents disturbs urodele limb regeneration. Journal of Experimental Zoology, 1979. 209(3): pp. 377–86.
205. Zhao, M., J.V. Forrester, and C.D. McCaig, A small, physiological electric field orients cell division. Proceedings of the National Academy of Sciences of the United States of America, 1999. 96(9): pp. 4942–6.
206. Borgens, R.B., et al., An imposed oscillating electrical field improves the recovery of function in neurologically complete paraplegic dogs. J Neurotrauma, 1999. 16(7): pp. 639–57.
207. Borgens, R.B., Electrically mediated regeneration and guidance of adult mammalian spinal axons into polymeric channels. Neuroscience, 1999. 91(1): pp. 251–64.
208. Borgens, R.B., et al., An imposed oscillating electrical field improves the recovery of function in neurologically complete paraplegic dogs. [erratum appears in J Neurotrauma 2000 Aug;17(8):727.] Journal of Neurotrauma, 1999. 16(7): pp. 639–57.
209. Jenrow, K.A., C.H. Smith, and A.R. Liboff, Weak extremely-low-frequency magnetic field-induced regeneration anomalies in the planarian Dugesia tigrina. Bioelectromagnetics, 1996. 17(6): pp. 467–74.
210. Jenrow, K.A., C.H. Smith, and A.R. Liboff, Weak extremely-low-frequency magnetic fields and regeneration in the planarian Dugesia tigrina. Bioelectromagnetics, 1995. 16(2): pp. 106–12.
211. Levin, M. and S. Ernst, DC magnetic field effects on early sea urchin development. Bioelectromagnetics, 1997. 18(3): pp. 255–263.
212. Misakian, M., et al., Biological, physical, and electrical parameters for in vitro studies with ELF magnetic and electric fields: a primer. Bioelectromagnetics, 1993. Suppl 2: pp. 1–73.
213. Chen, Y.H., et al., KCNQ1 gain-of-function mutation in familial atrial fibrillation. Science, 2003. 299(5604): pp. 251–4.
214. Herbert, E., et al., KCNQ1 gene mutations and the respective genotype-phenotype correlations in the long QT syndrome. Med Sci Monit, 2002. 8(10): pp. RA240–8.
215. Seebohm, G., et al., Identification of specific pore residues mediating KCNQ1 inactivation. A novel mechanism for long QT syndrome. Journal of Biological Chemistry, 2001. 276(17): pp. 13600–5.
216. Gillespie, J.I. and S. McHanwell, Measurement of intra-embryonic pH during the early stages of development in the chick embryo. Cell & Tissue Research, 1987. 247(2): pp. 445–51.
217. Gillespie, J.I. and J.R. Greenwell, Changes in intracellular pH and pH regulating mechanisms in somitic cells of the early chick embryo: a study using fluorescent pH-sensitive dye. Journal of Physiology, 1988. 405: pp. 385–95.
218. Grandin, N. and M. Charbonneau, Intracellular pH and the increase in protein synthesis accompanying activation of Xenopus eggs. Biology of the Cell, 1989. 67(3): pp. 321–30.
219. Grandin, N. and M. Charbonneau, An increase in the intracellular pH of fertilized eggs of Xenopus laevis is associated with inhibition of protein and DNA syntheses and followed by an arrest of embryonic development. Experimental Cell Research, 1989. 185(2): pp. 436–52.

220. Lee, S.C. and R.A. Steinhardt, Observations on intracellular pH during cleavage of eggs of Xenopus laevis. Journal of Cell Biology, 1981. 91(2 Pt 1): pp. 414–19.
221. Lee, S.C. and R.A. Steinhardt, pH changes associated with meiotic maturation in oocytes of Xenopus laevis. Developmental Biology, 1981. 85(2): pp. 358–69.
222. Turin, L. and A.E. Warner, Intracellular pH in early Xenopus embryos: its effect on current flow between blastomeres. Journal of Physiology, 1980. 300: pp. 489–504.
223. Webb, D.J. and R. Nuccitelli, Direct measurement of intracellular pH changes in Xenopus eggs at fertilization and cleavage. Journal of Cell Biology, 1981. 91(2 Pt 1): pp. 562–7.
224. Webb, D.J. and R. Nuccitelli, A comparative study of the membrane potential from before fertilization through early cleavage in two frogs, Rana pipiens and Xenopus laevis. Comparative Biochemistry & Physiology A—Comparative Physiology, 1985. 82(1): pp. 35–42.
225. Nuccitelli, R., et al., 31P NMR reveals increased intracellular pH after fertilization in Xenopus eggs. Proceedings of the National Academy of Sciences of the United States of America, 1981. 78(7): pp. 4421–5.
226. Grandin, N. and M. Charbonneau, Changes in intracellular free calcium activity in Xenopus eggs following imposed intracellular pH changes using weak acids and weak bases. Biochimica et Biophysica Acta, 1991. 1091(2): pp. 242–50.
227. Uzman, J.A., et al., The role of intracellular alkalinization in the establishment of anterior neural fate in Xenopus. Developmental Biology, 1998. 193(1): pp. 10–20.
228. Moolenaar, W.H., L.G. Tertoolen, and S.W. de Laat, Phorbol ester and diacylglycerol mimic growth factors in raising cytoplasmic pH. Nature, 1984. 312(5992): pp. 371–4.
229. Harvey, W. and N. Nelson, V-ATPases. Journal of Experimental Biology, 1992. 172.
230. Harvey, W., Physiology of V-ATPases, in V-ATPases, W. Harvey and N. Nelson (Eds.). 1992. pp. 1–17.
231. Masuda, C.A. and M. Montero-Lomeli, An NH2-terminal deleted plasma membrane H+-ATPase is a dominant negative mutant and is sequestered in endoplasmic reticulum derived structures. Biochemistry & Cell Biology, 2000. 78(1): pp. 51–8.
232. Soleimani, M., et al., Localization of the Na+/H+ exchanger isoform NHE-3 in rabbit and canine kidney. Biochim Biophys Acta, 1994. 1195(1): pp. 89–95.
233. Counillon, L. and J. Pouyssegur, The expanding family of eucaryotic Na(+)/H(+) exchangers. J Biol Chem, 2000. 275(1): pp. 1–4.
234. Goodenough, D. and L. Musil, Gap junctions and tissue business: problems and strategies for developing specific functional reagents. Journal of Cell Science—Supplement, 1993. 17: pp. 133–8.
235. Lo, C.W., Genes, gene knockouts, and mutations in the analysis of gap junctions. Developmental Genetics, 1999. 24(1–2): pp. 1–4.
236. Maestrini, E., et al., A missense mutation in connexin26, D66H, causes mutilating keratoderma with sensorineural deafness (Vohwinkel's syndrome) in three unrelated families. Human Molecular Genetics, 1999. 8(7): pp. 1237–43.
237. Loewenstein, W.R. and B. Rose, The cell-cell channel in the control of growth. Seminars in Cell Biology, 1992. 3(1): pp. 59–79.
238. Li, G. and M. Herlyn, Dynamics of intercellular communication during melanoma development. Molecular Medicine Today, 2000. 6(4): pp. 163–9.
239. Omori, Y., et al., Involvement of gap junctions in tumor suppression: analysis of genetically-manipulated mice. Mutation Research, 2001. 477(1–2): pp. 191–6.
240. Krutovskikh, V. and H. Yamasaki, The role of gap junctional intercellular communication (GJIC) disorders in experimental and human carcinogenesis. Histology and Histopathology, 1997. 12: pp. 761–768.

241. Yamasaki, H., et al., Intercellular communication and carcinogenesis. Mutation Research, 1995. 333(1–2): pp. 181–8.
242. Pitts, J.D., M.E. Finbow, and E. Kam, Junctional communication and cellular differentiation. British Journal of Cancer—Supplement, 1988. 9: pp. 52–7.
243. Omori, Y. and H. Yamasaki, Mutated connexin43 proteins inhibit rat glioma cell growth suppression mediated by wild-type connexin43 in a dominant-negative manner. International Journal of Cancer, 1998. 78(4): pp. 446–53.
244. Mehta, P.P., et al., Incorporation of the gene for a cell-cell channel protein into transformed cells leads to normalization of growth. Journal of Membrane Biology, 1991. 124(3): pp. 207–25.
245. Rose, B., P.P. Mehta, and W.R. Loewenstein, Gap-junction protein gene suppresses tumorigenicity. Carcinogenesis, 1993. 14(5): pp. 1073–5.
246. Hellmann, P., et al., Transfection with different connexin genes alters growth and differentiation of human choriocarcinoma cells. Experimental Cell Research, 1999. 246(2): pp. 480–90.
247. Lo, C.W., The role of gap junction membrane channels in development. Journal of Bioenergetics & Biomembranes, 1996. 28(4): pp. 379–85.
248. Warner, A., Interactions between growth factors and gap junctional communication in developing systems. Novartis Foundation Symposium, 1999. 219: pp. 60–72.
249. Levin, M., Isolation and community: the role of gap junctional communication in embryonic patterning. Journal of Membrane Biology, 2001. 185(3): pp. 177–192.
250. Goodenough, D.A., J.A. Goliger, and D.L. Paul, Connexins, connexons, and intercellular communication. Annual Review of Biochemistry, 1996. 65: pp. 475–502.
251. Bruzzone, R., et al., Pannexins, a family of gap junction proteins expressed in brain. Proceedings of the National Academy of Sciences of the United States of America, 2003. 100(23): pp. 13644–9.
252. Phelan, P. and T.A. Starich, Innexins get into the gap. Bioessays, 2001. 23(5): pp. 388–96.
253. Finbow, M.E., M. Harrison, and P. Jones, Ductin—a proton pump component, a gap junction channel and a neurotransmitter release channel. Bioessays, 1995. 17(3): pp. 247–55.
254. Finbow, M.E. and J.D. Pitts, Is the gap junction channel—the connexon—made of connexin or ductin? Journal of Cell Science, 1993. 106(Pt 2): pp. 463–71.
255. Bruzzone, R. and D.A. Goodenough, Gap junctions: ductin or connexins—which component is the critical one? Bioessays, 1995. 17(8): pp. 744–5.
256. Falk, M., Genetic tags for labelling live cells: gap junctions and beyond. Trends Cell Biol, 2002. 12(9): pp. 399–404.
257. Segretain, D. and M.M. Falk, Regulation of connexin biosynthesis, assembly, gap junction formation, and removal. Biochim Biophys Acta, 2004. 1662(1–2): pp. 3–21.
258. Flagg-Newton, J.L. and W.R. Loewenstein, Asymmetrically permeable membrane channels in cell junction. Science, 1980. 207(4432): pp. 771–3.
259. Robinson, S.R., et al., Unidirectional coupling of gap junctions between neuroglia. Science, 1993. 262(5136): pp. 1072–4.
260. Suchyna, T.M., et al., Different ionic selectivities for connexins 26 and 32 produce rectifying gap junction channels. Biophysical Journal, 1999. 77(6): pp. 2968–87.
261. Oh, S., et al., Molecular determinants of electrical rectification of single channel conductance in gap junctions formed by connexins 26 and 32. J Gen Physiol, 1999. 114(3): pp. 339–64.
262. Guthrie, S., L. Turin, and A. Warner, Patterns of junctional communication during development of the early amphibian embryo. Development, 1988. 103(4): pp. 769–83.

263. Lee, S.H., et al., Human epileptic astrocytes exhibit increased gap junction coupling. Glia, 1995. 15(2): pp. 195–202.
264. Pappas, C.A., M.G. Rioult, and B.R. Ransom, Octanol, a gap junction uncoupling agent, changes intracellular [H+] in rat astrocytes. Glia, 1996. 16(1): pp. 7–15.
265. Suadicani, S.O., et al., Gap junction channels coordinate the propagation of intercellular Ca2+ signals generated by P2Y receptor activation. Glia, 2004. 48(3): pp. 217–29.
266. Bedner, P., et al., A method to determine the relative cAMP permeability of connexin channels. Exp Cell Res, 2003. 291(1): pp. 25–35.
267. Braet, K., et al., Photoliberating inositol-1,4,5-trisphosphate triggers ATP release that is blocked by the connexin mimetic peptide gap 26. Cell Calcium, 2003. 33(1): pp. 37–48.
268. Meda, P., Probing the function of connexin channels in primary tissues. Methods, 2000. 20: pp. 232–244.
269. Paul, D., et al., Expression of a dominant negative inhibitor of intercellular communication in the early Xenopus embryo causes delamination and extrusion of cells. Development, 1995. 121(2): pp. 371–81.
270. Rouan, F., et al., Divergent effects of two sequence variants of GJB3 (G12D and R32W) on the function of connexin 31 in vitro. Exp Dermatol, 2003. 12(2): pp. 191–197.
271. Diestel, S., et al., Exchange of serine residues 263 and 266 reduces the function of mouse gap junction protein connexin31 and exhibits a dominant-negative effect on the wild-type protein in HeLa cells. Exp Cell Res, 2004. 294(2): pp. 446–57.
272. Diestel, S., et al., Expression of a connexin31 mutation causing erythrokeratodermia variabilis is lethal for HeLa cells. Biochem Biophys Res Commun, 2002. 296(3): pp. 721–8.
273. Kudo, T., et al., Transgenic expression of a dominant-negative connexin26 causes degeneration of the organ of Corti and non-syndromic deafness. Hum Mol Genet, 2003. 12(9): pp. 995–1004.
274. Mire, P., J. Nasse, and S. Venable-Thibodeaux, Gap junctional communication in the vibration-sensitive response of sea anemones. Hearing Research, 2000. 144(1–2): pp. 109–23.
275. Brooks, R.A. and R.I. Woodruff, Calmodulin transmitted through gap junctions stimulates endocytic incorporation of yolk precursors in insect oocytes. Dev Biol, 2004. 271(2): pp. 339–49.
276. Bruzzone, R., et al., Pharmacological properties of homomeric and heteromeric pannexin hemichannels expressed in Xenopus oocytes. J Neurochem, 2005. 92(5): pp. 1033–43.
277. Muller, A., et al., Increase in gap junction conductance by an antiarrhythmic peptide. European Journal of Pharmacology, 1997. 327(1): pp. 65–72.
278. Blackman, C.F., et al., Physiological levels of melatonin enhance gap junction communication in primary cultures of mouse hepatocytes. Cell Biol Toxicol, 2001. 17(1): pp. 1–9.
279. Webb, R.J., et al., Follicle-stimulating hormone induces a gap junction-dependent dynamic change in [cAMP] and protein kinase a in mammalian oocytes. Dev Biol, 2002. 246(2): pp. 441–54.
280. Burnside, A.S. and P. Collas, Induction of Oct-3/4 expression in somatic cells by gap junction-mediated cAMP signaling from blastomeres. Eur J Cell Biol, 2002. 81(11): pp. 585–91.
281. Stains, J.P., et al., Gap junctional communication modulates gene transcription by altering the recruitment of Sp1 and Sp3 to connexin-response elements in osteoblast promoters. J Biol Chem, 2003. 278(27): pp. 24377–87.

282. Stains, J.P. and R. Civitelli, Gap junctions regulate extracellular signal-regulated kinase (ERK) signaling to affect gene transcription. J Mol. Biol. Cell, 2004.
283. Bassett, C.A., et al., Panel discussion: To what extent can electrical stimulation be used in the treatment of human disorders? Annals of the New York Academy of Sciences, 1974. 238: pp. 586–93.
284. Shapiro, S., et al., Oscillating field stimulation for complete spinal cord injury in humans: a phase 1 trial. J Neurosurg Spine, 2005. 2(1): pp. 3–10.
285. Gilbert, S.F., Developmental biology. 7th ed. 2003, Sunderland, MA: Sinauer Associates, Inc.
286. Fischbarg, J. and F.P. Diecke, A mathematical model of electrolyte and fluid transport across corneal endothelium. J Membr Biol, 2005. 203(1): pp. 41–56.
287. Hotary, K.B. and K.R. Robinson, Endogenous electrical currents and voltage gradients in Xenopus embryos and the consequences of their disruption. Developmental Biology, 1994. 166(2): p. 789–800.
288. Shi, R. and R.B. Borgens, Three-dimensional gradients of voltage during development of the nervous system as invisible coordinates for the establishment of embryonic pattern. Developmental Dynamics, 1995. 202(2): p. 101–114.
289. Gruler, H. and R. Nuccitelli, Neural crest cell galvanotaxis: new data and a novel approach to the analysis of both galvanotaxis and chemotaxis. Cell Motility & the Cytoskeleton, 1991. 19(2): p. 121–133.
290. Nuccitelli, R. and C.A. Erickson, Embryonic cell motility can be guided by physiological electric fields. Experimental Cell Research, 1983. 147(1): p. 195–201.

APPENDIX A

SEQUENTIAL HIERARCHICAL TREATMENT PROTOCOL - SHTEP SCREEN

CALCIUM

DRUGS (Increasing Specificity) Drug is inhibitor unless otherwise indicated		TARGETS (Ion Translocators) Translocator (aliases)
STEP 1	Cd^{2+}; Co^{2+}; BAPTA; EDTA; EGTA; Bradykinin; A23187; Ruthenium Red	**CALCIUM IONS (Ca^{2+})**
STEP 2	Calcicludine; Flunarizine; Lamotrigine; Lanthanum; Loperamide HCl; Riluzole.	Calcium Channels (CaC)
STEP 3	Amlodipine; Allethrin; Amiodarone; (R)-+-Bay K 8644; Benidipine; Benzothiazepines; Cilnidipine; Darodipine; Dihydropyridines; Diltiazem; Diphenyl-butylpiperidine; DPBP (Diphenylbutylpiperidine); Felodipine; FPL 64176; Gabapentin; HMR-1556; Isradipine; Mibefradil; Modipine; Nifedipine; Nilvadipine; Nisoldipine; Riodipine; Verapamil; Niguldipine HCl; Nimodipine; Nitrendipine; Phenylalkylamines; Riodipine; Ryosidine; Sevoflurane; SR 33805 Oxalate. *ENHANCERS*: (L)-+-Bay K 8644; (±)-Bay K 8644.	Voltage-Sensitive Ca Channels (Cav)
STEP 4	ω-Agatoxin IIIA; Amlodipine; Barnidipine; Benidipine; Calciseptine; Cilnidipine; ω-Conotoxin MVIIC; Diphenylbutylpiperidine; Mibefradil; Nicardipine.	L Type (CaCL); Cav1.1 - 1.4
STEP 5		

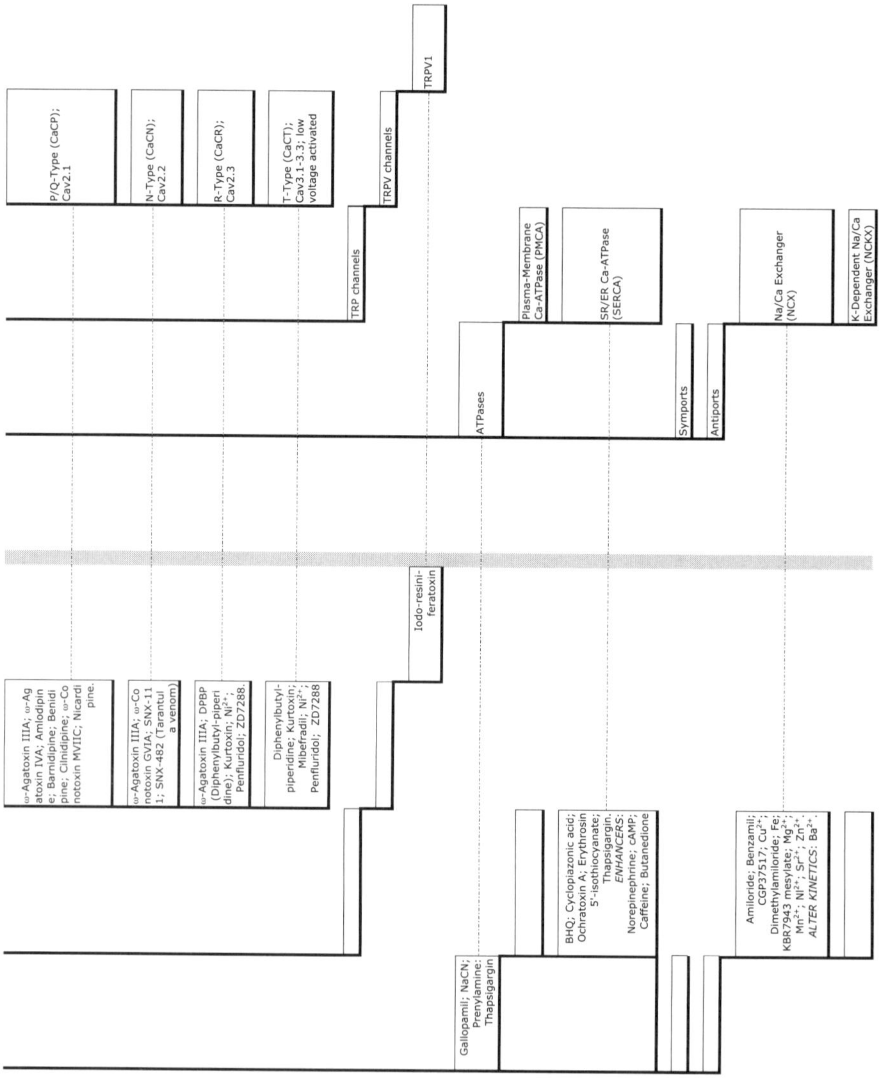
P/Q-Type (CaCP); Cav2.1
N-Type (CaCN); Cav2.2
R-Type (CaCR); Cav2.3
T-Type (CaCT); Cav3.1-3.3; low voltage activated
TRP channels
TRPV channels
TRPV1
ATPases
Plasma-Membrane Ca-ATPase (PMCA)
SR/ER Ca-ATPase (SERCA)
Symports
Antiports
Na/Ca Exchanger (NCX)
K-Dependent Na/Ca Exchanger (NCKX)
ω-Agatoxin IIIA; ω-Agatoxin IVA; Amlodipine; Barnidipine; Benidipine; Cilnidipine; ω-Conotoxin MVIIC; Nicardipine.
ω-Agatoxin IIIA; ω-Conotoxin GVIA; SNX-111; SNX-482 (Tarantula venom)
ω-Agatoxin IIIA; DPBP (Diphenylbutyl-piperidine); Kurtoxin; Ni^{2+}; Penfluridol; ZD7288.
Diphenylbutyl-piperidine; Kurtoxin; Mibefradil; Ni^{2+}; Penfluridol; ZD7288
Iodo-resini-feratoxin
Gallopamil; NaCN; Prenylamine; Thapsigargin
BHQ; Cyclopiazonic acid; Ochratoxin A; Erythrosin 5'-isothiocyanate; Thapsigargin. *ENHANCERS*: Norepinephrine; cAMP; Caffeine; Butanedione
Amiloride; Benzamil; CGP37517; Cu^{2+}; Dimethylamiloride; Fe; KBR7943 mesylate; Mg^{2+}; Mn^{2+}; Ni^{2+}; Sr^{2+}; Zn^{2+}. *ALTER KINETICS*: Ba^{2+}.

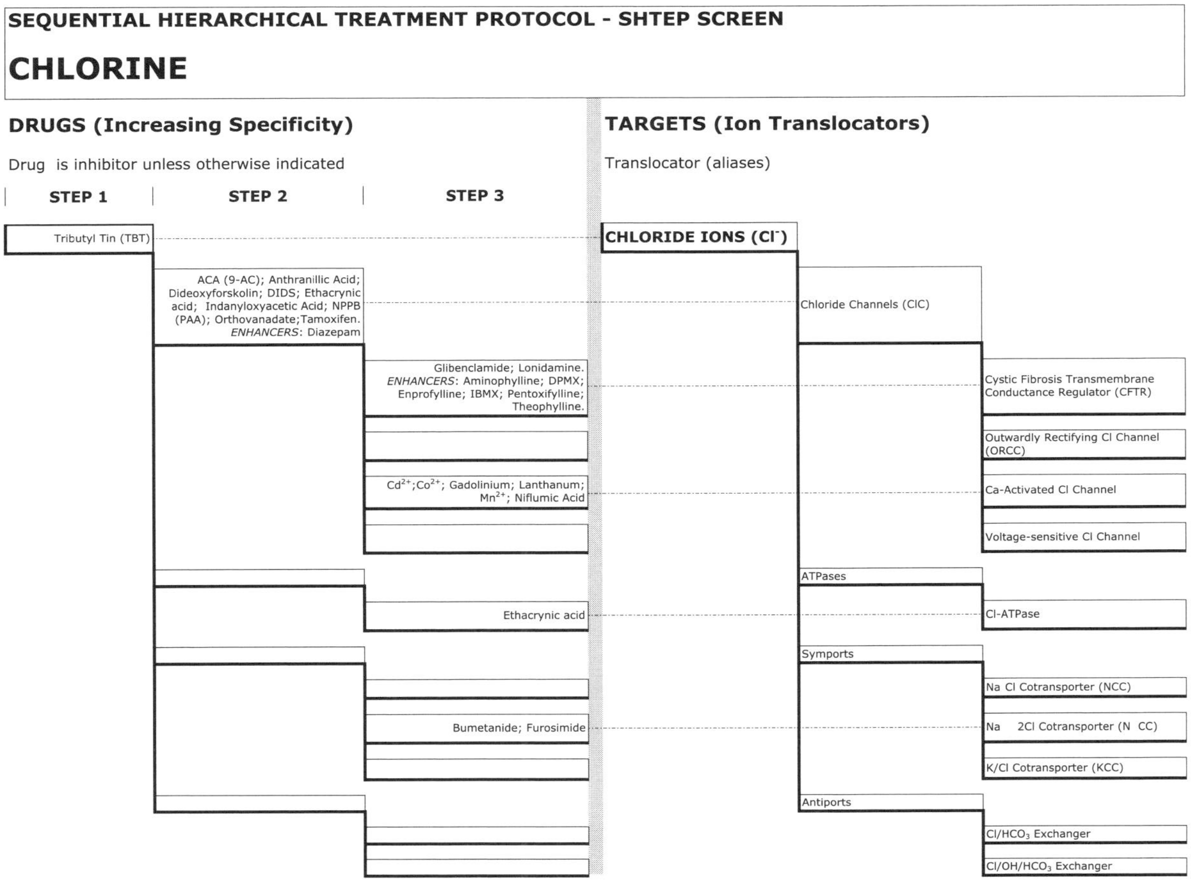
SEQUENTIAL HIERARCHICAL TREATMENT PROTOCOL - SHTEP SCREEN
CHLORINE
DRUGS (Increasing Specificity)
Drug is inhibitor unless otherwise indicated
STEP 1
STEP 2
STEP 3
Tributyl Tin (TBT)
ACA (9-AC); Anthranillic Acid; Dideoxyforskolin; DIDS; Ethacrynic acid; Indanyloxyacetic Acid; NPPB (PAA); Orthovanadate;Tamoxifen. ENHANCERS: Diazepam
Glibenclamide; Lonidamine. ENHANCERS: Aminophylline; DPMX; Enprofylline; IBMX; Pentoxifylline; Theophylline.
Cd2+;Co2+; Gadolinium; Lanthanum; Mn2+; Niflumic Acid
Ethacrynic acid
Bumetanide; Furosimide
TARGETS (Ion Translocators)
Translocator (aliases)
CHLORIDE IONS (Cl-)
Chloride Channels (ClC)
Cystic Fibrosis Transmembrane Conductance Regulator (CFTR)
Outwardly Rectifying Cl Channel (ORCC)
Ca-Activated Cl Channel
Voltage-sensitive Cl Channel
ATPases
Cl-ATPase
Symports
Na Cl Cotransporter (NCC)
Na 2Cl Cotransporter (N CC)
K/Cl Cotransporter (KCC)
Antiports
Cl/HCO3 Exchanger
Cl/OH/HCO3 Exchanger

SEQUENTIAL HIERARCHICAL TREATMENT PROTOCOL - SHTEP SCREEN

HYDROGEN

DRUGS (Increasing Specificity)

Drug is inhibitor unless otherwise indicated

STEP 1 | **STEP 2** | **STEP 3**

TARGETS (Ion Translocators)

Translocator (aliases)

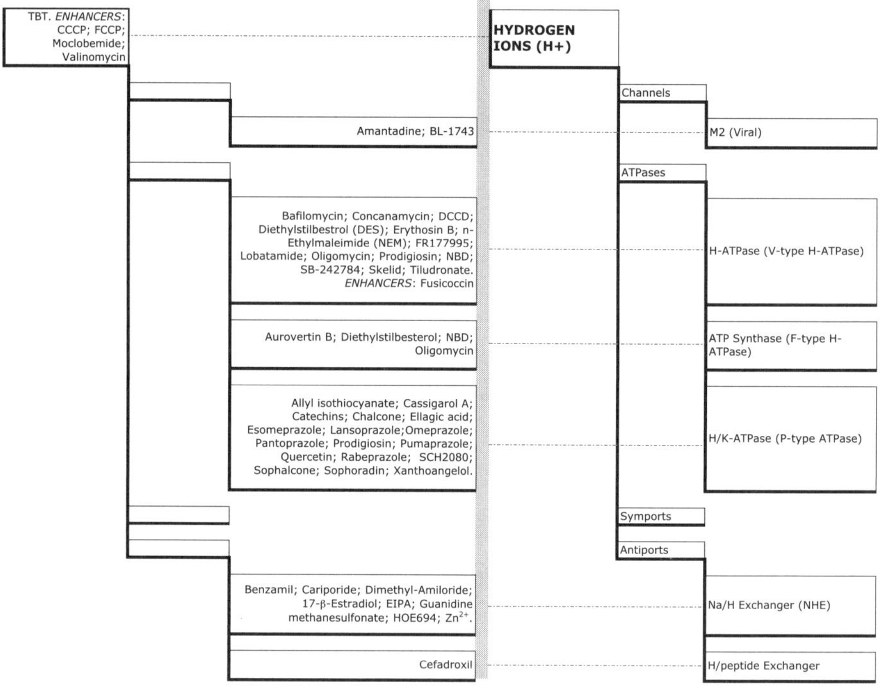

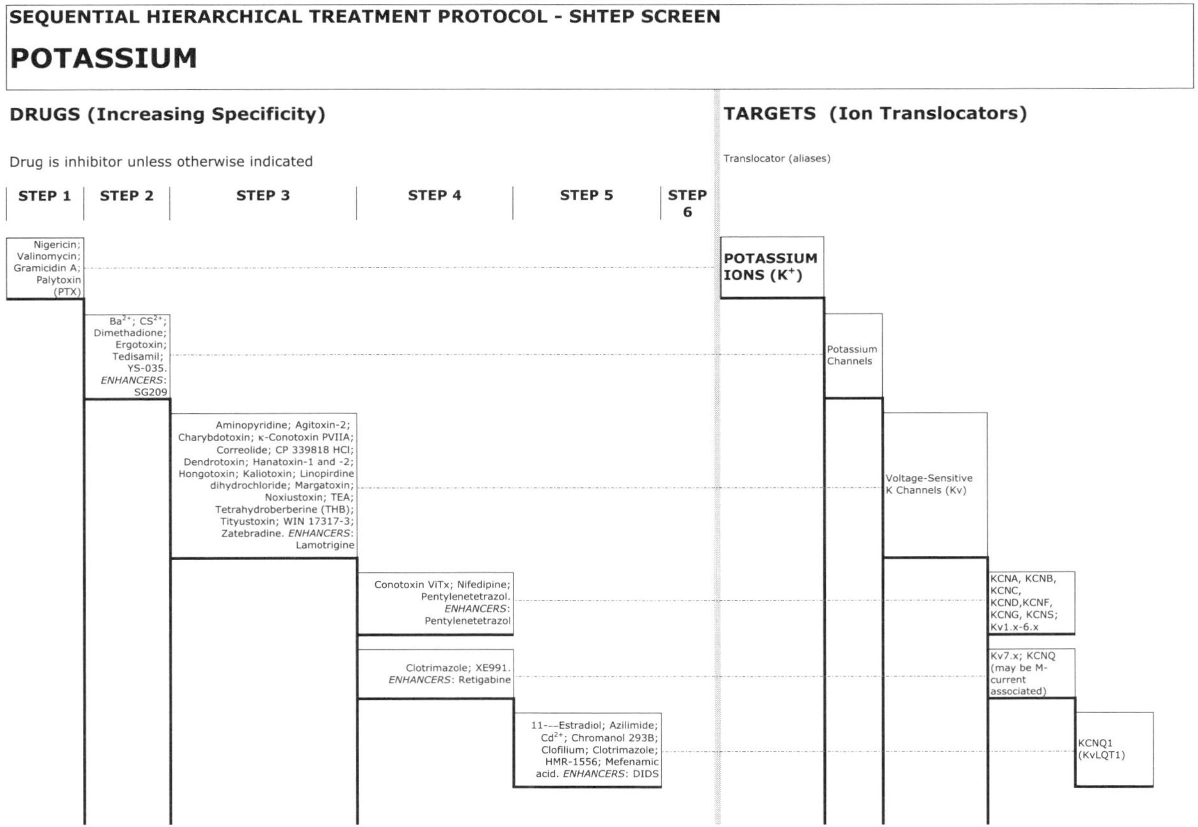
SEQUENTIAL HIERARCHICAL TREATMENT PROTOCOL - SHTEP SCREEN
POTASSIUM
DRUGS (Increasing Specificity)
Drug is inhibitor unless otherwise indicated
STEP 1
STEP 2
STEP 3
STEP 4
STEP 5
STEP 6
Nigericin; Valinomycin; Gramicidin A; Palytoxin (PTX)
Ba2+; CS2+; Dimethadione; Ergotoxin; Tedisamil; YS-035. ENHANCERS: SG209
Aminopyridine; Agitoxin-2; Charybdotoxin; κ-Conotoxin PVIIA; Correolide; CP 339818 HCl; Dendrotoxin; Hanatoxin-1 and -2; Hongotoxin; Kaliotoxin; Linopirdine dihydrochloride; Margatoxin; Noxiustoxin; TEA; Tetrahydroberberine (THB); Tityustoxin; WIN 17317-3; Zatebradine. ENHANCERS: Lamotrigine
Conotoxin ViTx; Nifedipine; Pentylenetetrazol. ENHANCERS: Pentylenetetrazol
Clotrimazole; XE991. ENHANCERS: Retigabine
11-–-Estradiol; Azilimide; Cd2+; Chromanol 293B; Clofilium; Clotrimazole; HMR-1556; Mefenamic acid. ENHANCERS: DIDS
TARGETS (Ion Translocators)
Translocator (aliases)
POTASSIUM IONS (K+)
Potassium Channels
Voltage-Sensitive K Channels (Kv)
KCNA, KCNB, KCNC, KCND,KCNF, KCNG, KCNS; Kv1.x-6.x
Kv7.x; KCNQ (may be M-current associated)
KCNQ1 (KvLQT1)

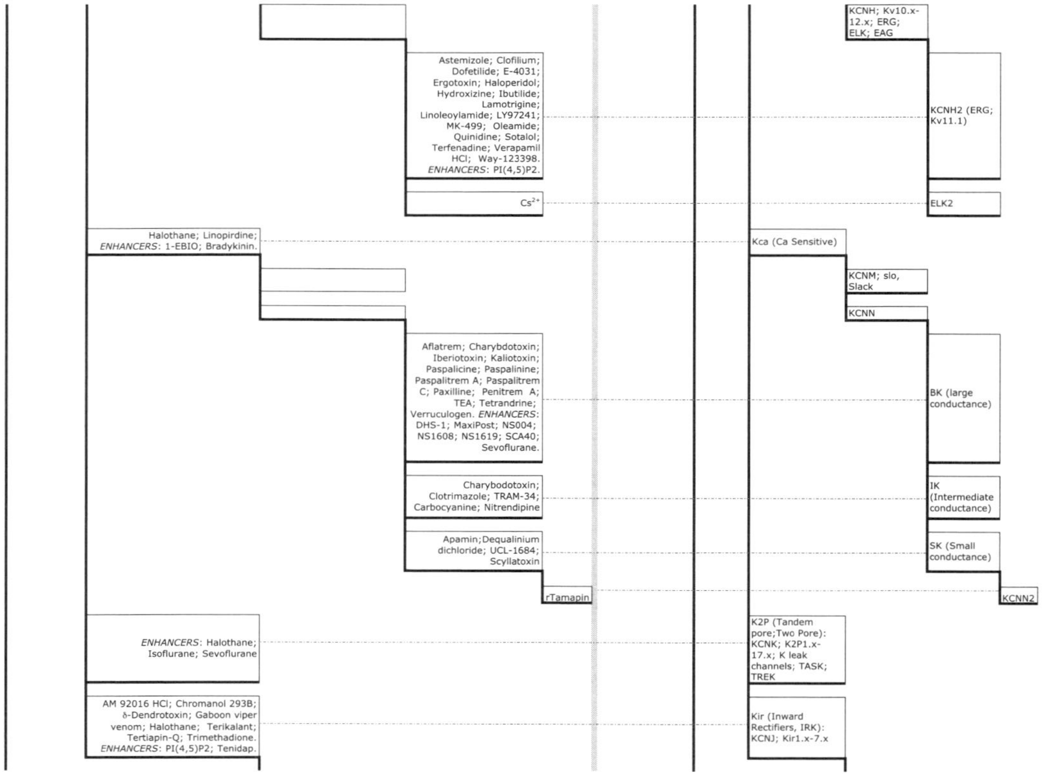
KCNH; Kv10.x-12.x; ERG; ELK; EAG
KCNH2 (ERG; Kv11.1)
Astemizole; Clofilium; Dofetilide; E-4031; Ergotoxin; Haloperidol; Hydroxizine; Ibutilide; Lamotrigine; Linoleoylamide; LY97241; MK-499; Oleamide; Quinidine; Sotalol; Terfenadine; Verapamil HCl; Way-123398. ENHANCERS: PI(4,5)P2.
ELK2
Cs2+
Kca (Ca Sensitive)
Halothane; Linopirdine; ENHANCERS: 1-EBIO; Bradykinin.
KCNM; slo, Slack
KCNN
BK (large conductance)
Aflatrem; Charybdotoxin; Iberiotoxin; Kaliotoxin; Paspalicine; Paspalinine; Paspalitrem A; Paspalitrem C; Paxilline; Penitrem A; TEA; Tetrandrine; Verruculogen. ENHANCERS: DHS-1; MaxiPost; NS004; NS1608; NS1619; SCA40; Sevoflurane.
IK (Intermediate conductance)
Charybodotoxin; Clotrimazole; TRAM-34; Carbocyanine; Nitrendipine
SK (Small conductance)
Apamin;Dequalinium dichloride; UCL-1684; Scyllatoxin
KCNN2
rTamapin
K2P (Tandem pore;Two Pore): KCNK; K2P1.x-17.x; K leak channels; TASK; TREK
ENHANCERS: Halothane; Isoflurane; Sevoflurane
Kir (Inward Rectifiers, IRK): KCNJ; Kir1.x-7.x
AM 92016 HCl; Chromanol 293B; δ-Dendrotoxin; Gaboon viper venom; Halothane; Terikalant; Tertiapin-Q; Trimethadione. ENHANCERS: PI(4,5)P2; Tenidap.

	KATP (ATP sensitive): Kir6.x (note: form active channels in combination with sulfonylurea receptors (SUR).	Prostaglandin J2; Alinidine; Aprikalim; Benzopyran; Bimakalim; BMS-180448; BRL 55843; Bupivacaine; Chloroquine; Chlor-propamide; Ciclazindol; Decanoic acid; Dibu-caine; Efaroxan; Fenofibrate; Gliben-clamide; Glipizide; Guanethidine; HMR-1098; Hydroxydeca-noate; Lidocaine; Mefloquine; Meglitinide; Mepivacaine; Mexiletine; Nataglinide; Phento-lamine; PHM; PI(4,5)P2; PKF-217-744; Procaine; Propanolol; Quinine; Repaglinide; Tetracaine; TMB-8; ZM 181,037; Tolazamide; Tolbut-amide; Troglitazone; U-37883. *ENHANCERS*: Cromakalim; Cyano-guanidine; Diazoxide; Levcromakalim; Minoxidil; Nicorandil; P1075; Pentylene-tetrazol; Pinacidil; Rimakalim; RP 66471; SG209; Y-26762; ZD 6169; ZM 226600.
ATPases	K-ATPase	
	Na/K-ATPase	Cu^{2+}; Fe^{2+}; Oligomycin; Ouabain; Palytoxin (PTX); Strophanthidin; Vanadate.
	H/K-ATPase	Allyl isothiocyanate; Cassigarol A; Catechins; Chalcone; Ellagic acid; Esomeprazole; Lansoprazole;Omeprazole; Pantoprazole; Prodigiosin; Pumaprazole; Quercetin; Rabeprazole; SCH2080; Sophalcone; Sophoradin; Xanthoangelol.
Symports	Na/K/2Cl co-transporter (NKCC)	Bumetanide; Furosimide
	K/Cl co-transporter	
Antiports		

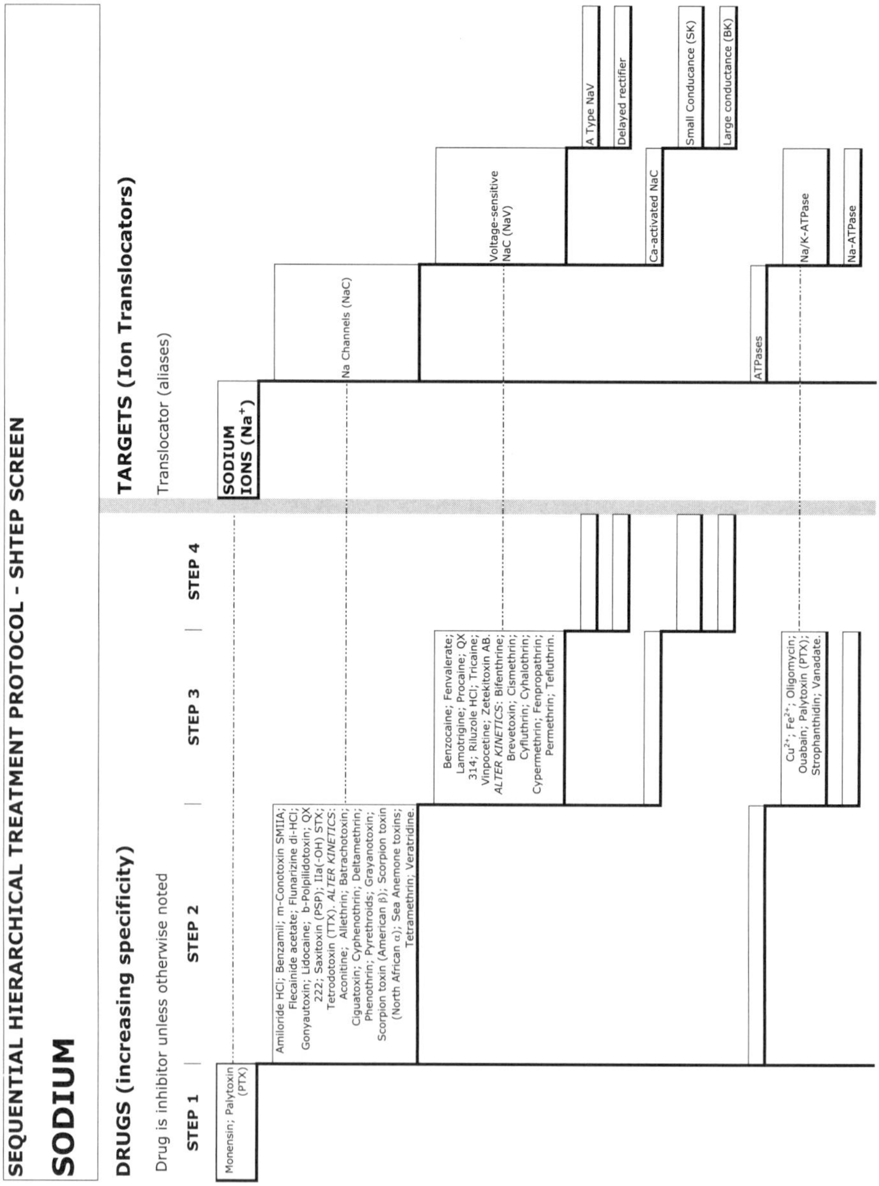
SEQUENTIAL HIERARCHICAL TREATMENT PROTOCOL - SHTEP SCREEN
SODIUM
DRUGS (increasing specificity)
Drug is inhibitor unless otherwise noted
STEP 1
STEP 2
STEP 3
STEP 4
Monensin; Palytoxin (PTX)
Amiloride HCl; Benzamil; m-Conotoxin SMIIA; Flecainide acetate; Flunarizine di-HCl; Gonyautoxin; Lidocaine; b-Polpilidotoxin; QX 222; Saxitoxin (PSP); IIa(-OH) STX; Tetrodotoxin (TTX). ALTER KINETICS: Aconitine; Allethrin; Batrachotoxin; Ciguatoxin; Cyphenothrin; Deltamethrin; Phenothrin; Pyrethroids; Grayanotoxin; Scorpion toxin (American β); Scorpion toxin (North African α); Sea Anemone toxins; Tetramethrin; Veratridine.
Benzocaine; Fenvalerate; Lamotrigine; Procaine; QX 314; Riluzole HCl; Tricaine; Vinpocetine; Zetekitoxin AB. ALTER KINETICS: Bifenthrine; Brevetoxin; Cismethrin; Cyfluthrin; Cyhalothrin; Cypermethrin; Fenpropathrin; Permethrin; Tefluthrin.
Cu2+; Fe2+; Oligomycin; Ouabain; Palytoxin (PTX); Strophanthidin; Vanadate.
TARGETS (Ion Translocators)
Translocator (aliases)
SODIUM IONS (Na+)
Na Channels (NaC)
Voltage-sensitive NaC (NaV)
A Type NaV
Delayed rectifier
Ca-activated NaC
Small Conducance (SK)
Large conductance (BK)
ATPases
Na/K-ATPase
Na-ATPase

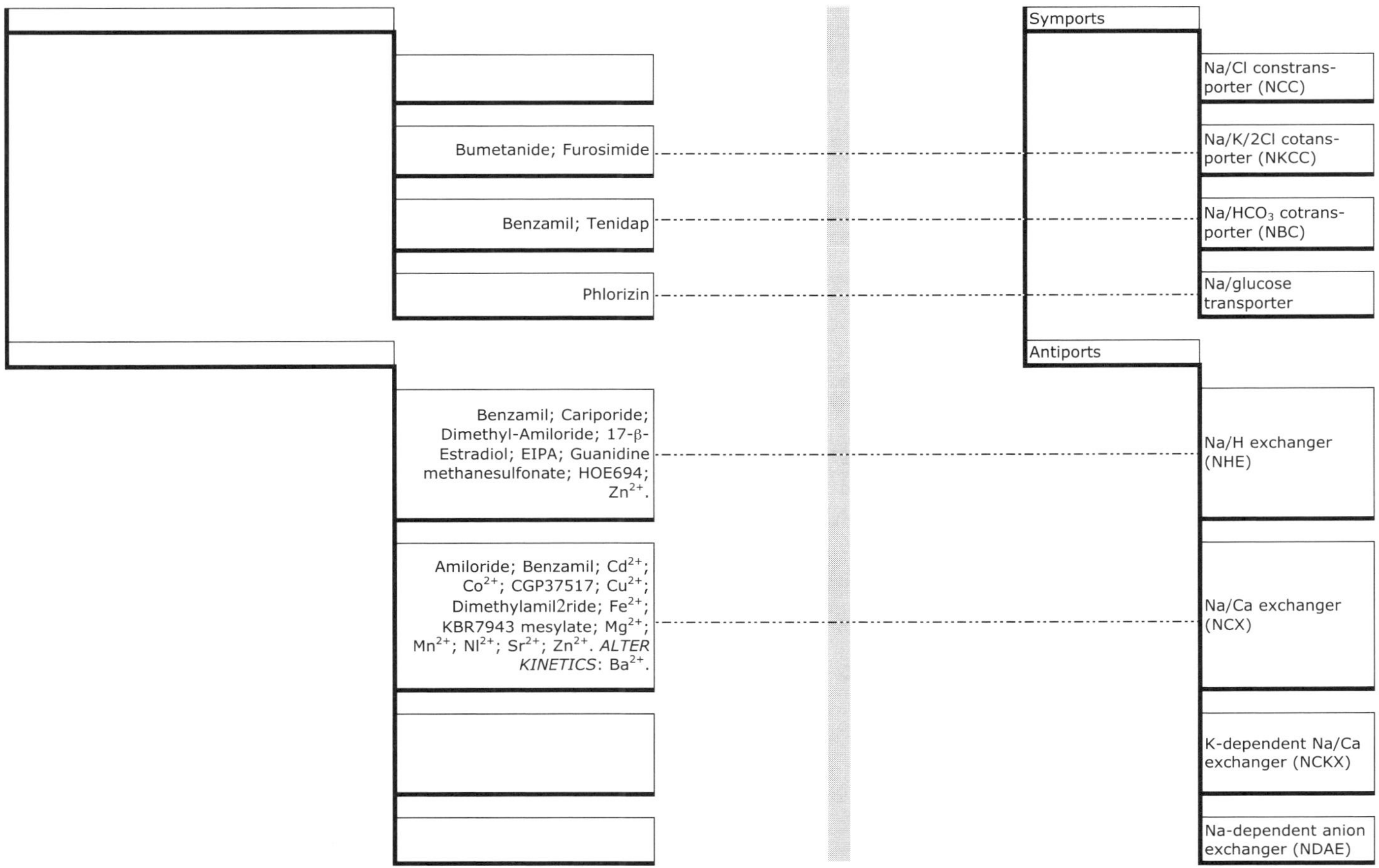
Symports
Na/Cl constrans-porter (NCC)
Na/K/2Cl cotans-porter (NKCC)
Bumetanide; Furosimide
Na/HCO_3 cotrans-porter (NBC)
Benzamil; Tenidap
Na/glucose transporter
Phlorizin
Antiports
Na/H exchanger (NHE)
Benzamil; Cariporide; Dimethyl-Amiloride; 17-β-Estradiol; EIPA; Guanidine methanesulfonate; HOE694; Zn^{2+}.
Na/Ca exchanger (NCX)
Amiloride; Benzamil; Cd^{2+}; Co^{2+}; CGP37517; Cu^{2+}; Dimethylamil2ride; Fe^{2+}; KBR7943 mesylate; Mg^{2+}; Mn^{2+}; NI^{2+}; Sr^{2+}; Zn^{2+}. *ALTER KINETICS*: Ba^{2+}.
K-dependent Na/Ca exchanger (NCKX)
Na-dependent anion exchanger (NDAE)

SEQUENTIAL HIERARCHICAL TREATMENT PROTOCOL – SHTEP SCREEN

OTHER

Drugs Drug is inhibitor unless otherwise indicated	Targets
Cu^{2+}	Aquaporins
Trifluoroperazine	Calmodulin
Acetazolamide; Dichlorphenamide; Ethoxyzolamide; Methazolamide.	Carbonic Anhydrase
Carbenoxolone;18-β-Glycyrrhetinic acid; Heptanol; Lindane; TAC-101	Gap Junctions (GJ)
Amitriptyline; Chloroquine; $CuCl_2$; EPA; HOCl; Sodium Caprate	Tight Junctions
Streptolysin O.	Membrane integrity
Diltiazem; Nifedipine; Tetracaine; LY83583	CNG (cyclic nucleotide gated) channels (monovalent cations, Ca^{2+})
Cs^+; ZD7288; Ivabradine; Zatebradine; Alinidine	HCN (hyperpolarisation-activated cyclic-nucleotide gated) channels (Na, K)
Gadolinium	Mechanosensitive channels
TPEN	Zn^{2+}, Cu^{2+}, Fe^{2+} (ignores Ca^{2+} and Mg^{2+})

APPENDIX B

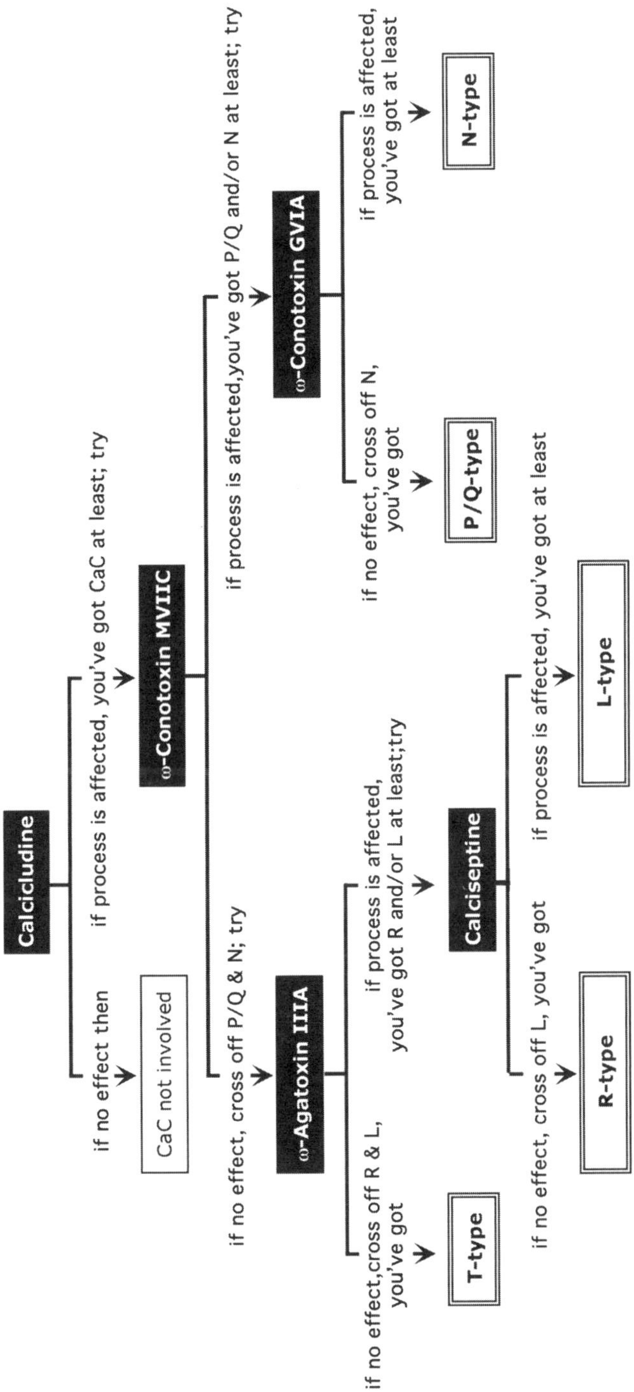
Calcicludine
if process is affected, you've got CaC at least; try
ω-Conotoxin MVIIC
if no effect then
CaC not involved
if process is affected,you've got P/Q and/or N at least; try
ω-Conotoxin GVIA
if process is affected, you've got at least
N-type
if no effect, cross off N, you've got
P/Q-type
if no effect, cross off P/Q & N; try
ω-Agatoxin IIIA
if process is affected, you've got R and/or L at least;try
Calciseptine
if process is affected, you've got at least
L-type
if no effect, cross off L, you've got
R-type
if no effect,cross off R & L, you've got
T-type

APPENDIX C

SEQUENTIAL HIERARCHICAL TREATMENT PROTOCOL – SHTEP SCREEN

ALPHABETICAL LISTING OF DRUGS

with information and references

Drugs with known multiple targets are indicated by a double line to the left of the name. *Always* confirm all targets for any drug before use.

Drug	Ion	Target	Action	Comments	Reference
1-EBIO	K	Ca-activated K channels	Pos		Adeagbo 1999
A23187	Ca	Membrane	Neg	ionophore, also transports Mn	Boot and Hilten 1996
A23187 (4-Bromo-A23187)	Ca	Membrane	Neg	ionophore, also transports Mn, avail. Caged	Boot and Hilten 1996
ACA (9-ACA)	Cl	Cl channels	Neg	strong on PKA activated cardiac I_{Cl}	Hui and Chen 1994
Acetazolamide	H, HCO_3	Carbonic Anhydrase	Neg	CA catalyzes hydration of CO_2 to H and HCO_3	Lindskog 1997
Aconitine	Na	Na channel kinetics	Pos	persistent activation	Lazdunski et al. 1985
Aflatrem	K	Ca-activated K channels, BK	Neg		Knaus et al 1994
Agatoxin IIIA (ω-Agatoxin IIIA)	Ca	L-, N-, P/Q- and R-type Ca channels	Neg		Yan & Adams 2000; Ertel et al 1994
Agatoxin IVA (ω-Agatoxin IVA)	Ca	P/Q-type Ca channels	Neg		Bowersox et al 1995
Agitoxin-2	K	Kv channels	Neg		Lipkind and Fozzard 1997
Alinidine	K	K-ATP channels	Neg		Challinor-Rogers et al 1997
Allethrin	Ca	Cav channels	Neg		Hildebrand et al 2004
Allethrin	Na	Na channel kinetics	Pos	a pyrethroid, alters kinetics	Narahashi et al 1992
Allyl isothiocyanate	H,K	H/K-ATPase	Neg		Takeguchi et al 1983
AM 92016 HCl	K	K inward rectifier	Neg	blocks time dependent delayed KIR current	Katsuki et al 2005
Amantadine	H	M2 channels	Neg	also blocks Ach receptor	Warnick et al 1982
Amiloride	Na,Ca	Na/Ca exchanger	Neg		Jurkowitz et al, 1983

Amiloride	H, Na	Na/H exchanger (NHE)	Neg		Harris and Fliegel 1999
Amiloride HCl	Na	Na channels			Kleyman et al 1988
Aminophylline	Cl	CFTR	Pos	enhance Cl efflux	Chappe et al 1998
Aminopyridine (4-Aminopyridine; 4-AP)	K	Kv channels	Neg		Zhang et al 1998
Amiodarone	Ca	Cav channels	Neg		Hancox 1997
Amitriptyline		tight junctions	Neg		Dahlin et al 1999
Amlodipine	Ca	Cav channels	Neg		Furukawa et al 1999
Amlodipine	Ca	N- and P/Q-type Ca channels	Neg		Furukawa et al 1999
Anthranilic acid (AA)	Cl	Cl channels	Neg		Landry et al 1987
Apamin	K	Ca-activated K channels, SK	Neg	potent, highly specific	Adeagbo 1999
Aprikalim	K	K-ATP channels	Pos		Russ et al 2003
Astemizole	K	KCNH2 (ERG)	Neg		Suessbrich et al 1996
Aurovertin B	H	ATP Synthase (F-type H-ATPase)	Neg	Mitochondrial H-ATPase	Weber and Senior 1998
Azimilide	K	KCNQ1 (KvLQT1)	Neg		Busch et al 1997
Ba^{2+}	K	K channels	Neg		Lock and Valverde 2000
Ba^{2+}	Na,Ca	Na/Ca exchanger	Neg	changes kinetics	Trac et al 1997
Bafilomycin	H	H-ATPase (V-type H-ATPase)	Neg		Gagliardi et al 1999
BAPTA	Ca	Aqueous compartments	Neg	chelator (avail. Permeable)	Turetta et al 2002
Barnidipine	Ca	N- and P/Q-type Ca channels	Neg		Furukawa et al 1999
Batrachotoxin	Na	Na channel kinetics	Pos	longer opening	Lazdunski et al. 1985
Bay K 8644 [(R)-(+) isomer]	Ca	Cav channels	Neg	Opposite effect of S isomer	Franckowiak et al 1985
Bay K 8644 [(S)-(+) isomer]	Ca	Cav channels	Pos	Opposite effect of R isomer	Franckowiak et al 1985
Benidipine	Ca	Cav channels	Neg		Furukawa et al 1999
Benidipine	Ca	N- and P/Q-type Ca channels	Neg		Furukawa et al 1999

Benzamil	Na	Na channels	Neg	Amiloride sensitive Na channels	Bell et al 1997
Benzamil	Na,Ca	Na/Ca exchanger	Neg		Jurkowitz et al, 1983
Benzamil	Na,H	Na/H exchanger (NHE)	Neg		Jurkowitz et al, 1983
Benzamil	Na	Na/HCO3 cotransporter	Neg		Ducoudret et al 2001
Benzocaine	Na	NaV channels	Neg	slow-blocking	Wang et al 1998
Benzopyran	K	K-ATP channels	Pos		Russ et al 2003
Benzothiazepines	Ca	Cav channels	Neg	Family of CaCL blockers	Striessnig 1999
BHQ	Ca	SERCA	Neg		Gukovskaya et al 2000
Bifenthrin	Na	NaV channels	Neg	a pyrethroid, alters kinetics	Choi and Soderlund 2005
Bimakalim	K	K-ATP channels	Pos		Mizumura et al 1995
BL-1743	H	M2 channels	Neg		Tu et al 1996
BMS-180448	K	K-ATP channels	Pos		Atwal 1994
Bradykinin	Ca	Ca stores (internal)	Neg		Schuster et al 2003
Bradykinin	K	Ca-activated K channels	Pos		Schuster et al 2003
Brevetoxin	Na	NaV channels	Neg	red tide toxin; Brevenal inhibits	Bourdelais et al 2004
BRL 55834	K	K-ATP channels	Pos		Arch et al 1994
Bumetanide	Na,K,Cl	Na/K/Cl cotransporter	Neg		Landry et al 1987
Bupivacaine	K	K-ATP channels	Neg		Yoneda et al 1993
Butanedione monoxime (2,3-Butanedione monoxime)	Ca	SERCA	Pos		Phillips and Altschuld 1996
Caffeine	Ca	SERCA	Neg	Ca from SR increased	Bers and Bridge 1989
Calciseptine	Ca	L-type Ca channels	Neg		Kuroda et al 1992
cAMP	Ca	SERCA	Pos	Ca from SR increased	Ulianich et al 2004
Carbenoxolone		Gap Junctions	Neg		Winmill and Hedrick 2003
Carbocyanine	K	Ca-activated K channels, IK	Neg		Simons 1976
Cariporide (HOE642)	Na,H	Na/H antiporter, (NHE)	Neg		Mathur and Karmazyn 1997
Cassigarol A	H,K	H/K-ATPase	Neg		Murukami et al 1992a

Catechins	H,K	H/K-ATPase	Neg		Murukami et al 1992a
CCCP	H	Membrane	Neg	ionophore	Holoubek et al 2003
Cd^{2+}	Cl	Ca-activated Cl channels	Neg		Tokimasa and North 1996
Cd^{2+}	K	KCNQ1 (KvLQT1)	Neg	also may bind minK	Tapper and George 2001
Cd^{2+}	Ca	L- and N-Type Ca channels	Neg		Boulton and O'Shaughnessy 1991
Cd^{2+}	Na,Ca	Na/Ca exchanger	Pos/Neg	Cd substitutes for Ca	Le et al 2005
Cefadroxil	H	H/peptide transport	Neg		Nozawa et al 2003
CGP 37157	Na,Ca	Na/Ca exchanger	Neg	mitochondrial	Cox et al 1993
Chalcone	H,K	H/K-ATPase	Neg		Murukami et al 1990
Charybdotoxin	K	Ca-activated K channels, BK	Neg		Borsotto 2001
Charybdotoxin	K	Ca-activated K channels, IK	Neg		Borsotto 2001
Charybdotoxin	K	Kv channels	Neg		Lipkind and Fozzard 1997
Chloroquine	K	K-ATP channels	Neg	Interact with Kir6.2	Gribble et al 2000
Chloroquine		tight junctions		increase number	Okanlawon and Dym 1996
Chlorpropamide	K	K-ATP channels	Neg		Smoak 1999
Chromanol 293B	K	KCNQ1 (KvLQT1)	Neg	3R,4S enantiomer is stronger	Yang et al 2000
Ciclazindol	K	K-ATP channels	Neg		McKay et al 2000
Ciguatoxin	Na	Na channel kinetics	Pos	increase permeability, complex polyether, red tide toxin	Lazdunski et al, 1985
Cilnidipine	Ca	Cav channels	Neg		Furukawa et al 1999
Cilnidipine	Ca	N- and P/Q-type Ca channels	Neg		Furukawa et al 1999
Cismethrin	Na	NaV channels	Neg	a pyrethroid, alters kinetics	Choi and Soderlund 2005

Clofilium	K	KCNH2 (ERG)	Neg	IC_{50} voltage dependent	Suessbrich et al 1996
Clofilium	K	KCNQ1 (KvLQT1)	Neg		Lock and Valverde 2000
Clotrimazole	K	Ca-activated K channels, IK	Neg		Brugnara 1997
Clotrimazole	K	KCNQ1 (KvLQT1)	Neg		Unsold et al 2000
Co^{2+}	Cl	Ca-activated Cl channels	Neg		Tokimasa and North 1996
Co^{2+}	Ca	L- and N-Type Ca channels	Neg		Boulton and O'Shaughnessy 1991
Co^{2+}	Na/Ca	Na/Ca exchanger	Neg		Bers and Bridge 1989
Concanamycin	H	H-ATPase (V-type H-ATPase)	Neg		Drose and Altendorf 1997
Conotoxin GVIA (ω-Conotoxin GVIA)	Ca	N-type Ca channels	Neg	reversible block	Bernstein and Jones 2001
Conotoxin MVIIC (ω-Conotoxin MVIIC)	Ca	N- and P/Q-type Ca channels	Neg	wide spectrum	McDonough et al 1996
Conotoxin PVIIA (κ-Conotoxin PVIIA)	K	Kv channels	Neg		Scanlon et al 1997
Conotoxin SMIIIA (μ-Conotoxin SMIIIA)	Na	Na channels	Neg	pore blocker	West et al 2002
Conotoxin ViTx	K	K channels	Neg	Kv 1.1 and 1.3, not 1.2	Kauferstein et al 2003
Correolide	K	Kv channels	Neg		Vianna-Jorge et al 2000
CP 339818 HCl	K	Kv channels	Neg	binds to C-type inactivated state	Nguyen et al 1996
Cromakalim	K	K-ATP channels	Pos	levcromakalim is active enantiomer	Dunne et al 1989
Cs^{+}	K	K channels	Neg		Rouzaire-Dubois and Dubois 1998
Cs^{+}	K	Kv channels	Neg		Lin et al 2001
Cu^{2+}		Aquaporin-3	Neg		Zelenina et al 2004
Cu^{2+}	Na,Ca	Na/Ca exchanger	Neg		Benders et al 1994
Cu^{2+}	Na,K	Na/K-ATPase	Neg		Benders et al 1994
Cu^{2+}		tight junctions	Pos	increases TEER	Ferruzza et al 2002

CyanoGuanidine	K	K-ATP channels	Pos		Russ et al 2003
Cyclopiazonic Acid	Ca	SERCA	Neg	reversible	Soler et al 1998
Cyfluthrin	Na	NaV channels	Neg	a pyrethroid, alters kinetics	Choi and Soderlund 2005
Cyhalothrin	Na	NaV channels	Neg	a pyrethroid, alters kinetics	Choi and Soderlund 2005
Cypermethrin	Na	NaV channels	Neg	a pyrethroid, alters kinetics	Choi and Soderlund 2005
Cyphenothrin	Na	Na channel kinetics	Neg	a pyrethroid, alters kinetics	Narahashi 1982
Darodipine	Ca	Cav channels	Neg		Matucci et al 1997
DCCD	H	H-ATPase (V-type H-ATPase)	Neg		Finbow et al 1993
Decanoic acid	K	K-ATP channels	Neg		Miyoshi et al 1996
Decarbamylsaxitoxin	Na	Na channels	Neg	synthetic, related to saxitoxin	Schantz 1985
Deltamethrin	Na	Na channel kinetics	Neg	a pyrethroid, alters kinetics	Narahashi et al 1992
Dendrotoxin (α Dendrotoxin)	K	Kv channels	Neg		Reid et al 1999
Dendrotoxin (β Dendrotoxin)	K	Kv channels	Neg		Zygmunt et al 1997
Dendrotoxin (δ Dendrotoxin)	K	K inward rectifier	Neg		Awan and Dolly 1991
Dequalinium dichloride	K	Ca-activated K channels, SK	Neg	potent, highly specific	Dunn 1994
DHS-1 (Dehydrosoyasaponin-1)	K	Ca-activated K channels, BK	Pos		Giangiacomo et al 1998
Diazepam	Cl	Cl channels	Pos		Kourie 1998
Diazoxide	K	K-ATP channels	Pos		Gribble et al 1997
Dibucaine	K	K-ATP channels	Neg		Yoneda et al 1993
Dichlorphenamide	H, HCO_3	Carbonic Anhydrase	Neg	CA catalyzes hydration of CO_2 to H and HCO_5	Lindskog 1997
Dideoxyforskolin	Cl	Cl channels	Neg	stretch activated Cl channels	Lock and Valverde 2000
DIDS	Cl	Cl channels	Neg		Landry et al 1987
DIDS	K	KCNQ1 (KvLQT1)	Pos	activates (when heteromeric)	Busch et al 1997

Diethylstilbestrol (DES)	H	ATP Synthase (F-type H-ATPase)	Neg		Strid et al 1988
Diethylstilbestrol (DES)	H	H-ATPase (V-type H-ATPase)	Neg		Van Der heyden and Docampo 2000
Dihydropyridines	Ca	Cav channels	Neg	Family of CaCl blockers	Yamaguchi et al 2003
Diltiazem (D-cis-Diltiazem-HCl)	Ca	Cav channels	Neg	antagonist, highly specific; a benzothiazepine	Budde 2001
Diltiazem (L-cis-diltiazem)	Ca	CNG channels			Hart et al 2003
Diltiazem (L-cis-diltiazem)	K	Cav channels	Neg	D-diltiazem is Neg control	Kraus et al 1998
Dimethadione	K	K channels	Neg		Azarbayjani and Danielsson 2002
Dimethylamiloride	Na,Ca	Na/Ca exchanger	Neg	more potent than amiloride	Cui et al 2002
Dimethylamiloride	Na,H	Na/H antiporter, (NHE)	Neg	more potent than amiloride	Bonnet, et al. 2000
Diphenylbutylpiperidine (DPBP)	Ca	L- and T-type Ca Channels	Neg		Enyeart et al 1990
Dofetilide	K	KCNH2 (ERG)	Neg		Becchetti et al 2002
DPMX	Cl	CFTR	Pos	enhance Cl efflux	Chappe et al 1998
E-4031	K	KCNH2 (ERG)	Neg		Liu and Wu 2003
EDTA	Ca	Aqueous compartments	Neg	chelator (avail. Photoactivatable)	Christensen et al 2003
Efaroxan	K	K-ATP channels	Neg		Chan et al 1991
EGTA	Ca	Aqueous compartments	Neg	chelator (available labeled and caged)	Miller and Smith 1984
Eicosapentanoic acid (EPA)		tight junctions	Neg		Usami et al 2001
EIPA (Ethylisopropyl amiloride)	Na,H	Na/H antiporter, (NHE)	Neg		Bagnis et al 2001
Ellagic acid	H,K	H/K-ATPase	Neg		Murukami et al 1991a
Englitazone	K	K-ATP channels	Neg		McKay et al 2000
Enprofylline	Cl	CFTR	Pos	enhance Cl efflux	Chappe et al 1998
Ergotoxin	K	KCNH2 (ERG)	Neg		Frenal et al 2004
Erythrosin 5'-isothiocyanate	Ca	SERCA	Neg		Mueller et al 2004
Erythrosine B	H	H-ATPase (V-type H-ATPase)	Neg		Gibbon and Kropf 1993

Esomeprazole	H,K	H/K-ATPase	Neg		Koch et al 2004
Estradiol (11-β-Estradiol)	K	KCNQ1 (KvLQT1)	Neg		Busch et al 1997
Estradiol (17-β-Estradiol)	Na,H	NHE	Neg		Kaloyianni et al 2005
Ethacrynic acid	Cl	Cl channels	Neg		Landry et al 1987
Ethacrynic acid	Cl	Cl-ATPase	Neg		Inagaki et al 1996
Ethoxyzolamide	H	Carbonic Anhydrase	Neg		Furla et al 2000
FCCP	H	Membrane	Neg	ionophore	Park et al 2002
Fe^{2+}	Na,Ca	Na/Ca exchanger	Neg		Kaplan et al 1997
Fe^{2+}	Na,K	Na/K-ATPase	Neg		Kaplan et al 1997
Felodipine	Ca	Cav channels	Neg		Furukawa et al 1999
Fenofibrate	K	K-ATP channels	Neg		Shimomura et al 2004
Fenpropathrin	NA	NaV channels	Neg	a pyrethroid, alters kinetics	Choi and Soderlund 2005
Fenvalerate	Na	NaV channels	Neg		Choi and Soderlund 2005
Flecainide acetate	Na	Na channels	Neg	blocks open channels	Singh et al 1984
Flunarizine dihydrochloride	Ca	Ca channels	Neg	also blocks Na channels	Pauwels et al 1991
Flunarizine dihydrochloride	Na	Na channels	Neg	also blocks Ca channels	Pauwels et al 1991
FPL 64176	Ca	Cav channels	Pos	40X more potent than Bay K 8644	Rampe et al 1993
FR177995	H	H-ATPase (V-type H-ATPase)	Neg		Farina and Gagliardi 2002
Furosemide	Na,K,Cl	Na/K/Cl cotransporter	Neg		O'Grady et al 1987
Fusicoccin	H,K	H-ATPase (V-type H-ATPase)	Pos	effect inhibited by phenylarsine oxide	Olivari et al 2000
Gabapentin HCl	Ca	Cav channels	Neg	binds to site associated with CaVs	Fink et al 2000
Gaboon viper venom	K	K inward rectifier	Neg		Busso et al 1988
Gadolinium	Ca	Ca-activated Cl channels	Neg		Tokimasa and North 1996
Gadolinium		Mechano-sensitive channels	Neg		Krasznai et al 2003
Gallopamil	Ca	Ca-ATPase	Neg		Mitrovic et al 1998
Glibenclamide	Ca	CFTR	Neg		Chappe et al 1998

Glibenclamide (Glyburide)	K	K-ATP channels	Neg		Hambrock et al 2004
Glipizide	K	K-ATP channels	Neg		Zunkler et al 1988
Glycyrrhetinic acid (18-b-Glycyrrhet nic acid)	GJC	Gap Junctions	Neg		Davidson & Baumgarten 1988
Gonyautoxin II	Na	Na channels	Neg	natural, related to saxitoxin	Shimizu 1985
Gonyautoxin III	Na	Na channels	Neg	natural, related to saxitoxin	Shimizu 1985
Gramicidin A	K	Membrane	Neg	creates K channels	Levitt et al 1978
Grayanotoxin	Na	Na channel kinetics	Pos	persistent activation	Lazdunski et al. 1985
Guanethidine	K	K-ATP channels	Neg		Fabiani & Story 1996
Guanidine methanesulfonate	Na,H	NHE	Neg		Weichert et al 1997
Haloperidol	K	KCNH2 (ERG)	Neg		Becchetti et al 2002
Halothane	K	Ca-activated K channels	Neg		Hashiguchi-Ikeda et al 2003
Halothane	K	K inward rectifier	Neg		Milovic et al 2004
Halothane	K	K Tandem Pore channels	Pos		Patel et al 1999
Hanatoxin-1	K	Kv channels	Neg		Shiau et al 2003
Hanatoxin-2	K	Kv channels	Neg		Shiau et al 2003
Heptanol	GJC	Gap Junctions	Neg		Weingart and Bukauskas 1998
HMR-1098	K	K-ATP channels	Neg		Sato et al. 2000
HMR-1556	K	KCNQ1 (KvLQT1)	Neg		Gogelein et al 2000
HMR-1556	Ca	Cav channels	Neg		Gogelein et al 2000
HOCl		tight junctions	Neg		Guo et al 1996
HOE642 (cariporide)	Na,H	NHE	Neg		Weichert et al 1997
HOE694	Na,H	NHE	Neg		Bagnis et al 2001
Hongotoxin	K	Kv channels	Neg		Pragl et al 2002
Hydroxydecanoate (5-Hydroxydecanoate)	K	K-ATP channels (mitochondrial)	Neg		Sato et al 2000
Hydroxyzine	K	KCNH2 (ERG)	NEG		Becchetti et al 2002
Iberiotoxin	K	Ca-activated K channels, BK	Neg		Borsotto 2001
IBMX	Cl	CFTR	Pos	enhance Cl efflux	Chappe et al 1998

Ibutilide	K	KCNH2 (ERG)	Neg		Perry et al 2004
indanyloxyacetic acid (IAA)	Cl	Cl channels	Neg		Landry et al 1987
Iodo-resiniferatoxin	Ca	TRPV1	Neg		Lee et al 2005
Ionomycin	Ca	Membrane		ionophore, more potent than A23187, also transports Pb	Kauffman et al 1980
Isoflurane	K	K Tandem Pore channels	Pos		Patel et al 1999
Isradipine	Ca	Cav channels	Neg		Ruegg & Hof 1990
Kaliotoxin	K	Ca-activated K channels, BK	Neg		Crest et al 1992
Kaliotoxin	K	Kv channels	Neg		Lipkind and Fozzard 1997
KB-R7943 mesylate	Na,Ca	Na/Ca exchanger	Neg	potent, selective	Watano and Kimura 1998
Kurtoxin	Ca	T-type Ca channels	Neg	inhibits activation gating	Sidach and Mintz 2002
Lamotrigine	Ca	Ca channels			Stefani et al 1997
Lamotrigine	K	KCNH2 (ERG)	Neg		Daniellson et al 2005
Lamotrigine	K	Kv channels	Pos		Zona et al 2002
Lamotrigine	Na	NaV channels	Neg		Zona et al 2002
Lansoprazole	H,K	H/K-ATPase	Neg		Besancon et al 1997
Lanthanum	Ca	Ca channels	Neg		Friedman et al 1998
Lanthanum	Cl	Ca-activated Cl channels	Neg		Tokimasa and North 1996
Levocromakalim	K	K-ATP channels	Pos	see also Cromakalim	Noack et al 1992
Lidocaine	K	K-ATP channels	Neg		Yoneda et al 1993
Lidocaine n-ethyl bromide	Na	Na channels	Neg		Tan and Saint 2000
Lindane		Gap Junctions	Neg		Guan et al 1995
Linoleoylamide	K	KCNH2 (ERG)	Neg		Liu and Wu 2003
Linopirdine	K	Ca-activated K channels	Neg		Schnee and Brown 1998
Linopirdine (specific to 2/3)	K	KCNQ2/3	Neg		Higashida et al 2000
Linopirdine dihydrochloride	K	Kv channels	Neg		Schnee and Brown 1998
Lobatamide C	H	H-ATPase (V-type H-ATPase)	Neg		Shen et al 2002

Lobatamide C A15	H	H-ATPase (V-type H-ATPase)	Neg		Shen et al 2002
Lobatamide C A16	H	H-ATPase (V-type H-ATPase)	Neg		Shen et al 2002
Lobatamide C A6	H	H-ATPase (V-type H-ATPase)	Neg		Shen et al 2002
Lonidamine	Cl	CFTR	Neg		Gong et al 2002
Loperamide	Ca	Cav channels	Neg	Hivh-voltage activated, NMDA	Church et al 1994
LY83583	Ca	CNG channels	Neg		Leinders-Zufall 1997
LY97241	K	KCNH2 (ERG)	Neg	more potent analog of clofilium	Suessbrich et al 1996
Margatoxin	K	Kv channels	Neg		Pragl et al 2002
MaxiPost	K	Ca-activated K channels, BK	Pos		Kiesewetter et al 2002
Mefenamic acid	K	KCNQ1 (KvLQT1)	Pos	activates (when heteromeric)	Busch et al 1997
Mefloquine	K	K-ATP channels	Neg	Interact with Kir6.2	Gribble et al 2000
Meglitinide	K	K-ATP channels	Neg		Gribble et al 1997
Mepivacaine	K	K-ATP channels	Neg		Yoneda et al 1993
Methazolamide	H, HCO3	Carbonic Anhydrase	Neg	CA catalyzes hydration of CO_2 to H and HCO_4	Lindskog 1997
Mexiletine	K	K-ATP channels	Neg		Yoneda et al 1993
Mg^{2+}	Na,Ca	Na/Ca exchanger	Neg		Smith et al 1987
Mibefradil	Ca	L- and T-type Ca channels	Neg		Mocanu et al 1999
Minoxidil	K	K-ATP channels	Pos		Merck Index **12** 6290; Russ et al 2003
MK-499	K	KCNH2 (ERG)	Neg		Baskin and Lynch 1994
Mn^{2+}	Cl	Ca-activated Cl channels	Neg		Tokimasa and North 1996
Mn^{2+}	Ca	Na/Ca exchanger	Neg		Frame and Milanick 1991

Mn^{2+}	Na	Na/Ca exchanger	Neg		Frame and Milanick 1991
Moclobemide	H	Aqueous compartments	Pos	lowers cytplasmic pH	Bonnet, et al. 2000
Monensin	Na	Membrane	Neg	ionophore	Richard et al 1989
Nateglinide	K	K-ATP channels	Neg		Hu et al 2000
NBD (chloronitrobenzoxadiazole)	H	ATP Synthase (F-type H-ATPase)	Neg		Orriss et al 1998
NBD (chloronitrobenzoxadiazole)	H	H-ATPase (V-type H-ATPase)	Neg		Benderra et al 1998
Neosaxitoxin	Na	Na channels	Neg	natural, related to saxitoxin, pH sensitive	Hu and Kao 1991
N-Ethylmaleimide (NEM)	H	H-ATPase (V-type H-ATPase)	Neg		Forgac 1999
Nicardipine	Ca	L-type Ca channels	Neg		Cui et al 2002
Nicardipine	Ca	N- and P/Q-type Ca channels	Neg		Furukawa et al 1999
$NiCl_2$	Na,Ca	Na/Ca exchanger	Neg		Dong et al 2003
$NiCl_2$	Ca	T-type Ca channels	Neg		Boulton and O'Shaughnessy 1991
Nicorandil	K	K-ATP channels (mitochondrial)	Pos	see also SG 209	Sato et al 2000
Nifedipine		KCNG channels	Neg		Lin et al 2001
Nifedipine	Ca	Cav channels	Neg		Furukawa et al 1999
Niflumic acid	Cl	Ca-activated Cl channels	Neg		Lee et al 2005
Nigericin	K	Membrane		ionophore	Horiuchi and Iwata 1983
Niguldipine HCl [(R)(-) isomer]	Ca	Cav channels	Neg	less active than S isomer	Hollt et al 1992
Niguldipine HCl [(S)(+) isomer]	Ca	Cav channels	Neg	more active than R isomer	Hollt et al 1992
Nilvadipine	Ca	Cav channels	Neg		Furukawa et al 1999
Nimodipine	Ca	Cav channels	Neg		Batuecas et al 1998
Nisoldipine	Ca	Cav channels	Neg		Mitchell et al 1993
Nitrendipine	K	Ca-activated K channels, IK	Neg		Merck Index 12 6667; Watanabe et al 1995

Nitrendipine	Ca	Cav channels	Neg		Merck Index 12 6667; Watanabe et al 1995
Norepinephrine	Ca	Ca-ATPase	Pos	Ca from SR increased	Kowarski et al 1985
Noxiustoxin	K	Kv channels	Neg		Gurrola and Possani 1995
NPPB (PAA)	Cl	Cl channels	Neg	very potent	Kirkup et al 1996
NS004	K	Ca-activated K channels, BK	Pos		Debska et al 2003
NS1608	K	Ca-activated K channels, BK	Pos		Hu et al 1995
NS1619	K	Ca-activated K channels, BK	Pos		Debska et al 2003
Ochratoxin A	Ca	Ca-ATPase	Neg	Endoplasmic reticulum	Chong and Rahimtula 1992
Oleamide	K	KCNH2 (ERG)	Neg		Liu and Wu 2003
Oligomycin	H	ATP Synthase (F-type H-ATPase)	Neg		Schultz et al 1996
Oligomycin	H	H-ATPase (V-type H-ATPase)	Neg		Perlin et al 1985
Oligomycin	Na,K	Na/K-ATPase	Neg		Arato-Oshima et al 1996
Omeprazole	H,K	H/K-ATPase	Neg		Besancon et al 1997
Orthovanadate	Cl	Cl transport	Neg		Candia and Podos 1981
Ouabain	Na,K	Na/K-ATPase	Neg		Croyle et al 1997
P1075	K	K-ATP channels (sarcolemmal)	Pos		Sato et al. 2000
PAA (NPPB)	Cl	Cl channels	Neg	very potent	Kirkup et al 1996
Palytoxin (PTX)	Na,K	Na/K-ATPase	Neg	converts Na/K-ATPase to pore	Tosteson 2003
Pantoprazole	H	H/K-ATPase	Neg		Besancon et al 1997
Paspalicine	K	Ca-activated K channels, BK	Neg		Knaus et al 1994
Paspalinine	K	Ca-activated K channels, BK	Neg		Knaus et al 1994

Paspalitrem A	K	Ca-activated K channels, BK	Neg		Knaus et al 1994
Paspalitrem C	K	Ca-activated K channels, BK	Neg		Knaus et al 1994
Paxilline	K	Ca-activated K channels, BK	Neg		Bilmen et al 2002
Penfluridol	Ca	T-type Ca channels			Enyeart et al 1992
Penitrem A	K	Ca-activated K channels, BK	Neg		Adeagbo 1999
Pentoxifylline	Cl	CFTR	Pos	enhance Cl efflux	Chappe et al 1998
Pentylenetetrazol	K	K-ATP channels	Pos		Klocker et al 1996
Pentylenetetrazol	K	Kv channels	Pos/Neg		Madeja et al 1997
Permethrin	Na	NaV channels	Neg	a pyrethroid, alters kinetics	Choi and Soderlund 2005
Phenothrin	Na	Na channel kinetics	Neg	a pyrethroid, alters kinetics	Narahashi 1982
Phentolamine	K	K-ATP channels	Neg		Proks and Ashcroft 1997
Phenylalkylamines	Ca	Cav channels	Neg	Family of CaCL blockers	Striessnig 1999
PHM (phentolamine mesylate)	K	K-ATP channels	Neg		Proks and Ashcroft 1997
PI(4,5)P2	K	K inward rectifier	Pos	not Kir6.x	Bian et al 2001
PI(4,5)P2	K	K-ATP channels	Neg	Reduce sensitivity to ATP	Day et al 1999
PI(4,5)P2	K	KCNH2 (ERG)	pos		Bian et al 2001
Pinacidil	K	K-ATP channels	Pos	Mg-ATPase in medium required	Gribble et al 1997
PKF 217-744	K	K-ATP channels	Pos		Weiss et al 2002
Pompilidotoxin (β-Pompilidotoxin; wasp toxin)	Na	Na channels	Pos	slows inactivation	Konno et al 1998
Prenylamine	Ca	Ca-ATPase	Neg		Lamers et al 1985
Procaine	K	K-ATP channels	Neg		Yoneda et al 1993
Procaine	Na	NaV channels	Neg	slow-blocking	Ribeiro and Costa 2003
Prodigiosin	H	H/K-ATPase	Neg		Matsuya et al 2000

Prodigiosin	H	H-ATPase (V-type H-ATPase)	Neg	uncouples by H+/Cl- symport	Ohkuma et al 1998
Propranolol	K	K-ATP channels	Neg		Yoneda et al 1993
Prostaglandin J2 (15-deoxy-Delta12,14-Prostaglandin J2)	K	K-ATP channels	Neg		Shimomura et al 2004
Pumaprazole	H	H/K-ATPase	Neg		Kromer et al 2000
Pyrethroids	Na	Na channel kinetics		modify closing	Lazdunski et al, 1985
Quercetin	H,K	H/K-ATPase	neg		Murakami et al 1992b
Quinidine	K	KCNH2 (ERG)	Neg		Becchetti et al 2002
Quinine	K	K-ATP channels	Neg	Interact with Kir6.2	Gribble et al 2000
QX 222	Na	Na channels	Neg		Cuevas and Adams 1994
QX 314	Na	NaV channels	Neg	impermeable	Alreja and Aghajanian 1994
Rabeprazole	H,K	H/K-ATPase	Neg		Besancon et al 1997
repaglinide	K	K-ATP channels	Neg		Hu et al 2000
Retigabine	K	KCNQ	Pos		Tatulian et al 2001
Rilmakalim	K	K-ATP channels	Pos		Kocic et al 2004
Riluzole	Ca	Ca channels	Neg		Stefani et al 1997
Riluzole HCl	Na	Na channels	Neg		Stefani et al 1997
Riodipine (Ryosidine)	Ca	Cav channels	Neg	highly specific	Kryzhanovskii et al 1992
RP 66471	K	K-ATP channels	Pos		Szewczyk et al 1993
rTamapin	K	KCNN2	Neg		Pedarzani et al 2002
Ruthenium red	Ca	Ca translocators (many)	Neg	blocks mitochondrial Ca flux,CaV & RyR; capsaicin antagonist	Bernardi et al 1984
Saxitoxin (IIα (-OH) STX)	Na	Na channels	Neg	synthetic, related to saxitoxin	Hu and Kao 1991
Saxitoxin (PSP)	Na	Na channels	Neg	pore blocker, many analogs	Catterall et al 1985
Saxitoxinol (α-saxitoxinol)	Na	Na channels	Neg	synthetic, related to saxitoxin	Schantz 1985

Saxitoxinol (β-saxitoxinol)	Na	Na channels	Neg	synthetic, related to saxitoxin	Schantz 1985
SB-242784	H	H-ATPase (V-type H-ATPase)	Neg		Farina and Gagliardi 2002
SCA40	K	Ca-activated K channels, BK	Pos		Michel et al 1993
SCH28080	H,K	H/K-ATPase	Neg		Vagin et al 2002
Scorpion toxins (American β)	Na	Na channel kinetics	Pos	enhance activation	Lazdunski et al, 1985
Scorpion toxins (North African α)	Na	Na channel kinetics	Pos	slow inactivation	Lazdunski et al, 1985
Scyllatoxin	K	Ca-activated K channels, SK	Neg		Wittekindt et al 2004
Sea Anemone toxins	Na	Na channel kinetics	Pos	slow inactivation	Lazdunski et al, 1985
Sevoflurane	K	Ca-activated K channels, BK	Pos		Kehl et al 2002
Sevoflurane	K	K Tandem Pore channels	Pos		Liu et al 2004
Sevoflurane	Ca	Cav channels	Neg		Huneke et al 2004
SG 209	K	K channels	Pos	analog of Nicorandil	Ishibashi et al 1991
SNX-111	Ca	N-type Ca channels	Neg		Bowersox et al 1997
SNX-482	Ca	R-type Ca channels	Neg	from tarantula venom	Newcomb et al 1998
sodium caprate		tight junctions	Neg		Soderholm et al 1998
Sophalcone	H,K	H/K-ATPase	Neg		Murukami et al 1991b
Sophoradin	H,K	H/K-ATPase	Neg		Murukami et al 1991b
Sotalol	K	KCNH2 (ERG)	Neg		Kupershmidt et al 2003
SR 33805 oxalate	Ca	Cav channels	Neg	potent, binds to unusual site	Chatelain et al 1993
Sr^{2+}	Na,Ca	Na/Ca exchanger	Neg		Richard et al 1989
Streptolysin O		Membrane	Neg	makes holes	Palmer 1998
Strophanthidin	Na,K	Na/K-ATPase	Neg		Eisner et al 1987
TAC-101		Gap Junctions	Pos		Miyaguchi et al 2001
Tamoxifen	Cl	Cl channels	Neg	stretch activated Cl channels	Lock and Valverde 2000
TEA (Tetraethylammonium)	K	Ca-activated K channels, BK	Neg	permeable	Borsotto 2001

TEA (Tetraethylammonium)	K	KCNQ	Neg	permeable	Hadley et al 2000
TEA (Tetraethylammonium)	K	Kv channels	Neg	permeable	Lipkind and Fozzard 1997
Tedisamil	K	K channels	Neg		Mitrovic et al 1998
Tefluthrin	Na	NaV channels	Neg	a pyrethroid, alters kinetics	Choi and Soderlund 2005
Tenidap	K	K inward rectifier	Pos	Kir2.3	Liu et al 2002
Tenidap	Na	Na/HCO3 co-transporter	Neg		Ducoudret et al 2001
Terfenadine	K	KCNH2 (ERG)	Neg		Suessbrich et al 1996
Terikalant	K	K inward rectifier	Neg		Jurkiewicz et al 1996
Tertiapin-Q (bee venom toxin)	K	K inward rectifier	Neg	high affinity to ROMK1 and GIRK1/4	Jin et al 1999
Tetracaine		CNG channels	Neg		Ghatpande et al 2003
Tetracaine	K	K-ATP channels	Neg		Yoneda et al 1993
Tetrahydroberberine (THB)	K	Kv channels	Neg		Wu and Jin 1997
Tetramethrin	Na	Na channel kinetics		a pyrethroid	Lazdunski et al, 1985
Tetrandrine	K	Ca-activated K channels, BK	Neg		Wu et al 2000
Tetrodotoxin (TTX)	Na	Na channels	Neg	very specific, reversible, pore blocker	Catterall et al 1985
Thapsigargin	Ca	SERCA	Neg	potent, avail. Fluorescent	Procida et al 1998
Theophylline	Cl	CFTR	Pos	enhance Cl efflux	Chappe et al 1998
Tiludronate (skelid)	H	H-ATPase (V-type H-ATPase)	Neg		David et al 1996
Tityustoxin-Kα	K	Kv channels	Neg		Rodrigues et al 2003
TMB-8	K	K-ATP channels	Neg		Szewczyk et al 1992
Tolazamide	K	K-ATP channels	Neg		Szabo and Salzman 1996
Tolbutamide	K	K-ATP channels	Neg		Gribble et al 1997
TPEN	Zn,Cu,Fe	Other	Neg	chelator, ignores Ca & Mg	Shumaker et al 1998
TRAM-34	K	Ca-activated K channels, IK	Neg		Wulff et al 2000
Tributyl Tin (TBT)	Cl,OH	Membrane		ionophore	Matsuya et al 2000
Tricaine	Na	NaV channels	Neg		Wang et al 1994

Trifluorperazine	Ca	Calmodulin	Neg		Bouquin et al 1995
Trimethadione	K	K inward rectifier	Neg	Ikr	Azarbayjani and Danielsson 2002
Troglitazone	K	K-ATP channels	Neg		Shimomura et al 2004
U-37883	K	K-ATP channels	Neg		Meisheri et al 1995
U-37883A	K	K-ATP channels	Neg		Meisheri et al 1995
UCL 1684	K	Ca-activated K channels, SK	Neg	highly potent	Malik-Hall et al 2000
Valinomycin	H,K	Membrane		ionophore	Carmeliet and Lieberman 1975
Vanadate	Na,K	Na/K-ATPase	Neg		Bond and Hudgins 1982
Verapamil HCl	K	KCNH2 (ERG)	Neg		Becchetti et al 2002
Verapamil HCl	Ca	Cav channels	Neg	Isoproterenol rescues	Budde 2001
Verapamil (methoxy-Verapamil)	Ca	Cav channels	Neg		Klockner and Isenberg 1991
Veratridine	Na	Na channel kinetics	Pos	persistent activation	Lazdunski et al, 1985
Verruculogen	K	Ca-activated K channels, BK	Neg		Knaus et al 1994
Vinpocetine	Na	NaV channels	Neg	also inhibits PDE1	Molnar and Erdo 1995
Way-123398	K	KCNH2 (ERG)	Neg		Becchetti et al 2002
WIN 17317-3	K	Kv channels	Neg		Hill et al 1995
Xanthoangelol	H,K	H/K-ATPase	Neg		Murukami et al 1990
XE 991 dihydrochloride	K	KCNQ1 (KvLQT1), KCNQ1 (KvLQT1)+minK, KCNQ2+3	Neg	less potent against KCNQ1/minK	Zaczek et al 1998
Y-26762	K	K-ATP channels	Pos		Yoneda et al 1993
YS-035	K	K channels	Neg	also a Ca^{2+} blocker	Berger et al 1991
Zatebradine	K	Kv channels	Neg	Kv1.5	Valenzuela et al 1996
ZD6169	K	K-ATP channels	Pos		Ohnmacht et al 1996
ZD7288	K	HCN channels	Neg		Felix et al 2003
ZD7288	Na	HCN channels	Neg		Felix et al 2003
ZD7288	Ca	T-type Ca channels	Neg		Felix et al 2003
Zetekitoxin AB	Na	NaV channels	Neg	analog of Saxitoxin, extremely potent	Yotsu-Yamashita et al 2004

ZM 181,037	K	K-ATP channels	Neg		Kau et al 1994
ZM 226600 (Anilide tertiary carbinols)	K	K-ATP channels	Pos		Grant et al 1994
Zn	H, Na	NHE	Neg		Kaloyianni et al 2005
$ZnCl_2$	Na,Ca	Na/Ca exchanger	Neg		Colvin 1998

REFERENCES

Abriel, H., U. Katz, and P. Kucera (1994). Ion Transport across the Early Chick Embryo: Ii. Characterization and pH Sensitivity of the Transembryonic Short-Circuit Current. *Journal of Membrane Biology* **141**(2): 159–166.

Adeagbo, A. S. (1999). 1-Ethyl-2-Benzimidazolinone Stimulates Endothelial K(Ca) Channels and Nitric Oxide Formation in Rat Mesenteric Vessels. *Eur J Pharmacol* **379**(2–3): 151–159.

Alreja, M. and G. K. Aghajanian (1994). Qx-314 Blocks the Potassium but Not the Sodium-Dependent Component of the Opiate Response in Locus Coeruleus Neurons. *Brain Res* **639**(2): 320–324.

Arato-Oshima, T., H. Matsui, A. Wakizaka, and H. Homareda (1996). Mechanism Responsible for Oligomycin-Induced Occlusion of Na+ within Na/K-Atpase. *J Biol Chem* **271**(41): 25604–25610.

Arch, J. R., N. E. Bowring, and D. R. Buckle (1994). Evaluation of the Novel Potassium Channel Activator Brl 55834 as an Inhaled Bronchodilator in Guinea-Pigs and Rats: Comparison with Levcromakalim and Salbutamol. *Pulm Pharmacol* **7**(2): 121–128.

Atwal, K. S. (1994). Pharmacology and Structure-Activity Relationships for Katp Modulators: Tissue-Selective Katp Openers. *Journal of Cardiovascular Pharmacology* **24 Suppl 4**: S12–17.

Awan, K. A. and J. O. Dolly (1991). K+ Channel Sub-Types in Rat Brain: Characteristic Locations Revealed Using Beta-Bungarotoxin, Alpha- and Delta-Dendrotoxins. *Neuroscience* **40**(1): 29–39.

Azarbayjani, F. and B. R. Danielsson (2002). Embryonic Arrhythmia by Inhibition of Herg Channels: A Common Hypoxia-Related Teratogenic Mechanism for Antiepileptic Drugs? *Epilepsia* **43**(5): 457–468.

Bagnis, C., M. Marsolais, D. Biemesderfer, R. Laprade, and S. Breton (2001). Na+/H+-Exchange Activity and Immunolocalization of Nhe3 in Rat Epididymis. *Am J Physiol Renal Physiol* **280**(3): F426–436.

Baskin, E. P. and J. J. Lynch, Jr. (1994). Comparative Effects of Increased Extracellular Potassium and Pacing Frequency on the Class Iii Activities of Methanesulfonanilide Ikr Blockers Dofetilide, D-Sotalol, E-4031, and Mk-499. *J Cardiovasc Pharmacol* **24**(2): 199–208.

Batuecas, A., R. Pereira, C. Centeno, J. A. Pulido, M. Hernandez, A. Bollati, E. Bogonez, and J. Satrustegui (1998). Effects of Chronic Nimodipine on Working Memory of Old Rats in Relation to Defects in Synaptosomal Calcium Homeostasis. *Eur J Pharmacol* **350**(2–3): 141–150.

Becchetti, A., M. De Fusco, O. Crociani, A. Cherubini, R. Restano-Cassulini, M. Lecchi, A. Masi, A. Arcangeli, G. Casari, and E. Wanke (2002). The Functional Properties of the Human Ether-a-Go-Go-Like (Helk2) K+ Channel. *Eur J Neurosci* **16**(3): 415–428.

Beil, W., K. F. Sewing, and W. Kromer (1999). Basic Aspects of Selectivity of Pantoprazole and Its Pharmacological Actions. *Drugs Today (Barc)* **35**(10): 753–764.

Bell, S. M., C. M. Schreiner, E. Resnick, C. Vorhees, and W. J. Scott, Jr. (1997). Exacerbation of Acetazolamide Teratogenesis by Amiloride and Its Analogs Active against Na+/H+ Exchangers and Na+ Channels. *Reproductive Toxicology* **11**(6): 823–831.

Benderra, Z., H. Morjani, A. Trussardi, and M. Manfait (1998). Characterization of H+-Atpase-Dependent Activity of Multidrug Resistance-Associated Protein in Homoharringtonine-Resistant Human Leukemic K562 Cells. *Leukemia* **12**(10): 1539–1544.

Benders, A. A., J. Li, R. A. Lock, R. J. Bindels, S. E. Bonga, and J. H. Veerkamp (1994). Copper Toxicity in Cultured Human Skeletal Muscle Cells: The Involvement of Na+/K(+)-Atpase and the Na+/Ca(2+)-Exchanger. *Pflugers Arch* **428**(5–6): 461–467.

Berger, F., U. Borchard, D. Hafner, T. Kammer, and T. Weis (1991). Inhibition of Potassium Outward Currents and Pacemaker Current in Sheep Cardiac Purkinje Fibres by the Verapamil Derivative Ys 035. *Naunyn Schmiedebergs Arch Pharmacol* **344**(6): 653–661.

Bernardi, P., V. Paradisi, T. Pozzan, and G. F. Azzone (1984). Pathway for Uncoupler-Induced Calcium Efflux in Rat Liver Mitochondria: Inhibition by Ruthenium Red. *Biochemistry* **23**(8): 1645–1651.

Bernstein, G. M. and O. T. Jones (2001). Mapping N-Type Calcium Channel Distributions with Ω-Conotoxins. *Ion Channel Localization: Methods and Protocols*. A. N. Lopatin and C. G. Nichols. Totowa, NJ, Humana Press.

Bers, D. M. and J. H. Bridge (1989). Relaxation of Rabbit Ventricular Muscle by Na-Ca Exchange and Sarcoplasmic Reticulum Calcium Pump. Ryanodine and Voltage Sensitivity. *Circ Res* **65**(2): 334–342.

Besancon, M., A. Simon, G. Sachs, and J. M. Shin (1997). Sites of Reaction of the Gastric H,K-Atpase with Extracytoplasmic Thiol Reagents. *Journal of Biological Chemistry* **272**(36): 22438–22446.

Bian, J., J. Cui, and T. V. McDonald (2001). Herg K(+) Channel Activity Is Regulated by Changes in Phosphatidyl Inositol 4,5-Bisphosphate. *Circ Res* **89**(12): 1168–1176.

Bilmen, J. G., L. L. Wootton, and F. Michelangeli (2002). The Mechanism of Inhibition of the Sarco/Endoplasmic Reticulum Ca2+ Atpase by Paxilline. *Arch Biochem Biophys* **406**(1): 55–64.

Bond, G. H. and P. M. Hudgins (1982). Low-Affinity Na+ Sites on (Na+ +K+)-Atpase Modulate Inhibition of Na+-Atpase Activity by Vanadate. *Biochim Biophys Acta* **687**(2): 310–314.

Bonnet, U., T. Leniger, and M. Wiemann (2000). Moclobemide Reduces Intracellular pH and Neuronal Activity of Ca3 Neurones in Guinea-Pig Hippocampal Slices-Implication for Its Neuroprotective Properties. *Neuropharmacology* **39**(11): 2067–2074.

Boot, J. H. and J. A. Van Hilten (1996). The Use of the Divalent Calcium-Ionophore A23187 as a Biochemical Tool in Pharmacological and in Vitro Toxicological Studies. *Cell Struct Funct* **21**(2): 97–99.

Borsotto, M. (2001). Small Conductance Calcium-Activated Potassium Channels in Rat Brain. *Ion Channel Localization: Methods and Protocols*. A. N. Lopatin and C. G. Nichols. Totowa, NJ, Humana Press.

Boulton, C. L. and C. T. O'Shaughnessy (1991). The Effect of Calcium Channel Antagonists on Spontaneous and Evoked Epileptiform Activity in the Rat Neocortex in Vitro. *Eur J Neurosci* **3**(10): 992–1000.

Bouquin, N., M. Tempete, I. B. Holland, and S. J. Seror (1995). Resistance to Trifluoroperazine, a Calmodulin Inhibitor, Maps to the Fabd Locus in Escherichia Coli. *Mol Gen Genet* **246**(5): 628–637.

Bourdelais, A. J., S. Campbell, H. Jacocks, J. Naar, J. L. Wright, J. Carsi, and D. G. Baden (2004). Brevenal Is a Natural Inhibitor of Brevetoxin Action in Sodium Channel Receptor Binding Assays. *Cell Mol Neurobiol* **24**(4): 553–563.

Bowersox, S., J. Mandema, K. Tarczy-Hornoch, G. Miljanich, and R. R. Luther (1997). Pharmacokinetics of Snx-111, a Selective N-Type Calcium Channel Blocker, in Rats and Cynomolgus Monkeys. *Drug Metab Dispos* **25**(3): 379–383.

Bowersox, S. S., G. P. Miljanich, Y. Sugiura, C. Li, L. Nadasdi, B. B. Hoffman, J. Ramachandran, and C. P. Ko (1995). Differential Blockade of Voltage-Sensitive Calcium Channels at the Mouse Neuromuscular Junction by Novel Omega-Conopeptides and Omega-Agatoxin-Iva. *J Pharmacol Exp Ther* **273**(1): 248–256.

Brugnara, C. (1997). Erythrocyte Membrane Transport Physiology. *Current Opinion in Hematology* **4**(2): 122–127.

Budde, T. (2001). Fluorescent Calcium Antagonists. *Ion Channel Localization: Methods and Protocols*. A. N. Lopatin and C. G. Nichols. Totowa, NJ, Humana Press.

Busch, A. E., G. L. Busch, E. Ford, H. Suessbrich, H. J. Lang, R. Greger, K. Kunzelmann, B. Attali, and W. Stuhmer (1997). The Role of the Isk Protein in the Specific Pharmacological Properties of the Iks Channel Complex. *British Journal of Pharmacology* **122**(2): 187–189.

Busso, C., E. Camino, L. Cedrini, and D. Lovisolo (1988). The Effects of Gaboon Viper (Bitis Gabonica) Venom on Voltage-Clamped Single Heart Cells. *Toxicon* **26**(6): 559–570.

Candia, O. A. and S. M. Podos (1981). Inhibition of Active Transport of Chloride and Sodium by Vanadate in the Cornea. *Invest Ophthalmol Vis Sci* **20**(6): 733–737.

Carmeliet, E. E. and M. Lieberman (1975). Increase of Potassium Flux by Valinomycin in Embryonic Chick Heart. *Pflugers Arch* **358**(3): 243–257.

Catterall, W., A., J. W. Schmidt, D. J. Messner, and D. J. Feller (1985). *Structure and Biosynthesis of Neuronal Sodium Channels*. Tetrodotoxin, Saxitoxin, and the Molecular Biology of the Sodium Channel, New York, The New York Academy of Sciences.

Catterall, W. A. (1980). Neurotoxins That Act on Voltage-Sensitive Sodium Channels in Excitable Membranes. *Annual Review of Pharmacology & Toxicology* **20**: 15–43.

Challinor-Rogers, J. L., F. L. Rosenfeldt, X. J. Du, and G. A. McPherson (1997). Antiischemic and Antiarrhythmic Activities of Some Novel Alinidine Analogs in the Rat Heart. *J Cardiovasc Pharmacol* **29**(4): 499–507.

Chan, S. L., M. J. Dunne, M. R. Stillings, and N. G. Morgan (1991). The Alpha 2-Adrenoceptor Antagonist Efaroxan Modulates K+Atp Channels in Insulin-Secreting Cells. *Eur J Pharmacol* **204**(1): 41–48.

Chappe, V., Y. Mettey, J. M. Vierfond, J. W. Hanrahan, M. Gola, B. Verrier, and F. Becq (1998). Structural Basis for Specificity and Potency of Xanthine Derivatives as Activators of the Cftr Chloride Channel. *Br J Pharmacol* **123**(4): 683–693.

Chatelain, P., M. Clinet, P. Polster, B. Christophe and A. S. Manning (1993). In Vitro Characterization of a Novel Ca2+ Entry Blocker: Sr 33805. *Eur J Pharmacol* **246**(3): 181–193.

Choi, J. S. and D. M. Soderlund (2005). Structure-Activity Relationships for the Action of 11 Pyrethroid Insecticides on Rat Na(V)1.8 Sodium Channels Expressed in Xenopus Oocytes. *Toxicol Appl Pharmacol*.

Chong, X. and A. D. Rahimtula (1992). Alterations in Atp-Dependent Calcium Uptake by Rat Renal Cortex Microsomes Following Ochratoxin a Administration in Vivo or Addition in Vitro. *Biochem Pharmacol* **44**(7): 1401–1409.

Christensen, T., D. M. Gooden, J. E. Kung, and E. J. Toone (2003). Additivity and the Physical Basis of Multivalency Effects: A Thermodynamic Investigation of the Calcium Edta Interaction. *J Am Chem Soc* **125**(24): 7357–7366.

Church, J., E. J. Fletcher, K. Abdel-Hamid, and J. F. MacDonald (1994). Loperamide Blocks High-Voltage-Activated Calcium Channels and N-Methyl-D-Aspartate-Evoked Responses in Rat and Mouse Cultured Hippocampal Pyramidal Neurons. *Mol Pharmacol* **45**(4): 747–757.

Colvin, R. A. (1998). Zinc Inhibits Ca2+ Transport by Rat Brain Na+/Ca2+ Exchanger. *Neuroreport* **9**(13): 3091–3096.

Cox, D. A., L. Conforti, N. Sperelakis, and M. A. Matlib (1993). Selectivity of Inhibition of Na(+)-Ca2+ Exchange of Heart Mitochondria by Benzothiazepine Cgp-37157. *J Cardiovasc Pharmacol* **21**(4): 595–599.

Crest, M., G. Jacquet, M. Gola, H. Zerrouk, A. Benslimane, H. Rochat, P. Mansuelle, and M. F. Martin-Eauclaire (1992). Kaliotoxin, a Novel Peptidyl Inhibitor of Neuronal Bk-Type Ca(2+)-Activated K+ Channels Characterized from Androctonus Mauretanicus Mauretanicus Venom. *J Biol Chem* **267**(3): 1640–1647.

Croyle, M. L., A. L. Woo, and J. B. Lingrel (1997). Extensive Random Mutagenesis Analysis of the Na+/K+-Atpase Alpha Subunit Identifies Known and Previously Unidentified Amino Acid Residues That Alter Ouabain Sensitivity — Implications for Ouabain Binding. *European Journal of Biochemistry* **248**(2): 488–495.

Cuevas, J. and D. J. Adams (1994). Local Anaesthetic Blockade of Neuronal Nicotinic Ach Receptor-Channels in Rat Parasympathetic Ganglion Cells. *Br J Pharmacol* **111**(3): 663–672.

Cui, X. L., H. Z. Chen, D. M. Wu, and B. W. Wu (2002). Enhancement of Ca(2+) Transients and Contraction of Single Ventricular Myocytes of Rats by 5-(N,N-Dimethyl) Amiloride. *Sheng Li Xue Bao* **54**(3): 219–224.

Dahlin, K. L., K. Bohlin, J. Strindlund, A. Ryrfeldt, and I. A. Cotgreave (1999). Amitriptyline-Induced Loss of Tight Junction Integrity in a Human Endothelial — Smooth Muscle Cell Bi-Layer Model. *Toxicology* **136**(1): 1–13.

Danielsson, B. R., K. Lansdell, L. Patmore, and T. Tomson (2005). Effects of the Antiepileptic Drugs Lamotrigine, Topiramate and Gabapentin on Herg Potassium Currents. *Epilepsy Res* **63**(1): 17–25.

David, P., H. Nguyen, A. Barbier, and R. Baron (1996). The Bisphosphonate Tiludronate Is a Potent Inhibitor of the Osteoclast Vacuolar H+-Atpase. *Journal of Bone and Mineral Research* **11**(10): 1498–.

Davidson, J. and I. Baumgarten (1988). Glycyrrhetinic Acid Derivatives: A Novel Class of Inhibitors of Gap-Junctional Intercellular Communication. Structure-Activity Relationships. *Journal of Pharmacology & Experimental Therapeutics* **246**(3): 1104–1107.

Day, Y. J., Z. Gao, P. C. Tan, and J. Linden (1999). Atp Sensitive Potassium Channel and Myocardial Preconditioning. *Acta Anaesthesiologica Sinica* **37**(3): 121–131.

Debska, G., A. Kicinska, J. Dobrucki, B. Dworakowska, E. Nurowska, J. Skalska, K. Dolowy, and A. Szewczyk (2003). Large-Conductance K+ Channel Openers Ns1619 and Ns004 as Inhibitors of Mitochondrial Function in Glioma Cells. *Biochem Pharmacol* **65**(11): 1827–1834.

Dong, J. W., H. F. Zhu, and Z. N. Zhou (2003). Inhibitors of Na(+)/H(+) and Na(+)/Ca(2+) Exchange Depress Intracellular Calcium Elevation Induced by Ischemia/Reperfusion in Rat Cardiac Myocytes. *Sheng Li Xue Bao* **55**(3): 245–250.

Drose, S. and K. Altendorf (1997). Bafilomycins and Concanamycins as Inhibitors of V-Atpases and P-Atpases. *Journal of Experimental Biology* **200**(Pt 1): 1–8.

Ducoudret, O., A. Diakov, S. Muller-Berger, M. F. Romero, and E. Fromter (2001). The Renal Na-Hco3-Cotransporter Expressed in Xenopus Laevis Oocytes: Inhibition by Tenidap and Benzamil and Effect of Temperature on Transport Rate and Stoichiometry. *Pflugers Arch* **442**(5): 709–717.

Dunn, P. M. (1994). Dequalinium, a Selective Blocker of the Slow Afterhyperpolarization in Rat Sympathetic Neurones in Culture. *Eur J Pharmacol* **252**(2): 189–194.

Dunne, M. J., D. I. Yule, D. V. Gallacher, and O. H. Petersen (1989). Cromakalim (Brl 34915) and Diazoxide Activate Atp-Regulated Potassium Channels in Insulin-Secreting Cells. *Pflugers Arch* **414 Suppl 1**: S154–155.

Eisner, D. A., M. Valdeolmillos, and S. Wray (1987). The Effects of Membrane Potential on Active and Passive Sodium Transport in Xenopus Oocytes. *Journal of Physiology* **385**: 643–659.

Enyeart, J. J., B. A. Biagi, R. N. Day, S. S. Sheu, and R. A. Maurer (1990). Blockade of Low and High Threshold Ca2+ Channels by Diphenylbutylpiperidine Antipsychotics Linked to Inhibition of Prolactin Gene Expression. *J Biol Chem* **265**(27): 16373–16379.

Enyeart, J. J., B. A. Biagi, and B. Mlinar (1992). Preferential Block of T-Type Calcium Channels by Neuroleptics in Neural Crest-Derived Rat and Human C Cell Lines. *Mol Pharmacol* **42**(2): 364–372.

Ertel, E. A., V. A. Warren, M. E. Adams, P. R. Griffin, C. J. Cohen, and M. M. Smith (1994). Type Iii Omega-Agatoxins: A Family of Probes for Similar Binding Sites on L- and N-Type Calcium Channels. *Biochemistry* **33**(17): 5098–5108.

Fabiani, M. E. and D. F. Story (1996). Inhibition of Sympathetic Noradrenergic Transmission by Guanabenz and Guanethidine in Rat Isolated Mesenteric Artery: Involvement of Neuronal Potassium Channels. *Pharmacol Res* **33**(3): 171–180.

Farina, C. and S. Gagliardi (2002). Selective Inhibition of Osteoclast Vacuolar H(+)-Atpase. *Curr Pharm Des* **8**(23): 2033–2048.

Felix, R., A. Sandoval, D. Sanchez, J. C. Gomora, J. L. De la Vega-Beltran, C. L. Trevino, and A. Darszon (2003). Zd7288 Inhibits Low-Threshold Ca(2+) Channel Activity and Regulates Sperm Function. *Biochem Biophys Res Commun* **311**(1): 187–192.

Ferruzza, S., M. Scacchi, M. L. Scarino, and Y. Sambuy (2002). Iron and Copper Alter Tight Junction Permeability in Human Intestinal Caco-2 Cells by Distinct Mechanisms. *Toxicol In Vitro* **16**(4): 399–404.

Finbow, M. E., S. John, E. Kam, D. K. Apps, and J. D. Pitts (1993). Disposition and Orientation of Ductin (Dccd-Reactive Vacuolar H(+)-Atpase Subunit) in Mammalian Membrane Complexes. *Experimental Cell Research* **207**(2): 261–270.

Fink, K., W. Meder, D. J. Dooley, and M. Gothert (2000). Inhibition of Neuronal Ca(2+) Influx by Gabapentin and Subsequent Reduction of Neurotransmitter Release from Rat Neocortical Slices. *Br J Pharmacol* **130**(4): 900–906.

Forgac, M. (1999). The Vacuolar H+-Atpase of Clathrin-Coated Vesicles Is Reversibly Inhibited by S-Nitrosoglutathione. *Journal of Biological Chemistry* **274**(3): 1301–1305.

Frame, M. D. and M. A. Milanick (1991). Mn and Cd Transport by the Na-Ca Exchanger of Ferret Red Blood Cells. *Am J Physiol* **261**(3 Pt 1): C467–475.

Franckowiak, G., M. Bechem, M. Schramm, and G. Thomas (1985). The Optical Isomers of the 1,4-Dihydropyridine Bay K 8644 Show Opposite Effects on Ca Channels. *Eur J Pharmacol* **114**(2): 223–226.

Frenal, K., C. Q. Xu, N. Wolff, K. Wecker, G. B. Gurrola, S. Y. Zhu, C. W. Chi, L. D. Possani, J. Tytgat, and M. Delepierre (2004). Exploring Structural Features of the Interaction between the Scorpion Toxincnergl and Erg K+ Channels. *Proteins* **56**(2): 367–375.

Friedman, H., S. Meir, I. Rosenberger, A. H. Halevy, P. B. Kaufman, and S. Philosoph-Hadas (1998). Inhibition of the Gravitropic Response of Snapdragon Spikes by the Calcium-Channel Blocker Lanthanum Chloride. *Plant Physiol* **118**(2): 483–492.

Fujisawa, S., K. Ono, and T. Iijima (2000). Time-Dependent Block of the Slowly Activating Delayed Rectifier K(+) Current by Chromanol 293b in Guinea-Pig Ventricular Cells. *Br J Pharmacol* **129**(5): 1007–1013.

Furla, P., D. Allemand, and M. N. Orsenigo (2000). Involvement of H(+)-Atpase and Carbonic Anhydrase in Inorganic Carbon Uptake for Endosymbiont Photosynthesis. *Am J Physiol Regul Integr Comp Physiol* **278**(4): R870–881.

Furukawa, T., T. Yamakawa, T. Midera, T. Sagawa, Y. Mori, and T. Nukada (1999). Selectivities of Dihydropyridine Derivatives in Blocking Ca(2+) Channel Subtypes Expressed in Xenopus Oocytes. *J Pharmacol Exp Ther* **291**(2): 464–473.

Gagliardi, S., M. Rees, and C. Farina (1999). Chemistry and Structure Activity Relationships of Bafilomycin A1, a Potent and Selective Inhibitor of the Vacuolar H+-Atpase. *Curr Med Chem* **6**(12): 1197–1212.

Ghatpande, A. S., R. Uma, and J. W. Karpen (2003). A Multiply Charged Tetracaine Derivative Blocks Cyclic Nucleotide-Gated Channels at Subnanomolar Concentrations. *Biochemistry* **42**(2): 265–270.

Giangiacomo, K. M., A. Kamassah, G. Harris, and O. B. McManus (1998). Mechanism of Maxi-K Channel Activation by Dehydrosoyasaponin-I. *J Gen Physiol* **112**(4): 485–501.

Gibbon, B. C. and D. L. Kropf (1993). Intracellular pH and Its Regulation in Pelvetia Zygotes. *Dev Biol* **157**(1): 259–268.

Gogelein, H., A. Bruggemann, U. Gerlach, J. Brendel, and A. E. Busch (2000). Inhibition of Iks Channels by Hmr 1556. *Naunyn Schmiedebergs Arch Pharmacol* **362**(6): 480–488.

Gong, X., S. M. Burbridge, A. C. Lewis, P. Y. Wong, and P. Linsdell (2002). Mechanism of Lonidamine Inhibition of the Cftr Chloride Channel. *Br J Pharmacol* **137**(6): 928–936.

Grant, T. L., C. J. Ohnmacht, and B. B. Howe (1994). Anilide Tertiary Carbinols: A Novel Series of K+ Channel Openers. *Trends Pharmacol Sci* **15**(11): 402–404.

Gribble, F. M. and F. M. Ashcroft (2000). Sulfonylurea Sensitivity of Adenosine Triphosphate-Sensitive Potassium Channels from Beta Cells and Extrapancreatic Tissues. *Metabolism: Clinical & Experimental* **49**(10 Suppl 2): 3–6.

Gribble, F. M., R. Ashfield, C. Ammala, and F. M. Ashcroft (1997). Properties of Cloned Atp-Sensitive K+ Currents Expressed in Xenopus Oocytes. *J Physiol* **498 (Pt 1)**: 87–98.

Gribble, F. M., T. M. Davis, C. E. Higham, A. Clark, and F. M. Ashcroft (2000). The Antimalarial Agent Mefloquine Inhibits Atp-Sensitive K-Channels. *British Journal of Pharmacology* **131**(4): 756–760.

Grunze, H., J. von Wegerer, R. W. Greene, and J. Walden (1998). Modulation of Calcium and Potassium Currents by Lamotrigine. *Neuropsychobiology* **38**(3): 131–138.

Guan, X., W. J. Bonney, and R. J. Ruch (1995). Changes in Gap Junction Permeability, Gap Junction Number, and Connexin43 Expression in Lindane-Treated Rat Liver Epithelial Cells. *Toxicol Appl Pharmacol* **130**(1): 79–86.

Gukovskaya, A. S., S. Gukovsky, and S. J. Pandol (2000). Endoplasmic Reticulum Ca(2+)-Atpase Inhibitors Stimulate Membrane Guanylate Cyclase in Pancreatic Acinar Cells. *Am J Physiol Cell Physiol* **278**(2): C363–371.

Guo, Y., M. Krumwiede, J. G. White, and O. D. Wangensteen (1996). Hocl Effects on the Tight Junctions of Rabbit Tracheal Epithelium. *Am J Physiol* **270**(2 Pt 1): L224–231.

Gurrola, G. B. and L. D. Possani (1995). Structural and Functional Features of Noxiustoxin: A K+ Channel Blocker. *Biochem Mol Biol Int* **37**(3): 527–535.

Hadley, J. K., M. Noda, A. A. Selyanko, I. C. Wood, F. C. Abogadie, and D. A. Brown (2000). Differential Tetraethylammonium Sensitivity of Kcnq1-4 Potassium Channels. *Br J Pharmacol* **129**(3): 413–415.

Hambrock, A., T. Kayar, D. Stumpp, and H. Osswald (2004). Effect of Two Amino Acids in Tm17 of Sulfonylurea Receptor Sur1 on the Binding of Atp-Sensitive K+ Channel Modulators. *Diabetes* **53 Suppl 3**: S128–134.

Hancox, J. C. (1997). Amiodarone Blocks L-Type Calcium Current in Single Myocytes Isolated from the Rabbit Atrioventricular Node. *General Pharmacology* **29**(3): 429–435.

Harris, C. and L. Fliegel (1999). Amiloride and the Na(+)/H(+) Exchanger Protein: Mechanism and Significance of Inhibition of the Na(+)/H(+) Exchanger (Review). *International Journal of Molecular Medicine* **3**(3): 315–321.

Hart, J., M. F. Wilkinson, M. E. Kelly, and S. Barnes (2003). Inhibitory Action of Diltiazem on Voltage-Gated Calcium Channels in Cone Photoreceptors. *Exp Eye Res* **76**(5): 597–604.

Hashiguchi-Ikeda, M., T. Namba, T. M. Ishii, T. Hisano, and K. Fukuda (2003). Halothane Inhibits an Intermediate Conductance Ca2+-Activated K+ Channel by Acting at the Extracellular Side of the Ionic Pore. *Anesthesiology* **99**(6): 1340–1345.

Higashida, H., D. A. Brown, and J. Robbins (2000). Both Linopirdine- and Way123,398-Sensitive Components of I K(M,Ng) Are Modulated by Cyclic Adp Ribose in Ng108-15 Cells. *Pflugers Arch* **441**(2–3): 228–234.

Hildebrand, M. E., J. E. McRory, T. P. Snutch, and A. Stea (2004). Mammalian Voltage-Gated Calcium Channels Are Potently Blocked by the Pyrethroid Insecticide Allethrin. *J Pharmacol Exp Ther* **308**(3): 805–813.

Hill, R. J., A. M. Grant, W. Volberg, L. Rapp, C. Faltynek, D. Miller, K. Pagani, E. Baizman, S. Wang, J. W. Guiles, and et al. (1995). Win 17317-3: Novel Nonpeptide Antagonist of Voltage-Activated K+ Channels in Human T Lymphocytes. *Mol Pharmacol* **48**(1): 98–104.

Hollt, V., M. Kouba, M. Dietel, and G. Vogt (1992). Stereoisomers of Calcium Antagonists Which Differ Markedly in Their Potencies as Calcium Blockers Are Equally Effective in Modulating Drug Transport by P-Glycoprotein. *Biochem Pharmacol* **43**(12): 2601–2608.

Holoubek, A., J. Vecer, M. Opekarova, and K. Sigler (2003). Ratiometric Fluorescence Measurements of Membrane Potential Generated by Yeast Plasma Membrane H(+)-Atpase Reconstituted into Vesicles. *Biochim Biophys Acta* **1609**(1): 71–79.

Horiuchi, M. and S. Iwata (1983). Effect of Nigericin on Distribution of Sodium, Potassium and Calcium Ion in Rabbit Lens. *Exp Eye Res* **37**(5): 439–445.

Hu, S., H. S. Kim, and C. A. Fink (1995). Differential Effects of the Bkca Channel Openers Ns004 and Ns1608 in Porcine Coronary Arterial Cells. *Eur J Pharmacol* **294**(1): 357–360.

Hu, S., S. Wang, B. Fanelli, P. A. Bell, B. E. Dunning, S. Geisse, R. Schmitz, and B. R. Boettcher (2000). Pancreatic Beta-Cell K(Atp) Channel Activity and Membrane-Binding Studies with Nateglinide: A Comparison with Sulfonylureas and Repaglinide. *Journal of Pharmacology & Experimental Therapeutics* **293**(2): 444–452.

Hu, S. L. and C. Y. Kao (1991). Interactions of Neosaxitoxin with the Sodium Channel of the Frog Skeletal Muscle Fiber. *J Gen Physiol* **97**(3): 561–578.

Hui, C. S. and W. Chen (1994). Origin of Delayed Outward Ionic Current in Charge Movement Traces from Frog Skeletal Muscle. *J Physiol* **479 (Pt 1)**: 109–125.

Huneke, R., D. Zitzelsberger, J. Fassl, E. Jungling, S. Brose, W. Buhre, R. Rossaint, and A. Luckhoff (2004). Temperature-Independent Inhibition of L-Type Calcium Currents by Halothane and Sevoflurane in Human Atrial Cardiomyocytes. *Anesthesiology* **101**(2): 409–416.

Inagaki, C., M. Hara, and X. T. Zeng (1996). A Cl- Pump in Rat Brain Neurons. *J Exp Zool* **275**(4): 262–268.

Ishibashi, T., M. Hamaguchi, and S. Imai (1991). 2-Nicotinamidoethyl Acetate (Sg-209) Is a Potassium Channel Opener: Structure Activity Relationship among Nicorandil Derivatives. *Naunyn Schmiedebergs Arch Pharmacol* **344**(2): 235–239.

Jin, W., A. M. Klem, J. H. Lewis, and Z. Lu (1999). Mechanisms of Inward-Rectifier K+ Channel Inhibition by Tertiapin-Q. *Biochemistry* **38**(43): 14294–14301.

Jurkiewicz, N. K., J. Wang, B. Fermini, M. C. Sanguinetti, and J. J. Salata (1996). Mechanism of Action Potential Prolongation by Rp 58866 and Its Active Enantiomer, Terikalant. Block of the Rapidly Activating Delayed Rectifier K+ Current, Ikr. *Circulation* **94**(11): 2938–2946.

Jurkowitz, M. S., R. A. Altschuld, G. P. Brierley, and E. J. Cragoe, Jr. (1983). Inhibition of Na+-Dependent Ca2+ Efflux from Heart Mitochondria by Amiloride Analogues. *FEBS Lett* **162**(2): 262–265.

Kaloyianni, M., R. Stamatiou, and S. Dailianis (2005). Zinc and 17beta-Estradiol Induce Modifications in Na(+)/H(+) Exchanger and Pyruvate Kinase Activity through Protein Kinase C in Isolated Mantle/Gonad Cells of Mytilus Galloprovincialis. *Comp Biochem Physiol C Toxicol Pharmacol.*

Kang, J. J., K. S. Hsu, and S. Y. Lin-Shiau (1994). Effects of Bipyridylium Compounds on Calcium Release from Triadic Vesicles Isolated from Rabbit Skeletal Muscle. *Br J Pharmacol* **112**(4): 1216–1222.

Kaplan, P., M. Matejovicova, and V. Mezesova (1997). Iron-Induced Inhibition of Na+, K(+)-Atpase and Na+/Ca2+ Exchanger in Synaptosomes: Protection by the Pyridoindole Stobadine. *Neurochem Res* **22**(12): 1523–1529.

Katsuki, H., A. Shinohara, S. Fujimoto, T. Kume, and A. Akaike (2005). Tetraethylammonium Exacerbates Ischemic Neuronal Injury in Rat Cerebrocortical Slice Cultures. *Eur J Pharmacol* **508**(1–3): 85–91.

Kau, S. T., P. Zografos, M. L. Do, T. J. Halterman, M. W. McConville, C. L. Yochim, S. Trivedi, B. B. Howe, and J. H. Li (1994). Characterization of Atp-Sensitive Potassium Channel-Blocking Activity of Zeneca Zm181,037, a Eukalemic Diuretic. *Pharmacology* **49**(4): 238–248.

Kauferstein, S., I. Huys, H. Lamthanh, R. Stocklin, F. Sotto, A. Menez, J. Tytgat, and D. Mebs (2003). A Novel Conotoxin Inhibiting Vertebrate Voltage-Sensitive Potassium Channels. *Toxicon* **42**(1): 43–52.

Kauffman, R. F., R. W. Taylor, and D. R. Pfeiffer (1980). Cation Transport and Specificity of Ionomycin. Comparison with Ionophore A23187 in Rat Liver Mitochondria. *J Biol Chem* **255**(7): 2735–2739.

Kehl, F., J. G. Krolikowski, J. P. Tessmer, P. S. Pagel, D. C. Warltier, and J. R. Kersten (2002). Increases in Coronary Collateral Blood Flow Produced by Sevoflurane Are Mediated by Calcium-Activated Potassium (Bkca) Channels in Vivo. *Anesthesiology* **97**(3): 725–731.

Kiesewetter, D. O., E. M. Jagoda, J. E. Starrett, Jr., V. K. Gribkoff, P. Hewawasam, N. Srinivas, D. Salazar, and W. C. Eckelman (2002). Radiochemical Synthesis and Biodistribution of a Novel Maxi-K Potassium Channel Opener. *Nucl Med Biol* **29**(1): 55–59.

Kirkup, A. J., G. Edwards, and A. H. Weston (1996). Investigation of the Effects of 5-Nitro-2-(3-Phenylpropylamino)-Benzoic Acid (Nppb) on Membrane Currents in Rat Portal Vein. *Br J Pharmacol* **117**(1): 175–183.

Kleyman, T. R. and E. J. Cragoe, Jr. (1988). Amiloride and Its Analogs as Tools in the Study of Ion Transport. *J Membr Biol* **105**(1): 1–21.

Klocker, N., U. Musshoff, M. Madeja, and E. J. Speckmann (1996). Activation of Atp-Sensitive Potassium Channels in Follicle-Enclosed Xenopus Oocytes by the Epileptogenic Agent Pentylenetetrazol. *Pflugers Arch* **431**(5): 736–740.

Klockner, U. and G. Isenberg (1991). Myocytes Isolated from Porcine Coronary Arteries: Reduction of Currents through L-Type Ca-Channels by Verapamil-Type Ca-Antagonists. *J Physiol Pharmacol* **42**(2): 163–179.

Knaus, H. G., O. B. McManus, S. H. Lee, W. A. Schmalhofer, M. Garcia-Calvo, L. M. Helms, M. Sanchez, K. Giangiacomo, J. P. Reuben, A. B. Smith, 3rd, and et al. (1994). Tremorgenic Indole Alkaloids Potently Inhibit Smooth Muscle High-Conductance Calcium-Activated Potassium Channels. *Biochemistry* **33**(19): 5819–5828.

Koch, T. R., A. Petro, M. Darrabie, and E. C. Opara (2004). Effect of the H, K-Atpase Inhibitor, Esomeprazole Magnesium, on Gut Total Antioxidant Capacity in Mice. *J Nutr Biochem* **15**(9): 522–526.

Kocic, I., Y. Hirano, and M. Hiraoka (2004). Hypotonic Stress Increases Efficacy of Rilmakalim, but Not Pinacidil, to Activate Atp-Sensitive K(+) Current in Guinea Pig Ventricular Myocytes. *J Pharmacol Sci* **95**(2): 189–195.

Konno, K., M. Hisada, Y. Itagaki, H. Naoki, N. Kawai, A. Miwa, T. Yasuhara, and H. Takayama (1998). Isolation and Structure of Pompilidotoxins, Novel Peptide Neurotoxins in Solitary Wasp Venoms. *Biochem Biophys Res Commun* **250**(3): 612–616.

Kourie, J. I. (1998). Effects of Atp-Sensitive Potassium Channel Regulators on Chloride Channels in the Sarcoplasmic Reticulum Vesicles from Rabbit Skeletal Muscle. *J Membr Biol* **164**(1): 47–58.

Kowarski, D., H. Shuman, A. P. Somlyo, and A. V. Somlyo (1985). Calcium Release by Noradrenaline from Central Sarcoplasmic Reticulum in Rabbit Main Pulmonary Artery Smooth Muscle. *J Physiol* **366**: 153–175.

Krasznai, Z., M. Morisawa, Z. T. Krasznai, S. Morisawa, K. Inaba, Z. K. Bazsane, B. Rubovszky, B. Bodnar, A. Borsos, and T. Marian (2003). Gadolinium, a Mechano-Sensitive Channel Blocker, Inhibits Osmosis-Initiated Motility of Sea- and Freshwater Fish Sperm, but Does Not Affect Human or Ascidian Sperm Motility. *Cell Motil Cytoskeleton* **55**(4): 232–243.

Kraus, R. L., S. Hering, M. Grabner, D. Ostler, and J. Striessnig (1998). Molecular Mechanism of Diltiazem Interaction with L-Type Ca2+ Channels. *J Biol Chem* **273**(42): 27205–27212.

Kromer, W., S. Postius, and R. Riedel (2000). Animal Pharmacology of Reversible Antagonism of the Gastric Acid Pump, Compared to Standard Antisecretory Principles. *Pharmacology* **60**(4): 179–187.

Kryzhanovskii, G. N., M. N. Karpova, O. Pankov, E. M. Abramova, and I. Abrosimov (1992). [the Antiepileptic Effects of Sodium Valproate and the Calcium Antagonist Riodipine When Used Jointly in a Model of Generalized Korazol-Induced Epileptic Activity]. *Biull Eksp Biol Med* **114**(10): 376–378.

Kupershmidt, S., I. C. Yang, K. Hayashi, J. Wei, S. Chanthaphaychith, C. I. Petersen, D. C. Johns, A. L. George, Jr., D. M. Roden, and J. R. Balser (2003). The Ikr Drug Response Is Modulated by Kcr1 in Transfected Cardiac and Noncardiac Cell Lines. *Faseb J* **17**(15): 2263–2265.

Kuroda, H., Y. N. Chen, T. X. Watanabe, T. Kimura, and S. Sakakibara (1992). Solution Synthesis of Calciseptine, an L-Type Specific Calcium Channel Blocker. *Pept Res* **5**(5): 265–268.

la Cour, M., A. Baekgaard, and T. Zeuthen (1997). Antibodies against a Furosemide Binding Protein from Ehrlich Ascites Tumour Cells Inhibit Na+, K+, Cl- Co-Transport in Frog Retinal Pigment Epithelium. *Acta Ophthalmol Scand* **75**(4): 405–408.

Lamers, J. M., K. J. Cysouw, and P. D. Verdouw (1985). Slow Calcium Channel Blockers and Calmodulin. Effect of Felodipine, Nifedipine, Prenylamine and Bepridil on Cardiac Sarcolemmal Calcium Pumping Atpase. *Biochem Pharmacol* **34**(21): 3837–3843.

Landry, D. W., M. Reitman, E. J. Cragoe, and Q. Al-Awqati (1987). Epithelial Chloride Channel. Development of Inhibitory Ligands. *Journal of General Physiology* **90**(6): 779–798.

Lazdunski, M., C. Frelin, J. Barhanin, A. Lombet, H. Meiri, D. Pauron, G. Romey, A. Schmid, H. Schweitz, P. Vigne, and H. P. M. Vijverberg (1985). *Polypeptide Toxins as Tools to Study Voltage-Sensitive Na^+ Channels*. Tetrodotoxin, Saxitoxin, and the Molecular Biology of the Sodium Channel, New York, The New York Academy of Sciences.

Le, H. D., A. Omelchenko, L. V. Hryshko, A. Uliyanova, M. Condrescu, and J. P. Reeves (2005). Allosteric Activation of Sodium-Calcium Exchange by Picomolar Concentrations of Cadmium. *J Physiol* **563**(Pt 1): 105–117.

Lee, M. G., D. Macglashan Jr, and B. J. Undem (2005). Role of Chloride Channels in Bradykinin-Induced Guinea Pig Airway Vagal C-Fiber Activation. *J Physiol.*

Leinders-Zufall, T., M. N. Rand, G. M. Shepherd, C. A. Greer, and F. Zufall (1997). Calcium Entry through Cyclic Nucleotide-Gated Channels in Individual Cilia of Olfactory Receptor Cells: Spatiotemporal Dynamics. *J Neurosci* **17**(11): 4136–4148.

Levitt, D. G., S. R. Elias, and J. M. Hautman (1978). Number of Water Molecules Coupled to the Transport of Sodium, Potassium and Hydrogen Ions Via Gramicidin, Nonactin or Valinomycin. *Biochim Biophys Acta* **512**(2): 436–451.

Lin, S., Z. Wang, and D. Fedida (2001). Influence of Permeating Ions on Kv1.5 Channel Block by Nifedipine. *Am J Physiol Heart Circ Physiol* **280**(3): H1160–1172.

Lindskog, S. (1997). Structure and Mechanism of Carbonic Anhydrase. *Pharmacology & Therapeutics* **74**(1): 1–20.

Lipkind, G. M. and H. A. Fozzard (1997). A Model of Scorpion Toxin Binding to Voltage-Gated K+ Channels. *J Membr Biol* **158**(3): 187–196.

Liu, C., J. D. Au, H. L. Zou, J. F. Cotten, and C. S. Yost (2004). Potent Activation of the Human Tandem Pore Domain K Channel Tresk with Clinical Concentrations of Volatile Anesthetics. *Anesth Analg* **99**(6): 1715–1722, table of contents.

Liu, Y., D. Liu, D. Printzenhoff, M. J. Coghlan, R. Harris, and D. S. Krafte (2002). Tenidap, a Novel Anti-Inflammatory Agent, Is an Opener of the Inwardly Rectifying K+ Channel Hkir2.3. *Eur J Pharmacol* **435**(2–3): 153–160.

Liu, Y. C. and S. N. Wu (2003). Block of Erg Current by Linoleoylamide, a Sleep-Inducing Agent, in Pituitary Gh3 Cells. *Eur J Pharmacol* **458**(1–2): 37–47.

Lock, H. and M. A. Valverde (2000). Contribution of the Isk (Mink) Potassium Channel Subunit to Regulatory Volume Decrease in Murine Tracheal Epithelial Cells. [Erratum Appears in J Biol Chem 2001 Mar 23;276(12):9582.]. *Journal of Biological Chemistry* **275**(45): 34849–34852.

Madeja, M., U. Musshoff, and E. J. Speckmann (1997). Diversity of Potassium Channels Contributing to Differences in Brain Area-Specific Seizure Susceptibility: Sensitivity of Different Potassium Channels to the Epileptogenic Agent Pentylenetetrazol. *Eur J Neurosci* **9**(2): 390–395.

Malik-Hall, M., C. R. Ganellin, D. Galanakis, and D. H. Jenkinson (2000). Compounds That Block Both Intermediate-Conductance (Ik(Ca)) and Small-Conductance (Sk(Ca)) Calcium-Activated Potassium Channels. *Br J Pharmacol* **129**(7): 1431–1438.

Mathur, S. and M. Karmazyn (1997). Interaction between Anesthetics and the Sodium-Hydrogen Exchange Inhibitor Hoe 642 (Cariporide) in Ischemic and Reperfused Rat Hearts. *Anesthesiology* **87**(6): 1460–1469.

Matsuya, H., M. Okamoto, T. Ochi, A. Nishikawa, S. Shimizu, T. Kataoka, K. Nagai, H. H. Wasserman, and S. Ohkuma (2000). Reversible and Potent Uncoupling of Hog Gastric (H(+)+K(+))-Atpase by Prodigiosins. *Biochemical Pharmacology* **60**(12): 1855–1863.

Matucci, R., M. F. Ottaviani, M. Barbieri, E. Cerbai, and A. Mugelli (1997). Protective Effect of Darodipine, a Calcium Antagonist, on Rat Cardiomyocytes against Oxygen Radical-Mediated Injury. *Br J Pharmacol* **122**(7): 1353–1360.

McDonough, S. I., K. J. Swartz, I. M. Mintz, L. M. Boland, and B. P. Bean (1996). Inhibition of Calcium Channels in Rat Central and Peripheral Neurons by Omega-Conotoxin Mviic. *J Neurosci* **16**(8): 2612–2623.

McKay, N. G., J. M. Kinsella, C. M. Campbell, and M. L. Ashford (2000). Sensitivity of Kir6.2-Sur1 Currents, in the Absence and Presence of Sodium Azide, to the K(Atp) Channel Inhibitors, Ciclazindol and Englitazone. *Br J Pharmacol* **130**(4): 857–866.

Meisheri, K., M. Fosset, S. Humphrey, and M. Lazdunski (1995). Receptor Binding Characterization in Kidney Membrane of [3h]U-37883, a Novel Atp-Sensitive K+ Channel Blocker with Diuretic/Natriuretic Properties. *Mol Pharmacol* **47**(1): 155–163.

Michel, A., F. Laurent, J. Bompart, K. Hadj-Kaddour, J. P. Chapat, M. Boucard, and P. A. Bonnet (1993). Cardiovascular Effects of Sca40, a Novel Potassium Channel Opener, in Rats. *Br J Pharmacol* **110**(3): 1031–1036.

Miller, D. J. and G. L. Smith (1984). Egta Purity and the Buffering of Calcium Ions in Physiological Solutions. *Am J Physiol* **246**(1 Pt 1): C160–166.

Milovic, S., B. Steinecker-Frohnwieser, W. Schreibmayer, and L. G. Weigl (2004). The Sensitivity of G Protein-Activated K+ Channels toward Halothane Is Essentially Determined by the C Terminus. *J Biol Chem* **279**(33): 34240–34249.

Mitchell, J., W. Frishman, and M. Heiman (1993). Nisoldipine: A New Dihydropyridine Calcium-Channel Blocker. *J Clin Pharmacol* **33**(1): 46–52.

Mitrovic, V., E. Oehm, J. Thormann, H. Pitschner, and W. Haberbosch (1998). Comparison of the Potassium Channel Blocker Tedisamil with the Beta-Adrenoceptor Blocker Esmolol and the Calcium Antagonist Gallopamil in Patients with Coronary Artery Disease. *Clin Cardiol* **21**(7): 492–502.

Miyaguchi, T., K. Nomata, M. Noguchi, J. Watanabe, H. Satoh, and H. Kanetake (2001). Tac-101, a Novel Retinobenzoic-Acid Derivative, Enhances Gap Junctional Intercellular Communication among Renal Epithelial Cells Treated with Renal Carcinogens. *Anticancer Res* **21**(6A): 4025–4030.

Miyoshi, S., T. Miyazaki, K. Moritani, and S. Ogawa (1996). Different Responses of Epicardium and Endocardium to Katp Channel Modulators During Regional Ischemia. *American Journal of Physiology* **271**(1 Pt 2): H140–147.

Mizumura, T., K. Nithipatikom, and G. J. Gross (1995). Bimakalim, an Atp-Sensitive Potassium Channel Opener, Mimics the Effects of Ischemic Preconditioning to Reduce Infarct Size, Adenosine Release, and Neutrophil Function in Dogs. *Circulation* **92**(5): 1236–1245.

Mocanu, M. M., S. Gadgil, D. M. Yellon, and G. F. Baxter (1999). Mibefradil, a T-Type and L-Type Calcium Channel Blocker, Limits Infarct Size through a Glibenclamide-Sensitive Mechanism. *Cardiovasc Drugs Ther* **13**(2): 115–122.

Molnar, P. and S. L. Erdo (1995). Vinpocetine Is as Potent as Phenytoin to Block Voltage-Gated Na+ Channels in Rat Cortical Neurons. *Eur J Pharmacol* **273**(3): 303–306.

Mueller, B., M. Zhao, I. V. Negrashov, R. Bennett, and D. D. Thomas (2004). Serca Structural Dynamics Induced by Atp and Calcium. *Biochemistry* **43**(40): 12846–12854.

Murakami, S., I. Arai, M. Muramatsu, S. Otomo, K. Baba, T. Kido, and M. Kozawa (1992a). Inhibition of Gastric H+,K(+)-Atpase and Acid Secretion by Cassigarol a, a Polyphenol from Cassia Garrettiana Craib. *Biochem Pharmacol* **44**(1): 33–37.

Murakami, S., Y. Isobe, H. Kijima, H. Nagai, M. Muramatu, and S. Otomo (1991a). Inhibition of Gastric H+, K(+)-Atpase and Acid Secretion by Ellagic Acid. *Planta Med* **57**(4): 305–308.

Murakami, S., H. Kijima, Y. Isobe, M. Muramatsu, H. Aihara, S. Otomo, K. Baba, and M. Kozawa (1990). Inhibition of Gastric H+, K(+)-Atpase by Chalcone Derivatives, Xanthoangelol and 4-Hydroxyderricin, from Angelica Keiskei Koidzumi. *J Pharm Pharmacol* **42**(10): 723–726.

Murakami, S., M. Muramatsu, H. Aihara, and S. Otomo (1991b). Inhibition of Gastric H+,K(+)-Atpase by the Anti-Ulcer Agent, Sofalcone. *Biochem Pharmacol* **42**(7): 1447–1451.

Murakami, S., M. Muramatsu, and S. Otomo (1992b). Inhibition of Gastric H+, K(+)-Atpase by Quercetin. *J Enzyme Inhib* **5**(4): 293–298.

Narahashi, T. (1982). Cellular and Molecular Mechanisms of Action of Insecticides: Neurophysiological Approach. *Neurobehav Toxicol Teratol* **4**(6): 753–758.

Narahashi, T., J. M. Frey, K. S. Ginsburg, and M. L. Roy (1992). Sodium and Gaba-Activated Channels as the Targets of Pyrethroids and Cyclodienes. *Toxicol Lett* **64–65 Spec No**: 429–436.

Newcomb, R., B. Szoke, A. Palma, G. Wang, X. Chen, W. Hopkins, R. Cong, J. Miller, L. Urge, K. Tarczy-Hornoch, J. A. Loo, D. J. Dooley, L. Nadasdi, R. W. Tsien, J. Lemos, and G. Miljanich (1998). Selective Peptide Antagonist of the Class E Calcium Channel from the Venom of the Tarantula Hysterocrates Gigas. *Biochemistry* **37**(44): 15353–15362.

Nguyen, A., J. C. Kath, D. C. Hanson, M. S. Biggers, P. C. Canniff, C. B. Donovan, R. J. Mather, M. J. Bruns, H. Rauer, J. Aiyar, A. Lepple-Wienhues, G. A. Gutman, S. Grissmer, M. D. Cahalan, and K. G. Chandy (1996). Novel Nonpeptide Agents Potently Block the C-Type Inactivated Conformation of Kv1.3 and Suppress T Cell Activation. *Mol Pharmacol* **50**(6): 1672–1679.

Noack, T., G. Edwards, P. Deitmer, and A. H. Weston (1992). Potassium Channel Modulation in Rat Portal Vein by Atp Depletion: A Comparison with the Effects of Levcromakalim (Brl 38227). *Br J Pharmacol* **107**(4): 945–955.

Nozawa, T., H. Toyobuku, D. Kobayashi, K. Kuruma, A. Tsuji, and I. Tamai (2003). Enhanced Intestinal Absorption of Drugs by Activation of Peptide Transporter Pept1 Using Proton-Releasing Polymer. *J Pharm Sci* **92**(11): 2208–2216.

O'Grady, S. M., H. C. Palfrey, and M. Field (1987). Na-K-2cl Cotransport in Winter Flounder Intestine and Bovine Kidney Outer Medulla: [3h] Bumetanide Binding and Effects of Furosemide Analogues. *J Membr Biol* **96**(1): 11–18.

Ohkuma, S., T. Sato, M. Okamoto, H. Matsuya, K. Arai, T. Kataoka, K. Nagai, and H. H. Wasserman (1998). Prodigiosins Uncouple Lysosomal Vacuolar-Type Atpase through Promotion of H+/Cl- Symport. *Biochemical Journal* **334**(Pt 3): 731–741.

Ohnmacht, C. J., K. Russell, J. R. Empfield, C. A. Frank, K. H. Gibson, D. R. Mayhugh, F. M. McLaren, H. S. Shapiro, F. J. Brown, D. A. Trainor, C. Ceccarelli, M. M. Lin, B. B. Masek, J. M. Forst, R. J. Harris, J. M. Hulsizer, J. J. Lewis, S. M. Silverman, R. W. Smith, P. J. Warwick, S. T. Kau, A. L. Chun, T. L. Grant, B. B. Howe, K. L. Neilson, and et al. (1996). N-Aryl-3,3,3-Trifluoro-2-Hydroxy-2-Methylpropanamides: Katp Potassium Channel Openers. Modifications on the Western Region. *J Med Chem* **39**(23): 4592–4601.

Okanlawon, A. and M. Dym (1996). Effect of Chloroquine on the Formation of Tight Junctions in Cultured Immature Rat Sertoli Cells. *J Androl* **17**(3): 249–255.

Olivari, C., C. Albumi, M. C. Pugliarello, and M. I. De Michelis (2000). Phenylarsine Oxide Inhibits the Fusicoccin-Induced Activation of Plasma Membrane H(+)-Atpase. *Plant Physiology* **122**(2): 463–470.

Orriss, G. L., A. G. Leslie, K. Braig, and J. E. Walker (1998). Bovine F1-Atpase Covalently Inhibited with 4-Chloro-7-Nitrobenzofurazan: The Structure Provides Further Support for a Rotary Catalytic Mechanism. *Structure* **6**(7): 831–837.

Palmer, M. (1998). Assembly Mechanism of the Oligomeric Streptolysin O Pore: The Early Membrane Lesion Is Lined by a Free Edge of the Lipid Membrane and Is Extended Gradually During Oligomerization. *EMBO Journal* **17**(6): 1598–1605.

Park, K. S., I. Jo, K. Pak, S. W. Bae, H. Rhim, S. H. Suh, J. Park, H. Zhu, I. So, and K. W. Kim (2002). Fccp Depolarizes Plasma Membrane Potential by Activating Proton and Na+ Currents in Bovine Aortic Endothelial Cells. *Pflugers Arch* **443**(3): 344–352.

Patel, A. J., E. Honore, F. Lesage, M. Fink, G. Romey, and M. Lazdunski (1999). Inhalational Anesthetics Activate Two-Pore-Domain Background K+ Channels. *Nat Neurosci* **2**(5): 422–426.

Pauwels, P. J., J. E. Leysen, and P. A. Janssen (1991). Ca++ and Na+ Channels Involved in Neuronal Cell Death. Protection by Flunarizine. *Life Sci* **48**(20): 1881–1893.

Pedarzani, P., D. D'Hoedt, K. B. Doorty, J. D. Wadsworth, J. S. Joseph, K. Jeyaseelan, R. M. Kini, S. V. Gadre, S. M. Sapatnekar, M. Stocker, and P. N. Strong (2002). Tamapin, a Venom Peptide from the Indian Red Scorpion (Mesobuthus Tamulus) That Targets Small Conductance Ca2+-Activated K+ Channels and Afterhyperpolarization Currents in Central Neurons. *J Biol Chem* **277**(48): 46101–46109.

Perlin, D. S., L. R. Latchney, and A. E. Senior (1985). Inhibition of Escherichia Coli H+-Atpase by Venturicidin, Oligomycin and Ossamycin. *Biochim Biophys Acta* **807**(3): 238–244.

Perry, M., M. J. de Groot, R. Helliwell, D. Leishman, M. Tristani-Firouzi, M. C. Sanguinetti, and J. Mitcheson (2004). Structural Determinants of Herg Channel Block by Clofilium and Ibutilide. *Mol Pharmacol* **66**(2): 240–249.

Phillips, R. M. and R. A. Altschuld (1996). 2,3-Butanedione 2–Monoxime (Bdm) Induces Calcium Release from Canine Cardiac Sarcoplasmic Reticulum. *Biochem Biophys Res Commun* **229**(1): 154–157.

Pragl, B., A. Koschak, M. Trieb, G. Obermair, W. A. Kaufmann, U. Gerster, E. Blanc, C. Hahn, H. Prinz, G. Schutz, H. Darbon, H. J. Gruber, and H. G. Knaus (2002). Synthesis, Characterization, and Application of Cy-Dye- and Alexa-Dye-Labeled Hongotoxin(1) Analogues. The First High Affinity Fluorescence Probes for Voltage-Gated K+ Channels. *Bioconjug Chem* **13**(3): 416–425.

Price, P. A., H. H. June, J. R. Buckley, and M. K. Williamson (2002). Sb 242784, a Selective Inhibitor of the Osteoclastic V-H+Atpase, Inhibits Arterial Calcification in the Rat. *Circ Res* **91**(6): 547–552.

Procida, K., C. Caspersen, H. Kromann, S. B. Christensen, and M. Treiman (1998). Acta, a Fluorescent Analogue of Thapsigargin, Is a Potent Inhibitor and a Conformational Probe of Skeletal Muscle Ca2+-Atpase. *FEBS Lett* **439**(1–2): 127–132.

Proks, P. and F. M. Ashcroft (1997). Phentolamine Block of Katp Channels Is Mediated by Kir6.2. *Proc Natl Acad Sci U S A* **94**(21): 11716–11720.

Puscas, I. (1999). Omeprazole Has a Dual Mechanism of Action: It Inhibits Both H(+)K(+)Atpase and Gastric Mucosa Carbonic Anhydrase Enzyme in Humans (in Vitro and in Vivo Experiments). *Journal of Pharmacology & Experimental Therapeutics* **290**(2): 530–534.

Rampe, D., B. Anderson, V. Rapien-Pryor, T. Li, and R. C. Dage (1993). Comparison of the in Vitro and in Vivo Cardiovascular Effects of Two Structurally Distinct Ca++ Channel Activators, Bay K 8644 and Fpl 64176. *J Pharmacol Exp Ther* **265**(3): 1125–1130.

Reid, P. F., O. Pongs, and J. O. Dolly (1992). Cloning of a Bovine Voltage-Gated K+ Channel Gene Utilising Partial Amino Acid Sequence of a Dendrotoxin-Binding Protein from Brain Cortex. *FEBS Lett* **302**(1): 31–34.

Ribeiro, M. A. and P. F. Costa (2003). The Sensitivity of Sodium Channels in Immature and Mature Rat Ca1 Neurones to the Local Anaesthetics Procaine and Lidocaine. *Brain Res Dev Brain Res* **146**(1–2): 59–70.

Richard, S., D. Potreau, P. Charnet, G. Raymond, and J. Nargeot (1989). Are Ba2+ and Sr2+ Ions Transported by the Na+-Ca2+ Exchanger in Frog Atrial Cells? *J Mol Cell Cardiol* **21**(9): 865–875.

Rodrigues, A. R., E. C. Arantes, F. Monje, W. Stuhmer, and W. A. Varanda (2003). Tityustoxin-K(Alpha) Blockade of the Voltage-Gated Potassium Channel Kv1.3. *Br J Pharmacol* **139**(6): 1180–1186.

Rouzaire-Dubois, B. and J. M. Dubois (1998). K+ Channel Block-Induced Mammalian Neuroblastoma Cell Swelling: A Possible Mechanism to Influence Proliferation. *Journal of Physiology* **510**(Pt 1): 93–102.

Ruegg, U. T. and R. P. Hof (1990). Pharmacology of the Calcium Antagonist Isradipine. *Drugs* **40 Suppl 2**: 3–9.

Russ, U., U. Lange, C. Loffler-Walz, A. Hambrock, and U. Quast (2003). Binding and Effect of K Atp Channel Openers in the Absence of Mg2+. *Br J Pharmacol* **139**(2): 368–380.

Sato, T., N. Sasaki, J. Seharaseyon, B. O'Rourke, and E. Marban (2000). Selective Pharmacological Agents Implicate Mitochondrial but Not Sarcolemmal K(Atp) Channels in Ischemic Cardioprotection. *Circulation* **101**(20): 2418–2423.

Scanlon, M. J., D. Naranjo, L. Thomas, P. F. Alewood, R. J. Lewis, and D. J. Craik (1997). Solution Structure and Proposed Binding Mechanism of a Novel Potassium Channel Toxin Kappa-Conotoxin Pviia. *Structure* **5**(12): 1585–1597.

Schantz, E. J. (1985). *Chemistry and Biology of Saxitoxin and Related Toxins*. Tetrodotoxin, Saxitoxin, and the Molecular Biology of the Sodium Channel, New York, The New York Academy of Sciences.

Schnee, M. E. and B. S. Brown (1998). Selectivity of Linopirdine (Dup 996), a Neurotransmitter Release Enhancer, in Blocking Voltage-Dependent and Calcium-Activated Potassium Currents in Hippocampal Neurons. *J Pharmacol Exp Ther* **286**(2): 709–717.

Schultz, B. D., R. J. Bridges, and R. A. Frizzell (1996). Lack of Conventional Atpase Properties in Cftr Chloride Channel Gating. *J Membr Biol* **151**(1): 63–75.

Schuster, A., J. L. Beny, and J. J. Meister (2003). Modelling the Electrophysiological Endothelial Cell Response to Bradykinin. *Eur Biophys J* **32**(4): 370–380.

Shen, R., C. T. Lin, E. J. Bowman, B. J. Bowman, and J. A. Porco, Jr. (2002). Synthesis and V-Atpase Inhibition of Simplified Lobatamide Analogues. *Organic Letters* **4**(18): 3103–3106.

Shiau, Y. S., P. T. Huang, H. H. Liou, Y. C. Liaw, Y. Y. Shiau, and K. L. Lou (2003). Structural Basis of Binding and Inhibition of Novel Tarantula Toxins in Mammalian Voltage-Dependent Potassium Channels. *Chem Res Toxicol* **16**(10): 1217–1225.

Shimizu, Y. (1985). *Chemistry and Biochemistry of Saxitoxin Analogues and Tetrodotoxin*. Tetrodotoxin, Saxitoxin, and the Molecular Biology of the Sodium Channel, New York, The New York Academy of Sciences.

Shimomura, K., H. Shimizu, M. Ikeda, S. Okada, M. Kakei, S. Matsumoto, and M. Mori (2004). Fenofibrate, Troglitazone, and 15-Deoxy-Delta12,14-Prostaglandin J2 Close Katp Channels and Induce Insulin Secretion. *J Pharmacol Exp Ther* **310**(3): 1273–1280.

Shumaker, D. K., L. R. Vann, M. W. Goldberg, T. D. Allen, and K. L. Wilson (1998). Tpen, a Zn2+/Fe2+ Chelator with Low Affinity for Ca2+, Inhibits Lamin Assembly, Destabilizes Nuclear Architecture and May Independently Protect Nuclei from Apoptosis in Vitro. *Cell Calcium* **23**(2–3): 151–164.

Sidach, S. S. and I. M. Mintz (2002). Kurtoxin, a Gating Modifier of Neuronal High- and Low-Threshold Ca Channels. *J Neurosci* **22**(6): 2023–2034.

Simons, T. J. (1976). Carbocyanine Dyes Inhibit Ca-Dependent K Efflux from Human Red Cell Ghosts. *Nature* **264**(5585): 467–469.

Singh, B. N., K. Nademanee, M. A. Josephson, N. Ikeda, N. Venkatesh, and R. Kannan (1984). The Electrophysiology and Pharmacology of Verapamil, Flecainide, and Amiodarone: Correlations with Clinical Effects and Antiarrhythmic Actions. *Ann N Y Acad Sci* **432**: 210–235.

Smith, J. B., E. J. Cragoe, Jr., and L. Smith (1987). Na+/Ca2+ Antiport in Cultured Arterial Smooth Muscle Cells. Inhibition by Magnesium and Other Divalent Cations. *J Biol Chem* **262**(25): 11988–11994.

Smoak, I. W. (1999). Cromakalim: Embryonic Effects and Reduction of Tolbutamide-Induced Dysmorphogenesis in Vitro. *Teratology* **60**(5): 260–264.

Soderholm, J. D., H. Oman, L. Blomquist, J. Veen, T. Lindmark, and G. Olaison (1998). Reversible Increase in Tight Junction Permeability to Macromolecules in Rat Ileal Mucosa in Vitro by Sodium Caprate, a Constituent of Milk Fat. *Dig Dis Sci* **43**(7): 1547–1552.

Soler, F., F. Plenge-Tellechea, I. Fortea, and F. Fernandez-Belda (1998). Cyclopiazonic Acid Effect on Ca2+-Dependent Conformational States of the Sarcoplasmic Reticulum Atpase. Implication for the Enzyme Turnover. *Biochemistry* **37**(12): 4266–4274.

Stefani, A., F. Spadoni, and G. Bernardi (1997). Differential Inhibition by Riluzole, Lamotrigine, and Phenytoin of Sodium and Calcium Currents in Cortical Neurons: Implications for Neuroprotective Strategies. *Exp Neurol* **147**(1): 115–122.

Strid, A., P. Nyren, and M. Baltscheffsky (1988). Diethylstilbestrol. Interactions with Membranes and Proteins and the Different Effects Upon Ca2+- and Mg2+-Dependent Activities of the F1-Atpase from Rhodospirillum Rubrum. *Eur J Biochem* **176**(2): 281–285.

Striessnig, J. (1999). Pharmacology, Structure and Function of Cardiac L-Type Ca(2+) Channels. *Cell Physiol Biochem* **9**(4–5): 242–269.

Suessbrich, H., S. Waldegger, F. Lang, and A. E. Busch (1996). Blockade of Herg Channels Expressed in Xenopus Oocytes by the Histamine Receptor Antagonists Terfenadine and Astemizole. *FEBS Lett* **385**(1–2): 77–80.

Sutko, J. L., J. A. Airey, W. Welch, and L. Ruest (1997). The Pharmacology of Ryanodine and Related Compounds. *Pharmacol Rev* **49**(1): 53–98.

Szabo, C. and A. L. Salzman (1996). Inhibition of Atp-Activated Potassium Channels Exerts Pressor Effects and Improves Survival in a Rat Model of Severe Hemorrhagic Shock. *Shock* **5**(6): 391–394.

Szewczyk, A., J. R. De Weille, and M. Lazdunski (1992). Tmb-8 (8-(N,N-Diethylamino) Octyl-3,4,5-Trimethoxybenzoate) Inhibits the Atp-Sensitive K+ Channel. *Eur J Pharmacol* **226**(2): 175–177.

Szewczyk, A., B. Mikolajek, S. Pikula, and M. J. Nalecz (1993). Potassium Channel Openers Induce Mitochondrial Matrix Volume Changes Via Activation of Atp-Sensitive K+ Channel. *Pol J Pharmacol* **45**(4): 437–443.

Takeguchi, N., Y. Nishimura, T. Watanabe, Y. Mori, and M. Morii (1983). A Pungent Ingredient of Mustard, Allylisothiocyanate, Inhibits (H+ + K+)-Atpase. *Biochem Biophys Res Commun* **112**(2): 464–468.

Tan, J. H. and D. A. Saint (2000). Interaction of Lidocaine with the Cardiac Sodium Channel: Effects of Low Extracellular pH Are Consistent with an External Blocking Site. *Life Sci* **67**(22): 2759–2766.

Tapper, A. R. and A. L. George, Jr. (2001). Location and Orientation of Mink within the I(Ks) Potassium Channel Complex. *Journal of Biological Chemistry* **276**(41): 38249–38254.

Tatulian, L., P. Delmas, F. C. Abogadie, and D. A. Brown (2001). Activation of Expressed Kcnq Potassium Currents and Native Neuronal M-Type Potassium Currents by the Anti-Convulsant Drug Retigabine. *J Neurosci* **21**(15): 5535–5545.

Tokimasa, T. and R. A. North (1996). Effects of Barium, Lanthanum and Gadolinium on Endogenous Chloride and Potassium Currents in Xenopus Oocytes. *J Physiol* **496 (Pt 3)**: 677–686.

Tosteson, M. T., J. Thomas, J. Arnadottir, and D. C. Tosteson (2003). Effects of Palytoxin on Cation Occlusion and Phosphorylation of the (Na+,K+)-Atpase. *J Membr Biol* **192**(3): 181–189.

Trac, M., C. Dyck, M. Hnatowich, A. Omelchenko, and L. V. Hryshko (1997). Transport and Regulation of the Cardiac Na(+)-Ca2+ Exchanger, Ncx1. Comparison between Ca2+ and Ba2+. *J Gen Physiol* **109**(3): 361–369.

Tu, Q., L. H. Pinto, G. Luo, M. A. Shaughnessy, D. Mullaney, S. Kurtz, M. Krystal, and R. A. Lamb (1996). Characterization of Inhibition of M2 Ion Channel Activity by Bl-1743, an Inhibitor of Influenza a Virus. *Journal of Virology* **70**(7): 4246–4252.

Turetta, L., E. Bazzan, K. Bertagno, E. Musacchio, and R. Deana (2002). Role of Ca(2+) and Protein Kinase C in the Serotonin (5-Ht) Transport in Human Platelets. *Cell Calcium* **31**(5): 235–244.

Ulianich, L., A. Secondo, S. De Micheli, S. Treglia, F. Pacifico, D. Liguoro, F. Moscato, S. Marsigliante, L. Annunziato, S. Formisano, E. Consiglio, and B. Di Jeso (2004). Tsh/Camp up-Regulate Sarco/Endoplasmic Reticulum Ca2+-Atpases Expression and Activity in Pc Cl3 Thyroid Cells. *Eur J Endocrinol* **150**(6): 851–861.

Unsold, B., G. Kerst, H. Brousos, M. Hubner, R. Schreiber, R. Nitschke, R. Greger, and M. Bleich (2000). Kcne1 Reverses the Response of the Human K+ Channel Kcnq1 to Cytosolic pH Changes and Alters Its Pharmacology and Sensitivity to Temperature. *Pflugers Archiv - European Journal of Physiology* **441**(2–3): 368–378.

Usami, M., K. Muraki, M. Iwamoto, A. Ohata, E. Matsushita, and A. Miki (2001). Effect of Eicosapentaenoic Acid (Epa) on Tight Junction Permeability in Intestinal Monolayer Cells. *Clin Nutr* **20**(4): 351–359.

Vagin, O., S. Denevich, K. Munson, and G. Sachs (2002). Sch28080, a K+-Competitive Inhibitor of the Gastric H,K-Atpase, Binds near the M5-6 Luminal Loop, Preventing K+ Access to the Ion Binding Domain. *Biochemistry* **41**(42): 12755–12762.

Valenzuela, C., E. Delpon, L. Franqueza, P. Gay, O. Perez, J. Tamargo, and D. J. Snyders (1996). Class Iii Antiarrhythmic Effects of Zatebradine. Time-, State-, Use-, and Voltage-Dependent Block of Hkv1.5 Channels. *Circulation* **94**(3): 562–570.

Van Der Heyden, N., and R. Docampo (2000). Intracellular pH in Mammalian Stages of Trypanosoma Cruzi Is K+-Dependent and Regulated by H+-Atpases. *Mol Biochem Parasitol* **105**(2): 237–251.

Vianna-Jorge, R., C. F. Oliveira, M. L. Garcia, G. J. Kaczorowski, and G. Suarez-Kurtz (2000). Correolide, a nor-Triterpenoid Blocker of Shaker-Type Kv1 Channels Elicits Twitches in Guinea-Pig Ileum by Stimulating the Enteric Nervous System and Enhancing Neurotransmitter Release. *Br J Pharmacol* **131**(4): 772–778.

Wang, G. K., W. M. Mok, and S. Y. Wang (1994). Charged Tetracaine as an Inactivation Enhancer in Batrachotoxin-Modified Na+ Channels. *Biophys J* **67**(5): 1851–1860.

Wang, G. K., C. Quan, and S. Wang (1998). A Common Local Anesthetic Receptor for Benzocaine and Etidocaine in Voltage-Gated Mu1 Na+ Channels. *Pflugers Arch* **435**(2): 293–302.

Warnick, J. E., M. A. Maleque, N. Bakry, A. T. Eldefrawi, and E. X. Albuquerque (1982). Structure-Activity Relationships of Amantadine. I. Interaction of the N-Alkyl Analogues with the Ionic Channels of the Nicotinic Acetylcholine Receptor and Electrically Excitable Membrane. *Mol Pharmacol* **22**(1): 82–93.

Watanabe, H., M. Nishio, M. Miura, and T. Iijima (1995). L-Type Ca Channel Block by Highly Hydrophilic Dihydropyridines in Single Ventricular Cells of Guinea-Pig Hearts. *J Mol Cell Cardiol* **27**(6): 1271–1279.

Watano, T. and J. Kimura (1998). Calcium-Dependent Inhibition of the Sodium-Calcium Exchange Current by Kb-R7943. *Can J Cardiol* **14**(2): 259–262.

Weber, J. and A. E. Senior (1998). Effects of the Inhibitors Azide, Dicyclohexylcarbodiimide, and Aurovertin on Nucleotide Binding to the Three F1-Atpase Catalytic Sites Measured Using Specific Tryptophan Probes. *Journal of Biological Chemistry* **273**(50): 33210–33215.

Weichert, A., S. Faber, H. W. Jansen, W. Scholz, and H. J. Lang (1997). Synthesis of the Highly Selective Na+/H+ Exchange Inhibitors Cariporide Mesilate and (3-Methanesulfonyl-4-Piperidino-Benzoyl) Guanidine Methanesulfonate. *Arzneimittelforschung* **47**(11): 1204–1207.

Weingart, R. and F. F. Bukauskas (1998). Long-Chain N-Alkanols and Arachidonic Acid Interfere with the Vm-Sensitive Gating Mechanism of Gap Junction Channels. *Pflugers Arch* **435**(2): 310–319.

Weiss, R., M. Mevissen, D. S. Hauser, G. Scholtysik, C. J. Portier, B. Walter, U. E. Studer, and H. Danuser (2002). Inhibition of Human and Pig Ureter Motility in Vitro and in Vivo by the K(+) Channel Openers Pkf 217–744b and Nicorandil. *J Pharmacol Exp Ther* **302**(2): 651–658.

West, P. J., G. Bulaj, J. E. Garrett, B. M. Olivera, and D. Yoshikami (2002). Mu-Conotoxin Smiiia, a Potent Inhibitor of Tetrodotoxin-Resistant Sodium Channels in Amphibian Sympathetic and Sensory Neurons. *Biochemistry* **41**(51): 15388–15393.

Winmill, R. E. and M. S. Hedrick (2003). Gap Junction Blockade with Carbenoxolone Differentially Affects Fictive Breathing in Larval and Adult Bullfrogs. *Respir Physiol Neurobiol* **138**(2–3): 239–251.

Wittekindt, O. H., V. Visan, H. Tomita, F. Imtiaz, J. J. Gargus, F. Lehmann-Horn, S. Grissmer, and D. J. Morris-Rosendahl (2004). An Apamin- and Scyllatoxin-Insensitive Isoform of the Human Sk3 Channel. *Mol Pharmacol* **65**(3): 788–801.

Wu, J. and G. Z. Jin (1997). Tetrahydroberberine Blocks Membrane K+ Channels Underlying Its Inhibition of Intracellular Message-Mediated Outward Currents in Acutely Dissociated Ca1 Neurons from Rat Hippocampus. *Brain Res* **775**(1–2): 214–218.

Wu, S. N., H. F. Li, and Y. C. Lo (2000). Characterization of Tetrandrine-Induced Inhibition of Large-Conductance Calcium-Activated Potassium Channels in a Human Endothelial Cell Line (Huv-Ec-C). *J Pharmacol Exp Ther* **292**(1): 188–195.

Wulff, H., M. J. Miller, W. Hansel, S. Grissmer, M. D. Cahalan, and K. G. Chandy (2000). Design of a Potent and Selective Inhibitor of the Intermediate-Conductance Ca2+-Activated K+ Channel, Ikca1: A Potential Immunosuppressant. *Proc Natl Acad Sci U S A* **97**(14): 8151–8156.

Yamaguchi, S., B. S. Zhorov, K. Yoshioka, T. Nagao, H. Ichijo, and S. Adachi-Akahane (2003). Key Roles of Phe1112 and Ser1115 in the Pore-Forming Iiis5-S6 Linker of L-Type Ca2+ Channel Alpha1c Subunit (Cav 1.2) in Binding of Dihydropyridines and Action of Ca2+ Channel Agonists. *Mol Pharmacol* **64**(2): 235–248.

Yan, L. and M. E. Adams (2000). The Spider Toxin Omega-Aga Iiia Defines a High Affinity Site on Neuronal High Voltage-Activated Calcium Channels. *J Biol Chem* **275**(28): 21309–21316.

Yang, I. C., M. W. Scherz, A. Bahinski, P. B. Bennett, and K. T. Murray (2000). Stereoselective Interactions of the Enantiomers of Chromanol 293b with Human Voltage-Gated Potassium Channels. *J Pharmacol Exp Ther* **294**(3): 955–962.

Yoneda, I., H. Sakuta, K. Okamoto, and Y. Watanabe (1993). Effects of Local Anesthetics and Related Drugs on Endogenous Glibenclamide-Sensitive K+ Channels in Xenopus Oocytes. *Eur J Pharmacol* **247**(3): 267–272.

Yotsu-Yamashita, M., Y. H. Kim, S. C. Dudley, Jr., G. Choudhary, A. Pfahnl, Y. Oshima, and J. W. Daly (2004). The Structure of Zetekitoxin Ab, a Saxitoxin Analog from the Panamanian Golden Frog Atelopus Zeteki: A Potent Sodium-Channel Blocker. *Proc Natl Acad Sci U S A* **101**(13): 4346–4351.

Zaczek, R., R. J. Chorvat, J. A. Saye, M. E. Pierdomenico, C. M. Maciag, A. R. Logue, B. N. Fisher, D. H. Rominger, and R. A. Earl (1998). Two New Potent Neurotransmitter Release Enhancers, 10,10-Bis(4-Pyridinylmethyl)-9(10h)-Anthracenone and 10,10-Bis(2-Fluoro-4-Pyridinylmethyl)-9(10h)-Anthracenone: Comparison to Linopirdine. *J Pharmacol Exp Ther* **285**(2): 724–730.

Zelenina, M., S. Tritto, A. A. Bondar, S. Zelenin, and A. Aperia (2004). Copper Inhibits the Water and Glycerol Permeability of Aquaporin-3. *J Biol Chem* **279**(50): 51939–51943.

Zhang, H., B. Zhu, J. A. Yao, and G. N. Tseng (1998). Differential Effects of S6 Mutations on Binding of Quinidine and 4-Aminopyridine to Rat Isoform of Kv1.4: Common Site but Different Factors in Determining Blockers' Binding Affinity. *J Pharmacol Exp Ther* **287**(1): 332–343.

Zhao, F., P. Li, S. R. Chen, C. F. Louis, and B. R. Fruen (2001). Dantrolene Inhibition of Ryanodine Receptor Ca2+ Release Channels. Molecular Mechanism and Isoform Selectivity. *J Biol Chem* **276**(17): 13810–13816.

Zona, C., V. Tancredi, P. Longone, G. D'Arcangelo, M. D'Antuono, M. Manfredi, and M. Avoli (2002). Neocortical Potassium Currents Are Enhanced by the Antiepileptic Drug Lamotrigine. *Epilepsia* **43**(7): 685–690.

Zunkler, B. J., S. Lenzen, K. Manner, U. Panten, and G. Trube (1988). Concentration-Dependent Effects of Tolbutamide, Meglitinide, Glipizide, Glibenclamide and Diazoxide on Atp-Regulated K+ Currents in Pancreatic B-Cells. *Naunyn Schmiedebergs Arch Pharmacol* **337**(2): 225–230.

Zygmunt, P. M., G. Edwards, A. H. Weston, B. Larsson, and E. D. Hogestatt (1997). Involvement of Voltage-Dependent Potassium Channels in the Edhf-Mediated Relaxation of Rat Hepatic Artery. *Br J Pharmacol* **121**(1): 141–149.

Section III

Transcriptional Regulation of Target Genes

10 Expression Profiling in *Xenopus* Embryos

Curtis Altmann

CONTENTS

I. INTRODUCTION

Since its introduction in the 1990s [1, 2], microarray technology has become an extensively applied approach for studying gene expression in a wide variety of fields. Many of the initial studies focused on expression studies in tumors and cell lines as well as the development of the technology itself. The application of the technology to the study of *Drosophila* metamorphosis [3] initiated the flow of publications using the technology to understand embryonic development in more than a dozen systems.

The purpose of this chapter is to present an overview of microarray options and general approaches to consider when applying this technology to questions of embryonic development and signaling. It is assumed that the readers of this chapter are not in the business of preparing the arrays themselves but rather are embryologists/developmental biologists interested in applying this powerful technology to their model system. Thus, the preparation of microarrays, which has been well covered in numerous articles and books [4–6], will not be discussed in detail. Instead, issues relating to the selection of array format (e.g., oligonucleotide versus cDNA), types of experimental approaches, preparation of sample, analysis of array data, and independent confirmation of results will be examined. With regard to detailed methods, protocols that have been used successfully for *Xenopus* are provided at the end of the chapter and can be readily adapted to other species. For those expecting to do only a few arrays, the general recommendation is to purchase and use kits and reagents from one of the established suppliers. Although the discussion is focused on the application of array technologies to *Xenopus*, many of the comments will be applicable to other systems.

Though only 5 years have passed since the publication of the first application of microarrays to the study of metazoan development, an increasing number of studies have applied the technology to a variety of vertebrate (including zebrafish [7–9], *Xenopus* [10–16], chicken [17–20], medaka [21, 22]) and invertebrate embryonic systems (including *Drosophila* [23–26] and *C. elegans* [27–30]. The wide availability of array technologies for mouse and human has resulted in many published studies of embryonic development. Among the explored areas are studies of oocyte development, implantation and hatching [31–36]; toxicity and birth defects [37–40] and organ development [41–44]; gene targeting [44–46]; and stem cells [47], to name but a few. An approach that will likely gain in popularity combines gene targeting, fluorescein-activated cell sorting, and microarrays [44, 48, 49]. Though the disruption and isolation of GFP expressing cell populations is likely to effect gene expression, these methods will certainly allow a more detailed description of the molecular signatures of various cell and tissue types during normal development.

In *Xenopus*, two general approaches have been applied: whole embryo and tissue explant analysis [10, 11, 13–16]. An advantage of whole embryo studies is the presence of the full complement of regulatory interactions. In contrast, the relative homogeneity sought in explant studies (for example, ectodermal explants) means the loss of interactions from other germ layers and the potential restriction of cell fate. As will be noted below, the presence of heterogeneity can create both false negative and false positive expression changes. Thus, the selection of the approach has significant and direct consequences on the outcome.

II. GOALS OF EXPRESSION PROFILING IN EMBRYOS

A major goal of expression profiling in embryos is the identification and characterization of gene regulatory networks (GRNs) [50–52]. Critical to this goal is the identification of downstream targets of a given signaling pathway and the distinction between "direct" and "indirect" targets and regulators. Here, we define direct regulators and targets as those that activate target genes in the absence of protein translation, whereas secondary regulators and targets require translation of an additional factor. Thus, direct regulators can include all components of a growth factor signaling pathway as well as direct transcription regulators [50]. Direct transcriptional regulators act through cis-regulatory elements and act as either activators or repressors of transcription. An important goal in defining and characterizing a GRN is the identification of genes involved in integrating signaling pathways. Although expression profiling does not allow the identification of such important regulators, comparisons between the sets of genes activated by two pathways allow the identification of commonly regulated genes, which may be the putative targets of growth factor signaling "cross talk" [53].

The criteria for demonstrating a direct target of transcriptional regulation has been rigorously and clearly defined [50], though meeting these criteria requires significantly more work than identifying potential regulatory targets available through microarray analysis. The first criterion involves the demonstration that the target genes are perturbed in a manner consistent with the effector. Here, expression profiling can be used by generating both activating and repressing forms of the transcription factor using fusions with VP16 activation domains and Engrailed repressor domains [54]. The second criterion requires that the regulatory target be expressed in a spatial and temporal manner consistent with the transcriptional regulator. In some cases, this might entail being a member of a syn-expression group [12, 55–59]; however, complete overlap is not an absolute requirement. Finally, to be considered a *bona fide* direct target, the regulatory factor must be shown to physically interact with the regulatory region of the target gene.

Immediate regulators include the direct activators of transcription defined above as well as various components of the growth factor signaling pathway. Here, expression profiling can define both direct and secondary targets using inhibitors of protein translation such as cycloheximide. Although a powerful approach, this largely limits the distinction between these targets to very early gastrula stages of embryonic development [60, 61]. For transcription factors, this limitation can potentially be circumvented by the preparation of hormone-inducible constructs [62] such that the translation inhibitor can be added at later stages simultaneously with the hormone inducer. Given that many growth factor signaling pathways act through multiple transcription factors, methods to block or activate the pathway that are independent of translation are necessary to distinguish immediate and secondary targets using microarray methods. One approach is the use of inhibiting or activating small molecules such as SU5402, an inhibitor specific to FGF receptor 1 [13]. The application of such pharmacologic agents shows great promise for the analysis of signaling pathways.

III. TYPES OF ARRAYS

Microarrays are prepared from either oligonucleotides or amplified cDNAs. Many of the earliest array studies employed spotted PCR products amplified from cDNA clones either from random libraries or from unique clone libraries available commercially. Oligonucleotide arrays have become available for a growing number of organisms and are limited only by the availability of EST /or genomic sequence. Affymetrix Gene Chips (www.affymetrix.com) are both widely known and well established, but a number of other options are available. A major difference between the Gene Chip technology and other alternatives is that the latter are probed using two or more labeled samples.

IV. OLIGONUCLEOTIDE ARRAYS

The advantage of oligonucleotide-based methods is that they avoid the problems and labor associated with the preparation, amplification, and printing (for *in situ* synthesized oligonucleotides) of the array [63]. In addition, oligonucleotides can be designed that distinguish different splice forms, closely related homologues, and single nucleotide polymorphisms. Spotted arrays (e.g., those that are printed from synthesized oligonucleotides) employ oligonucleotides that are from 40 to 80 nucleotides in length. Shorter oligonucleotides are often used for detecting single nucleotide polymorphisms. Oligonucleotide probes are sensitive to hybridization temperature and washing stringency and may require more development to use effectively. Operon Biotechnologies provides a variety predesigned oligonucleotide sets from a number of species including *Xenopus*, zebrafish, *Drosophila*, and *C. elegans*. Many oligonucleotide synthesis companies will aid in the design and production of oligonucleotides for arraying.

In situ synthesized arrays employ oligonucleotides prepared directly on the chip (Table 10.1). Affymetrix uses a photolithographic masking procedure to produce multiple 25 nucleotide probes (11–16 per transcript) on chips containing more than 1 million features (and growing as feature size decreases). Agilent (www.agilent.com) uses ink-jet technology to print 22,000–44,000 oligonucleotides base by base on a slide. Nimblegen (www.nimblegen.com) uses a maskless technology to generate oligonucleotides and, together with Agilent, offers the ability to prepare custom-designed oligonucleotide arrays.

V. cDNA ARRAYS

In contrast, cDNA arrays consist of PCR-amplified double-stranded hybridization targets that are typically 100s to 1000s of bases long. Due to the greater length, they are unable to distinguish subtle differences between sequences and are more subject to cross-hybridization between closely related homologues. Although the length of the target limits its ability to distinguish small differences, it allows for robust hybridization (i.e., higher sensitivity) and is relatively easy to use. A large number of protocols (as well as commercially prepared kits) are available online, making

TABLE 10.1
Suppliers of Oligonucleotide Arrays

Supplier	Species	Oligo	Target Number	Link/Notes
Nimblegen	Zebrafish, *Xenopus*, Chicken, *C. elegans*, *Drosophila*	24–150mer	382,000–32,889	http://www.nimblegen.com/
Affymetrix	*Xenopus*, Chicken, *Drosophila*, Zebrafish, *C. elegans*	25mer	14,400	
			28,000	
			18,500	
			14,900	
			22,500	
Agilent	*Xenopus*, Zebrafish, *C. elegans*, Chicken	25–60	44,000	www.agilent.com

this technology accessible to many labs (provided the actual arrays are already available). If the cDNA array is made in-house, the advantage of having a cDNA clone available for further analysis such as *in situ* hybridization is significant.

VI. EXPERIMENTAL DESIGN

A critical decision in the application of expression profiling methods concerns the experimental design. There are a number of reviews that cover this topic in detail [64–67]. As with all well-designed experiments, microarrays require a clearly defined question or hypothesis. Often, this is simply, "What is the set of genes activated by gene/treatment X?" Second, systematic and experimental errors such as dye-specific effects, print tip-dependent variation (for printed cDNA and oligonucleotide arrays), and signal amplification should be minimized. A sufficient number of arrays and replicates should be used to allow simple statistical analysis of the experiments (discussed in greater detail below). Finally, with publication of the results in mind, the experimental design must comply with the Minimum Information About a Microarray Experiment (MIAME) standard (http://www.mged.org/Workgroups/MIAME/miame.html) developed by the Microarray Gene Expression Data (MGED) Society (www.mged.org) [68–71]. Not only do many journals require compliance with this standard, the main public repositories of array data (ArrayExpress at the EBI and GEO at NCBI, CIBEX in Japan) are designed to accept, hold, and distribute MIAME-compliant microarray data.

VII. NUMBERS OF ARRAYS AND REPLICATES

How many arrays are necessary, and how many replicates of each spot are needed (if this is under the control of the experimenter)? The answer to this question depends on the overall goals of the researcher. Although not statistically reliable, a single-array experiment can yield a set of prospective regulated genes, many of which can be independently tested and confirmed. Thus, the effect of increasing numbers of array hybridizations, spot replicates, and various loop and reference experimental designs is to affect the statistical significance of the regulated gene set and determines the type of statistical methods that can be applied. Recent reviews by Churchill [72] and Dobbin et al. [73] address many of the critical issues related to experimental design for array analysis.

The key issues that affect the decision of the number of replicates are cost and sample availability. For commercial offerings, the cost of the arrays themselves can be high, whereas the costs of preparing the probe can exceed the cost of the arrays. For Affymetrix Gene Chips, a single paired experiment (control and test sample) requires two arrays, as they are based on single dyes. In contrast, printed arrays can be considerably less expensive to both purchase and use, as up to three dyes can be used [64]. Churchill [72] breaks down the experimental design into three layers: biological replicates (e.g., the treated and untreated samples from independent animals/experiments/pools), the technical replicates (e.g., multiple samples from a single treatment, dye swap samples), and duplication of array elements. Studies in *Xenopus* benefit from the relative ease of isolating various samples and the variety and ease of experimental manipulation. Thus, to minimize the effects of biological variation and allow the application of statistical approaches, at least three paired samples should be prepared on separate days. From each of the six samples, a single amino-allyl labeled cDNA should be prepared, divided into two aliquots, and inverse dye labeled (i.e., dye swap). This generates a total of 12 labeled probes that can then be hybridized to an array.

VIII. DYE SWAP

The dye swap approach involves preparing two differentially labeled probes from each of two samples and is a simple and effective design for the direct comparison of two samples. The purpose of dye swap experiments is to minimize the effects of dye-specific bias, which is an intensity difference due to factors other than differences in gene expression in the sample [74–76]. Dobbin et al. define four types of dye bias: 1. Dye bias that is the same for all genes, causing higher overall fluorescence in one image. 2. Dye bias that depends on spot intensity. 3. Dye bias associated with a subset of genes. 4. Dye bias that varies with the gene and sample [74]. Possible sources of dye bias include differences in incorporation during RT reactions in direct labeling approaches or by differences in quantum efficiencies and stability of the dyes in indirect labeling methods (see Methods section). Martin-Magniette and co-workers have suggested a statistical model and an index to evaluate the magnitude of this effect [75] and conclude that the bias can increase the false negative rate. Proper normalization eliminates dye bias across all spots (type 1), whereas Loess normalization methods [77] eliminate intensity-dependent bias (type 2) but not type 3 and type 4 dye bias. The dye swap approach does eliminate type 3 bias, as the samples are no longer confounded by association with the dye. Type 4 bias, which

Dobbin et al. call gene- and sample-specific dye bias, is not easily eliminated by current methods.

IX. SAMPLE PREPARATION: RNA ISOLATION AND LABELING

There are two general methods for total RNA isolation in common use: guanidinium-acid phenol methods [78] and SDS-proteinase K-based methods [79, 80]. For speed and simplicity, we use guanidinium-based methods, though alternative methods are likely to work as well (see Methods section). In the course of following up array results with real-time SYBR Green™-based PCR methods, we have noted the presence of an inhibitor that co-purifies with RNA from *Xenopus* embryos when using a guanidinium-based approach. This inhibitor cannot be removed from the RNA by phenol extraction or alcohol precipitation but can be eliminated by a LiCl precipitation (CRA, personal observation). Although this does not appear to affect the reverse transcription reaction and, therefore, is not likely to interfere with microarray probe preparation, we have nonetheless incorporated a LiCl precipitation step in our RNA isolation protocols.

Original array protocols employed a direct labeling method using dye-conjugated dNTPs (typically Cy3™ or Cy5™). This method is particularly vulnerable to dye-specific bias largely due to differences in incorporation. Most recently, indirect methods that employ amino-allyl-conjugated dNTPs have become the method of choice for labeling microarray samples, as they avoid differences in the enzymatic incorporation of dye during reverse transcription (see Methods section). In addition, as the amino-allyl-labeled sample can be divided prior to coupling of the fluorescent dye, variation in the sample preparation is decreased (though differential bleaching and quantum effects still apply). Finally, the availability of a number of dyes with characteristics suitable for microarray scanning allows for three or four color experiments to be performed. Experimental designs incorporating three color hybridizations are more efficient (e.g., require fewer arrays), though they come at the expense of a small loss of dynamic range [64].

In situations where the amount of sample is limiting, various amplification strategies have been developed that seek to either amplify the hybridized signal [81–84] or amplify the amount of sample. Probably the most popular is the antisense amplification approach originally developed by Eberwine and co-workers [85, 86]. The essence of this method is the preparation of a double-stranded template that incorporates a bacteriophage RNA polymerase promoter, which is then used to prepare antisense RNA. Under appropriate conditions, there is a linear accumulation of RNA (e.g., acid precipitable material) as the DNA-dependent RNA polymerase undergoes multiple rounds of transcription. From this, amplification of up to 2000X can be achieved in a single round, whereas incorporation of a second round of amplification can yield up to 10^6X amplification [87]. Of course, such power comes with a cost: Although the overall amplification is linear, and even the amplification of a single gene may be linear, there is reason to suspect that the amplification of different genes is not proportional [88]. In principle, for example, shorter genes may amplify more rapidly than longer genes within a sample, whereas the total number of template (e.g., transcripts) and the length distribution will affect comparisons across samples.

A number of studies have attempted to address this issue directly by comparing array results obtained by amplified and unamplified methods [88–103]. The overall conclusion is that there is reasonable agreement between the two approaches, with a range of correlation values between expression ratios ranging from 0.63–0.92 [97, 100, 102]. When unamplified samples prepared separately and hybridized to different batches of slide were compared by Zhao et al., they observe a correlation coefficient of 0.95 [102]. Similarly, when amplified materials are compared against each other, a correlation coefficient of 0.90 is observed. This indicates that both methods generate internally consistent results but have a greater degree of difference between each other. But consider that even at a correlation of 0.95, a 10,000-gene array would yield a significant number of false negatives and positives. Importantly, although the analyses are somewhat informative, they do not address the higher order moments of statistical distributions, skewness (whether the distribution is symmetric), kurtosis (the distribution of the data about the mean), and skedasticity (which addresses whether the variances for each gene are different). Thus, we recommend that this method be used and interpreted with these differences in mind and avoided if sufficient quantities of sample can be obtained.

X. SAMPLE SOURCES

For studies in embryos, two general types of sample are generally considered: whole embryo and tissue explants. The advantage of whole embryo experiments is that the full range of factors normally expressed in the embryo are available, thus all the components necessary in a given inductive interaction or signaling pathway are present. At the same time, the presence of the wide variety of tissue types results in a higher degree of complexity and the potential to mask regulated genes (e.g., false negatives). Similar complications arise in studies of various tumor samples, as any given tumor will have varying contributions of non-tumor cells such as vascular endothelium, smooth muscle, and other cell types [104, 105]. To address this complication, a variety of methods have been applied that generally decrease the size of the sample [106], including Laser Capture Microdissection (LCM) [107–110] and Fluorescence Activated Cell Sorting (FACS) [111–113].

The tissue explant offers the advantage of decreasing the heterogeneity of the sample and eliminating potentially competing signaling. Using such an approach, one can examine the neural induction pathway in the absence of mesoderm formation [114]. At the same time, the absence of other tissues may lead to the loss of regulatory factors induced by a separate pathway. For neural induction, this could be the loss of caudalizing signals such as FGF signaling [115, 116] or factors required for competence [117, 118]. Finally, whereas the tissue explant may decrease potential heterogeneity, it is well established that the animal pole explant (or animal cap, the most commonly used explant type in *Xenopus*) has a pre-existing dorsal-ventral axis that will generate heterogeneity [119]. Moreover, for perturbations involving microinjection, the distribution of the agent (either RNA or morpholino) is nonhomogeneous, which can generate concentration-dependent differences in responsive cells [120].

XI. TIME AND SPACE

The study of development in the embryo is critically concerned with the temporal and spatial changes that occur. At the gene expression level, this awareness is reflected in gene expression studies in whole embryos [121, 122] and the application of tissue-specific markers. For microarrays, beyond the obvious limitation to studying the range of time during which one's tissue/organ/cell type is induced/specified/differentiated, there are additional experimental limitations to consider. For some studies, it may be of great interest to identify the direct target of a signaling pathway. As noted above, inclusion of cycloheximide allows induced transcripts that do not require new protein synthesis to be identified, though this limits the analysis to earlier stages of development [60, 61]. Such an approach has been used successfully in the identification by microarray of direct targets of BMP signaling [14] or in the follow-up experiments of microarray-identified genes regulated by VegT [16]. The identification of direct targets facilitates an analysis of promoter elements in conjunction with genome sequence.

In many instances, the cycloheximide approaches described above are not readily applicable, as the pathway or process under study occurs later in development (e.g., neural patterning and kidney development) or requires the synthesis of additional factors. One approach applicable to nuclear transcription factors is the generation of steroid-inducible fusion proteins [62]. Such constructs have been used successfully to dissect the temporal requirements of transcription factor activity [123]. Similarly, a temporal dissection using microarray technology can be readily performed. To identify potential direct targets, the inducing steroid would be applied together with the translation inhibitor cycloheximide. Alternatively, during the design of the study, one might consider the post-hoc testing of regulated targets, perhaps based on the overall number of targets that are expected from the array analysis. Thus, if one expects 100–200 regulated genes (both direct and indirect), it might be reasonable to distinguish the direct targets by an alternative method such as quantitative RT-PCR based methods (see Section XVII).

XII. ORGANIZING AND ANALYZING AND YOUR PROJECT

Before addressing the various issues involved after the preparation and hybridization of the samples, it is important to discuss some of the software options available for both analyzing and organizing your data [124, 125]. Depending on the scientific goals and size of the group, purchasing a commercial package might be appropriate; however, many of these commercial packages do not come cheaply. In contrast, there are a number of excellent open source and freely available packages that implement the same feature set as the commercial packages. These open source packages do not come with a technical support hotline but often have discussion groups and help available. In addition, as new methods are developed and published, they are often incorporated into the various updates, allowing the new methods to be applied to your existing data. Many of the packages incorporate the open source statistical language R (www.r-project.org). For those with advanced computer skills, the largely command line-driven package BioConductor (www.bioconductor.org) may be appropriate, though for some tools, graphical interfaces have been developed. Alternatively, the National Cancer Institute Biometric Research Branch (NCI BRB) offers an Excel-based package called BRB Arraytools (http://linus.nci.nih.gov/BRB-ArrayTools.html) that incorporates a variety of analysis using the R statistical language.

XIII. SCANNING AND DATA EXTRACTION

Once the hybridization is completed, an image is acquired using one of a number of commercial scanners. For all arrays, regardless of the format, the raw data is in fact a 16-bit greyscale image, usually at 5–10 μm resolution. Thus, taking a high-quality picture is essential and determines the results of all downstream analysis. For each scanner, there are two parameters that are varied: laser power and photomultiplier tube (PMT) gain. The overall goal is to obtain an image with the brightest spots (e.g., highest expressed genes) at or near saturation. To minimize photobleaching of the fluorescent signal, lower laser settings can be selected and the PMT gain increased to obtain the correct exposure. Some scanners acquire multiple channels simultaneously, whereas others perform separate scans, but one objective is to obtain images from each channel that have balanced signals. In most cases, the majority of the spots on the array will not be differentially expressed, and therefore histograms of pixel intensity and visual inspection can be useful in balancing the various channels. As discussed below, computational methods are available for normalizing the signals (e.g., balancing the channels), which suggests that maximizing the dynamic range and limiting pixel saturation should take precedence.

The next step in data acquisition is spot recognition and data extraction, and a variety of software tools, both free and commercial, are available for this task (see http://ihome.cuhk.edu.hk/~b400559/arraysoft_image.html [126]). In the first step, the spots are recognized (gridded) and distinguished from background noise and various artifacts (segmentation). It is at this stage that an area of the image is selected that encloses the signal (often a circle, though some software can recognize different shapes), while the pixels adjacent to the enclosure are defined as background. The software next sums the intensity for each pixel to generate a unitless number, a pixel volume for each channel (e.g., dye) that is related to the amount of a given message. As all downstream analysis depends on the data extraction step, it is critical that the user carefully examine the array image spot by spot and identify failed targets, spurious signals such as dust, and regions of high background. Most software packages allow such problem areas to be flagged for exclusion or special attention. The end result of the process is essentially a tab-delimited file, one row for each spot and one column for the location, area, intensity background, and so on.

XIV. NORMALIZATION

For most biologists, getting through the previous steps is relatively straightforward, as many of the tasks are essentially similar to established methods and approaches. Still, most biologists and embryologists have never had to address the type and amount of data generated in an array analysis and may feel overwhelmed at this point. Fortunately, the scientific community has responded to this flood of data by training a cadre of specialists and developing software tools. This section will address the issue of normalization, in which systematic errors are removed and the fluorescent intensities are balanced between the image channels. A complete discussion of the various issues for both normalization and statistical analysis is beyond the scope of this chapter and is best left to the experts [127–130]. A common theme that runs through the growing numbers of studies and reviews is the desirability of collaboration between a statistician and the biologist. Critical to such a successful collaboration is the incorporation of a proper experimental design such as those outlined and referenced above, which will allow the application of established methods. Many of the software tools available (both free and commercial) incorporate well-accepted statistical analysis methods but are unlikely to replace expert advice.

The first task in the analysis of array data is normalization, both within an array (between channels in multidye experiments) and between multiple arrays. Within the array, systematic variation in signal is introduced not only by the labeling method and the characteristics of the fluorescent dyes, but also in the preparation and printing of the arrays themselves [77, 131]. Early efforts addressing the normalization problem utilized a global approach [132, 133], a set of housekeeping genes [2, 134–137], and more recently universal reference RNAs have been developed for normalizing both within and between arrays [138–140]. In the global approach, one channel was multiplied by a single factor so that the average log ratio of the two channels is equal to 0. Although the global approach may still be applicable (assuming that the number of up- and down-regulated genes are roughly equivalent), it should be used with caution. The latter two methods have become less favored as the variability of housekeeping genes has been recognized [136], and the exogenous controls do not adequately account for dye bias effects. Current best practice implements a non-linear normalization method that takes into account signal intensity and may include a spatial normalization; it may also account for differences arising from the physical printing process (pin normalization). The most commonly applied method is the locally weighted scatterplot smoothing (Lowess aka Loess smoothing) [77, 134, 141], which is implemented in many of the commercial and public domain analysis packages.

Finally, adjustments are necessary to match the distribution of log ratios across multiple arrays, and the method used depends on the overall experimental design. The starting point for normalizing across multiple arrays is a set of properly normalized arrays, as outlined above. Arrays employing a reference sample can be scaled using this common reference [138–140, 142–144]. Scale normalization involves adjusting the log ratios of expression, M ($M = \log_2 R - \log_2 G$), such that each array has the same median absolute deviation [77, 145]. In general, the normalization procedures for within slide normalization can be applied across multiple slides as well. The goal of this normalization adjustment is to prevent any slide(s) from having undue weight when averaging across multiple slides [77, 146]. As with many of the methods described in this chapter, integrated solutions are available in the variety of software packages available [124, 125].

XV. CLUSTERING, GROUPING, AND STATISTICAL ANALYSIS

Once the primary data are collected and normalized both within and across arrays, there are a number of higher order analyses that are often performed. The following discussion presents the basic issues and choices available and provides references to comprehensive reviews for each area. The overall questions that are addressed at this stage of analysis are (1) Which genes are regulated, or more specifically, which genes are statistically different between samples? (2) Are genes or experiments related to each other, or can they be grouped together (e.g., clustered)? (3) Can a regulated pathway or gene network be identified? (4) Do regulated genes share common genomic regulatory elements? In the subsequent sections, an exhaustive list of available approaches is neither provided nor attempted, as the available tools are constantly in flux. Rather, the various programs and links are provided as a starting point for analysis and to provide examples of the types of analysis that might be of interest. An excellent resource for programs and databases can be found yearly in *Nucleic Acids Research* [147], whereas Y. F. Leung maintains a well-organized list of analysis software (http://ihome.cuhk.edu.hk/~b400559/arraysoft.html).

Initial array experiments identified differentially expressed genes based on a fixed fold change cutoff [148, 149], whereas current methods employ variations of statistical tests. There are a number of approaches, and the methods are extensively covered elsewhere [128, 129, 150, 151]. As noted above, collaboration with a statistician can be of significant value. The goals of the statistical tests are to identify the regulated genes and assign a confidence value while minimizing both false negative and false positive results. A variety of software packages are available that incorporate standard statistical tests for both two-color and single-color analysis. These packages include four common tests for comparing two conditions: (1) the global and gene-specific t-test statistics [152], (2) the significance of microarrays (SAM) test statistic [153], (3) the regularized t-test statistic [154], and (4) the B statistic [155]. The second part of the statistical analysis is the determination of the significance and often includes permutation analysis in which the data is shuffled to generate the null condition (e.g., no regulated genes). For experiments in which there are more than two conditions, a simple calculation of expression ratio is no longer possible. Statistical analysis of experiments of this type involves analysis of variance (ANOVA), in which the relative expression is compared [133, 150, 156].

For some experiments, the relationship of expression between the samples and the genes is of interest, and a variety of methods discover patterns within the data. These methods have been applied to time course experiments in which the co-regulated genes are sought to identify markers of specific tissue types, more often in cancer [157–165]. The methods are unsupervised in that they do not incorporate any additional knowledge and they simplify the data, often by providing structure.

One set of approaches (principal component analysis, singular value decomposition, multidimensional scaling) reduces the data to a small number of variables that can be readily visualized [166]. As a large number of variables are reduced into two to three dimensions, such analysis may both reveal and obscure important relationships. The failure to capture known differences within the data (based on the design, for example) should be considered carefully.

A popular unsupervised method of data analysis is clustering, and there are numerous approaches and software solutions available [167–169]. The two most common methods are hierarchical and centroid clustering. With hierarchical clustering, all genes (and samples) start in an individual cluster and are joined sequentially by a measure of similarity. This joining generates a tree structure that is familiar to most biologists. Centroid clustering groups the data to a set of defined groups that are specified by the experimenter. In this method, the data is grouped iteratively within these groups based on some function. The best known of the centroid approaches is the k-means method, which seeks to minimize the average distance of the samples from the centroid. The Self Organizing Map (SOM) is another centroid method where the samples influence the location of the neighboring clusters [168]. Critically, there is no right method or set of parameters for clustering [170]. This point is well illustrated by Leung and Cavalieri, in which a variety of clustering parameters are applied to reveal different underlying relationships within the data [171]. There is considerable debate as to how to evaluate clustering results, as well [166, 172, 173]. Thus, we recommend that the researcher apply a variety of clustering approaches using a number of parameters and closely examine the results.

The co-identification of genes involved in a specific pathway can yield useful information as to the molecular mechanism and is implicit in the interpretation of clustering approaches. The principle underlying an examination at the pathway level is existence of predefined classes of genes. A wide variety of classifications exist with a growing number of public resources, including (1) the Gene Ontology (GO) database [174, 175], (2) *Kyoto Encyclopedia for Genes and Genomes* (KEGG) [176, 177], (3) *Encyclopedia of Metabolic Pathways MetaCyc* [178], (4) Reactome [179], and (5) Biocarta (http://www.biocarta.com/). *Nucleic Acids Research* each year publishes an online resource of publicly available biology databases that includes summaries [147] and is an invaluable resource.

Although identifying a pathway might be done by visual inspection, more comprehensive methods that incorporate a variety of annotations are being developed. Although there are still a large number of uncharacterized (and therefore incompletely annotated) genes, by combining the clustering approaches discussed above with pathway mapping, one can potentially identify new members of a pathway that can be tested. Unfortunately for *Xenopus* and other model organisms, essentially all of the tools are designed for analysis in mouse or human. Thus, a key requirement for performing such analysis is the ability to translate the gene identities between organisms. Although NCBI provides some tools to do this task (e.g., Homologene, Homologous Unigene), they do not handle batch methods well. For *Xenopus*, we have developed a system to perform such batch translations as part of a larger EST analysis project (XenDb) [180].

With the appropriate gene identifiers, a variety of informatic methods are available for mapping array data to pathways [147, 181]. One example is Gene Microarray Pathway Profiler (GenMAPP), which is a program for displaying a pathway overlaid with color-coded array expression information [182]. The software includes MAPPBuilder, which allows pathways to be built and existing pathways to be modified. A second program, MAPPFinder, integrates gene expression data with GO terms and determines which pathways have statistically significant numbers of regulated genes [183]. The output can then be displayed in GenMAPP. The Database for Annotation, Visualization, and Integrated Discovery (DAVID [184]) brings together data from the KEGG, GO, PFAM, and other databases and searches for common functional categories using EASE [185]. A limitation of these approaches is that they currently do not incorporate statistical tests and significance information [181]. This limitation is beginning to be addressed by a variety of groups; however, these are focused on the yeast system and include Pathway Processor [186], GeneMerge [187], and GeneXpress [188], which integrates cluster analysis, promoter analysis, and biological process data. Clearly, researchers should be prepared to reanalyze their data periodically as more tools are developed.

A final area of interest is the identification of common regulatory elements that may control the coordinated expression of the gene set. As with previous questions, there are a growing number of tools available, but these tools are primarily directed towards mouse and human analysis. The completion of draft sequences for *Xenopus tropicalis* (http://genome.jgi-psf.org/Xentr4/Xentr4.home.html), zebrafish (http://www. sanger.ac.uk/Projects/D_rerio/), chicken [189], and others (http://www.ensembl.org/index.html) allows the identification of conserved regulatory sequences to be identified. rVista provides precomputed alignments for multiple genomes integrated with transcription factor binding information [190, 191] and identifies evolutionary conserved regions (ECRs). The availability of multiple genomes allows for phylogenetic analysis of regulatory sequences [192–194], suggesting that examining the set of sequences identified by array among multiple species may allow improved detection of important regulatory elements. A general discussion of computational methods to identify regulatory elements has been recently reviewed [195].

The first step to a regulatory element analysis is the extraction of the appropriate sequences from the genome. A number of batch extraction tools are available for this (again, targeted to mouse and human) such as BEARR [196], EZ-Retrieve [197], TRED [198], and others. The result of such searches provides the set of promoter elements in FASTA format for subsequent analysis.

With the appropriate promoter elements in hand, there are two general approaches for examining the promoter: identifying known transcription factor binding sites and using the sequences to identify conserved motifs (i.e., motif discovery). The BEARR and TRED programs provide tools to search for defined binding sites. TRED uses the JASPAR transcription factor database [199] or can search for a user-inputted motif, whereas BEARR requires consensus binding sites to be provided (e.g., TRANSFAC [200, 201]). The Promoter Analysis and Interaction Network Toolset (PAINT) program uses a gene identifier list (mouse, human, rat) to fetch the promoter regions, followed by a factor binding site search using the TRANSFAC database [202]. One might be interested in searching for over-represented transcription factor binding sites using OTFBS [203].

The various motif discovery methods employ either word counting methods [204–208] or probabilistic sequence models [209–215]. For word counting-based methods, the selection of the appropriate background model (e.g., nonrelated, nonregulated sequences) is important to identify and minimize false positive motifs [216–219]. Multiple Expectation-maximization for Motif Elicitation (MEME) [220] allows one to discover conserved regions from groups of related sequences using an iterative model. MELINA integrates multiple search methods (CONSENSUS, MEME, GIBBS sampler, CORESEARCH) into a single front end [221]. POBO combines promoter searching within co-regulated genes with bootstrapping methods to detect over- and under-represented motifs [222]. CONFAC allows hundreds of genes to be searched and incorporates a comparison between mouse and human genomes and statistical testing [223]. As transcription factor binding sites are often found grouped, efforts are being focused on identifying motifs that are separated by small gaps [205, 208, 224]. A variety of sites incorporate multiple tools to aid in the acquisition and analysis of regulatory sequences [225–228]. In summary, the growth of genomic sequence from multiple organisms coupled with a growing set of analysis tools should allow the identification and testing of regulatory regions controlling the co-expression of genes identified in microarray assays.

XVI. CONFIRMATION AND FOLLOW-UP ANALYSIS OF MICROARRAYS

Although the major desired outcome of the various sections above is to accurately identify all of the regulated genes in one's pathway or system, one will still remain with both false negative and false positive targets. With a proper experimental design and statistical analysis, one can have reasonable confidence in an expected number of false results. By definition, however, one cannot identify which array targets they are. Thus, independent confirmation of array results is critical. In addition, there may be additional questions about the observed regulation that will need to be addressed. On a positive note, the array results have normally narrowed the set of regulated genes to a few hundred or less. Unfortunately, many of the follow-up and confirmatory experiments are not as amenable to high-throughput methods.

With a potential regulated set of genes in hand, one should address the following questions: (1) Can the genes be confirmed independently? (2) Are the genes expressed in a temporally and spatially consistent manner *in vivo*? (3) What role do the confirmed genes play in the process under study? This last question is obviously the most critical to the researcher interested in understanding signaling pathways in the embryo. The answer to this final question distinguishes the array paper providing a list of genes from that providing insight into function. It is in some ways unfortunate that microarray studies have been categorized as functional genomics, as there is little direct information regarding function obtained from an array study.

XVII. INDEPENDENT CONFIRMATION OF ARRAY RESULTS

There are a number of methods that can be used to demonstrate gene regulation independently of the array study. The initial method often is the identification of previously identified genes within the array data, and is referred to as *in silico* analysis based on literature and other database searches. An excellent resource is the Pubgene database (www.pubgene.org), which maps known relationships extracted from the public literature. It is important to keep in mind that all known genes may not be identified in an array experiment. Such false negative targets are both expected and predicted. First, a gene may not be present on the array or may not be expressed at high enough levels to be detected with confidence. Alternatively, the target (either oligonucleotide or cDNA) may cross hybridize with a related gene, which can mask the true signal or may not hybridize efficiently, or the hybridization and washing may have been suboptimal [229–231]. Therefore, there are a number of reasons why all known genes may not be identified, and the inability to detect some of these control genes should not result in the dismissal of the array experiment by itself.

The second and most popular method to confirm array results is the use of real-time quantitative PCR (qPCR) methods in its various forms [232–234]. Among the advantages of this method are sensitivity and scalability, as well as its apparent simplicity. On the last point, however, significant care must be exercised, as there are a number of technical issues that the researcher needs to address. Just as with microarray analysis, the amount of starting material needs to be normalized (either by calculation or by design). Specificity must also be addressed both in the choice of qPCR methods and in the design and use of the oligonucleotide primers. Finally, there are important issues regarding the appropriate mathematical treatment of the qPCR data. The following section introduces these various issues, discusses general solutions, and provides references to detailed methods. For extensive discussions on the application of this technique, the gene quantification website is a good resource (http://www.gene-quantification.info/). A number of reviews highlight the major issues that need to be considered for effective application of this method [232, 235–237].

The underlying principle of this method is the measurement during each cycle of PCR amplification of the amount of product using fluorescence. In one chemistry, the amount of product is detected using an intercalating dye (typically, SYBR Green) that fluoresces brightly when bound to double-stranded DNA. A number of alternative chemistries employ a hybridization probe, where a fluorescent signal (from a donor dye) is initially absorbed by a second dye (quencher) by Fluorescence Resonance Energy Transfer (FRET). During PCR, the fluorescence signal is revealed either by the hydrolysis of the oligonucleotide and separation of the donor and quenching dyes (TaqMan™) or by the separation of the donor and quenching dyes by hybridization (Scorpion™, Sunrise™). The FRET-based methods offer a greater degree of specificity than dye-based methods, as they employ a third primer that hybridizes to the product. Dye-based methods are flexible and are considerably less expensive but can be error prone due to the nonspecific binding of the dye. The following discussion addresses the central issues related to the analysis and interpretation of real-time qPCR data.

As noted above, the raw fluorescence data obtained by the two chemistries are essentially similar and reflect four phases of amplification: a linear ground phase (lag phase), an early exponential phase, an exponential phase, and a plateau phase (similar to bacterial growth curves) [238]. The fluorescence measured during the ground phase defines a baseline, and the fractional cycle when the signal rises above this background level is known as the threshold cycle (C_t) or crossing point (C_p) and is related to the amount of starting material. Thus, a comparison of the C_t values obtained for any given product in each of two samples provides a simple measure of relative expression. In theory, one can determine the initial concentration of the unknown by generating a standard curve using serial dilutions of known standards (this is known as absolute quantitation). Alternatively, one can use an internal or external reference sample to calibrate (normalize) two samples and determine an expression ratio (known as relative quantitation). In most cases, relative quantitative methods are used for the confirmation of microarray data.

The first step in the application of qPCR to the confirmation of array data is the preparation of cDNA by reverse transcription (RT). During the preparation of the cDNA during reverse transcription, equivalent amounts of RNA should be used to help minimize initial differences. The quality of the RNA can play a significant role in qPCR analysis, generally increasing the estimated amount of a template but not proportionally for each gene [235, 236]. The quality can be reliably assessed using the LabChip™ (Agilent 2100), which also provides quantitation and requires minimal amounts of sample. Limited amounts of RNA can also be readily measured using a fluorimeter (for example, the TD700 by Turner Designs) or directly using a Nanodrop spectrophotometer that can accurately measure 1 μl of sample. Although the RT and PCR reactions can be combined in a single tube, in most cases a large number of genes will be examined in each sample; therefore, a two-step method where the RT reaction is performed separately is preferred.

Finally, the choice between random primers and Oligo-dT for RT priming can affect the performance of the cDNA in PCR reactions [239, 240]. The use of target-specific primers, which have been suggested to provide the greatest specificity and sensitivity [240], are not practical as this approach requires separate synthesis reactions for each target. Random oligonucleotide priming can lead to multiple priming events for an mRNA target and produces cDNA from ribosomal targets, and the cDNA produced may be linear over a narrower range [236, 239]. Priming with Oligo-dT is more specific than random priming, though it is more sensitive to the quality of the RNA and secondary structure, which might block the progression of the transcriptase. In our hands, we have not observed significant differences between either priming method and generally employ random hexamers. As the majority of the assays performed compare the expression of a gene between two or more samples and not the expression of genes within a sample, differences in copy numbers generated due to priming effects [239] are likely to be similar in magnitude in each sample.

As noted above, using equivalent amounts of high-quality RNA in the RT reaction is important for comparing expression between two samples, but it does not guarantee equivalence between the samples. Most RT-qPCR approaches include a normalization procedure to adjust for potential differences introduced during the RT reaction (or due to intrinsic differences in the RNA). Although microarray data in general no longer uses the concept of housekeeping genes for normalization, RT-qPCR often relies on the expression of a nonregulated gene (or genes) for both normalization and reference. A growing body of evidence suggests that using a single reference gene is likely to be inappropriate [241–243]. To minimize the potential variation due to a single gene, we have developed a set of 12 primer pairs that are expressed at relatively consistent levels in the *Xenopus* embryo during development (presented in Table 10.2). In the appropriate selection of such primer sets, the method of Vandesompele et al. is useful for identifying the most stable genes for use as a control [241–250]. Alternatively, one might determine the number of cells in the initial sample and demonstrate that the selected genes do not vary with respect to genome copy number. Whereas in most cases this may be appropriate, this result can be skewed by a differential abundance of dividing cells.

Following the identification of nonregulated genes, a qPCR analysis is performed to determine the appropriate scalar factor for normalization and applied either by using appropriate dilutions of cDNA in the test analysis or by correcting the C_t values following qPCR.

$$\frac{N_A}{N_B} = K_{RS} \frac{\eta_B \left(1+E_B\right)^{Ct_B-1}}{\eta_A \left(1+E_A\right)^{Ct_A-1}}$$

where N = RT yield; C_t = cycle threshold; E = PCR efficiency; K_{RS} = relative qPCR sensitivity; and η = initial number of molecules.

The second major issue faced in qPCR-based mRNA quantitation has been recognized and addressed by a variety of researchers: PCR efficiency and calculation of mRNA differences [251–255]. The concept of efficiency refers to the fact that the PCR product accumulates at a rate less than the doubling at each cycle as predicted by PCR. Efficiency can be determined directly from the slope of a dilution series [256–258] or estimated by increases in the fluorescence during cycling [253, 259–262]. Alternatively, the fluorescence data can be estimated by fitting the data to mathematical models of PCR amplification [253, 263]. To aid in this process, a number of software solutions have been made available, including REST [258], Q-Gene [264], Linreg [260], and DART-PCR [255]. Irrespective of the method selected, it is important to take PCR efficiency into account to get accurate results. This can be illustrated by considering the equation controlling the relative amounts of product N in two samples, A and B [257]. K_{RS} reflects the difference in the probes' fluorescence and binding efficiencies in the two assays, and N is the number of target DNA molecules present at threshold. Importantly, the impact of differences in efficiency (E) are inflated by the exponential of the threshold value, and larger effects are observed for products that are represented at lower initial copy numbers (e.g., higher threshold cycle numbers).

TABLE 10.2
Real-Time PCR Primers for Sample Normalization

Name	Unigene	Gene Name	Length	Upstream Primer/Downstream Primer	Melt	DimerMelt
Elongation factor 1 alpha, oocyte form	X.23	eef1a-o1	230	TGGTGTTGGAGAATTTGAGG CAGATATTGGCACAAATGGAA	82.1°C	67.4°C
Alpha Tubulin	Xl.33489	Cicn03	245	CTACCCCCGTATCCATTTCC ACTGGATAGTGCGCTTGGTT	84.3°C	n/d
Ribosomal protein L13a	Xl.36182	rpl13a-prov	239	CTGGCAACTTCTACCGCAAC AGCAGCAGGAACCACCATAC	85.3°C	70.3°C
Beta-Actin	Xl.4138	ActB	250	GTCATGGACTCAGGTGATGGT ATGAAGAAGAGGCAGCTGTGG	84.2°C	73.7°C
Ornithine decarboxylase 1	Xl.4223	odc1-prov	245	AGCCTCCTTGGGAGTGAGA GTCGCCAAGATCAGCAACA	82.4°C	75.6°C
Beta-Tubulin at 56D	Xl.7933	betatub56d-prov	239	GCACCAGCTCGTAGAGAACA CAGGCATGAAGAAGTGCAGA	85.6°C	n/d
Glyceraldehyde-3-phosphate dehydrogenase	Xl.995	gapd-prov	197	TGGTTCACGCCAACTACAAA CAAGTCATCACCGTCTTCCA	85.5°C	n/d
Histone H2A	Xl.31145	HIST1H2AH	165	TGGTCGTGTTCATCGTCTGT CCTGGGGATAATACGGGTCT	84.5°C	n/d
TAF12 RNA polymerase II, TATA box binding protein	xl.6374	taf12-A-prov	170	CCTCATACCCCAAAACAACG ATGGCTGGTTCCTGTTTGAC	83.8°C	73.4°C
Succinate dehydrogenase complex, subunit A, flavoprotein (Fp)	xl.1718	sdha-prov	173	GTGGAATACTTTGCCCTGGA CCAGTGCAGGTATGAGCAGA	81.9°C	n/d
Tyrosine 3-monooxygenase/tryptophan 5 monooxygenase activation protein	xl.4573	ywhaz-prov	193	AGATATCGTGGGCCAGTCAC AAGCTCAGCAATGGCTTCAT	82.6°C	70.6°C
Ubiquitin C	xl.29333	ubc-prov	215	TCAAGGCCAAAATCCAAGAC ATCGCTAAGCTCAACCTCAA	82.3°C	78.2°C
Histone H4 Histone 1, H4l	Xl.48317	hist1h4l-prov	175	AAGCGCATCTCAGGTCTCAT CTCCGAAGCCGTAGAGAGTG	85.4°C	68°C

n/d = Not detected/generated.

XVIII. SPATIAL AND TEMPORAL ANALYSIS OF GENE EXPRESSION OF ARRAY GENES

A critical issue to consider when addressing the relevance of the regulated genes identified by microarray is the whether the genes are expressed appropriately to be involved in the process being examined. Given the repeated use of various factors and pathways in diverse developmental functions, it is possible that true regulated genes may be identified but outside the context under consideration. This issue has been well understood in the field and is addressed by examining gene expression *in situ*, often in whole mount assays. For array experiments, both the untreated and treated embryos are important, as the normal pattern of expression can reveal a potential role in the normal process, wherease the expression in Gain of Function (GOF) or Loss of Function (LOF) embryos can confirm the array regulation by either ectopic or decreased expression. Such misexpression studies may also provide insight into the mechanisms of the gene or pathway in question, as the observed phenotypes may in fact result secondarily from a regulated gene.

XIX. FUNCTIONAL TESTING OF REGULATED GENES IDENTIFIED BY ARRAY ANALYSIS

For most researchers, the overall research objective is to determine not only the set of regulated genes but also their role in the process in question. Thus, the limited ability to pursue confirmed regulated genes following a high-throughput screen can be frustrating. In mouse, experiments examining the function of a given gene involve considerable expense, labor, and time. High-throughput loss of function using RNAi is available for *C. elegans* [265–267], though rapid gain of function methods have not been described. In *Drosophila*, transgenic flies can be generated using the GAL4-UAS system in about 2–3 months, and RNAi studies can be performed in whole embryos using microinjection. In *Xenopus*, the microinjection approach provides a reasonably rapid method to assess gene function using gain and loss of function approaches [62, 120]. Similar approaches are possible in zebrafish.

In vertebrate embryos, gain and loss of function approaches allow the functional significance of a regulated gene identified by microarray to be assessed. In its simplest form, gain of function of a regulated gene can be assayed for its ability to recapitulate some or all of the activities known for the regulator by which it was identified. Similarly, by loss of function (both in the presence and absence of the regulator), the necessity of the gene can be examined. The ability of the regulated gene to cooperate in the embryonic pathway may be assessed by co-injection with subthreshold levels of the regulator. Thus, an additional factor, not normally present but rather co-induced by the original treatment, may be present to complement a gene's activity.

The array results allow a more direct approach to the question of complementation of co-induced activities. As noted above, clustering of array experiments and mapping of regulated genes to functional categories (such as Gene Ontologies) can suggest potential molecular mechanisms for gene activities. Such mapping can also suggest functional experiments in which a set of genes (linked by pathway, ontology, or cluster) are pooled and co-expressed by microinjection. As with the approaches described above, this can be done in the presence or absence of subthreshold levels of other gene activities. Similar pooled approaches have been successful in a variety of functional screens using random pools. Among one of the largest groups of genes identified in array screens are the set of conserved genes with no known function. These genes can be pooled and assayed as noted above. The fact that they were identified by a known treatment or gene activity immediately suggests potential phenotypes that might be sought.

XX. CONCLUSIONS AND SUMMARY

This chapter attempts to bring together a wide range of issues to be considered in the contemplation and implementation of a genomic approach to embryonic signaling and development. In each section, we have attempted to provide references to comprehensive treatments of each topic and highlight the major points to be considered. A clear hypothesis or question combined with adherence to established methods is critical to the success of the experiments and can significantly simplify the analysis of the large data set. International cooperation between both academic and commercial groups (MGED) has defined standards for sharing of data required by a growing number of journals (MIAME) [70]. Issues relating to the preparation of sample through initial data extraction are considered, and preferred approaches recommended. Higher order analysis of array data is discussed, including various statistical approaches and methods for clustering the data. It is suggested that scientists establish working relationships with statisticians and other experts, as needed. Finally, the adoption of a well-supported database system is recommended to not only enable tracking and organization of the data, but also take advantage of new approaches in data analysis that get regularly incorporated.

With the array data fully analyzed, normalized, clustered, and categorized, the researcher faces the task of independently confirming and assessing the functional significance of the results. The current method of choice for array confirmation is the real-time PCR approach due to its sensitivity and scalability. As with microarrays themselves, there remain a number of outstanding issues that should be carefully addressed when using this method. These issues are highlighted, and references to more extensive discussions of qPCR methods are provided. Finally, some ideas and approaches to following up the array results are suggested.

METHOD 1: RNA ISOLATION BY METHOD OF SAACHI AND CHOMCYNSKI

(For 100 animal caps, 15–20 embryos)

1. Add 600 μl of Solution D* and homogenize.
2. Add 60 μl of 2 M sodium acetate (pH 4.0) and mix.
3. Add 600 μl of acid-phenol and mix.
4. Add 120 μl of Chroloform/IAA and mix.
5. Put on ice for 15 minutes.

6. Centrifuge 14,000 rpm for 20 minutes at 4°C.
7. Transfer the supernatants to new tubes (if solution did not separate into two phases after centrifuging, add 120 μl of Chroloform/IAA and repeat steps 4 to 7).
8. Add 600 μl of Chroloform/IAA and mix.
9. Centrifuge 14,000 rpm for 20 minutes at 4°C.
10. Transfer the supernatants (usually 600 μl) to new tubes.
11. Add same volume of 2-propanol as the supernatants.
12. Shake tubes gently.
13. Leave at –20°C for 1 hour or longer.
14. Centrifuge 14,000 rpm at 4°C for 20 minutes.
15. Discard the sup.
16. Dissolve the pellet by 50 μl of DEPC -water (incubate the tube at 65°C for 5 to 10 minutes to dissolve).
17. Add 200 μl of 5 M LiCl.
18. Leave at –20°C for 12 hours or longer.
19. Centrifuge 14,000 rpm at 4°C for 20 minutes and discard the sup.
20. Add 1 ml of 75% EtOH.
21. Centrifuge 14,000 rpm at 4°C for 5 minutes and discard the sup.
22. Add 50 μl of DEPC-water and incubate the tube at 65°C for 5 to 10 minutes to dissolve.
23. Use 1 μl of RNA solution to measure O.D.
24. Add 7 μl of 3 M sodium acetate and mix.
25. Add 170 μl of 100% EtOH and mix.
26. Leave at –80°C for 1 hour or longer.
27. Centrifuge 14,000 rpm at 4°C for 20 minutes and discard the sup.
28. Dissolve the RNA by DEPC-water with a concentration greater than 4 μg/μl.

Solution D*

Stock solution: Mix 29.3 ml DEPC-water, 1.76 ml 0.75 M sodium citrate, pH 7.0, and 2.64 ml of 10% Sarkosyl. Add 25 g guanidine thiocyanate and stir at 60–65°C to dissolve. Store up to 3 months at room temperature.

Working solution: Add 350 μl 2-mercaptoethanol to 50 ml stock solution above. Store up to 1 month at room temperature.

PROBE PREPARATION, HYBRIDIZATION, AND WASHING

Method of Tetsuya Koide 06.3.28 Edition

Reverse Transcription (Oligo(dT) and Aminoallyl)

1. Do ethanol precipitation of 50 μg of total RNA and resuspend the RNA in 27 μl of DEPC-water.

2. Combine:

Total RNA of interest (50 μg)	27 μl
T_{20}VN (20 μM)	5.0 μl

3. Incubate at 70°C for 10 minutes.
4. Chill on ice for 2 minutes.
5. Combine:

5X RT buffer	10 μl
0.1 M DTT	2.5 μl
Labeling dNTP mix (20X)	2.5 μl
RNase inhibitor (Promega)	0.5 μl
Superscript III	2.5 μl
Total volume	50 μl

6. Incubate at 50°C for 2 hours.

Labeling dNTP Mix 20X 100 μl	Stock	Volumn	[Final]
DATP	100 mM	10 μl	10 mM
DGTP	100 mM	10 μl	10 mM
DCTP	100 mM	10 μl	10 mM
DTTP	100 mM	2.0 μl	2 mM
Amino allyl dUTP	64.2 mM	12.46 μl	8 mM

PURIFICATION OF PROBE (USING QIA QUICK COLUMN)

1. Add 5 μl of 2.5 M NaOH to degrade RNA.
2. Incubate at 65°C for 45 minutes.
3. Spin briefly.
4. Add 25 μl of 2 M Hepes to neutralize reaction.
5. Add 500 μl of buffer PB into QIA Quick Column, add cDNA to this column, and mix well.
6. Centrifuge for 1 minute. Save the flow-through and pour back to the same column.
7. Centrifuge for 1 minute and discard the flow-through.
8. Add 750 μl of wash buffer.
9. Centrifuge for 1 minute and discard the flow-through.
10. Add 750 μl of wash buffer.
11. Centrifuge for 1 minute and discard the flow-through.
12. Centrifuge for 1 minute to remove residual wash buffer.
13. Add 30 μl of water (pH 7–8.5) to the center of the column (elution 1).
14. Incubate for 1 minute and centrifuge for 1 minute.
15. Add 30 μl of water (pH 7–8.5) to the center of the column (elution 2).
16. Incubate for 1 minute and centrifuge for 1 minute.
17. Air-dry cDNA solution completely by SpeedVac with heater on. (Option: use of Microcon YM-30.)
18. Resuspend the cDNA with 15 μl of water. (If you have two samples, you can decrease the volume to 7.5 μl.)

19. Open an aliquote of Cy dye (or Alexa dyes), add 15 μl of 0.1 M $NaHCO_3$ (pH 9.0), and combine Cy dye and cDNA solution (1:1).
20. Incubate the mixture for 1 hour at room temperature in the dark.

PREHYBRIDIZATION OF GLASS SLIDE

1. Prepare the prehybridization solution.

Final conc.	For 50 ml solution
5X SSC	12.5 ml of 20X SSC
1% SDS	0.5 ml of 10% SDS
1% BSA	0.5 g of BSA (Sigma)

2. Incubate prehybridization solution at 42°C.
3. Preparation of slideglass:
   Mark the area of DNA spots on the side of the spots with a diamond pen and the date.
   Steam the side of the spots above the hot water quickly and dry on the heat block at about 80°C immediately.
   UV-crosslink the slides with the 1500 μJ of the energy.
4. Prehybridize slides at 42°C for more than 1 hour in a waterbath.
5. After prehybridization, wash slides in mQ water in glass box five times each for 10 seconds and wash in isopropanol in plastic slidecase once for 10 seconds.
6. Spin to remove residual isopropanol at 1000 rpm for 2 minutes at room temperature. Store slideglasses in blackbox.

PURIFICATION AND HYBRIDIZATION OF PROBE

1. Add 15 (or 7.5) μl of 4 M hydroxylamine hydrochloride and incubate the mixture for 15 minutes at room temperature in the dark.
2. Purify probe same as previously on Qia Quick Column.
3. Add 30 μl of buffer EB. Incubate for 1 minute at room temperature. Centrifuge for 1 minute.
4. Add 30 μl of buffer EB. Incubate for 1 minute at room temperature. Centrifuge for 1 minute, save residual solution from the inside of column.
5. Check the quality of probe with UV spectrometer.
6. Concentrate the eluted probe to 18 μl using Microcon-30.
7. Prepare 2X hybridization mix as follows:

20X SSC	17 μL
10 μg/μl yeast tRNA	5 μL
Water	27 μl
10% SDS	1 μl
Probe	17 μl

8. Mix with 17 µl of 2X hybridization buffer.
9. Heat the probe for 5 minutes at 95°C.
10. Spin 2 minutes at room temperature.
11. Apply 32 µl of hybridization solution.
12. Place the side of the DNA spots to cover slip so as not to put any bubbles. Flip it quickly.
13. Apply 13 µl of 3X SSC on the two points of the slideholder and on the three points of the slide.
14. Close the slideholder and wrap it with aluminum foil.
15. Place a ring weight on the slideholder and sink it in water bath at 65°C.
16. Incubate overnight.

WASHING (USE VIGOROUS SHAKING)

1. Prewarm the solution 1 at 42°C overnight.
2. 2X SSC, 0.1% SDS for 5 minutes at 42°C (60 ml of 20X SSC, 6 ml of 10% SDS, water 534 ml).
3. 0.1X SSC, 0.1% SDS for 10 minutes at room temperature (2 ml of 20X SSC, 4 ml of 10% SDS, water 394 ml).
4. 0.1X SSC for 1 minute at room temperature four times (8 ml of 20X SSC, water 1592 ml).
5. Spin the slides to remove the residual solution at 700 rpm for 2 minutes at room temperature.
6. Keep in dark till scanning (perform as soon as possible).

REFERENCES

1. Schena M, Shalon D, Davis RW, Brown PO: Quantitative monitoring of gene expression patterns with a complementary DNA microarray [see comments]. *Science* 1995, 270: 467–470.
2. Fodor SP, Read JL, Pirrung MC, Stryer L, Lu AT, Solas D: Light-directed, spatially addressable parallel chemical synthesis. *Science* 1991, 251: 767–773.
3. White KP, Rifkin SA, Hurban P, Hogness DS: Microarray analysis of Drosophila development during metamorphosis. *Science* 1999, 286: 2179–2184.
4. Schena M: *DNA Microarrays*. Oxford: Oxford University Press; 1999.
5. Schena M: *Microarray Analysis*. Hoboken, NJ: Wiley-Liss; 2003.
6. *DNA Microarrays: A Molecular Cloning Manual*. Bowtell, D. and Sambrook, J. (eds) Cold Spring Harbor, NY: Cold Spring Harbor Laboratory Press; 2003.
7. Pichler FB, Black MA, Williams LC, Love DR: Design, normalization, and analysis of spotted microarray data. *Methods Cell Biol* 2004, 77: 521–543.
8. Qian F, Zhen F, Ong C, Jin SW, Meng SH, Stainier DY et al.: Microarray analysis of zebrafish cloche mutant using amplified cDNA and identification of potential downstream target genes. *Dev Dyn* 2005, 233: 1163–1172.
9. Stickney HL, Schmutz J, Woods IG, Holtzer CC, Dickson MC, Kelly PD et al.: Rapid mapping of zebrafish mutations with SNPs and oligonucleotide microarrays. *Genome Res* 2002, 12: 1929–1934.

10. Altmann CR, Bell E, Sczyrba A, Pun J, Bekiranov S, Gaasterland T et al.: Microarray-based analysis of early development in Xenopus laevis. *Dev Biol* 2001, 236: 64–75.
11. Arima K, Shiotsugu J, Niu R, Khandpur R, Martinez M, Shin Y et al.: Global analysis of RAR-responsive genes in the Xenopus neurula using cDNA microarrays. *Dev Dyn* 2005, 232: 414–431.
12. Baldessari D, Shin Y, Krebs O, Konig R, Koide T, Vinayagam A et al.: Global gene expression profiling and cluster analysis in Xenopus laevis. *Mech Dev* 2005, 122: 441–475.
13. Chung HA, Hyodo-Miura J, Kitayama A, Terasaka C, Nagamune T, Ueno N: Screening of FGF target genes in Xenopus by microarray: temporal dissection of the signalling pathway using a chemical inhibitor. *Genes Cells* 2004, 9: 749–761.
14. Peiffer DA, von Bubnoff A, Shin Y, Kitayama A, Mochii M, Ueno N et al.: A Xenopus DNA microarray approach to identify novel direct BMP target genes involved in early embryonic development. *Dev Dyn* 2005, 232: 445–456.
15. Shin Y, Kitayama A, Koide T, Peiffer DA, Mochii M, Liao A et al.: Identification of neural genes using Xenopus DNA microarrays. *Dev Dyn* 2005, 232: 432–444.
16. Taverner NV, Kofron M, Shin Y, Kabitschke C, Gilchrist MJ, Wylie C et al.: Microarray-based identification of VegT targets in Xenopus. *Mech Dev* 2005, 122: 333–354.
17. Afrakhte M, Schultheiss TM: Construction and analysis of a subtracted library and microarray of cDNAs expressed specifically in chicken heart progenitor cells. *Dev Dyn* 2004, 230: 290–298.
18. Hackam AS, Bradford RL, Bakhru RN, Shah RM, Farkas R, Zack DJ et al.: Gene discovery in the embryonic chick retina. *Mol Vis* 2003, 9: 262–276.
19. Cogburn LA, Wang X, Carre W, Rejto L, Porter TE, Aggrey SE et al.: Systems-wide chicken DNA microarrays, gene expression profiling, and discovery of functional genes. *Poult Sci* 2003, 82: 939–951.
20. van Hemert S, Ebbelaar BH, Smits MA, Rebel JM: Generation of EST and microarray resources for functional genomic studies on chicken intestinal health. *Anim Biotechnol* 2003, 14: 133–143.
21. Kimura T, Jindo T, Narita T, Naruse K, Kobayashi D, Shin I et al.: Large-scale isolation of ESTs from medaka embryos and its application to medaka developmental genetics. *Mech Dev* 2004, 121: 915–932.
22. Katogi R, Nakatani Y, Shin-i T, Kohara Y, Inohaya K, Kudo A: Large-scale analysis of the genes involved in fin regeneration and blastema formation in the medaka, Oryzias latipes. *Mech Dev* 2004, 121: 861–872.
23. Stathopoulos A, Levine M: Whole-genome analysis of Drosophila gastrulation. *Curr Opin Genet Dev* 2004, 14: 477–484.
24. Gupta V, Oliver B: Drosophila microarray platforms. *Brief Funct Genomic Proteomic* 2003, 2: 97–105.
25. Montalta-He H, Reichert H: Impressive expressions: developing a systematic database of gene-expression patterns in Drosophila embryogenesis. *Genome Biol* 2003, 4: 205.
26. Stathopoulos A, Levine M: Whole-genome expression profiles identify gene batteries in Drosophila. *Dev Cell* 2002, 3: 464–465.
27. Lee J, Nam S, Hwang SB, Hong M, Kwon JY, Joeng KS et al.: Functional genomic approaches using the nematode Caenorhabditis elegans as a model system. *J Biochem Mol Biol* 2004, 37: 107–113.
28. Schlecht U, Primig M: Mining meiosis and gametogenesis with DNA microarrays. *Reproduction* 2003, 125: 447–456.

29. GuhaThakurta D, Palomar L, Stormo GD, Tedesco P, Johnson TE, Walker DW et al.: Identification of a novel cis-regulatory element involved in the heat shock response in Caenorhabditis elegans using microarray gene expression and computational methods. *Genome Res* 2002, 12: 701–712.
30. Kim SK, Lund J, Kiraly M, Duke K, Jiang M, Stuart JM et al.: A gene expression map for Caenorhabditis elegans. *Science* 2001, 293: 2087–2092.
31. Chen HW, Chen JJ, Yu SL, Li HN, Yang PC, Su CM et al.: Transcriptome analysis in blastocyst hatching by cDNA microarray. *Hum Reprod* 2005, 20: 2492–2501.
32. Wang QT, Piotrowska K, Ciemerych MA, Milenkovic L, Scott MP, Davis RW et al.: A genome-wide study of gene activity reveals developmental signaling pathways in the preimplantation mouse embryo. *Dev Cell* 2004, 6: 133–144.
33. Wang S, Cowan CA, Chipperfield H, Powers RD: Gene expression in the preimplantation embryo: in-vitro developmental changes. *Reprod Biomed Online* 2005, 10: 607–616.
34. Kawamura K, Kawamura N, Mulders SM, Sollewijn G, Hsueh AJ: Ovarian brain-derived neurotrophic factor (BDNF) promotes the development of oocytes into preimplantation embryos. *Proc Natl Acad Sci U S A* 2005, 102: 9206–9211.
35. Maekawa M, Yamamoto T, Tanoue T, Yuasa Y, Chisaka O, Nishida E: Requirement of the MAP kinase signaling pathways for mouse preimplantation development. *Development* 2005, 132: 1773–1783.
36. Tanaka TS, Jaradat SA, Lim MK, Kargul GJ, Wang X, Grahovac MJ et al.: Genome-wide expression profiling of mid-gestation placenta and embryo using a 15,000 mouse developmental cDNA microarray. *Proc Natl Acad Sci U S A* 2000, 97: 9127–9132.
37. Afshari CA, Nuwaysir EF, Barrett JC: Application of complementary DNA microarray technology to carcinogen identification, toxicology, and drug safety evaluation. *Cancer Res* 1999, 59: 4759–4760.
38. Mikheeva S, Barrier M, Little SA, Beyer R, Mikheev AM, Kerr MK et al.: Alterations in gene expression induced in day-9 mouse embryos exposed to hyperthermia (HS) or 4-hydroperoxycyclophosphamide (4CP): analysis using cDNA microarrays. *Toxicol Sci* 2004, 79: 345–359.
39. Nemeth KA, Singh AV, Knudsen TB: Searching for biomarkers of developmental toxicity with microarrays: normal eye morphogenesis in rodent embryos. *Toxicol Appl Pharmacol* 2005, 206: 219–228.
40. Sulik KK: Genesis of alcohol-induced craniofacial dysmorphism. *Exp Biol Med (Maywood)* 2005, 230: 366–375.
41. Chauhan BK, Reed NA, Zhang W, Duncan MK, Kilimann MW, Cvekl A: Identification of genes downstream of Pax6 in the mouse lens using cDNA microarrays. *J Biol Chem* 2002, 277: 11539–11548.
42. Schwab K, Patterson LT, Aronow BJ, Luckas R, Liang HC, Potter SS: A catalogue of gene expression in the developing kidney. *Kidney Int* 2003, 64: 1588–1604.
43. Park YK, Franklin JL, Settle SH, Levy SE, Chung E, Jeyakumar LH et al.: Gene expression profile analysis of mouse colon embryonic development. *Genesis* 2005, 41: 1–12.
44. Takasato M, Osafune K, Matsumoto Y, Kataoka Y, Yoshida N, Meguro H et al.: Identification of kidney mesenchymal genes by a combination of microarray analysis and Sall1-GFP knockin mice. *Mech Dev* 2004, 121: 547–557.
45. Vasseur S, Hoffmeister A, Garcia-Montero A, Barthet M, Saint-Michel L, Berthezene P et al.: Mice with targeted disruption of p8 gene show increased sensitivity to lipopolysaccharide and DNA microarray analysis of livers reveals an aberrant gene expression response. *BMC Gastroenterol* 2003, 3: 25.

46. Wang G, Zhang J, Moskophidis D, Mivechi NF: Targeted disruption of the heat shock transcription factor (hsf)-2 gene results in increased embryonic lethality, neuronal defects, and reduced spermatogenesis. *Genesis* 2003, 36: 48–61.
47. Kelly DL, Rizzino A: DNA microarray analyses of genes regulated during the differentiation of embryonic stem cells. *Mol Reprod Dev* 2000, 56: 113–123.
48. He X, Gonzalez V, Tsang A, Thompson J, Tsang TC, Harris DT: Differential gene expression profiling of CD34+ CD133+ umbilical cord blood hematopoietic stem progenitor cells. *Stem Cells Dev* 2005, 14: 188–198.
49. Bouchard M, Grote D, Craven SE, Sun Q, Steinlein P, Busslinger M: Identification of Pax2-regulated genes by expression profiling of the mid-hindbrain organizer region. *dev* 2005, 132: 2633–2643.
50. Koide T, Hayata T, Cho KW: Xenopus as a model system to study transcriptional regulatory networks. *Proc Natl Acad Sci U S A* 2005, 102: 4943–4948.
51. Levine M, Davidson EH: Gene regulatory networks for development. *Proc Natl Acad Sci U S A* 2005, 102: 4936–4942.
52. Davidson EH, Rast JP, Oliveri P, Ransick A, Calestani C, Yuh CH et al.: A genomic regulatory network for development. *Science* 2002, 295: 1669–1678.
53. Miyazono K, Maeda S, Imamura T: BMP receptor signaling: Transcriptional targets, regulation of signals, and signaling cross-talk. *Cytokine Growth Factor Rev* 2005, 16: 251–263.
54. Kessler DS: Siamois is required for formation of Spemann's organizer. *Proc Natl Acad Sci U S A* 1997, 94: 13017–13022.
55. Gawantka V, Pollet N, Delius H, Vingron M, Pfister R, Nitsch R et al.: Gene expression screening in xenopus identifies molecular pathways, predicts gene function and provides a global view of embryonic patterning [In Process Citation]. *Mech Dev* 1998, 77: 95–141.
56. Pollet N, Muncke N, Verbeek B, Li Y, Fenger U, Delius H et al.: An atlas of differential gene expression during early Xenopus embryogenesis. *Mech Dev* 2005, 122: 365–439.
57. Pollet N, Delius H, Niehrs C: In situ analysis of gene expression in Xenopus embryos. *C R Biol* 2003, 326: 1011–1017.
58. Scholpp S, Groth C, Lohs C, Lardelli M, Brand M: Zebrafish fgfr1 is a member of the fgf8 synexpression group and is required for fgf8 signalling at the midbrain-hindbrain boundary. *Dev Genes Evol* 2004, 214: 285–295.
59. Niehrs C, Pollet N: Synexpression groups in eukaryotes. *Nature* 1999, 402: 483–487.
60. Cascio S, Gurdon JB: The initiation of new gene transcription during Xenopus gastrulation requires immediately preceding protein synthesis. *dev* 1987, 100: 297–305.
61. Grainger RM, Gurdon JB: Loss of competence in amphibian induction can take place in single nondividing cells. *Proc Natl Acad Sci U S A* 1989, 86: 1900–1904.
62. Kolm PJ, Sive HL: Efficient hormone-inducible protein function in Xenopus laevis. *Dev Biol* 1995, 171: 267–272.
63. Wang HY, Malek RL, Kwitek AE, Greene AS, Luu TV, Behbahani B et al.: Assessing unmodified 70-mer oligonucleotide probe performance on glass-slide microarrays. *Genome Biol* 2003, 4: R5.
64. Woo Y, Krueger W, Kaur A, Churchill G: Experimental design for three-color and four-color gene expression microarrays. *Bioinformatics* 2005, 21 Suppl 1: i459–i467.
65. Kerr MK, Churchill GA: Experimental design for gene expression microarrays. *Biostatistics* 2001, 2: 183–201.

66. Churchill GA: Fundamentals of experimental design for cDNA microarrays. *Nat Genet* 2002, 32 Suppl.: 490–495.
67. Kerr MK, Churchill GA: Statistical design and the analysis of gene expression microarray data. *Genet Res* 2001, 77: 123–128.
68. Ball CA, Sherlock G, Parkinson H, Rocca-Sera P, Brooksbank C, Causton HC et al.: Standards for microarray data. *Science* 2002, 298: 539.
69. Ball CA, Brazma A, Causton H, Chervitz S, Edgar R, Hingamp P et al.: Submission of microarray data to public repositories. *PLoS Biol* 2004, 2: E317.
70. Brazma A, Hingamp P, Quackenbush J, Sherlock G, Spellman P, Stoeckert C et al.: Minimum information about a microarray experiment (MIAME)—toward standards for microarray data. *Nat Genet* 2001, 29: 365–371.
71. Spellman PT, Miller M, Stewart J, Troup C, Sarkans U, Chervitz S et al.: Design and implementation of microarray gene expression markup language (MAGE-ML). *Genome Biol* 2002, 3: RESEARCH0046.
72. Churchill GA: Fundamentals of experimental design for cDNA microarrays. *Nat Genet* 2002, 32 Suppl.: 490–495.
73. Dobbin K, Shih JH, Simon R: Questions and answers on design of dual-label microarrays for identifying differentially expressed genes. *J Natl Cancer Inst* 2003, 95: 1362–1369.
74. Dobbin KK, Kawasaki ES, Petersen DW, Simon RM: Characterizing dye bias in microarray experiments. *Bioinformatics* 2005, 21: 2430–2437.
75. Martin-Magniette ML, Aubert J, Cabannes E, Daudin JJ: Evaluation of the gene-specific dye bias in cDNA microarray experiments. *Bioinformatics* 2005, 21: 1995–2000.
76. Rosenzweig BA, Pine PS, Domon OE, Morris SM, Chen JJ, Sistare FD: Dye bias correction in dual-labeled cDNA microarray gene expression measurements. *Environ Health Perspect* 2004, 112: 480–487.
77. Yang YH, Dudoit S, Luu P, Lin DM, Peng V, Ngai J et al.: Normalization for cDNA microarray data: a robust composite method addressing single and multiple slide systematic variation. *Nucleic Acids Res* 2002, 30: e15.
78. Chomczynski P, Sacchi N: Single-step method of RNA isolation by acid guanidinium thiocyanate-phenol-chloroform extraction. *Anal Biochem* 1987, 162: 156–159.
79. Hilz H, Wiegers U, Adamietz P: Stimulation of proteinase K action by denaturing agents: application to the isolation of nucleic acids and the degradation of 'masked' proteins. *Eur J Biochem* 1975, 56: 103–108.
80. Wiegers U, Hilz H: A new method using 'proteinase K' to prevent mRNA degradation during isolation from HeLa cells. *Biochem Biophys Res Commun* 1971, 44: 513–519.
81. Karsten SL, Van Deerlin VM, Sabatti C, Gill LH, Geschwind DH: An evaluation of tyramide signal amplification and archived fixed and frozen tissue in microarray gene expression analysis. *Nucleic Acids Res* 2002, 30: E4.
82. Badiee A, Eiken HG, Steen VM, Lovlie R: Evaluation of five different cDNA labeling methods for microarrays using spike controls. *BMC Biotechnol* 2003, 3: 23.
83. Manduchi E, Scearce LM, Brestelli JE, Grant GR, Kaestner KH, Stoeckert CJ, Jr.: Comparison of different labeling methods for two-channel high-density microarray experiments. *Physiol Genomics* 2002, 10: 169–179.
84. Stears RL, Getts RC, Gullans SR: A novel, sensitive detection system for high-density microarrays using dendrimer technology. *Physiol Genomics* 2000, 3: 93–99.
85. Eberwine J, Yeh H, Miyashiro K, Cao Y, Nair S, Finnell R et al.: Analysis of gene expression in single live neurons. *Proc Natl Acad Sci U S A* 1992, 89: 3010–3014.

86. Van Gelder RN, von Zastrow ME, Yool A, Dement WC, Barchas JD, Eberwine JH: Amplified RNA synthesized from limited quantities of heterogeneous cDNA. *Proc Natl Acad Sci U S A* 1990, 87: 1663–1667.
87. Phillips J, Eberwine JH: Antisense RNA amplification: a linear amplification method for analyzing the mRNA population from single living cells. *Methods* 1996, 10: 283–288.
88. Pabon C, Modrusan Z, Ruvolo MV, Coleman IM, Daniel S, Yue H et al.: Optimized T7 amplification system for microarray analysis. *BioTechniques* 2001, 31: 874–879.
89. Jenson SD, Robetorye RS, Bohling SD, Schumacher JA, Morgan JW, Lim MS et al.: Validation of cDNA microarray gene expression data obtained from linearly amplified RNA. *Mol Pathol* 2003, 56: 307–312.
90. Li J, Adams L, Schwartz SM, Bumgarner RE: RNA amplification, fidelity and reproducibility of expression profiling. *C R Biol* 2003, 326: 1021–1030.
91. Li Y, Ali S, Philip PA, Sarkar FH: Direct comparison of microarray gene expression profiles between non-amplification and a modified cDNA amplification procedure applicable for needle biopsy tissues. *Cancer Detect Prev* 2003, 27: 405–411.
92. McClintick JN, Jerome RE, Nicholson CR, Crabb DW, Edenberg HJ: Reproducibility of oligonucleotide arrays using small samples. *BMC Genomics* 2003, 4: 4.
93. Nygaard V, Loland A, Holden M, Langaas M, Rue H, Liu F et al.: Effects of mRNA amplification on gene expression ratios in cDNA experiments estimated by analysis of variance. *BMC Genomics* 2003, 4: 11.
94. Patel OV, Suchyta SP, Sipkovsky SS, Yao J, Ireland JJ, Coussens PM et al.: Validation and application of a high fidelity mRNA linear amplification procedure for profiling gene expression. *Vet Immunol Immunopathol* 2005, 105: 331–342.
95. Polacek DC, Passerini AG, Shi C, Francesco NM, Manduchi E, Grant GR et al.: Fidelity and enhanced sensitivity of differential transcription profiles following linear amplification of nanogram amounts of endothelial mRNA. *Physiol Genomics* 2003, 13: 147–156.
96. Puskas LG, Zvara A, Hackler L, Jr., Van Hummelen P: RNA amplification results in reproducible microarray data with slight ratio bias. *BioTechniques* 2002, 32: 1330–1334, 1336, 1338, 1340.
97. Rudnicki M, Eder S, Schratzberger G, Mayer B, Meyer TW, Tonko M et al.: Reliability of t7-based mRNA linear amplification validated by gene expression analysis of human kidney cells using cDNA microarrays. *Nephron Exp Nephrol* 2004, 97: e86–e95.
98. Saghizadeh M, Brown DJ, Tajbakhsh J, Chen Z, Kenney MC, Farber DB et al.: Evaluation of techniques using amplified nucleic acid probes for gene expression profiling. *Biomol Eng* 2003, 20: 97–106.
99. Schneider J, Buness A, Huber W, Volz J, Kioschis P, Hafner M et al.: Systematic analysis of T7 RNA polymerase based in vitro linear RNA amplification for use in microarray experiments. *BMC Genomics* 2004, 5: 29.
100. Stoyanova R, Upson JJ, Patriotis C, Ross EA, Henske EP, Datta K et al.: Use of RNA amplification in the optimal characterization of global gene expression using cDNA microarrays. *J Cell Physiol* 2004, 201: 359–365.
101. Wilson CL, Pepper SD, Hey Y, Miller CJ: Amplification protocols introduce systematic but reproducible errors into gene expression studies. *BioTechniques* 2004, 36: 498–506.
102. Zhao H, Hastie T, Whitfield ML, Borresen-Dale AL, Jeffrey SS: Optimization and evaluation of T7 based RNA linear amplification protocols for cDNA microarray analysis. *BMC Genomics* 2002, 3: 31.

103. Wadenback J, Clapham DH, Craig D, Sederoff R, Peter GF, von Arnold S et al.: Comparison of standard exponential and linear techniques to amplify small cDNA samples for microarrays. *BMC Genomics* 2005, 6: 61.
104. Ghosh D: Mixture models for assessing differential expression in complex tissues using microarray data. *Bioinformatics* 2004, 20: 1663–1669.
105. Francis P, Fernebro J, Eden P, Laurell A, Rydholm A, Domanski HA et al.: Intratumor versus intertumor heterogeneity in gene expression profiles of soft-tissue sarcomas. *Genes Chromosomes Cancer* 2005, 43: 302–308.
106. Glanzer JG, Haydon PG, Eberwine JH: Expression profile analysis of neurodegenerative disease: advances in specificity and resolution. *Neurochem Res* 2004, 29: 1161–1168.
107. Player A, Barrett JC, Kawasaki ES: Laser capture microdissection, microarrays and the precise definition of a cancer cell. *Expert Rev Mol Diagn* 2004, 4: 831–840.
108. Kamme F, Zhu J, Luo L, Yu J, Tran DT, Meurers B et al.: Single-cell laser-capture microdissection and RNA amplification. *Methods Mol Med* 2004, 99: 215–223.
109. Mizuarai S, Takahashi K, Kobayashi T, Kotani H: Advances in isolation and characterization of homogeneous cell populations using laser microdissection. *Histol Histopathol* 2005, 20: 139–146.
110. Rubin MA: Use of laser capture microdissection, cDNA microarrays, and tissue microarrays in advancing our understanding of prostate cancer. *J Pathol* 2001, 195: 80–86.
111. Fox RM, Von Stetina SE, Barlow SJ, Shaffer C, Olszewski KL, Moore JH et al.: A gene expression fingerprint of C. elegans embryonic motor neurons. *BMC Genomics* 2005, 6: 42.
112. Bryant Z, Subrahmanyan L, Tworoger M, LaTray L, Liu CR, Li MJ et al.: Characterization of differentially expressed genes in purified Drosophila follicle cells: toward a general strategy for cell type-specific developmental analysis. *Proc Natl Acad Sci U S A* 1999, 96: 5559–5564.
113. Bouchard M, Grote D, Craven SE, Sun Q, Steinlein P, Busslinger M: Identification of Pax2-regulated genes by expression profiling of the mid-hindbrain organizer region. *dev* 2005, 132: 2633–2643.
114. Altmann CR, Brivanlou AH: Neural patterning in the vertebrate embryo. *Int Rev Cytol* 2001, 203: 447–482.
115. Cox WG, Hemmati-Brivanlou A: Caudalization of neural fate by tissue recombination and bFGF. *dev* 1995, 121: 4349–4358.
116. Lamb TM, Harland RM: Fibroblast growth factor is a direct neural inducer, which combined with noggin generates anterior-posterior neural pattern. *dev* 1995, 121: 3627–3636.
117. Streit AC, Stern CD: Competence for neural induction: HGF/SF, HGF1/MSP and the c-Met receptor. *CIBA Found Symp* 1997, 212: 165-8, 155–165.
118. Rudel D, Sommer RJ: The evolution of developmental mechanisms. *Dev Biol* 2003, 264: 15–37.
119. Chen Y, Grunz H: The final determination of Xenopus ectoderm depends on intrinsic and external positional information. *Int J Dev Biol* 1997, 41: 525–528.
120. Vize PD, Melton DA, Hemmati-Brivanlou A, Harland RM: Assays for gene function in developing Xenopus embryos. *Methods Cell Biol* 1991, 36: 367–387.
121. Hemmati-Brivanlou A, Frank D, Bolce ME, Brown BD, Sive HL, Harland RM: Localization of specific mRNAs in Xenopus embryos by whole-mount in situ hybridization. *dev* 1990, 110: 325–330.

122. Harland RM: In situ hybridization: an improved whole-mount method for Xenopus embryos. *Methods Cell Biol* 1991, 36: 685–695.
123. Tada M, O'Reilly MA, Smith JC: Analysis of competence and of Brachyury autoinduction by use of hormone-inducible Xbra. *dev* 1997, 124: 2225–2234.
124. *The Analysis of Gene Expression Data: Methods and Software*. Springer; 2003.
125. Dudoit S, Yang YH: Bioconductor R packages for exploratory analysis and normalization of cDNA microarray data. In The Analysis of Gene Expression Data: Methods and Software. New York: Springer; 2002.
126. Yang YH, Buckley MJ, Dudoit S, Speed TP: Comparison of methods for image analysis on cDNA microarray data. Journal of Computational & Graphical Statistics 2002, 11: 108–136.
127. Dudoit S, Yang YH, Callow M, Speed T: Statistical methods for identifying differentially expressed genes in replicated cDNA microarray experiments. Computer program. 2000.
128. Speed TP: Statistical Analysis of Gene Expression Microarray Data. Boca Raton, FL: Chapman & Hall/CRC; 2003.
129. McLachlan GJ, Do KA, Ambroise C: Analyzing Microarray Gene Expression Data. Hoboken, NJ: Wiley-Interscience; 2004.
130. Cui X, Churchill GA: Statistical tests for differential expression in cDNA microarray experiments. Genome Biol 2003, 4: 210.
131. Newton MA, Kendziorski CM, Richmond CS, Blattner FR, Tsui KW: On differential variability of expression ratios: improving statistical inference about gene expression changes from microarray data. J Comput Biol 2001, 8: 37–52.
132. Chen Y, Dougherty ER, Bittner ML: Ratio-based decisions and the quantitative analysis of cDNA microarray images. J Biomed Optics 1997, 2: 364–374.
133. Kerr MK, Martin M, Churchill GA: Analysis of variance for gene expression microarray data. J Comput Biol 2000, 7: 819–837.
134. Tseng GC, Oh MK, Rohlin L, Liao JC, Wong WH: Issues in cDNA microarray analysis: quality filtering, channel normalization, models of variations and assessment of gene effects. Nucleic Acids Res 2001, 29: 2549–2557.
135. Tseng GC, Oh MK, Rohlin L, Liao JC, Wong WH: Issues in cDNA microarray analysis: quality filtering, channel normalization, models of variations and assessment of gene effects. Nucleic Acids Res 2001, 29: 2549–2557.
136. Khimani AH, Mhashilkar AM, Mikulskis A, O'Malley M, Liao J, Golenko EE et al.: Housekeeping genes in cancer: normalization of array data. BioTechniques 2005, 38: 739–745.
137. Warrington JA, Nair A, Mahadevappa M, Tsyganskaya M: Comparison of human adult and fetal expression and identification of 535 housekeeping/maintenance genes. Physiol Genomics 2000, 2: 143–147.
138. Park PJ, Cao YA, Lee SY, Kim JW, Chang MS, Hart R et al.: Current issues for DNA microarrays: platform comparison, double linear amplification, and universal RNA reference. J Biotechnol 2004, 112: 225–245.
139. Cronin M, Ghosh K, Sistare F, Quackenbush J, Vilker V, O'Connell C: Universal RNA reference materials for gene expression. Clin Chem 2004, 50: 1464–1471.
140. Novoradovskaya N, Whitfield ML, Basehore LS, Novoradovsky A, Pesich R, Usary J et al.: Universal Reference RNA as a standard for microarray experiments. BMC Genomics 2004, 5: 20.
141. Berger JA, Hautaniemi S, Jarvinen AK, Edgren H, Mitra SK, Astola J: Optimized LOWESS normalization parameter selection for DNA microarray data. BMC Bioinformatics 2004, 5: 194.

142. Dybkaer K, Zhou G, Iqbal J, Kelly D, Xiao L, Sherman S et al.: Suitability of stratagene reference RNA for analysis of lymphoid tissues. BioTechniques 2004, 37: 470–472, 474.
143. He XR, Zhang C, Patterson C: Universal mouse reference RNA derived from neonatal mice. BioTechniques 2004, 37: 464–468.
144. Belbin TJ, Gaspar J, Haigentz M, Perez-Soler R, Keller SM, Prystowsky MB et al.: Indirect measurements of differential gene expression with cDNA microarrays. BioTechniques 2004, 36: 310–314.
145. Smyth GK, Speed T: Normalization of cDNA microarray data. Methods 2003, 31: 265–273.
146. Quackenbush J: Microarray data normalization and transformation. Nat Genet 2002, 32: 496–501.
147. Galperin MY: The Molecular Biology Database Collection: 2005 update. Nucleic Acids Res 2005, 33: D5–24.
148. Schena M, Shalon D, Heller R, Chai A, Brown PO, Davis RW: Parallel human genome analysis: microarray-based expression monitoring of 1000 genes. Proc Natl Acad Sci U S A 1996, 93: 10614–10619.
149. DeRisi JL, Iyer VR, Brown PO: Exploring the metabolic and genetic control of gene expression on a genomic scale. Science 1997, 278: 680–686.
150. Cui X, Churchill GA: Statistical tests for differential expression in cDNA microarray experiments. Genome Biol 2003, 4: 210.
151. Lee MLT: Analysis of microarray gene expression data. Boston: Kluwer Academic Publishers; 2004.
152. Callow MJ, Dudoit S, Gong EL, Speed TP, Rubin EM: Microarray expression profiling identifies genes with altered expression in HDL-deficient mice. Genome Res 2000, 10: 2022–2029.
153. Tusher VG, Tibshirani R, Chu G: Significance analysis of microarrays applied to the ionizing radiation response. Proc Natl Acad Sci U S A 2001, 98: 5116–5121.
154. Baldi P, Long AD: A Bayesian framework for the analysis of microarray expression data: regularized t-test and statistical inferences of gene changes. Bioinformatics 2001, 17: 509–519.
155. Lonnstedt I, Speed T: Replicated microarray data. Statistica Sinica 2002, 12: 31.
156. Churchill GA: Using ANOVA to analyze microarray data. BioTechniques 2004, 37: 173–5, 177.
157. Cuperlovic-Culf M, Belacel N, Ouellette RJ: Determination of tumour marker genes from gene expression data. Drug Discov Today 2005, 10: 429–437.
158. Wadlow R, Ramaswamy S: DNA microarrays in clinical cancer research. Curr Mol Med 2005, 5: 111–120.
159. Meyerson M, Franklin WA, Kelley MJ: Molecular classification and molecular genetics of human lung cancers. Semin Oncol 2004, 31: 4–19.
160. Stremmel C, Wein A, Hohenberger W, Reingruber B: DNA microarrays: a new diagnostic tool and its implications in colorectal cancer. Int J Colorectal Dis 2002, 17: 131–136.
161. Haviv I, Campbell IG: DNA microarrays for assessing ovarian cancer gene expression. Mol Cell Endocrinol 2002, 191: 121–126.
162. Wang K, Gan L, Jeffery E, Gayle M, Gown AM, Skelly M et al.: Monitoring gene expression profile changes in ovarian carcinomas using cDNA microarray. Gene 1999, 229: 101–108.
163. Savage KJ, Gascoyne RD: Molecular signatures of lymphoma. Int J Hematol 2004, 80: 401–409.

164. Mora J, Gerald WL, Cheung NK: Evolving significance of prognostic markers associated with new treatment strategies in neuroblastoma. Cancer Lett 2003, 197: 119–124.
165. Mariadason JM, Augenlicht LH, Arango D: Microarray analysis in the clinical management of cancer. Hematol Oncol Clin North Am 2003, 17: 377–387.
166. Ben Hur A, Guyon I: Detecting stable clusters using principal component analysis. Methods Mol Biol 2003, 224: 159–182.
167. Quackenbush J: Computational analysis of microarray data. Nat Rev Genet 2001, 2: 418–427.
168. Tamayo P, Slonim D, Mesirov J, Zhu Q, Kitareewan S, Dmitrovsky E et al.: Interpreting patterns of gene expression with self-organizing maps: methods and application to hematopoietic differentiation. Proc Natl Acad Sci U S A 1999, 96: 2907–2912.
169. Sherlock G: Analysis of large-scale gene expression data Curr Opin Immunol 2000, 12: 201–205.
170. Garge NR, Page GP, Sprague AP, Gorman BS, Allison DB: Reproducible clusters from microarray research: whither? BMC Bioinformatics 2005, 6 Suppl 2:: S10.
171. Leung YF, Cavalieri D: Fundamentals of cDNA microarray data analysis. Trends Genet 2003, 19: 649–659.
172. McShane LM, Radmacher MD, Freidlin B, Yu R, Li MC, Simon R: Methods for assessing reproducibility of clustering patterns observed in analyses of microarray data. Bioinformatics 2002, 18: 1462–1469.
173. Yeung KY, Haynor DR, Ruzzo WL: Validating clustering for gene expression data. Bioinformatics 2001, 17: 309–318.
174. Ashburner M, Ball CA, Blake JA, Botstein D, Butler H, Cherry JM et al.: Gene ontology: tool for the unification of biology. The Gene Ontology Consortium. Nat Genet 2000, 25: 25–29.
175. Harris MA, Clark J, Ireland A, Lomax J, Ashburner M, Foulger R et al.: The Gene Ontology (GO) database and informatics resource. Nucleic Acids Res 2004, 32 Database issue: D258-D261.
176. Kanehisa M: The KEGG database. Novartis Found Symp 2002, 247: 91-101; discussion 101-103, 119-128, 244-252.
177. Kanehisa M, Goto S, Kawashima S, Okuno Y, Hattori M: The KEGG resource for deciphering the genome. Nucleic Acids Res 2004, 32: D277–D280.
178. Krieger CJ, Zhang P, Mueller LA, Wang A, Paley S, Arnaud M et al.: MetaCyc: a multiorganism database of metabolic pathways and enzymes. Nucleic Acids Res 2004, 32: D438–D442.
179. Joshi-Tope G, Gillespie M, Vastrik I, D'Eustachio P, Schmidt E, de Bono B et al.: Reactome: a knowledgebase of biological pathways. Nucleic Acids Res 2005, 33: D428–D432.
180. Sczyrba A, Beckstette M, Brivanlou AH, Giegerich R, Altmann CR: XenDB: full length cDNA prediction and cross species mapping in Xenopus laevis. BMC Genomics 2005, 6: 123.
181. Cavalieri D, De Filippo C: Bioinformatic methods for integrating whole-genome expression results into cellular networks. Drug Discov Today 2005, 10: 727–734.
182. Dahlquist KD, Salomonis N, Vranizan K, Lawlor SC, Conklin BR: GenMAPP, a new tool for viewing and analyzing microarray data on biological pathways. Nat Genet 2002, 31: 19–20.
183. Doniger SW, Salomonis N, Dahlquist KD, Vranizan K, Lawlor SC, Conklin BR: MAPPFinder: using Gene Ontology and GenMAPP to create a global gene-expression profile from microarray data. Genome Biol 2003, 4: R7.

184. Dennis G, Jr., Sherman BT, Hosack DA, Yang J, Gao W, Lane HC et al.: DAVID: Database for Annotation, Visualization, and Integrated Discovery. Genome Biol 2003, 4: 3.
185. Hosack DA, Dennis G, Jr., Sherman BT, Lane HC, Lempicki RA: Identifying biological themes within lists of genes with EASE. Genome Biol 2003, 4: R70.
186. Grosu P, Townsend JP, Hartl DL, Cavalieri D: Pathway Processor: a tool for integrating whole-genome expression results into metabolic networks. Genome Res 2002, 12: 1121–1126.
187. Castillo-Davis CI, Hartl DL: GeneMerge—post-genomic analysis, data mining, and hypothesis testing. Bioinformatics 2003, 19: 891–892.
188. Segal E, Friedman N, Koller D, Regev A: A module map showing conditional activity of expression modules in cancer. Nat Genet 2004, 36: 1090–1098.
189. Hillier LW, Miller W, Birney E, Warren W, Hardison RC, Ponting CP et al.: Sequence and comparative analysis of the chicken genome provide unique perspectives on vertebrate evolution. Nature 2004, 432: 695–716.
190. Loots GG, Ovcharenko I: rVISTA 2.0: evolutionary analysis of transcription factor binding sites. Nucleic Acids Res 2004, 32: W217–W221.
191. Loots GG, Ovcharenko I, Pachter L, Dubchak I, Rubin EM: rVista for comparative sequence-based discovery of functional transcription factor binding sites. Genome Res 2002, 12: 832–839.
192. Dickmeis T, Muller F: The identification and functional characterisation of conserved regulatory elements in developmental genes. Brief Funct Genomic Proteomic 2005, 3: 332–350.
193. Zhang Z, Gerstein M: Of mice and men: phylogenetic footprinting aids the discovery of regulatory elements. J Biol 2003, 2: 11.
194. Sandelin A, Wasserman WW, Lenhard B: ConSite: web-based prediction of regulatory elements using cross-species comparison. Nucleic Acids Res 2004, 32: W249–W252.
195. Rombauts S, Florquin K, Lescot M, Marchal K, Rouze P, van de PY: Computational approaches to identify promoters and cis-regulatory elements in plant genomes. Plant Physiol 2003, 132: 1162–1176.
196. Vega VB, Bangarusamy DK, Miller LD, Liu ET, Lin CY: BEARR: Batch Extraction and Analysis of cis-Regulatory Regions. Nucleic Acids Res 2004, 32: W257–W260.
197. Zhang H, Ramanathan Y, Soteropoulos P, Recce ML, Tolias PP: EZ-Retrieve: a web-server for batch retrieval of coordinate-specified human DNA sequences and underscoring putative transcription factor-binding sites. Nucleic Acids Res 2002, 30: e121.
198. Zhao F, Xuan Z, Liu L, Zhang MQ: TRED: a Transcriptional Regulatory Element Database and a platform for in silico gene regulation studies. Nucleic Acids Res 2005, 33: D103–D107.
199. Sandelin A, Alkema W, Engstrom P, Wasserman WW, Lenhard B: JASPAR: an open-access database for eukaryotic transcription factor binding profiles. Nucleic Acids Res 2004, 32: D91–D94.
200. Heinemeyer T, Wingender E, Reuter I, Hermjakob H, Kel AE, Kel OV et al.: Databases on transcriptional regulation: TRANSFAC, TRRD and COMPEL. Nucleic Acids Res 1998, 26: 362–367.
201. Wingender E, Dietze P, Karas H, Knuppel R: TRANSFAC: a database on transcription factors and their DNA binding sites. Nucleic Acids Res 1996, 24: 238–241.
202. Vadigepalli R, Chakravarthula P, Zak DE, Schwaber JS, Gonye GE: PAINT: a promoter analysis and interaction network generation tool for gene regulatory network identification. OMICS 2003, 7: 235–252.

203. Zheng J, Wu J, Sun Z: An approach to identify over-represented cis-elements in related sequences. Nucleic Acids Res 2003, 31: 1995–2005.
204. Jensen LJ, Knudsen S: Automatic discovery of regulatory patterns in promoter regions based on whole cell expression data and functional annotation. Bioinformatics 2000, 16: 326–333.
205. Vanet A, Marsan L, Labigne A, Sagot MF: Inferring regulatory elements from a whole genome. An analysis of Helicobacter pylori sigma(80) family of promoter signals. J Mol Biol 2000, 297: 335–353.
206. van Helden J: Metrics for comparing regulatory sequences on the basis of pattern counts. Bioinformatics 2004, 20: 399–406.
207. van Helden J: Regulatory sequence analysis tools. Nucleic Acids Res 2003, 31: 3593–3596.
208. van Helden J, Rios AF, Collado-Vides J: Discovering regulatory elements in non-coding sequences by analysis of spaced dyads. Nucleic Acids Res 2000, 28: 1808–1818.
209. Qin ZS, McCue LA, Thompson W, Mayerhofer L, Lawrence CE, Liu JS: Identification of co-regulated genes through Bayesian clustering of predicted regulatory binding sites. Nat Biotechnol 2003, 21: 435–439.
210. Bailey TL, Elkan C: Fitting a mixture model by expectation maximization to discover motifs in biopolymers. Proc Int Conf Intell Syst Mol Biol 1994, 2: 28–36.
211. Hughes JD, Estep PW, Tavazoie S, Church GM: Computational identification of cis-regulatory elements associated with groups of functionally related genes in Saccharomyces cerevisiae. J Mol Biol 2000, 296: 1205–1214.
212. Lawrence CE, Altschul SF, Boguski MS, Liu JS, Neuwald AF, Wootton JC: Detecting subtle sequence signals: a Gibbs sampling strategy for multiple alignment. Science 1993, 262: 208–214.
213. Liu JS, Lawrence CE: Bayesian inference on biopolymer models. Bioinformatics 1999, 15: 38–52.
214. Neuwald AF, Liu JS, Lawrence CE: Gibbs motif sampling: detection of bacterial outer membrane protein repeats. Protein Sci 1995, 4: 1618–1632.
215. Qin ZS, McCue LA, Thompson W, Mayerhofer L, Lawrence CE, Liu JS: Identification of co-regulated genes through Bayesian clustering of predicted regulatory binding sites. Nat Biotechnol 2003, 21: 435–439.
216. Thijs G, Lescot M, Marchal K, Rombauts S, De Moor B, Rouze P et al.: A higher-order background model improves the detection of promoter regulatory elements by Gibbs sampling. Bioinformatics 2001, 17: 1113–1122.
217. Robin S, Schbath S: Numerical comparison of several approximations of the word count distribution in random sequences. J Comput Biol 2001, 8: 349–359.
218. Schbath S: An overview on the distribution of word counts in Markov chains. J Comput Biol 2000, 7: 193–201.
219. Marsan L, Sagot MF: Algorithms for extracting structured motifs using a suffix tree with an application to promoter and regulatory site consensus identification. J Comput Biol 2000, 7: 345–362.
220. Grundy WN, Bailey TL, Elkan CP: ParaMEME: a parallel implementation and a web interface for a DNA and protein motif discovery tool. Comput Appl Biosci 1996, 12: 303–310.
221. Poluliakh N, Takagi T, Nakai K: Melina: motif extraction from promoter regions of potentially co-regulated genes. Bioinformatics 2003, 19: 423–424.
222. Kankainen M, Holm L: POBO, transcription factor binding site verification with bootstrapping. Nucleic Acids Res 2004, 32: W222–W229.

223. Karanam S, Moreno CS: CONFAC: automated application of comparative genomic promoter analysis to DNA microarray datasets. Nucleic Acids Res 2004, 32: W475–W484.
224. Sinha S, Tompa M: YMF: A program for discovery of novel transcription factor binding sites by statistical overrepresentation. Nucleic Acids Res 2003, 31: 3586–3588.
225. Liu Y, Wei L, Batzoglou S, Brutlag DL, Liu JS, Liu XS: A suite of web-based programs to search for transcriptional regulatory motifs. Nucleic Acids Res 2004, 32: W204–W207.
226. van Helden J: Regulatory sequence analysis tools. Nucleic Acids Res 2003, 31: 3593–3596.
227. Aerts S, Van Loo P, Thijs G, Mayer H, de Martin R, Moreau Y et al.: TOUCAN 2: the all-inclusive open source workbench for regulatory sequence analysis. Nucleic Acids Res 2005, 33: W393–W396.
228. Coessens B, Thijs G, Aerts S, Marchal K, De Smet F, Engelen K et al.: INCLUSive: A web portal and service registry for microarray and regulatory sequence analysis. Nucleic Acids Res 2003, 31: 3468–3470.
229. Sartor M, Schwanekamp J, Halbleib D, Mohamed I, Karyala S, Medvedovic M et al.: Microarray results improve significantly as hybridization approaches equilibrium. BioTechniques 2004, 36: 790–796.
230. Bhanot G, Louzoun Y, Zhu J, DeLisi C: The importance of thermodynamic equilibrium for high throughput gene expression arrays. Biophys J 2003, 84: 124–135.
231. Korkola JE, Estep AL, Pejavar S, DeVries S, Jensen R, Waldman FM: Optimizing stringency for expression microarrays. BioTechniques 2003, 35: 828–835.
232. Wong ML, Medrano JF: Real-time PCR for mRNA quantitation. BioTechniques 2005, 39: 75–85.
233. Higuchi R, Dollinger G, Walsh PS, Griffith R: Simultaneous amplification and detection of specific DNA sequences. Biotechnology (NY) 1992, 10: 413–417.
234. Higuchi R, Fockler C, Dollinger G, Watson R: Kinetic PCR analysis: real-time monitoring of DNA amplification reactions. Biotechnology (NY) 1993, 11: 1026–1030.
235. Huggett J, Dheda K, Bustin S, Zumla A: Real-time RT-PCR normalisation; strategies and considerations. Genes Immun 2005, 6: 279–284.
236. Bustin SA, Nolan T: Pitfalls of quantitative real-time reverse-transcription polymerase chain reaction. J Biomol Tech 2004, 15: 155–166.
237. Bustin SA: Real-time, fluorescence-based quantitative PCR: a snapshot of current procedures and preferences. Expert Rev Mol Diagn 2005, 5: 493–498.
238. Tichopad A, Dilger M, Schwarz G, Pfaffl MW: Standardized determination of real-time PCR efficiency from a single reaction set-up. Nucleic Acids Res 2003, 31: e122.
239. Zhang J, Byrne CD: Differential priming of RNA templates during cDNA synthesis markedly affects both accuracy and reproducibility of quantitative competitive reverse-transcriptase PCR. Biochem J 1999, 337: 231–241.
240. Lekanne Deprez RH, Fijnvandraat AC, Ruijter JM, Moorman AF: Sensitivity and accuracy of quantitative real-time polymerase chain reaction using SYBR green I depends on cDNA synthesis conditions. Anal Biochem 2002, 307: 63–69.
241. Dheda K, Huggett JF, Chang JS, Kim LU, Bustin SA, Johnson MA et al.: The implications of using an inappropriate reference gene for real-time reverse transcription PCR data normalization. Anal Biochem 2005, 344: 141–143.
242. Dheda K, Huggett JF, Bustin SA, Johnson MA, Rook G, Zumla A: Validation of housekeeping genes for normalizing RNA expression in real-time PCR. BioTechniques 2004, 37: 112–119.

243. Tricarico C, Pinzani P, Bianchi S, Paglierani M, Distante V, Pazzagli M et al.: Quantitative real-time reverse transcription polymerase chain reaction: normalization to rRNA or single housekeeping genes is inappropriate for human tissue biopsies. Anal Biochem 2002, 309: 293–300.
244. Vandesompele J, De Preter K, Pattyn F, Poppe B, Van Roy N, De Paepe A et al.: Accurate normalization of real-time quantitative RT-PCR data by geometric averaging of multiple internal control genes. Genome Biol 2002, 3: 0034.1–0034.11.
245. Aerts JL, Gonzales MI, Topalian SL: Selection of appropriate control genes to assess expression of tumor antigens using real-time RT-PCR. BioTechniques 2004, 36: 84–81.
246. de Kok JB, Roelofs RW, Giesendorf BA, Pennings JL, Waas ET, Feuth T et al.: Normalization of gene expression measurements in tumor tissues: comparison of 13 endogenous control genes. Lab Invest 2005, 85: 154–159.
247. Pfaffl MW, Tichopad A, Prgomet C, Neuvians TP: Determination of stable housekeeping genes, differentially regulated target genes and sample integrity: BestKeeper—Excel-based tool using pair-wise correlations. Biotechnol Lett 2004, 26: 509–515.
248. Andersen CL, Jensen JL, Orntoft TF: Normalization of real-time quantitative reverse transcription-PCR data: a model-based variance estimation approach to identify genes suited for normalization, applied to bladder and colon cancer data sets. Cancer Res 2004, 64: 5245–5250.
249. Zhang QJ, Chadderton A, Clark RL, Augustine-Rauch KA: Selection of normalizer genes in conducting relative gene expression analysis of embryos. Birth Defects Res A Clin Mol Teratol 2003, 67: 533–544.
250. Schmittgen TD, Zakrajsek BA: Effect of experimental treatment on housekeeping gene expression: validation by real-time, quantitative RT-PCR. J Biochem Biophys Methods 2000, 46: 69–81.
251. Livak KJ, Schmittgen TD: Analysis of relative gene expression data using real-time quantitative PCR and the 2(-Delta Delta C(T)) Method. Methods 2001, 25: 402–408.
252. Gentle A, Anastasopoulos F, McBrien NA: High-resolution semi-quantitative real-time PCR without the use of a standard curve. BioTechniques 2001, 31: 502, 504–506, 508.
253. Liu W, Saint DA: A new quantitative method of real time reverse transcription polymerase chain reaction assay based on simulation of polymerase chain reaction kinetics. Anal Biochem 2002, 302: 52–59.
254. Marino JH, Cook P, Miller KS: Accurate and statistically verified quantification of relative mRNA abundances using SYBR Green I and real-time RT-PCR. J Immunol Methods 2003, 283: 291–306.
255. Peirson SN, Butler JN, Foster RG: Experimental validation of novel and conventional approaches to quantitative real-time PCR data analysis. Nucleic Acids Res 2003, 31: e73.
256. Rasmussen R: Quantification on the LightCycler. In Rapid Cycle Real-time PCR, Methods and Applications. Meuer S, Wittwer C, Nakagawara K (Eds.). Heidelberg: Springer Press; 2001:21–34.
257. Stahlberg A, Aman P, Ridell B, Mostad P, Kubista M: Quantitative real-time PCR method for detection of B-lymphocyte monoclonality by comparison of kappa and lambda immunoglobulin light chain expression. Clin Chem 2003, 49: 51–59.
258. Pfaffl MW: A new mathematical model for relative quantification in real-time RT-PCR. Nucleic Acids Res 2001, 29: e45.

259. Pfaffl MW: Development and validation of an externally standardised quantitative Insulin like growth factor-1 (IGF-1) RT-PCR using LightCycler SYBR® Green I technology. In Rapid Cycle Real-time PCR, Methods and Applications. Meuer S, Wittwer C, Nakagawara K (Eds.). Heidelberg: Springer Press; 2001.
260. Ramakers C, Ruijter JM, Deprez RH, Moorman AF: Assumption-free analysis of quantitative real-time polymerase chain reaction (PCR) data. Neurosci Lett 2003, 339: 62–66.
261. Peccoud J, Jacob C: Theoretical uncertainty of measurements using quantitative polymerase chain reaction. Biophys J 1996, 71: 101–108.
262. Shiao YH: A new reverse transcription-polymerase chain reaction method for accurate quantification. BMC Biotechnol 2003, 3: 22.
263. Tichopad A, Dilger M, Schwarz G, Pfaffl MW: Standardized determination of real-time PCR efficiency from a single reaction set-up. Nucleic Acids Res 2003, 31: e122.
264. Muller PY, Janovjak H, Miserez AR, Dobbie Z: Processing of gene expression data generated by quantitative real-time RT-PCR. BioTechniques 2002, 32: 1372–1379.
265. Lee J, Nam S, Hwang SB, Hong M, Kwon JY, Joeng KS et al.: Functional genomic approaches using the nematode Caenorhabditis elegans as a model system. J Biochem Mol Biol 2004, 37: 107–113.
266. Brooks DR, Isaac RE: Functional genomics of parasitic worms: the dawn of a new era. Parasitol Int 2002, 51: 319–325.
267. Gunsalus KC, Piano F: RNAi as a tool to study cell biology: building the genome-phenome bridge. Curr Opin Cell Biol 2005, 17: 3–8.

11 Chromatin Immunoprecipitation for *In Vivo* Studies of Transcriptional Regulation during Development

Daniel R. Buchholz, Bindu Diana Paul, and Yun-Bo Shi

CONTENTS

I. INTRODUCTION

A. Importance of Detecting Protein/DNA Interactions in the Context of Chromatin *In Vivo*

Protein/DNA interactions form the molecular basis of transcriptional regulation in development and disease. Transcription is a complex process involving a large number of sequence-specific transcription factors and cofactors, and studies of the spatiotemporal nature of protein/DNA interactions and the consequences for chromatin remodeling are crucial for understanding the molecular mechanisms of transcriptional regulation *in vivo*. Previous techniques such as gel shift assays, though important for showing the capability of proteins to bind DNA *in vitro*, could not show whether these interactions actually take place in the context of chromatin in a cell or tissue, especially when considering tissue-specific gene regulation. The chromatin immunoprecipitation (ChIP) assay provides a snapshot of these critical protein/DNA interactions as they occur *in situ* (Kuo and Allis, 1999). The ChIP assay essentially involves formaldehyde crosslinking of proteins to each other and, importantly, to the site of DNA interaction. This crosslinking is followed by sonication to yield small fragments of chromatin, which are then immunoprecipitated using antibodies against protein components of interest to isolate pieces of DNA, bound to or associated with the protein. This DNA is then purified and analyzed by PCR-based methods.

B. Molecular Analysis of Gene Regulation, *In Vitro* vs. *In Vivo*

The ChIP assay was first developed in yeast and tissue culture cells and yielded many successes in understanding the role of transcription factors and cofactors in histone modifications and recruitment of basal transcription machinery (Kuo and Allis, 1999; Das et al., 2004). These studies provide a starting point for *in vivo* studies in molecular mechanisms of transcription during development because *in vitro* studies using cell lines, though derived from tissues, do not necessarily reflect normal cellular physiology, including gene regulation and transcription factor binding to promoters. Each cell type has its own set of transcribed genes and associated transcription factors, and even if the same gene is expressed in more than one cell type, there may be different molecular mechanisms regulating the gene given the presence of different or different levels of transcription factors and cofactors. More importantly, developmental or disease processes are not readily replicated in culture, making it necessary to analyze molecular mechanisms of gene regulation in animals *in vivo*.

C. Brief Summary of the Use of Chromatin Immunoprecipitation *In Vivo*

The ChIP assay has become a standard procedure in cell culture, as reflected by published overviews and methods (Damjanovski et al., 2002; Farnham and Weinmann, 2002; Spencer et al., 2003), and protocols are widely available from companies such as Upstate Biotechnology, Inc. Using this assay to study protein/DNA interactions in tissues presents added complications, when compared to cell culture, due to lack of single cell suspension and the fact that each tissue may have different requirements for fixing chromatin because of differences in composition of extracellular matrix and connective tissue. The protocols for tissues have been modified from protocols for cells (Parrizas et al., 2001; Wells and Farnham, 2002; Chaya and Zaret, 2004). These modifications have improved on methods for fixing chromatin with minimum disturbance of protein/DNA interactions while at the same time enabling isolation of fixed chromatin fragments from the tissue. These protocols for tissues converge with protocols for cells at the point of sonication after tissue nuclei are lysed with SDS.

D. Thyroid Hormone Receptor in Frog Metamorphosis as an Example

We have been using frog metamorphosis as a model to study the role of thyroid hormone (T3) receptors (TRs) in gene regulation and development (Sachs and Shi, 2000; Sachs et al., 2002; Buchholz et al., 2003; 2004; and 2005). Frog metamorphosis is a postembryonic process controlled by T3, which exerts its effects on various target tissues via binding to TRs (Shi, 1999). The expression of TRs in premetamorphic tadpoles before, as well as after, the start of T3 secretion into the blood motivated development of a dual-function model for the role of TR in development (Sachs et al., 2000; Buchholz et al., 2006). TR expression during premetamorphosis is hypothesized to function by recruiting corepressors to repress T3-regulated genes, allowing the tadpole to grow. During natural metamorphosis or T3-induced metamorphosis, TRs function as activators to induce expression of T3-regulated genes necessary for the larval to juvenile transition by recruiting coactivators to the promoters. By using the ChIP protocol described below, we have been able to investigate different aspects of the model, such as the binding of TR to DNA, the recruitment of transcriptional cofactors, and changes in histone acetylation at T3 target genes in whole tadpoles or specific tissues during development. The methods described here should be applicable to other tissues and developmental systems, as well.

II. Methodological Overview

A. Study of Transcription Factors

Tissue-specific gene regulation largely depends on the set of transcription factors expressed in the cells and their binding to promoter regions in chromatin. The ChIP assay can directly assess these protein/DNA interactions in a promoter-specific, tissue-specific, and hormone- or developmental stage-dependent fashion (Figure

11.1). We isolated fixed chromatin from intestines or tails of tadpoles for the ChIP assay and used quantitative PCR (qPCR) with primers to measure presence of promoters of widely expressed genes from chromatin immunoprecipitated with anti-TR antibodies. These experiments take advantage of genes that are expressed in most if not all cell types, because organs or biopsies from whole animals most often are composed of multiple tissues. Direct T3 response genes expressed in restricted tissues that make up a small fraction of material in the organ may not be detectable due to low signal. For example, the matrix metalloproteinase stromelysin 3 (ST3) is a highly T3-induced gene expressed exclusively in fibroblasts, which make up a minority of the cells of the intestine. Because ST3 is not expressed in epithelial cells, which make up most of the intestine, only a small fraction of the cells containing nuclei have TR bound to the ST3 promoter, making it difficult to study TR binding to this promoter.

B. Study of Cofactors

Using the ChIP assay to detect cofactor binding to a particular region of DNA includes problems similar to those encountered when studying transcription factors, resulting from tissue and promoter specificity, and poses additional difficulties. For any one transcription factor, there is likely to be a less than 1:1 ratio of a particular cofactor. For example, a number of corepressors seem to bind well to unliganded TR, with nuclear receptor corepressor (N-CoR) and silencing mediator for retinoic acid and thyroid hormone receptors (SMRT) as best documented (Jepsen and Rosenfeld, 2002; Yoon et al., 2003; Tomita et al., 2004). Thus, the binding of one particular corepressor presumably reduces the binding of the TR to a different corepressor. In the case of coactivators, the situation is even more complex because there are even more coactivators known to interact with TR under various *in vitro* conditions. An additional technical problem is the efficiency of crosslinking of cofactors to DNA, which will likely be less than that for transcription factors because the latter bind directly to DNA, whereas cofactors are at least one molecular interaction removed. Nevertheless, we have been able to show an increased recruitment of cofactor SRC3 to T3-responsive promoters in the intestine in the presence of T3, indicating the sensitivity and usefulness of *in vivo* ChIP assay for cofactors (Figure 11.1) (Paul et al., 2005).

C. Study of Histone Modifications

Because histones are bound directly to DNA, ChIP assay for histone modifications do not present the same problems as cofactors. Moreover, because of the relative abundance of histones, studies of histone modifications may not be as affected by tissue specificity as transcription factor-based studies (Figure 11.1c). With studying histone modifications, their relative abundance creates a potential problem, namely DNA region specificity. It is important to be able to conclude that any detectable modification is associated with the DNA region of interest. This suggests that the size of the DNA fragments cannot be too large to contain two or multiple regions

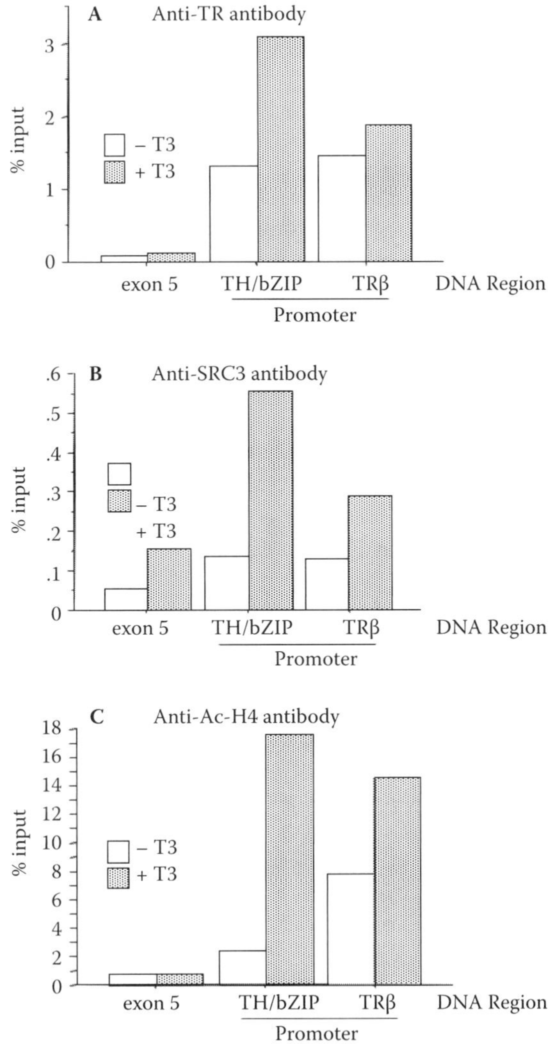

FIGURE 11.1 Using our *in vivo* model of frog metamorphosis, which totally depends on T3 and TR, we have shown hormone-dependent changes at T3-inducible promoters. Premetamorphic tadpoles before the beginning of endogenous T3 production were treated with 10 nM T3 for 2 days. ChIP assay was carried out on intestines from one to three tadpoles using antibodies for TR (a), SRC3 (b), and acetylated histone H4 (c). Notice the promoter-specific hormone-dependent changes of TR binding in (a) and hormone dependent SRC3 binding (b). Also, note the higher levels of histone acetylation in T3-treated animals. Also note that the ChIP signals, as a percentage of input DNA, were much higher for acetylated histone H4 and TR than for SRC3. This was likely due to the fact that TR and histones bind DNA directly, whereas SRC3 associate with DNA indirectly and compete for binding to TR with other coactivators. (Data from Paul et al., 2005; Buchholz et al., 2005; and unpublished observations.)

of modification, so that antibodies will not recognize modified histones outside the region of interest yet pull down the region of interest.

D. Cell Culture vs. Tissues, Frog vs. Mouse

The important differences between ChIP assay for cell culture and tissues is in preparation of the chromatin, whereas the subsequent immnoprecipitation steps and analysis are the same. In order to provide a snapshot of protein/DNA interactions, it is desirable to fix the cells or tissues with as little disturbance as possible. Because cells in culture are separate, they are quickly and efficiently fixed in a cell suspension with minimal disturbance. Therefore, chromatin isolation from blood cells can be similar to that for tissue culture cells. Direct fixation of tissues best preserves chromatin structure but creates problems in isolating chromatin relatively free of cellular debris. Fixation by perfusion of organs *in situ* is appropriate for tissues that are large enough and highly vascularized followed by mincing and then homogenizing in a dounce (Chaya and Zaret, 2004). Alternative methods are to isolate cell types from tissues (e.g., pancreatic islets) or to mince tissues first and then fix in formaldehyde before homogenization (Parrizas et al., 2001; Wells and Farnham, 2002).

In the current protocol, we homogenize tissues first to release cells and nuclei and then fix with formaldehyde. All these methods attempt to preserve, as closely as possible, the chromatin structure by fixing first before disrupting the nuclear membrane or altering the ionic strength found in the tissues. These methods can be used across organisms, such as frogs, mice, or tissue/cancer biopsies, but may have to be modified to accommodate different types of tissue. For example, organs high in connective tissue may require nuclear isolation first before fixation to avoid trapping the nuclei in the tissue, and embryos have little connective tissue but cannot be fixed *in situ* for lack of veins large enough for perfusion and so need to be minced first.

III. KEY ISSUES

A. Antibody

The antibody characteristics, namely, high specificity and titer, are desirable for ChIP assays be they from tissues or cell culture samples. However, these requirements may become more stringent in tissues for which it is difficult to get chromatin preparations free of cell debris that may exacerbate problems with high background from antibody cross-reactivity or in tissues where only some cell types express the gene of interest.

B. Important Controls

Reliability of results from ChIP assay in the evaluation of differences observed between samples is an important issue because of the many potential artifacts. Thus, control antibodies and DNA regions need to be included in the experimental design.

Differences between samples may be due to one batch of chromatin being more "sticky" than another as a result of a treatment, so that even if replication of the same treatment gives similar results, the results may not necessarily reflect real binding. To control for systematic differences in chromatin between treatments, a control antibody is necessary. The control antibody can be preimmune serum or immunized to an irrelevant protein and should give results that are clearly distinguishable from the antibody of interest.

Another issue in comparing across treatments is that different time points after a treatment or different developmental stages may constitute a different collection of cell types in a developing system. If widely different developmental stages are used, the ChIP assay would be documenting differences at a developmentally regulated promoter that may be due to changes in cell types or composition, rather than studying mechanisms of gene regulation within a given cell.

Comparison across DNA regions with the same antibody needs to be done with caution because the enhanceosome may be different at different promoters and thus represent different epitopes for the antibody. Such differences can lead to differences in immunoprecipitation efficiency and antibody cross-reactivity. Also, comparing across antibodies with the same DNA region needs to be done with the caveat that different antibodies may bind with different affinities. Finally, and of course, any result needs to be repeated from multiple chromatin preparations in order to represent true biological variability.

C. Quantitative PCR vs. PCR/Gel Electrophoresis

The advantages of quantitative PCR over PCR/gel elecrophoresis are both logistical and potentially critical. Quantitative PCR measures amplification during each cycle and compares samples in the exponential phase of the reaction that can be quantitatively compared using standards. In contrast, traditional PCR detects amplified products at the endpoint of the reaction, which may or may not reflect rank order differences in input among samples due to potential saturation of reaction components. In addition, qPCR obviates the need for post-PCR analysis. In many cases, the quantity of amplicons is too low to be detected by ethidium bromide staining after agarose gel electrophoresis, so visualization of PCR results requires Southern blotting followed by chemiluminescent detection or exposure of the dried gel to film if radioactive nucleotides are included in the PCR reaction.

A more important issue is in the case where two treatments using the same antibody and DNA region give the same results. It is not easy using conventional PCR to determine whether this similarity is due to similar binding levels or because both samples are at the background level. To distinguish these possibilities, one needs to use a control DNA region where the antibody is not expected to bind and compare the results with the DNA region of interest, remembering the cautions of comparing across DNA regions discussed above. When using qPCR, the specific product is not visualized, and thus, careful controls and calibration are needed to ensure the specificity of the signal detected, especially when SYBR Green or a similar detection method is used, because nonspecifically amplified DNA will contribute to the final signal.

IV. PROTOCOL

A. Animals and Treatment

Xenopus laevis tadpoles of different developmental stages (Nieuwkoop and Faber, 1994) were obtained from NASCO or Xenopus I, Inc. Stage 54 premetamorphic tadpoles at a density of 2–3 tadpoles per liter were treated with 10 nM T3 for 1–3 days at 18°C.

B. Reagents and Buffers

Stock materials:

37% formaldehyde solution
1 M Tris-HCl, pH 9.4 or 1 M glycine
1 M DTT, frozen in aliquots
0.2 M PMSF, in EtOH (half-life is 30 min in water)
Protease inhibitor tablet (Roche, Complete, Mini, EDTA-free)
2 ml all glass dounce homogenizer sets (Kontes Kimble) with pestles A (for initial homogenization) and B (for nuclei expulsion)
QIAquick PCR purification kit (Qiagen)

Working solutions:

0.6X phosphate buffered saline (PBS)
Nuclei extraction buffer (prepared fresh): 0.5% Triton X-100, 10 mM Tris-HCl, pH 7.5, 3 mM $CaCl_2$, 0.25 M sucrose, protease inhibitor tablet (1 tablet/20 μL), 1 mM DTT, and 0.2 mM PMSF
SDS lysis buffer: (Upstate, 1% SDS, 10 mM EDTA, 50 mM Tris-HCl, pH 8.1)
Salmon sperm DNA/protein A agarose (Upstate)
ChIP dilution buffer: (Upstate, 0.01% SDS, 1.1% Triton X-100, 1.2 mM EDTA, 16.7 mM Tris-HCl, pH 8.1, 167 mM NaCl. Add 1 mM DTT, 0.4 mM PMSF, and protease inhibitor tablet [1 tablet/20 μL] just before use.)
ChIP I low salt buffer: (0.1% SDS; 1% Triton X-100; 2 mM EDTA; 50 mM Hepes, pH 7.5; 150 mM NaCl)
ChIP II high salt buffer: (0.1% SDS; 1% Triton X-100; 2 mM EDTA; 50 mM Hepes, pH 7.5; 500 mM NaCl)
ChIP III LiCl wash buffer: (0.25 M LiCl; 0.5% NP-40; 0.5% sodium deoxycholate; 1 mM EDTA; 10 mM Tris-HCl, pH 8.0)
TE (10 mM Tris-HCl, 1 mM EDTA pH 8.0)
Elution buffer: 0.5% SDS, 0.1M NaHCO3 (Sigma), 25 μg/ml Proteinase K (Roche)

C. Chromatin Isolation from Tissue

1. Dissect tissue, up to 100 mg, from euthanized tadpoles and place in 1 mL nuclei extraction buffer in dounce on ice. We have tried whole

tadpole, tail, intestine, and liver. For small intestines, flush contents using 0.6X PBS in a syringe with a 27-gauge needle, and place in 1 mL nuclei extraction buffer in dounce on ice. Pooling one to three premetamorphic tadpole organs per treatment results in sufficient chromatin concentration without overwhelming capacity of the buffer and homogenization process.

2. Crush in a dounce homogenizer with 10–15 strokes using the large clearance pestle A for initial homogenization. See Note 1.
3. Pour homogenate into 1.7 ml eppendorf tubes and add 25 μL 37% formaldehyde (1% final concentration) to crosslink proteins and DNA. Rotate tubes at room temperature for 15 min.
4. Add 100 μL 1M Tris-HCl, pH 9.5 (or glycine to 0.125 M) and continue rotating for 5 min to stop crosslinking.
5. Centrifuge at 2000 g at 4°C for 2 min.
6. Discard supernatant, resuspend pellet in 1 mL nuclei extraction buffer, and transfer to the dounce tubes on ice.
7. Rehomogenize with 5–10 strokes using pestle B. Pestle B may be difficult to use with tissues high in connective tissue, so using pestle A again may be necessary.
8. Filter out the unhomogenized debris through a Falcon 100 μm cell strainer into fresh eppendorf tubes or 50 mL conical tubes and centrifuge at 2000 g at 4°C for 2 min.
9. Optional step. After filtering, the nuclei can be further purified by layering onto a 9 ml cushion of nuclei extraction buffer with 2.2 M sucrose rather than 0.25 M sucrose and then centrifuging the samples for 3 hr at 4°C at 55,000 rpm using an HB4 rotor and Sorvall Ultracentrifuge (Damjanovski et al., 2002). See Note 2.
10. Resuspend pellet in 200–300 μl SDS lysis buffer on ice.
11. Shear the chromatin to approximately 200 to 1000 bp fragments of DNA using a sonicator, while keeping the samples on ice throughout the process to avoid overheating. See Note 3.
12. Centrifuge the sonicated solution at 14,000 rpm in an eppendorf microfuge for 10 min at 4°C. Transfer the supernatant to fresh tubes and quantitate the DNA by measuring the absorbance at A260 using a spectrophotometer. See Note 4.
13. At this stage, the chromatin can be aliquotted, snap frozen in liquid nitrogen, and stored at –80°C. We have stored aliquots for many weeks without affecting the results.

D. Immunoprecipitation and DNA Purification

1. Remove a frozen aliquot and adjust the DNA to 100 ng/μl using the SDS lysis buffer. Then, dilute the samples to a final concentration of 10 ng/μl (down to 3–5 ng/μl also works) using the ChIP dilution buffer so the total volume is enough for 500 μL per immunoprecipitation. (For instance, when using three antibodies, take 200 μl of 100 ng/μl sample in SDS lysis

buffer and add 1800 μL of ChIP dilution buffer in 5 mL Falcon tubes so there is enough volume for each antibody and 20 μL for the input sample.)

2. Preclear the chromatin before dividing into separate tubes for antibodies using 60 μl of slurry of salmon sperm DNA/protein A agarose per 1 mL of chromatin solution for 30 min with rotation at 4°C to reduce background or nonspecific DNA/protein binding to the agarose beads. Use salmon sperm DNA/protein G agarose if using monoclonal antibodies.
3. Pellet the agarose beads by centrifugation at 1000 g for 2 min at 4°C.
4. Set up immunoprecipitation: Add 10–60 μl of slurry of salmon sperm DNA/protein A agarose into 1.7 mL eppendorf tubes, enough tubes for all antibodies and chromatin samples. Then, for each chromatin sample, add 500 μL of precleared sample into the same number of tubes as antibodies. (Between 0.5 and 1.0 ml of the chromatin solution with a total DNA concentration ranging from 3.5 to 10.0 μg can be used per immunoprecipitation.) Add the appropriate amount of antibody to each tube with 500 μL of sample and agarose. (We use between 5 and 40 μL of antibody per tube. Both antisera as well as purified antibodies can be used. See Note 5. A preimmune serum or irrelevant antibody should be used as a negative control. A no-antibody control can be used to troubleshoot high background.)
5. Pipet 20 μL of each chromatin sample into another 1.7 mL eppendorf tube for input controls and store at 4°C until de-crosslinking.
6. Incubate immunoprecipitation samples with rotation from 4 hr to overnight at 4°C.
7. After incubation, pellet the beads at 1000 g for 2 min at 4°C, remove supernant (as much as possible to avoid high background), and add 1 mL of ChIP I.
8. Rotate for 5–15 min at 4°C, then repeat wash (step 7) with 1 ml of ChIP II, III, and TE buffers.
9. After last wash with TE buffer and removal of supernatant, add 100 μL of elution buffer to the beads in each tube and rotate at 65°C for 6 hr to overnight to reverse crosslinks. Do not forget to include input samples at this step.
10. Purify DNA by phenol/chloroform extraction or using the QIAquick PCR purification kit (Qiagen).
11. Resuspend (for phenol/chloroform extraction) or elute DNA (for the Qiagen kit) in 40 μl of water or EB buffer (Qiagen, 10 mM Tris-HCl, pH 8.5); 4 μL can be used for each PCR reaction.

E. PCR/Gel Analysis and Quantitative PCR

Conventional PCR, using α-^{32}P-dNTP followed by polyacrylamide gel electrophoresis, or quantitative PCR can be employed for the DNA amplification and detection. For conventional PCR, for example, assemble a 20 μl reaction on ice as follows (the conditions for each primer set have to be determined empirically):

10X Taq buffer free of Mg	2 μl
25 mM $MgCl_2$	2 μl
Primer A	25 ng (0.1 μl)
Primer B	25 ng (0.1 μl)
25 mM dNTPs	0.1 μl
Taq polymerase (5 U/μl, promega)	0.1 μl
α-^{32}P-dCTP or ^{32}P -dATP	0.1 μl
H_2O	11.5 μl

PCR cycling: 95°C for 30 sec, 62°C for 30 sec, 72°C for 30 sec, 20–37 cycles depending on the intensity of signals, then 72°C for 5 min.

After PCR, add 4 μl of 5X DNA loading buffer to each tube. Separate on a 6% native polyacrylamide gel using 1X TBE. Load 5–8 μl of the PCR reaction per lane. Run the gel at 300 V for 40 min to 1 hr. Dry the gel on Whatman No. 3 filter paper and visualize by autoradiography. Care should be taken to avoid saturation of the PCR (especially for input DNA control) and autoradiography exposure so that any potential differences between samples can be identified, especially the input samples that are used to normalize (visually or using densitometry) the immunoprecipitation samples. Thus, it is worthwhile to try different numbers of PCR cycles to avoid saturation.

Quantitative PCR is used routinely in our lab for analysis of the ChIP assay. In our experience, qPCR yields the same rank order differences between samples as conventional PCR. We use promoter-specific primers and FAM (6-carboxyfluorescein)-labeled *Taq*-man probes, which are much preferred and increase the specificity of the reaction compared to SYBR Green, on an ABI 7000 (Applied Biosystems). Quantitation with *Taq*-man probes involves a fluorescent moiety attached to the probe that is quenched in the unbound state and gives a fluorescent signal when bound to the proper sequence in the PCR product. Thus, as the PCR cycle number increases, more probes are in the bound state to give more fluorescent signal. Each assay includes standards, a no-template control, a control sample, and the input and experimental samples. A standard curve is generated using six threefold serial dilutions from a standard (concentrated ChIP input DNA, enough made and frozen in aliquots, to last at least across all the assays in the experiment so different qPCR runs can be compared). The theoretical slope for the standard curve is –3.32. The initial concentration of the standard and dilutions are chosen to encompass the input and experimental samples so they fall within the standard curve. The no-template control is pure water to monitor PCR product contamination. The control sample is a known amount of DNA, e.g., 0.1 μg/μl tadpole genomic DNA, and is used to assess the consistency of the standard curve calculated from the standards across qPCR runs. The values for the input and experimental samples are calculated from the standard curve, and then percentage input is calculated for each experimental sample from the corresponding input sample.

We test three kinds of DNA regions in qPCR experiments. For us, the promoters of interest are thyroid hormone-regulated promoters containing T3 response elements (TREs), which are the regions that we amplify. We use two negative controls. First, a control promoter lacking a TRE is not expected to bind the transcription factor of interest, in our case thyroid hormone receptor, and is not expected to change across treatments. Second, a nonpromoter region at least 3000 base pairs away from the promoter region of interest, such as a downstream exon controlled by the promoter of interest, is used to identify potential high background from poor sonication, nonspecific antibody binding, incomplete washing after immunoprecipitation, or nonspecific binding from too much antibody. Also, by comparing the signal from this nonpromoter control region to that from the experimental DNA region, we can determine whether the antibody gives specific signal in the experimental DNA region (the TRE region in our case).

V. NOTES

Note 1: Examination using a hemocytometer after this step reveals debris from disrupted cells, intact nuclei, and few intact cells.

We have examined the homogenate after this step by treating the preparation with DAPI and examining an aliquot on a hemocytometer under a fluorescent compound microscope. We found large amounts of debris and that most cells are disrupted. Most of the cell debris is removed by washing in nuclei extraction buffer, followed by rehomogenization and filtration; most of the cell debris is removed, resulting in relatively pure nuclei. After the optional centrifugation through the high-sucrose buffer, the nuclei are difficult to distinguish from debris because they are shrunken and misshapen from the high ionic strength of the sucrose buffer, so that it is not clear if this step improves purity.

Note 2: The previous protocol purified the nuclei through a high-sucrose buffer (Damjanouski et al., 2002). This step is not necessary in our hands. In addition, in the previous protocol, DNA-protein crosslinking was done after nuclei isolation. To better preserve native DNA-protein interactions, we have tried to add formaldehyde during tissue homogenization or after homogenization but before nuclei isolation. In our hands, they all produced qualitatively similar ChIP results as the previous method, although we have noticed a higher signal using antibodies against modified histones by skipping the high-sucrose step.

Note 3: The sonication condition should be optimized depending on the sonicator used. To optimize the shearing conditions, set up a pilot experiment with different extents of sonication. Check the levels of sonication on a 1% agarose gel after de-crosslinking and DNA purification to ensure ~200–1000 base pair size. Immerse the microtip in the solution in a 1.7 mL eppendorf tube, keeping the tip quite close to the bottom and avoiding the walls of the tube. Lower sonication volumes lead to a higher shearing efficiency, so that changing the volume requires reoptimization. Avoid trapping air bubbles and emulsifying the sample during sonication, as this can compromise the efficiency of sonication. Under our conditions with a Branson sonifier 450 set at 30% duty cycle and output control 2, 10–12 cycles of 10-sec pulses with 10-sec cooling between pulses using a stepped microtip yielded optimal results.

Some workers include a restriction enzyme step as an additional means to fragment the DNA to reduce the proportion of larger sized fragments 1–2 kilobases (kb) in the chromatin mixture (Chaya and Zaret, 2004). This step is important for experiments where DNA regions of interest are relatively close to each other (within 2 kb) or for when the immunoprecipitated chromatin will be subsequently used on a microarray to identify unknown promoter targets. We have not tried to incorporate this step because our experiments have thus far involved only comparisons of transcription factor and cofactor binding to known promoters very distant from each other.

Note 4: If RNA concentration is too high and interferes with the DNA measurement, treat samples with RNAse, purify DNA, and then measure. We generally have very little RNA as evidenced by the lack of transfer RNAs in the gel for checking sonicated DNA. DNA yields vary depending upon tissue, where we usually get between 45 and 90 μg chromatin from tails and 60 and 180 μg from intestines with 30–50 mg of starting material.

Note 5: A titration needs to be performed for each antibody to identify the appropriate concentration. For example, in a pilot experiment, prepare six tubes with 500 μL of sample and add 5, 10, 15, 20, 40 μL to the first five tubes and add 40 μL of preimmune serum or an irrelevant antibody to the sixth tube. Complete the ChIP assay and choose the antibody amount that gives the highest signal from the experimental DNA region and the least background from a control DNA region.

VI. SHORT VERSION OF PROTOCOL

Chromatin isolation from tissue:

1. Place dissected tissue in 1 mL nuclei extraction buffer in dounce on ice.
2. Crush with 10–15 strokes using pestle A.
3. Transfer to 1.7 ml tubes, add 25 uL 37% formaldehyde, and rotate at room temperature for 15 min.
4. Add 100 μL 1 M Tris-HCl, pH 9.5 and rotate for another 5 min.
5. Centrifuge at 2000 g at 4°C for 2 min.
6. Resuspend pellet in 1 ml of nuclei extraction buffer, transfer to the dounce tubes on ice.
7. Rehomogenize with 5–10 strokes.
8. Filter through a Falcon 100 μm cell strainer and recentrifuge.
9. Resuspend pellet in 200–300 μl SDS lysis buffer on ice.
10. Sonicate.
11. Centrifuge at 14,000 rpm for 10 min at 4°C. Transfer the supernatant to fresh tubes and quantitate DNA.
12. Make frozen aliquots.

Immunoprecipitation and DNA purification:

1. Adjust the DNA to 100 ng/μl using the SDS lysis buffer. Then, dilute samples to 10 ng/μl with ChIP dilution buffer.
2. Preclear the chromatin using 60 μl of slurry of salmon sperm DNA/protein A agarose per 1 mL of chromatin solution for 30 min with rotation at 4°C.

3. Pellet the agarose beads by centrifugation at 1000 g for 2 min at 4°C.
4. Immunoprecipitation: Add 10–60 μl of slurry of salmon sperm DNA/protein A agarose to appropriate number of empty tubes. Aliquot 500 μL of precleared chromatin sample into one tube for each antibody. Add antibody.
5. Pipet 20 μL into another 1.7 mL eppendorf tube for input controls.
6. Incubate the samples with rotation from 4 hr to overnight at 4°C.
7. After incubation, pellet the beads at 1000 g for 2 min at 4°C, remove supernatant, and add 1 mL of ChIP I.
8. Rotate for 5–15 min at 4°C, then repeat wash (step 7) with 1 ml of ChIP II, III, and TE buffers.
9. After last wash, add 100 μL of elution buffer and rotate at 65°C for 6 hr to overnight. Do not forget to include input samples at this step.
10. Purify DNA by phenol/chloroform extraction or using the Qiaquick PCR purification kit (Qiagen).
11. Resuspend (for phenol/chloroform extraction) or elute DNA (for the Qiagen kit) in 40 μl of water or EB buffer (Qiagen, 10 mM Tris-HCl, pH 8.5).

REFERENCES

Buchholz, D.R., Hsia, S.-C.V., Fu, L., Shi, Y.-B., 2003. A dominant negative thyroid hormone receptor blocks amphibian metamorphosis by retaining corepressors at target genes. Mol. Cell. Biol. 23, 6750–6758.

Buchholz, D.R., Paul, B.D., Shi, Y.-B., 2005. Gene-specific changes in promoter occupancy by thyroid hormone receptor druing frog metamorphosis: Implications for gene regulation. J. Biol. Chem. 280, 41222–41228.

Buchholz, D.R., Tomita, A., Fu, L., Paul, B.D., Shi, Y.-B., 2004. Transgenic analysis reveals that thyroid hormone receptor is sufficient to mediate the thyroid hormone signal in frog metamorphosis. Mol. Cell. Biol. 24, 9026–9037.

Chaya, D., Zaret, K.S., 2004. Sequential chromatin immunoprecipitation from animal tissues. Methods Enzymol. 376, 361–372.

Damjanovski, S., Sachs, L.M., Shi, Y.-B., 2002. Function of thyroid hormone receptors during amphibian metamorphosis. Methods Mol. Biol. 202, 153–176.

Das, P.M., Ramachandran, K., vanWert, J., Singal, R., 2004. Chromatin immunoprecipitation assay. Biotechniques 37, 961–969.

Farnham, P.J., Weinmann, A.S., 2002. Identification of unknown target genes of human transcription factors using chromatin immunoprecipitation. Methods 26, 37–47.

Jepsen, K., Rosenfeld, M.G., 2002. Biological roles and mechanistic actions of co-repressor complexes. J. Cell Sci. 115, 689–698.

Kuo, M.-H., Allis, C.D., 1999. In vivo cross-linking and immunoprecipitation for studying dynamic protein: DNA associations in a chromatin environment. Methods 19, 425–433.

Nieuwkoop, P.D., Faber, J., 1994. Normal table of Xenopus laevis (Daudin). Garland Publishing, Inc., New York, p. 252.

Parrizas, M., Maestro, M.A., Boj, S.F., Paniagua, A., Casamitjana, R., Gomis, R., Rivera, F., Ferrer, J., 2001. Hepatic nuclear factor 1-alpha directs nucleosomal hyperacetylation to its tissue-specific transcriptional targets. Mol. Cell. Biol. 21, 3234–3243.

Paul, B.D., Fu, L., Buchholz, D.R., Shi, Y.-B., 2005. Coactivator recruitment is essential for liganded thyroid hormone receptor to initiate amphibian metamorphosis. Mol. Cell. Biol. 25, 5712–5724.

Sachs, L.M., Damjanovski, S., Jones, P.L., Li, Q., Amano, T., Ueda, S., Shi, Y.-B., Ishizuya-Oka, A., 2000. Dual functions of thyroid hormone receptors during *Xenopus* development. Comparative Biochemistry and Physiology, Part B 126, 199–211.

Sachs, L.M., Jones, P.L., Havis, E., Rouse, N., Demeneix, B.A., Shi, Y.-B., 2002. Nuclear receptor corepressor recruitment by unliganded thyroid hormone receptor in gene repression during *Xenopus laevis* development. Mol. Cell. Biol. 22, 8527–8538.

Sachs, L.M., Shi, Y.-B., 2000. Targeted chromatin binding and histone acetylation in vivo by thyroid hormone receptor during amphibian development. Proceedings of the National Academy of Sciences 97, 13138–13143.

Shi, Y.-B., 1999. Amphibian metamorphosis: From morphology to molecular biology. John Wiley and Sons, Inc., New York, pp. 1–288.

Spencer, V.A., Sun, J.-M., Li, L., Davie, J.R., 2003. Chromatin immunoprecipitation: a tool for studying histone acetylation and transcription factor binding. Methods 31, 67–75.

Tomita, A., Buchholz, D.R., Shi, Y.-B., 2004. Recruitment of N-CoR/SMRT-TBLR1 corepressor complexes by unliganded thyroid hormone receptor for gene repression during frog development. Mol. Cell. Biol. 24, 3337–3346.

Wells, J., Farnham, P.J., 2002. Characterizing transcription factor binding sites using formaldehyde crosslinking and immunoprecipitation. Methods 26, 48–56.

Yoon, H., Chan, D., Huang, Z., Li, J., Fondell, J., Qin, J., Wong, J.M., 2003. Purification and functional characterization of the human N-CoR complex: the roles of HDAC3, TBL1 and TBLR1. EMBO J. 22, 1336–1346.

Section IV

Emerging Strategies for the Analysis of Signaling in Development

12 Chemical Biology in Zebrafish Vascular Development

Joanne Chan and Thomas M. Roberts

CONTENTS

I. INTRODUCTION

A. Advantages of Using the Zebrafish for Chemical Biology

Advances in our understanding of biology and chemistry in the last 20 years have led to a new and exciting era for research and medicine as small molecule chemical inhibitors enter the clinical arena for the treatment of human diseases. Understanding the mechanism of disease and developing methods to inhibit selected proteins are the key elements in the success of molecularly targeted therapies. The development of Gleevec as a small molecule chemical inhibitor of the ABL tyrosine kinase is based on the discovery of BCR-ABL overactivity in chronic myelogenous leukemia.[1] As we look into the continued collaboration of chemistry and biology, signaling pathways and molecular disease targets may be revealed with unprecedented speed as new technologies are being developed. One recent development is the use of the zebrafish embryo as a model system for chemical biology.

The zebrafish, *Danio rerio*, is an attractive vertebrate organism for genetics and developmental biology, largely due to the transparency of the embryo and its rapid external development. This optical clarity, combined with the use of transgenic lines where specific cell types have been labeled with fluorescent proteins, has provided the researcher an opportunity to observe cell biology *in vivo*. As a vertebrate organism, the zebrafish also shares many organ systems with mammals, so that examination of complex cellular and molecular events leading up to organogenesis can be studied.[2, 3] In addition, the fecundity of the zebrafish allows the use of large sample sizes, which, in turn, improves statistical confidence in the experimental outcome. The cost of maintaining a large colony of zebrafish is at a fraction of the cost for an equivalent colony of mammalian model organisms, allowing more time and resources to be devoted to experimentation. Mutant zebrafish phenotypes, transgenic lines, molecular marker information, ESTs, and expression patterns are all available through the ZFIN website (http://zfin.org/cgi-bin/webdriver?MIval=aa-ZDB_home.apg). Zebrafish lines, molecular markers, and reagents can be obtained through the Zebrafish International Resource Center (http://zfin.org/zf_info/stckctr/stckctr.html). In addition, many established strategies for genetic screens, transgenesis, embryological assays, and molecular analyses are currently available in methodology books that provide a great resource for researchers interested in initiating zebrafish work.[4, 5]

B. Methods of Analysis Using the Zebrafish Embryo

In addition to visual analysis, the ease of molecular manipulations in the zebrafish has greatly facilitated the analysis of gene function *in vivo*. Overexpression or downregulation of a target protein by microinjection of mRNA or antisense morpholino frequently allows the researcher to examine the experimental results overnight or over the course of a few days.[6] Antisense morpholino technology can be used to target the translational start site of an mRNA or to block a splice junction of a pre-mRNA, to provide efficient functional knockdown of diverse targets.[6] For example, the use of a VEGF-A morpholino effectively inhibits blood vessel formation in the zebrafish embryo,[7] whereas overexpression of the VEGF-A ligand generates excessive and sometimes aberrant blood vessels.[8–10] Altering VEGF-A levels in the zebrafish embryo has been used in a number of studies to examine the components of this signaling pathway and its interactions with other molecules during vascular development.[9–12] Despite the usefulness of these experiments, both the RNA for overexpression and morpholino for downregulation of protein function have relatively short lives when they are microinjected into one-to four-cell stage embryos. In addition, these changed levels of VEGF signaling are induced at early embryonic stages that sometimes preclude the ability to examine their effects on later developmental events.

The ability to inhibit target gene function at later stages of development holds the greatest appeal for the use of chemical genetic approaches. In the last few years, the zebrafish has become a novel model system for chemical biology.[13, 14] As zebrafish genes and proteins are highly homologous to their mammalian counterparts, small molecule inhibitors designed for human use can also block the function of their zebrafish counterparts.[11, 15] The use of chemical inhibitors provides the unique advantage of facile temporal control over gene function that can complement molecular and genetic studies in the zebrafish. Compared with current animal models, the zebrafish appears extremely well suited for chemical biology. Its small size (1–5 mm in length during the first 5 days of development) and aquatic habitat allows easy addition of chemicals.[13]

The amount of compound needed for testing in this whole animal model system is relatively small. In the course of our study on chemical regulation of blood vessel formation in the zebrafish, we tested over 30,000 animals, with less than 5 mg of the compound in question for the entire study.[11] Forward chemical genetic screens can be performed in the zebrafish, as high-diversity compound libraries have been used to identify chemicals with biological activity in live embryos.[16, 17] In addition, selected compounds with high potency can be used in reverse chemical genetic studies, e.g., for the dissection of specific signaling pathways during vertebrate development.[11, 17] As an animal model system, zebrafish embryonic phenotypes can be exploited as a readout for the analysis of compound behavior, *in vivo* target validation, and the examination of adverse effects. Although chemical biology can be applied to many tissues and cell types in the zebrafish, in this review, we will discuss recent studies using chemical biology to study signaling pathways regulating blood vessel formation in the zebrafish embryo.

II. DEVELOPMENT OF THE ZEBRAFISH EMBRYONIC VASCULATURE

A. Zebrafish as a Model System for Vascular Biology

The transparent zebrafish embryo is particularly well suited for vascular biology studies. As a vertebrate organism, the initial patterning, differentiation, and migration of various cell types required for blood vessel formation closely resemble mammalian vasculogenesis and angiogenesis.[3, 18] The embryonic zebrafish circulatory system is perhaps one of the easiest systems to observe due to the transparency of the embryo during the first few days after fertilization. A detailed account of the developing vasculature during the first 7 days of development has been carefully described by Isogai et al.,[18] which provides a comprehensive guide for the phenotypic analysis. Assessment of vascular development is particularly easy in the transparent zebrafish embryo, as blood flow can be used as an indicator of vessel function. Blood vessel formation in the zebrafish share similarities with other vertebrates, as the processes of both vasculogenesis and angiogenesis also occur in the zebrafish embryo.[19–22]

The major axial vessels in the zebrafish, the dorsal aorta, and posterior cardinal vein form by vasculogenesis as angioblasts originating from the lateral plate mesoderm migrate towards the midline from 14 hours post fertilization (hpf) to 16 hpf to form a vascular cord that lumenizes over the next few hours.[21, 22] The zebrafish heart begins to beat at about 24 hpf; blood flow follows shortly afterwards, from about 24 hpf to 26 hpf (Figure 12.1).[3, 18, 23] By 48 hpf, intersegmental vessels are formed along the trunk by spouting angiogenesis.[18] Vascular complexity increases over the next few days as the zebrafish develops all the organ systems required to start feeding at about 5 days post fertilization (dpf). Although visual analysis can be very informative, a couple of simple techniques can also be used to provide pictorial records of the zebrafish vasculature (Figure 12.1). These include microangiography, where a fluorescent agent is injected into the circulation[24] (Figure 12.1b, d, e), and the detection of endogenous alkaline phosphatase activity found in vertebrate endothelial cells[10] (Figure 12.1d–f).

B. Transgenic Zebrafish Lines Facilitate Analysis of Cell Biology *In Vivo*

The generation of transgenic zebrafish lines with endothelial expression of fluorescent proteins has made it possible to observe endothelial cells in a living embryo. Several such lines have been reported, each using a vascular gene promoter to drive the expression of green or red fluorescent proteins (e.g., *Tg[Tie2:GFP]*[19]; *Tg[(VEGFR2/flk:G-RCFP]*[25]; *Tg[flk:EGFP]*[22]; *Tg[lmo2:EGFP]*[26]; and *Tg[lmo2:DsRed]*[26]). In particular, the fli-promoter driven-eGFP line,[12] *Tg[fli1:EGFP]y1*, has become a particularly useful tool for vascular biology studies, in part due to the high copy number of the transgene (great than 25 copies) and its availability through the Oregon stock center. The *Tg[fli1:EGFP]y1* line was used to examine endothelial cell behavior *in vivo* using time-lapse analysis.[27] This study revealed the how connections are formed between endothelial cells as they sprout from the major axial vessels. Expression of fluorescent proteins in the *Tg[fli1:EGFP]y1* line is not restricted to endothelial cells but is also expressed in nonvascular neural crest-derived tissues such as the developing cartilage of the jaw, to facilitate studies on the development of the endodermal pouches.[28] This line is extremely bright, and the endothelial-EGFP signal persists in blood vessels of the adult zebrafish, allowing examination of neovascularization during adult tail fin regeneration.[29, 30]

When used in combination with tissue- or organ-specific transgenic lines, cellular interactions between endothelial and adjacent cell types may be observed. A number of zebrafish transgenic lines where organs or tissues have been labeled with fluorescent tags have also been reported. Some examples include neuronal (e.g.,

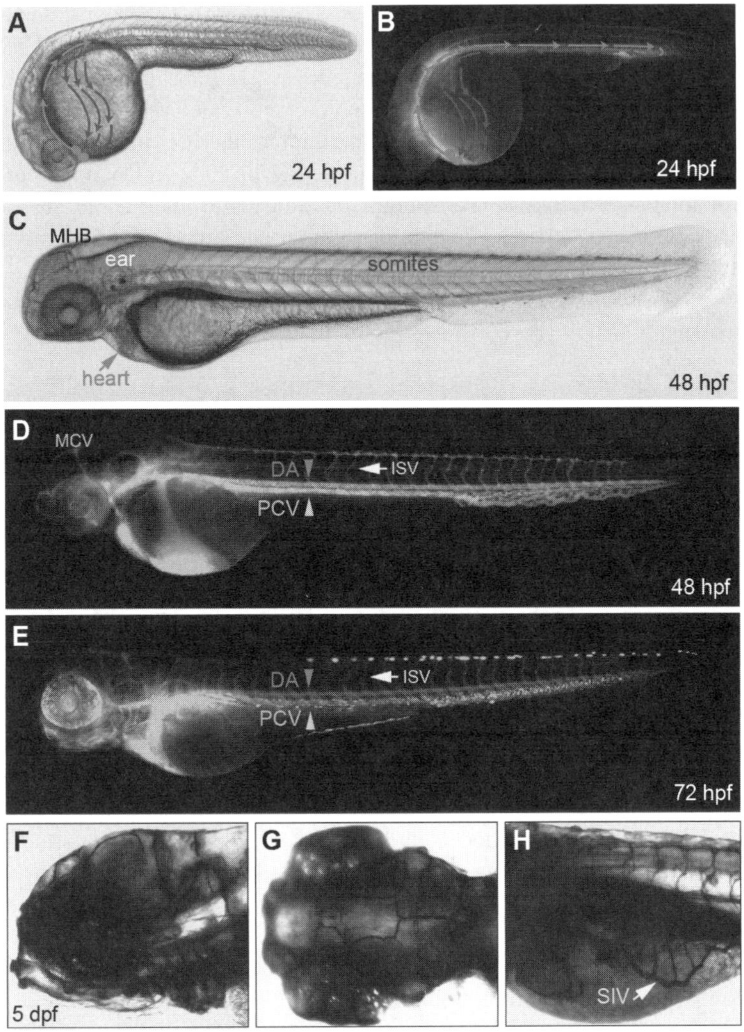

FIGURE 12.1 (See color insert following page 144.) Zebrafish vascular development during the first few days post fertilization using microangiography and detection of endogenous alkaline phosphatase activity in endothelial cells. Lateral views in all panels, anterior is to the left. **A.** Phase view of a zebrafish embryo at 24 hpf; the first circulatory loop is formed, the two-chambered heart begins to beat, and blood circulation can be observed from 24 hpf to 26 hpf. **B.** Microangiography in a 24 hpf embryo indicates simple circulatory loop. **C.** Phase view at 48 hpf, showing the locations for several landmarks such as the midbrain-hindbrain boundary (MHB), heart, ear, and somites. **D.** Microangiography of the same embryo as in **C**, at 48 hpf. Major vessels are labeled: DA (dorsal aorta), PCV (posterior cardinal vein), ISVs (intersegmental vessels), midcerebral vein (MCV). **E.** Microangiography of a zebrafish embryo at 72 hpf. Major vessels are labeled as above. **F, G, H.** Alkaline phosphatase staining of endogenous activity in endothelial cells in a zebrafish larva at 5 days post fertilization. **F, H.** Lateral views showing head and trunk vessels, SIV, subintestinal veins. **G.** Dorsal view of head vessels.

Tg[huC:GFP][31] as a pan-neural GFP line and *Tg[islet:EGFP]*[32] for cranial motoneuron neurons); heart muscle *Tg[cmlc:EGFP]*[33]; gut *Tg[nkx2.2a:eRGFP]:GFP]*[34]; and liver *Tg[L-FABP:EGFP]*[35] lines. As the methods for transgenesis have improved significantly over the last few years, many more vascular or tissue-specific promoter-driven lines are likely to be generated. For example, the use of the Tol2-transpose mediated transgenesis system has been reported to produce up to 50% founders.[36] The combined use of tissue and endothelial transgenic lines could significantly accelerate the speed of characterization of vascular and nonvascular phenotypes and facilitate studies on the cell-to-cell interactions in an intact animal.

C. Conserved Roles for Zebrafish and Mammalian Genes and Pathways

Of particular interest to biologists and chemists alike is the remarkable conservation of genes and pathways between zebrafish and mammals. For vascular biology, the master regulators for endothelial cell growth, migration, and survival are the vascular endothelial growth factors (VEGFs) and their receptors (VEGFRs). These angiogenic molecules control physiological and pathological angiogenesis in humans.[37] In addition to VEGF and its receptors, other proteins known to play important roles in vascular biology such as the angiopoietins and Tie receptors; members of the Notch signaling pathway; the arterial marker ephrinB2; and its cognate receptor ephB4 all have equivalent homologs in the zebrafish, with expression patterns or genetic data to demonstrate conserved roles in blood vessel formation.[20]

Indeed, the critical roles of shh, VEGF, and Notch in defining arterial-venous cell fate was determined in the zebrafish using genetic, molecular, and chemical inhibition.[38, 39] These important pathway interactions have been confirmed in mammalian studies using animal models and cell culture systems. Although null mutations in the mouse *shh* gene have not exhibited vascular defects, shh has been shown to induce the expression of *VEGF-A* isoforms in interstitial mesenchymal cells and in a whole animal hindlimb ischemia model.[40, 41] *Notch 1*, *Notch 4*, as well as the Notch ligand *Delta-like-4* have been shown in genetic studies in the mouse to affect arterial cell fate determination.[42–44] In addition, VEGF-A, but not the basic fibroblast growth factor, was also able to upregulate the expression of *Notch 1* and the *Delta-like-4* ligand expression in human arterial endothelial cells,[45] thus demonstrating strong conservation in these pathway interactions among vertebrate animals.

Intracellular signaling components also appear to be conserved. In mammalian studies, the *phospholipase C gamma 1* (*PLCγ1*) gene is an important effector downstream of the mouse VEGFR2/flk-1 receptor, as a point mutation at this site phenocopies the receptor null mutation.[46, 47] In addition, it has also been reported to be tyrosine phosphorylated by the VEGFR1/flt-1 receptor.[48, 49] In the zebrafish, two independent alleles have been reported, where deficiencies in *PLCγ1* gene function led to defects in cardiovascular function.[9, 50] A splicing defect in the *PLCg1*^{y10} allele[9] resulted in a high proportion of *PLCγ1* mRNAs bearing a stop codon, with a minor portion of wild-type mRNAs. Arterial-side defects observed in this mutant were proposed to be due to reduced VEGFR2/flk-1 signaling.[9] Positional cloning of the *dead beat* mutant, *ded*^{m582}, revealed a null mutation, also in the *PLCγ1* gene.[50] This mutant features impaired cardiac ventricular contractility, as well as the lack of a lumenized vasculature. In the *ded*^{m582} allele, a defect in VEGFR1/flt-1 signaling is implicated in the cardiac contractility defect.[50] Thus, the conserved use of homologous molecules and signaling pathways underscores the importance of the zebrafish as a model organism.

D. Targeting Blood Vessels for the Treatment of Disease

Abnormal blood vessels play important roles in many human diseases. Although some conditions are driven by overactive blood vessel formation, as in tumor angiogenesis, macular degeneration, and endometriosis, others are affected by poor perfusion because of partially or completely blocked vessels, as in atherosclerosis, or by arteriovenous malformations where improper connections between arteries and veins prevent proper blood flow.[37] The VEGF-A ligand and its receptors play crucial roles in physiological and tumor angiogenesis. For hyperactive vessels such as those found in tumors, blocking the ability of VEGF-A to interact with its receptors has been proven effective, as the anti-ligand monoclonal antibody, Avastin, appear to be able to neutralize its effects. Currently, Avastin is being used in combination with chemotherapy as the first line of defense against colorectal cancer.[51, 52]

The VEGFRs belong to a larger family of receptor tyrosine kinases (RTKs) that signal via a common mechanism. Ligand binding induces activation of its cytoplasmic tyrosine kinase domain and activation of effector molecules, which transduce the signal downstream from the receptor.[53] The successful clinical use of Gleevec as a small molecule inhibitor has demonstrated that blocking an overactive kinase can lead to cancer therapy.[1, 54] This, combined with the clinical use of Avastin, provides strong evidence that targeting the VEGF-VEGFR signaling pathway using small molecules might also yield therapeutic results. Effort from the pharmaceutical industry has generated a number of potent inhibitors, in the hopes of providing an orally available agent for anti-angiogenic therapy.[55] Several potent inhibitors have been identified with high affinity towards the VEGFRs. However, because RTKs share a similar intracellular enzymatic moiety, overlapping affinities are also found. Promising candidates are at various stages of clinical trials. Among these, PTK787/ZK222854, SU11248, and ZD6474 exhibit high affinity for the VEGFRs, with lower affinity for the related platelet-derived growth factor receptors (PDGFRs), or the epithelial growth factor receptors (EGFRs).[55, 56] In some cases, a broad-spectrum inhibitor might be useful in targeting both the cancer proper and its blood supply.

E. Zebrafish as a Model for Evaluation of Anti-angiogenic Compounds

Although small molecule inhibitors and the anti-angiogenic approach are both proven strategies, finding the most effective anti-angiogenic or anti-tumor compound with minimal adverse effects still poses a challenge. Current whole animal models for angiogenesis include the matrigel plug assay; the cornea micropocket assay; or tumor implants, typically done in small mammals such as the mouse, rat, or rabbit.[57]

Rodent tumor models are frequently used to test the effectiveness of an anti-angiogenic compound, using a reduction in tumor size and its vascularization as indicators, e.g., Wood et al.[58] We and others have shown that the zebrafish embryo can be used as an effective model for the evaluation of anti-angiogenic compounds.[11, 25, 59] The transparency of the zebrafish embryo permits easy evaluation of anti-angiogenic compounds by visual scoring for the presence or absence of blood vessels. The ability to screen through a large number of compounds using large sample sizes in a live vertebrate model system provides significant advantages over the use of cell culture or mammalian systems. The use of the zebrafish embryo can be a useful prescreen animal model for the evaluation of lead compounds before investment in more expensive and time-consuming mammalian model systems.

III. CHEMICAL ANALYSIS OF VASCULAR FUNCTION IN THE ZEBRAFISH

A. Precise Temporal and Dosage Regulation of Gene Function

The ability to introduce chemicals to the embryo medium at any point during the first few days of development has made it possible to develop and expand the use of chemical genetic approaches in the zebrafish. The use of highly specific small molecule inhibitors allows precise temporal control over gene function that facilitates the dissection of signaling pathways and developmental processes. So far, the use of chemical biology in zebrafish vascular development has improved our understanding of the VEGF signaling pathway and identified critical molecules in embryonic blood vessel formation.

The remarkable conservation between human and zebrafish versions of the VEGFR2 receptor (78% identity in kinase domain[11]) enabled us to use an inhibitor designed against the human VEGFRs, PTK787/ZK222584,[58, 60] to regulate blood vessel formation in the living zebrafish embryo.[11] We found this inhibitor to provide robust regulation of blood vessel formation without inducing adverse effects on other tissues by visual inspection. This small molecule kinase inhibitor was recently tested, along with 20 others, against 119 protein kinases and found to be one of the most selective.[61] It has high affinity for all three receptors of the VEGFR family and a 10-fold lower affinity towards the related PDGFR-β (e.g., for VEGFR2: IC_{50} is 42 nM; for PDGFR-β: IC_{50} is 490 nM[55]). By regulating inhibitor dosage and exposure time, this compound provided precise control over receptor function and blood vessel formation in the live zebrafish embryo.[11] Furthermore, particular sets of blood vessels can be targeted by the timing of inhibitor addition. When the VEGFR inhibitor is added early, before the onset of blood vessel formation, major vessels such as the dorsal aorta and the posterior cardinal vein fail to form. This effect is dose-dependent, with 100% inhibition of the dorsal aorta at a high inhibitor concentration of 5 μM. The micromolar level of inhibitor needed likely reflected the layers of cells within an embryo that limited cellular accessibility of the chemical inhibitor. The same inhibitor was effective at nanomolar levels when used on cells expressing the zebrafish VEGFR2 receptor.[11] The increased dosage needed for target inhibition in whole embryo has also been reported in other studies.[62]

B. Chemical Genetics Mimic Classical Genetic Studies

In classical genetic studies using *Drosophila* or *C. elegans*, modulation of gene dosage effects by the use of hypomorphic or temperature-sensitive alleles has provided valuable information on the components of a signaling pathway.[63–65] Using chemical genetic control of VEGFR function in place of a genetic mutation in the zebrafish, we were able to regulate the precise amount of target function in order to address questions about the receptor's signaling pathway *in vivo*. We were able to use the chemical VEGFR inhibitor from 24 hpf to 30 hpf, after the formation of the major axial vessels, to inhibit the formation of the intersegmental vessels, allowing about 30% of these vessels to form by 48 hpf. Under these conditions, we showed that upregulation of AKT/PKB activity was able to override the receptor block to allow about 80% of these vessels to form in the presence of the chemical inhibitor.[11] Only the activated, myristolyated form of AKT/PKB, but not the kinase-dead version, can provide functional bypass of VEGFR inhibition. This rescue experiment suggests that the signaling threshold required for endothelial cell function can be achieved by increasing the activity of a downstream effector in the pathway. It illustrates the use of chemical probes for the dissection of biological processes.

In addition, it demonstrates the effectiveness of chemical control of gene dosage in a vertebrate organism, mimicking classical genetic studies in flies and worms without the limitation of having a mutation in a particular gene. As many human genetic disorders result from compromised gene function, the ability to consistently generate a desired gene dosage in a vertebrate organism could provide insights into the signaling pathway defects relevant to human disease. Our study also suggests that other signaling pathways can be dissected once specific chemical inhibitors become available.

IV. FORWARD CHEMICAL GENETIC SCREENS IN THE ZEBRAFISH

A. Chemical Library Screening in Zebrafish

One major task in the post-genomic era is the functionalization of enormous amounts of sequence information, in order to define genes and gene function. The use of model organisms has been an essential part of this process, as gene functions defined in yeast, worms, flies, zebrafish, and mice have shown conserved roles in humans. Chemical biology provides another approach that can be used in conjunction with model organisms to expedite the dissection of signaling pathways. So far, chemical library screens for drug discovery have been carried out using *in vitro* and cell-based assays.[66, 67] Whole organism approaches for chemical screening is also possible using yeast, worms, flies, or zebrafish embryos.[14, 16] Of these, the zebrafish, as a vertebrate organism, affords an opportunity to investigate more complex processes such as organogenesis.[3] One of the major strengths of the zebrafish as a model system is its utility in genetic analysis. Many interesting zebrafish mutations have been identified through phenotypic or *in situ* expression pattern screens[68, 69]; however, the positional cloning of the affected gene is often very time-consuming. The use of specific chemical probes will be of great value in the characterization of mutant lines, with the potential for defining common signaling pathways that can provide candidates for the affected gene. In addition, biologically active compounds with minimal adverse effects may be identified that can lead to the development of important therapeutics.

Recently, high-diversity chemical libraries have become available through academic sources for screening. Thousands of transparent embryos can be generated each day for chemical library screening. A single person can screen through several hundred compounds per week. This type of visual screen has been used to screen chemical compound libraries to identify several highly specific chemicals affecting the nervous system, pigmentation, cardiovascular function, and the development of the zebrafish inner ear.[16] In large-scale chemical screens, three zebrafish embryos were placed into each well of a 96-well plate. Compounds were added to the embryo medium, and phenotypes were scored by visual analysis. Although the identification of the chemical target is a difficult task, it is possible to use chemical screens to reverse a particular phenotype. Thus, not knowing the exact protein target(s), chemicals can be identified in an unbiased manner with the goal of suppressing a specific defect or phenotype.

Peterson et al.[17] have recently performed a suppressor screen for compounds that can reverse the vascular defect in *gridlock(grl)* mutant embryos.[17] The *gridlockm145/m145* mutation is caused by a hypomorphic mutation in the *hey2* gene (also known as *HRT2*, *CHF1*, *HERP1*, and *HESR2*) and encodes a bHLH transcriptional repressor. This compromised *grl/hey2* gene function led to a short-circuiting of blood flow so that only the anterior portion of the zebrafish embryo is perfused.[24, 70, 71] Using this mutant line, a chemical suppressor screen was performed using 5000 compounds. Two structurally related compounds were identified, GS4012 and GS3999, that can generate a wild-type phenotype when added to $grl^{ml/145m145}$ homozygous mutants.[17] The molecular target for these chemicals has not yet been identified. As with genetic screens, the design of these chemical screens will play a crucial role in the successful identification of biologically interesting compounds.

B. Intersection of Genetics and Chemical Biology in Heart Development

The combined use of specific chemical probes and mutant zebrafish lines could provide a fast track to the elucidation of signaling pathways and biological processes. In the case of concentramide, a small molecule was identified that generated a highly specific patterning defect in the zebrafish embryo. Heart patterning was reversed so that the ventricle formed inside the atrium.[72, 73] This phenotype was also the distinguishing feature of the *heart-and-soul* mutant that resulted from defects in the atypical PKCλ gene (has^{m567}).[72, 73] In the $ha^{s}m^{567}$ mutant, the lack of normal PKCλ gene function resulted in cell polarity defects in many cell types in the embryo, whereas concentramide-induced defects were restricted to patterning of the anterior structures in the zebrafish, including the heart. Despite these differences, the chemical's target and the function of the PKCλ gene intersected at a point such that they both contributed to the proper patterning of the heart. Thus, in this case, information on the *has*^{m567} mutant has provided a specific phenotype that should facilitate the analysis of a biologically active chemical.

C. Combined Used of Multiple Approaches in the Zebrafish

Joining a wealth of mutant and transgenic lines and the ability to alter the levels of target proteins, chemical biology adds another set of tools for the zebrafish researcher. The use of chemical probes in the zebrafish is likely to accelerate our understanding of developmental process and signaling pathways. The ability to combine multiple approaches in a single vertebrate model system is a significant advantage for the zebrafish. One translational advantage to working with the zebrafish is that all zebrafish genes identified so far have human counterparts. A novel anti-angiogenic compound identified in a zebrafish screen can be immediately tested in mammalian systems to examine its effectiveness.

Studies in the zebrafish have already begun to use a combination of genetics, chemical inhibition, and the application of molecular approaches to examine signaling pathways. For instance, in vascular development, the role of genes controlling arterial cell fate was dissected using these combined approaches.[39] The chemical inhibitor, cyclopamine, blocks the function of smoothened, a receptor in the sonic hedgehog pathway; by doing so, cyclopamine essentially reproduces features of the genetic mutants deficient in shh signaling. The chemical inhibitor cyclopamine, and shh-deficient lines *sonic-you* (*syu*) or *you-too* (*yot*), were used to show that shh induces the expression of *VEGF-A* mRNA and functions upstream of genes in the formation of a vertebrate artery in the zebrafish embryo.[39] This relationship has been confirmed in mammalian studies, where shh was able to induce neovascularization in a hindlimb ischemia model and to induce the expression of *VEGF-A*, *Ang1*, and *Ang2* mRNAs in cell culture.[40, 41]

In the chemical suppressor study by Peterson et al.,[17] after the identification of two suppressors that can reverse the *grl*$^{m145/m145}$ phenotype, the authors further characterized the vascular-specific effects of these compounds. They discovered that the chemical suppressors induced an upregulation of *VEGF-A* mRNA levels in the zebrafish embryo. It turns out that *grl*$^{m145/m145}$ embryos have an endogenous reduction in *VEGF-A* mRNA that can be rescued by overexpression of this ligand. Thus, the analysis of the suppressor chemicals led to another genetic connection between the VEGF and the Notch signaling pathways, as the *gridlock/hey2* genes have been shown to be targets of Notch signaling.[74, 75] In this case, availability of the chemical suppressors helped reveal the signaling defects in this mutant line.

V. FUTURE DIRECTIONS FOR CHEMICAL BIOLOGY IN THE ZEBRAFISH

The future of chemical biology in the zebrafish certainly looks bright for both forward and reverse chemical approaches. In the forward approach, structurally diverse chemical libraries are now available through academic or commercial sources, so that chemical library screens can be performed by individual research labs to examine their favorite developmental process or signaling pathway. For reverse chemical genetics, more chemical compounds are becoming available for study. As a vertebrate model organism, the zebrafish is highly versatile, with the ability to accommodate a wide range of genetic, developmental, and chemical manipulations. As novel technologies are constantly being developed to improve its utility, the continued development of chemical approaches in the zebrafish will provide significant insights into gene function and signaling pathways relevant to developmental processes that might help understand their roles in human disease.

ACKNOWLEDGMENTS

We thank Robert E. Bolcome III and Kimberly L. Bellavance for preparation of the figure. In compliance with Harvard Medical School guidelines on possible conflict of interest, we disclose that one of the authors (T.M.R.) has consulting relationships with Upstate Biotechnology and Novartis Pharmaceuticals, Inc.

REFERENCES

1. Deininger, M., Buchdunger, E. & Druker, B. J. The development of imatinib as a therapeutic agent for chronic myeloid leukemia. *Blood* 105, 2640–53 (2005).
2. Kimmel, C. B., Ballard, W. W., Kimmel, S. R., Ullmann, B. & Schilling, T. F. Stages of embryonic development of the zebrafish. *Developmental Dynamics* 203, 253–310 (1995).
3. Thisse, C. & Zon, L. I. Organogenesis—heart and blood formation from the zebrafish point of view. *Science* 295, 457–62 (2002).
4. Nusslein-Volhard, C. D., R. *Zebrafish: A practical approach* (2002).
5. Westerfield, M. *The zebrafish book. Guide for the laboratory use of zebrafish (Danio rerio).* (Univ. of Oregon Press, Eugene, 1995).
6. Ekker, S. C. Morphants: a new systematic vertebrate functional genomics approach. *Yeast* 17, 302–306 (2000).
7. Nasevicius, A., Larson, J. & Ekker, S. C. Distinct requirements for zebrafish angiogenesis revealed by a VEGF-A morphant. *Yeast* 17, 294–301 (2000).
8. Liang, D. et al. The role of vascular endothelial growth factor (VEGF) in vasculogenesis, angiogenesis, and hematopoiesis in zebrafish development. *Mech Dev* 108, 29–43. (2001).
9. Lawson, N. D., Mugford, J. W., Diamond, B. A. & Weinstein, B. M. phospholipase C gamma-1 is required downstream of vascular endothelial growth factor during arterial development. *Genes Dev* 17, 1346–51 (2003).
10. Habeck, H. Analysis of zebrafish VEGF receptor mutant reveals specific disruption of angiogenesis. *Curr Biol* 12, 1405–1412 (2002).
11. Chan, J., Bayliss, P. E., Wood, J. M. & Roberts, T. M. Dissection of angiogenic signaling in zebrafish using a chemical genetic approach. *Cancer Cell* 1, 257–67 (2002).
12. Lawson, N. & Weinstein, B. *In vivo* imaging of embryonic vascular development using transgenic zebrafish. *Dev Biol* 248, 307–318 (2002).

13. MacRae, C. A. & Peterson, R. T. Zebrafish-based small molecule discovery. *Chem Biol* 10, 901–8 (2003).
14. Zon, L. I. & Peterson, R. T. *In vivo* drug discovery in the zebrafish. *Nat Rev Drug Discov* 4, 35–44 (2005).
15. Milan, D. J., Peterson, T. A., Ruskin, J. N., Peterson, R. T. & MacRae, C. A. Drugs that induce repolarization abnormalities cause bradycardia in zebrafish. *Circulation* 107, 1355–8 (2003).
16. Peterson, R. T., Link, B. A., Dowling, J. E. & Schreiber, S. L. Small molecule developmental screens reveal the logic and timing of vertebrate development. *Proc Natl Acad Sci U S A* 97, 12965–9 (2000).
17. Peterson, R. T. et al. Chemical suppression of a genetic mutation in a zebrafish model of aortic coarctation. *Nat Biotechnol* 22, 595–9 (2004).
18. Isogai, S., Horiguchi, M. & Weinstein, B. M. The vascular anatomy of the developing zebrafish: an atlas of embryonic and early larval development. *Dev Biol* 230, 278–301. (2001).
19. Motoike, T. et al. Universal GFP reporter for the study of vascular development. *Genesis* 28, 75–81. (2000).
20. Lawson, N. D. & Weinstein, B. M. Arteries and veins: making a difference with zebrafish. *Nat Rev Genet* 3, 674–82. (2002).
21. Torres-Vazquez, J., Kamei, M. & Weinstein, B. M. Molecular distinction between arteries and veins. *Cell Tissue Res* 314, 43–59 (2003).
22. Jin, S. W., Beis, D., Mitchell, T., Chen, J. N. & Stainier, D. Y. Cellular and molecular analyses of vascular tube and lumen formation in zebrafish. *Development* 132, 5199–209 (2005).
23. Stainier, D. Y. Zebrafish genetics and vertebrate heart formation. *Nat Rev Genet* 2, 39–48 (2001).
24. Weinstein, B. M., Stemple, D. L., Driever, W. & Fishman, M. C. Gridlock, a localized heritable vascular patterning defect in the zebrafish. *Nat Med* 1, 1143–7 (1995).
25. Cross, L. M., Cook, M. A., Lin, S., Chen, J. N. & Rubinstein, A. L. Rapid analysis of angiogenesis drugs in a live fluorescent zebrafish assay. *Arterioscler Thromb Vasc Biol* 23, 911–2 (2003).
26. Zhu, H. et al. Regulation of the lmo2 promoter during hematopoietic and vascular development in zebrafish. *Dev Biol* 281, 256–269 (2005).
27. Isogai, S., Lawson, N. D., Torrealday, S., Horiguchi, M. & Weinstein, B. M. Angiogenic network formation in the developing vertebrate trunk. *Development* 130, 5281–90 (2003).
28. Crump, J. G., Maves, L., Lawson, N. D., Weinstein, B. M. & Kimmel, C. B. An essential role for Fgfs in endodermal pouch formation influences later craniofacial skeletal patterning. *Development* 131, 5703–16 (2004).
29. Lawson, N. D. & Weinstein, B. M. In vivo imaging of embryonic vascular development using transgenic zebrafish. *Dev Biol* 248, 307–18 (2002).
30. Huang, C. C., Lawson, N. D., Weinstein, B. M. & Johnson, S. L. reg6 is required for branching morphogenesis during blood vessel regeneration in zebrafish caudal fins. *Dev Biol* 264, 263–74 (2003).
31. Park, H. C. et al. Analysis of upstream elements in the HuC promoter leads to the establishment of transgenic zebrafish with fluorescent neurons. *Dev Biol* 227, 279–93 (2000).
32. Higashijima, S., Hotta, Y. & Okamoto, H. Visualization of cranial motor neurons in live transgenic zebrafish expressing green fluorescent protein under the control of the islet-1 promoter/enhancer. *J Neurosci* 20, 206–18. (2000).

33. Huang, C. J., Tu, C. T., Hsiao, C. D., Hsieh, F. J. & Tsai, H. J. Germ-line transmission of a myocardium-specific GFP transgene reveals critical regulatory elements in the cardiac myosin light chain 2 promoter of zebrafish. *Dev Dyn* 228, 30–40 (2003).
34. Ng, A. N. et al. Formation of the digestive system in zebrafish: III. Intestinal epithelium morphogenesis. *Dev Biol* 286, 114–35 (2005).
35. Her, G. M., Chiang, C. C., Chen, W. Y. & Wu, J. L. *In vivo* studies of liver-type fatty acid binding protein (L-FABP) gene expression in liver of transgenic zebrafish (Danio rerio). *FEBS Lett* 538, 125–33 (2003).
36. Kawakami, K. et al. A transposon-mediated gene trap approach identifies developmentally regulated genes in zebrafish. *Dev Cell* 7, 133–44 (2004).
37. Ferrara, N., Gerber, H. P. & LeCouter, J. The biology of VEGF and its receptors. *Nat Med* 9, 669–76 (2003).
38. Lawson, N. D. et al. Notch signaling is required for arterial-venous differentiation during embryonic vascular development. *Development* 128, 3675–83 (2001).
39. Lawson, N. D., Vogel, A. M. & Weinstein, B. M. Sonic hedgehog and vascular endothelial growth factor act upstream of the Notch pathway during arterial endothelial differentiation. *Dev Cell* 3, 127–36. (2002).
40. Pola, R. et al. Postnatal recapitulation of embryonic hedgehog pathway in response to skeletal muscle ischemia. *Circulation* 108, 479–85 (2003).
41. Pola, R. et al. The morphogen Sonic hedgehog is an indirect angiogenic agent upregulating two families of angiogenic growth factors. *Nat Med* 7, 706–11 (2001).
42. Duarte, A. et al. Dosage-sensitive requirement for mouse Dll4 in artery development. *Genes Dev* 18, 2474–8 (2004).
43. Krebs, L. T. et al. Haploinsufficient lethality and formation of arteriovenous malformations in Notch pathway mutants. *Genes Dev* 18, 2469–73 (2004).
44. Gale, N. W. et al. Haploinsufficiency of delta-like 4 ligand results in embryonic lethality due to major defects in arterial and vascular development. *Proc Natl Acad Sci U S A* 101, 15949–54 (2004).
45. Liu, Z. J. et al. Regulation of Notch1 and Dll4 by vascular endothelial growth factor in arterial endothelial cells: implications for modulating arteriogenesis and angiogenesis. *Mol Cell Biol* 23, 14–25 (2003).
46. Takahashi, T., Yamaguchi, S., Chida, K. & Shibuya, M. A single autophosphorylation site on KDR/Flk-1 is essential for VEGF-A-dependent activation of PLC-gamma and DNA synthesis in vascular endothelial cells. *Embo J* 20, 2768–78 (2001).
47. Sakurai, Y., Ohgimoto, K., Kataoka, Y., Yoshida, N. & Shibuya, M. Essential role of Flk-1 (VEGF receptor 2) tyrosine residue 1173 in vasculogenesis in mice. *Proc Natl Acad Sci U S A* 102, 1076–81 (2005).
48. Ito, N. & Claesson-Welsh, L. Dual effects of heparin on VEGF binding to VEGF receptor-1 and transduction of biological responses. *Angiogenesis* 3, 159–66 (1999).
49. Ito, N., Huang, K. & Claesson-Welsh, L. Signal transduction by VEGF receptor-1 wild type and mutant proteins. *Cell Signal* 13, 849–54 (2001).
50. Rottbauer, W. et al. VEGF-PLCgamma1 pathway controls cardiac contractility in the embryonic heart. *Genes Dev* 19, 1624–1634 (2005).
51. Willett, C. G. et al. Direct evidence that the VEGF-specific antibody bevacizumab has antivascular effects in human rectal cancer. *Nat Med* 10, 145–7 (2004).
52. Ferrara, N., Hillan, K. J., Gerber, H. P. & Novotny, W. Discovery and development of bevacizumab, an anti-VEGF antibody for treating cancer. *Nat Rev Drug Discov* 3, 391–400 (2004).
53. Pawson, T. & Nash, P. Assembly of cell regulatory systems through protein interaction domains. *Science* 300, 445–52 (2003).

54. Druker, B. J. STI571 (Gleevec) as a paradigm for cancer therapy. *Trends Mol Med* 8, S14–8. (2002).
55. Manley, P. W. et al. Advances in the structural biology, design and clinical development of VEGF-R kinase inhibitors for the treatment of angiogenesis. *Biochim Biophys Acta* 1697, 17–27 (2004).
56. Arora, A. & Scholar, E. M. Role of tyrosine kinase inhibitors in cancer therapy. *J Pharmacol Exp Ther* (2005).
57. Hasan, J. et al. Quantitative angiogenesis assays in vivo—a review. *Angiogenesis* 7, 1–16 (2004).
58. Wood, J. M. et al. PTK787/ZK 222584, a novel and potent inhibitor of vascular endothelial growth factor receptor tyrosine kinases, impairs vascular endothelial growth factor-induced responses and tumor growth after oral administration. *Cancer Res* 60, 2178–2189 (2000).
59. Serbedzija, G. N., Flynn, E. & Willett, C. E. Zebrafish angiogenesis: a new model for drug screening. *Angiogenesis* 3, 353–9.
60. Bold, G. et al. New anilinophthalazines as potent and orally well absorbed inhibitors of the VEGF receptor tyrosine kinases useful as antagonists of tumor-driven angiogenesis. *J Med Chem* 43, 2310–2323 (2000).
61. Fabian, M. A. et al. A small molecule-kinase interaction map for clinical kinase inhibitors. *Nat Biotechnol* 23, 329–36 (2005).
62. Chan, J. & Serluca, F. C. Chemical approaches to angiogenesis. *Methods Cell Biol* 76, 475–87 (2004).
63. Rubin, G. M. Signal transduction and the fate of the R7 photoreceptor in Drosophila. *Trends Genet* 7, 372–7 (1991).
64. Sternberg, P. W. Intercellular signaling and signal transduction in C. elegans. *Annu Rev Genet* 27, 497–521 (1993).
65. Sternberg, P. W. & Horvitz, H. R. Signal transduction during C. elegans vulval induction. *Trends Genet* 7, 366–71 (1991).
66. Chanda, S. K. Fulfilling the promise: drug discovery in the post-genomic era. *Drug Discov. Today* 8, 168–174 (2003).
67. Drews, J. Drug discovery: a historical perspective. *Science* 287, 1960–1964 (2000).
68. Haffter, P. et al. The identification of genes with unique and essential functions in the development of the zebrafish, Danio rerio. *Development* 123, 1–36 (1996).
69. Patton, E. E. & Zon, L. I. The art and design of genetic screens: zebrafish. *Nat Rev Genet* 2, 956–66 (2001).
70. Zhong, T. P. Gridlock, an HLH gene required for assembly of the aorta in zebrafish. *Science* 287, 1820–1824 (2000).
71. Zhong, T. P. Gridlock signalling pathway fashions the first embryonic artery. *Nature* 414, 216–220 (2001).
72. Horne-Badovinac, S. et al. Positional cloning of heart and soul reveals multiple roles for PKC lambda in zebrafish organogenesis. *Curr Biol* 11, 1492–502 (2001).
73. Peterson, R. T., Mably, J. D., Chen, J. N. & Fishman, M. C. Convergence of distinct pathways to heart patterning revealed by the small molecule concentramide and the mutation heart-and-soul. *Curr Biol* 11, 1481–91 (2001).
74. Nakagawa, O. et al. Members of the HRT family of basic helix-loop-helix proteins act as transcriptional repressors downstream of Notch signaling. *Proc Natl Acad Sci U S A* 97, 13655–60 (2000).
75. Iso, T., Chung, G., Hamamori, Y. & Kedes, L. HERP1 is a cell type-specific primary target of Notch. *J Biol Chem* 277, 6598–607 (2002).

13 Investigating Gastrulation

Karen Symes

CONTENTS

I. INTRODUCTION

Gastrulation is a major morphogenetic event during early embryonic development, turning what is essentially a single-layered organism into a multilayered one. It forms the primary germ layers, the ectoderm, mesoderm, and endoderm, and establishes the basic body plan, which is elaborated on by successive developmental events. Gastrulation is driven by changes in cell shape, adhesion, and motility; however, this description fails to invoke the beauty of the process or the significance of bringing cells and tissues into contact with one another for the first time. In many embryos, most cells undergo an active form of movement, from intercalation to involution to migration to rotation and more, all coordinated with incredible precision. The variety and complexity of these movements and the resulting tissue reorganizations make gastrulation one of the most fascinating and demanding embryological processes to study. In recent years, significant progress has been made in identifying some of the molecular mechanisms of gastrulation; however, we are far from fully understanding how gastrulation is initiated and controlled, further adding to the allure of the process. When Lewis Wolpert said, "The most important event in your life is not birth, marriage, or death but gastrulation," he was not referring to the obsession of the researchers who work on gastrulation, but perhaps he should have been.

This chapter attempts to address the varied techniques used to understand the molecular basis of gastrulation and to draw attention to some of the potential pitfalls in performing the work. It is focused on vertebrates and, in particular, *Xenopus laevis*; however, the advantages of examining specific gastrulation movements in other embryos will also be discussed.

II. THE VARIED CELL MOTILITIES OF *XENOPUS* GASTRULATION

Detailed and excellent descriptions of the movements of gastrulation have been published elsewhere; for example, see the recent book edited by Claudio Stern, *Gastrulation: From Cells to Embryos* (1) and Ray Keller's chapter in *The Cellular Basis of Morphogenesis* edited by Leon Browder (2). Here, some of the major movements that will be discussed below are briefly outlined.

A. General Overview of Gastrulation

Prior to gastrulation, during the blastula stages, the *Xenopus* embryo is a sphere with a cavity called the blastocoel in the animal hemisphere (Figure 13.1a). At this stage, the embryo is essentially a single tissue layer, with the future ectoderm in the animal hemisphere, endoderm in the vegetal hemisphere, and mesoderm in a ring around the equator or marginal zone (Figure 13.1a). Gastrulation organizes these tissues into three layers, with the movement of the mesoderm and endoderm into the interior of the embryo and the expansion of the ectoderm over its surface (Figure 13.1c).

Gastrulation movements begin on the future dorsoanterior side of the embryo, and as gastrulation proceeds, they spread bilaterally until they are occurring around its entire circumference (3, 4). Internalization begins in a region called the involuting marginal zone (IMZ). These cells form a torus at the equator of the embryo and include both a superficial epithelial layer and an underlying region of mesenchymal cells (Figure 13.1b; 5, 6). The vegetal-most region of the IMZ turns inward, doubles back on itself, and moves toward the animal pole (Figure 13.1b; 6). Externally, this

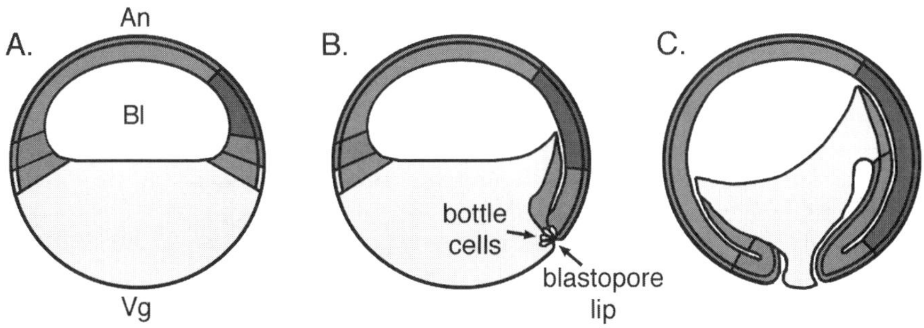

FIGURE 13.1 (See color insert following page 144.) *Xenopus* gastrulation. (A) Prior to gastrulation, at the blastula stage, the embryo is a sphere with a cavity called the blastocoel (Bl) in the animal hemisphere (An). At this stage, the prospective ectoderm (light blue) and neural ectoderm (dark blue) are located in the animal hemisphere, the prospective endoderm (yellow) is in the vegetal hemisphere (Vg), and the prospective mesoderm (red) forms a ring at the equator. (B) Shortly after the onset of gastrulation, stage 10+, the leading-edge mesoderm (light red) begins to migrate across the inner surface of the blastocoel roof, epiboly is already occurring in the ectoderm, and the endoderm is moving into the embryo through rotation. (C) At the mid-late gastrula stage, stage 11, convergent extension of the prospective notochord and somite tissue (dark red) has started, and gastrulation movements are occurring around the entire circumference of the embryo.

is seen as a localized indentation, the blastopore groove, that becomes a pigment line as epithelial cells, termed bottle cells due to their characteristic shape (Figure 13.1b; 7, 8), constrict apically and form the blastopore lip (stage 10; throughout, all stages are according to Reference 9). The involution of the IMZ is driven, at least in part, by the rotation movements of the vegetal endoderm that brings the IMZ cells into contact with the fibronectin-rich extracellular matrix of the inner surface of the blastocoel roof (10). The vanguard of the leading IMZ mesendoderm cells actively migrate across this matrix towards the animal pole (see Figure 13.2) to become the head mesoderm, liver, and anterior of the blood islands (11).

Early involution is accompanied by the thinning and vegetal expansion of the dorsoanterior IMZ and the neural tissue, which is situated animal to it (Figure 13.1b; 5, 6). This occurs through radial intercalation, in which deep mesenchymal cells intercalate or move between one another perpendicular to the plane of the tissue. Little is known about the molecular basis of radial intercalation in *Xenopus* except that during late gastrulation, these movements depend on the $\alpha5\beta1$ integrin-mediated interaction of the deep cells with the fibronectin-containing extracellular matrix at their basal surfaces (12). Recent evidence also suggests that the normal fibronectin fibril assembly, essential for the radial intercalation of these cells, is controlled by the planar cell polarity (PCP) pathway (13).

The next cells to move into the embryo will give rise to the trunk of the embryo including the future notochord and somites. The movement of these IMZ cells begins at the mid-gastrula stage (stage 10.5) and is characterized by mediolateral intercalation, the intercalation of cells within the plane of the tissue (Figure 13.3; 14–16). These movements, termed convergent extension, aid the involution of the

IMZ, lengthen the embryo in the anterior posterior direction and narrow it, and force the blastopore closed (5). Convergent extension of the IMZ is mirrored by the movements of the prospective neural ectoderm (5, 6) and accompanied by the overlying suprablastoporal endoderm that will eventually form the archenteron (gut) roof. Convergent extension of the IMZ and neural ectoderm extends these tissues over the future endoderm and, with a contribution from similarly moving tissues from around the circumference of the embryo (Figure 13.1c), closes the blastopore (stage 13; 3).

Prior to and throughout gastrulation, the animal cap is thinned and expanded in a process called epiboly. Epiboly is also the result of radial intercalation of deep cells, in addition to the cell division and spreading of the superficial cells within the plane of the tissue (17, 18). The expansion of the animal cap continues until the entire outer surface of the embryo is covered, forming the ectoderm. Thus, by the end of gastrulation, the ectoderm forms the outer layer, the endoderm forms the inner layer, and the mesoderm forms the layer between the two (stage 13).

B. Migration of the Future Head Mesoderm

The future head mesoderm of the embryo is formed by some of the first cells to move during gastrulation, the dorsoanterior mesendoderm cells of the IMZ. These cells actively migrate toward the animal pole across the extracellular matrix of the inner surface of the blastocoel roof. This matrix is rich in fibronectin that is organized into fibrils (19–21). The fibrils contribute to the directed migration of these cells, although they are not sufficient for this directed movement (22). The mesendoderm cells move as a sheet and are polarized with lamelliform protrusions extending in the direction of movement (Figure 13.2). Their trailing edges underlap one another in a distinct pattern that resembles shingles on a roof (23). The cells in contact with the blastocoel roof matrix, an interaction that is mediated by the integrin matrix receptor $\alpha5\beta1$ (12), actively migrate carrying the deeper layers along for the ride.

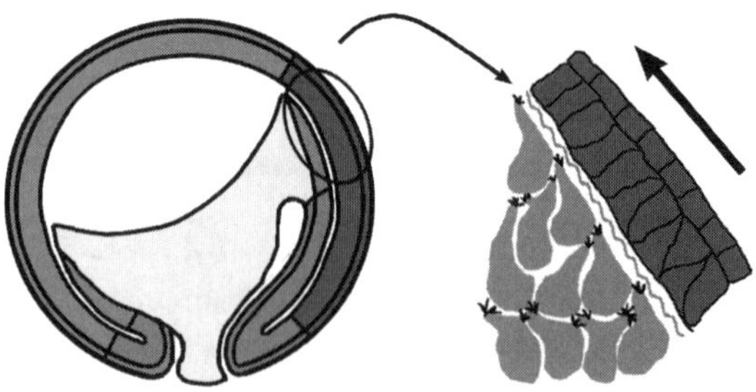

FIGURE 13.2 Mesendoderm migration. The dorsoanterior mesendoderm cells of the IMZ actively migrate towards the animal pole (in the direction of the arrow) across the extracellular matrix of the inner surface of the blastocoel roof. These cells move as a sheet and are polarized with lamelliform protrusions extending in the direction of movement and their trailing edges underlapping one another.

C. Convergent Extension

Convergent extension is an active, force generating mechanism that is the result of the biomechanical linkage of two processes, convergence of cells and extension of a tissue. These processes are not always linked, and one does not necessarily lead to the other. However, in gastrulation, the term implies this very specific process (3, 24).

Convergent extension is the result of the mediolateral intercalation of cells that converge towards the dorsoanterior midline, thickening and extending the tissues, and elongating the anterior posterior axis as cells move between one another (25–27). In *Xenopus*, two types of polarized cell motility drive convergent extension, characterized by the monopolar morphology of prospective neural cells (28, 29) and the bipolar morphology of intercalating mesoderm cells (26, 27).

In the prospective neural tissue, the medial cells are rounded and unbiased in their protrusive activity, and appear to ride passively on the developing notochord. In contrast, the more lateral cells have a monopolar protrusive activity that begins at the late gastrula stage (30). This polarity is biased in the direction of cell movement, towards the midline (30, 31).

In the mesoderm, convergent extension is driven by mediolateral intercalation behavior (known as MIB) that begins laterally and spreads toward the dorsoanterior midline (32). These movements result in the formation of successive arcs of cells, the vegetal alignment zone, that are thought to cause hoop stress, contributing to the involution of the tissue and closure of the blastopore (32). The mesoderm cells are initially rounded and multipolar, putting out lamelliform cytoplasmic protrusions around their entire circumference (Figure 13.3a; 26, 27). At the mid-gastrula stage (stage 10.5), these cells begin to polarize. The cells become bipolar by extending large lamelliform protrusions at their mediolateral ends (Figure 13.3b) as well as filiform protrusions in the animal-vegetal direction (26, 27). These lateral lamelliform protrusions contact adjacent cells (Figure 13.3b), creating traction and force that causes the cells to lengthen mediolaterally and enables them to move between one another. In the course of this process, a notochord-somite boundary forms and cells become captured at the boundaries (Figure 13.3c; 27). The result of these movements is the transformation of a short broad array of cells into a long thin one, extending the anterior-posterior axis while narrowing the embryo.

III. GROWTH FACTORS AND OTHER SIGNALS IMPORTANT FOR GASTRULATION

Although our knowledge of the molecular mechanisms that drive gastrulation has dramatically increased in the last 10 years, we are still a long way from understanding them. This in part is due to the fact that gastrulation immediately follows and often overlaps with other processes essential for normal embryonic development. The work is further complicated by the fact that embryos regularly reuse signaling pathways, and the identification of molecules important for gastrulation may be masked by their involvement in critical earlier embryonic processes. In addition, recent evidence suggests that divergent pathways downstream of a particular factor can distinguish

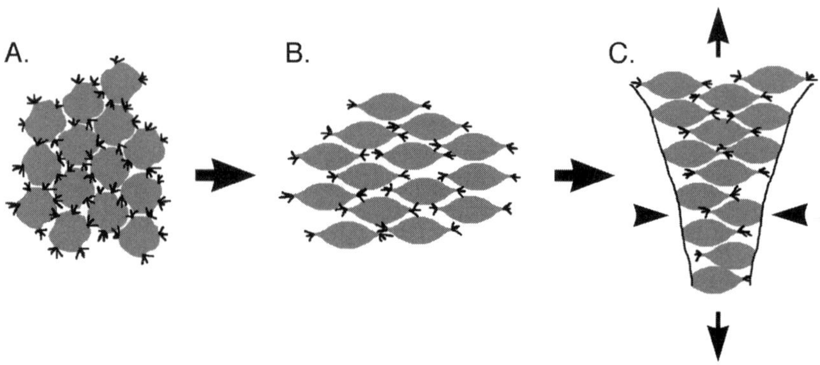

FIGURE 13.3 Convergent extension. (A) Prior to the onset of convergent extension, prospective mesoderm cells are rounded and multipolar, putting out lamelliform cytoplasmic protrusions around their entire circumference. (B) At the mid-gastrula stage (stage 10.5), they begin to polarize. The cells extend large lamelliform protrusions at their mediolateral ends that contact adjacent cells. This creates traction and force, which causes the cells to lengthen mediolaterally and enables them to move in between one another. (C) The notochord cells are captured at the forming notochord-somite boundary. These mediolateral intercalation movements cause the tissue to narrow (arrowheads) and extend (arrows).

mechanisms to specify tissues and control cell movements (for example, see References 33 and 34) or suppress one type of cell movement in favor of another (for example, see Reference 35). Thus, one of the major challenges in identifying the molecules that control gastrulation is to discriminate between the process of gastrulation and earlier events necessary for gastrulation to occur. In other words, obtaining a gastrulation defect does not necessarily mean that there is a defect in gastrulation. In this chapter, only molecules that have been shown to play a direct role in the process of gastrulation are considered.

A. Migration of the Head Mesoderm

To date, most work examining the molecular mechanisms of *Xenopus* gastrulation has focused on convergent extension; however, there is also some information on the less well-studied processes. For example, the directed migration of the prospective head mesoderm depends on platelet-derived growth factor (PDGF) signaling (36). Immediately prior to and during gastrulation, PDGF-A is expressed in the ectoderm, whereas its receptor, PDGFRα, is expressed in the mesoderm (37–39). Normally, prospective head mesoderm cells migrate as a sheet toward the animal pole across a matrix containing fibronectin fibrils. This directed movement can be recapitulated *in vitro* in an assay in which mesendoderm explants are plated on their natural substratum for migration, the blastocoel roof matrix (23, 36, 40, 41; see Figure 13.4).

In this assay, disruption of PDGFRα function in the mesoderm or PDGF-A in the blastocoel roof substratum does not inhibit migration per se, but results in movement that is randomized and no longer directed towards the animal pole (36). Overexpression of either the long or short form of PDGF-A suggests that PDGF-matrix binding is essential for directed movement (36). Consistent with this, overexpression of wild-type PDGF-A, or inhibition of PDGF-A *in vivo* abolishes the normal shingle arrangement of these cells, the frequency of their protrusive activity is reduced, and they are disoriented. These embryos also develop with reduced or absent head structures (36, 37). The network of fibronectin fibrils and rotation of the vegetal endoderm, however, are unaffected. The formation of a polarized fibronectin matrix on the blastocoel roof requires earlier instructive signals from the mesoderm and depends on activin and fibroblast growth factor (FGF) signaling in the blastocoel roof (22).

In zebrafish embryos, PDGF and its receptor are not spatially restricted during the gastrula stages (42, 43), and although inhibition of PDGF signaling in migrating mesendoderm cells causes a loss of polarization and differences in the number and type of cytoplasmic protrusions, the direction of migration is not altered (44). In contrast, the disruption of the PCP pathway causes the migrating mesendoderm cells to lose orientation, despite being polarized (45).

Phosphatidylinositol 3-kinase (PI3K) is a well-characterized effector of PDGF signaling and is important for a wide range of cell responses in a variety of cell types including cell division, chemotaxis, and membrane ruffling (for review, see Reference 46). In zebrafish and *Xenopus* embryos, respectively, PDGF signaling via PI3K is important for cell polarization, the formation of cytoplasmic protrusions, and cell spreading (44, 47). However, its role in directed migration is unresolved (36).

How the motility of the head mesoderm is controlled remains an important question. It seems likely, given its role in polarized matrix deposition (13), that the PCP pathway will be particularly important. Rho GTPases act downstream in both PCP and PDGF signaling pathways. Recent evidence suggests that they are important for the polarity and protrusive activity of migrating head mesoderm cells, but not for directed migration (48). They may also be important for remodeling the extracellular matrix. For example, in v-Abl transformed fibroblasts, c-Abl facilitates adhesion and cell spreading (36) that is dependent on an increase in fibronectin deposition, mediated by Rho-induced changes in the cytoskeleton (49). How these pathways converge and are integrated with signals from other factors, such as *Goosecoid* (*Gsc*), which promotes cell migration (50), and syndecan-2, which transmits left-right asymmetry to the migrating mesoderm (51, 52), however, is still to be determined.

B. Tissue Separation

The continued separation of tissues that are moving across one another is a prerequisite for gastrulation and necessary to keep cells in the involuting tissues from integrating with the overlying tissues. In *Xenopus*, the interface between the IMZ and the overlying ectoderm is called Brachet's Cleft. Brachet's Cleft is maintained by a separation behavior in the post involution tissue and a repulsive behavior in the overlying tissue that prevents their integration (53). This behavior appears to be established by Frizzled 7-dependent protein kinase C (PKC) signaling (54) and requires paraxial protocadherin (PAPC; 55). PAPC modulates Rho and c-Jun N-terminal kinase (JNK) activity, effectors of the planar cell polarity pathway (55). In addition, PAPC expression depends on Lim1 (56), which is also required for tissue separation in *Xenopus* embryos. Fibroblast growth factor and the paired-class homeodomain transcription factors Mix.1 and *Gsc* further regulate tissue separation (53). Disruption of *Mix.1, Gsc,* or FGF function in the marginal zone impedes normal separation behavior of the anterior mesoderm. In addition, modulation of cadherins may also be necessary for this process (53).

C. Epiboly

Despite its importance for the gastrulation of many organisms (reviewed in Reference 57), little is known about the molecular mechanisms of epiboly. In *Xenopus*, disruption of Xoom, a membrane-associated protein, inhibits epiboly (58, 59). In addition, recent evidence in the zebrafish suggests that regulation of the homeobox-containing Mix/Bix gene *mtx2* by the T-box gene *eomesodermin* (*eomes*) is important for epiboly (60). Repression of *eomes* or *mtx2* function inhibits epiboly movements without affecting tissue patterning. *Eomes* and *mtx2* appear to control different aspects of epiboly. *Eomes* is important for the initiation of radial intercalation, whereas *mtx2* acts later in the vegetal progression of the blastoderm margin. T-box genes have also been shown to be important for convergent extension during gastrulation (see below; for review, see Reference 61).

D. Convergent Extension

1. Fibroblast Growth Factor Signaling

Despite strong evidence that FGF is required for cell movement in gastrulating chick and mouse embryos (for review, see Reference 62), it has been difficult to identify the role of FGF in *Xenopus* gastrulation due to its importance in mesoderm induction (63–66) and maintenance (67, 68). Recently, however, Sprouty and Spred protein families have been shown to differentially modulate FGF signaling to separately control mesoderm patterning and morphogenesis (34, 69). In *Xenopus*, Xsproutys inhibit Ca^{2+} and PKCδ signaling and, unlike mammalian Sproutys, leave MAPK signaling intact (34, 69; see Chapter 4 in this volume by Sater and El-Hodiri). This inhibits convergent extension without affecting mesoderm induction and patterning (34, 69). In contrast, XSpreds inhibit MAPK activation and mesoderm specification, but not Ca2+ and PKCδ signaling (34). Both Ca^{2+} and PKCδ have been implicated in convergent extension (70, 71), with PKCd being required for Dishevelled function in non-canonical Wnt signaling (70).

Neurotrophin receptor homolog (NRH), an FGF target gene, is necessary for protrusive activity in marginal zone cells (72) and convergent extension (see below; 73). In addition, the maternal FGF receptor 1 (FGFR1), downstream signaling via MAPK, and its target genes are activated by *Xenopus* nodal related 3 (Xnr3), which is required for convergent extension (33). Xnr3 further regulates the expression of the T-box gene *brachyury* (Xbra) in the organizer at the gastrula stage (33). Xbra is required for convergent extension and actively inhibits migration by modulating cell adhesion to fibronectin (35). Both Xnr3 and NRH are also linked to Wnt signaling. Xnr3 is a direct target of the Wnt/β-catenin signaling pathway (74; see Chapter 2 in this volume by Kühl and Moon), whereas NRH can activate downstream effectors of the PCP pathway (73). This evidence further suggests that FGF, perhaps in combination with Wnt signaling, regulates the switch between tissue specification and morphogenesis, and that these two processes are distinct. How these signaling pathways are coordinated to control these different cell fates, however, remains to be determined.

2. Bone Morphogenetic Proteins

Like FGF, it has been difficult to establish a role for Bone Morphogenetic Proteins (BMPs) in *Xenopus* gastrulation due to their involvement in the early specification and patterning of the mesoderm (for review, see Reference 75). However, recent evidence in zebrafish embryos suggests that BMP may also play a role in modulating the switch between tissue specification and morphogenesis (76). Distinct cell shapes and movements are correlated with different levels of BMP activity in ventralized, chordin (chordino), and dorsalized smad-5 (somitabun) mutants. Low BMP levels allow mediolateral cell elongation and a dorsally biased intercalation, resulting in tissue extension (76). Conversely, high BMP activity promotes vegetal migration of mesoderm cells into the tailbud, increasing tail formation at the expense of head and trunk (76). These high levels of BMP limit convergence and extension by negatively regulating expression of wnt11 (silberblick) and wnt5a (pipetail) genes (76), which are required for convergent extension but not cell fate specification (see below; 77–79).

3. Wnt Signaling

The Wnt family of secreted glycoproteins play essential roles in embryonic development (for reviews see 62, 79–83), with different downstream signaling pathways that regulate tissue specification (canonical Wnt/β-catenin pathway; see Chapter 1 in this volume by Clements and Kimelman), and cell polarity and matrix assembly (non-canonical PCP and Wnt Ca^{2+} pathways; and Chapter 2 by Kühl and Moon, and Chapter 8 by Slusarski). During gastrulation, convergent extension is regulated, at least in part, by the non-canonical pathways. A brief overview of this role follows.

In *Xenopus* and zebrafish, the PCP pathway establishes cell polarity during convergent extension and is necessary for normal matrix assembly, independent of tissue specification (for reviews, see References 79 and 81). In vertebrates, this pathway involves Wnts and, in particular, wnt4, wnt5a, wnt7a, and wnt11 (see Chapter 2). Disruption of any of these Wnts in *Xenopus* development inhibits convergent extension, as does blocking certain frizzled (Fz) receptors, and their downstream effectors including Dishevelled, the Rho GTPases, and JNK (for reviews, see References 62 and 79). For example, disruption of wnt11 signaling inhibits convergent extension in embryos and activin-treated animal caps (84). Zebrafish mutants that lack wnt11 (silberblick) or glypican-4/6 (knypek), which is required to promote wnt11 signaling, have impaired cell polarity and convergent extension movements (77, 85). Dishevelled has also been shown to be necessary for cell polarization and protrusive activity during convergent extension in *Xenopus* (86), exerting these affects through Rho GTPases and JNK (87–90). In parallel to this pathway, Lim1 signals through PAPC, Rho GTPases, and JNK to coordinate the polarity of the intercalating mesoderm (56, 91). Activation of the Rho GTPase, Rac 1, by the hyaluronan synthesizing enzyme (Has2) has also been shown recently to control protrusive activity of cells that is necessary for dorsal convergence during zebrafish gastrulation (92).

Recent evidence indicates that the PCP pathway acts by regulating matrix deposition and, in particular, organized fibronectin fibril assembly, in addition to polarized cell morphology and motility (13). Overexpression of the PCP genes XStrabismus (Xstbm; Van Gogh), Xfz, or Xprickle causes defects in convergent extension that are correlated with the disorganized assembly of fibronectin fibrils on mesodermal tissue surfaces (13). Xfz and Xstbm are further required for the bipolar morphology of intercalating mesoderm cells (93).

A second non-canonical Wnt signaling pathway, the Wnt Ca^{2+} pathway, which involves intracellular calcium release and activation of protein kinase C (for review, see Reference 94 and Chapter 8), has also been shown to play a role in convergent extension (71). Activation of this pathway requires heterotrimeric G proteins (for review, see Reference 94). For example, $G\alpha_i$ and $G\alpha_t$ rescue convergent extension defects resulting from overexpression of wnt11/XFz7 (95). In addition, in the zebrafish, it was shown recently that inhibition of $G\alpha_{12}$ and $G\alpha_{13}$ disrupts convergent extension without altering tissue specification (96). Time-lapse analysis of these embryos revealed that the dorsal migration of mesoderm cells was slowed and their path of movement was more erratic than wild-type cells (96). The intercalation of prospective notochord cells was also impaired (96). $G\alpha_{12}$ and $G\alpha_{13}$ act independently of wnt11 signaling, but the involvement of the Wnt Ca^{2+} pathway remains to be determined (96).

The Wnt signaling pathways diverge at Dishevelled (see Chapters 1 and 2), and recent evidence suggests that cells can be diverted toward tissue specification or morphogenesis at this point (86, 97). For example, deletion mutations in Dishevelled revealed that DEP domain is more important for convergent extension and cell polarity than the DIX domain, which is necessary for Wnt/β-catenin signaling (86). In addition, the mammalian homolog of *Drosophila* Naked Cuticle (mNkd) interacts directly with Dishevelled and acts both to antagonize the canonical Wnt/β-catenin signaling pathway and to stimulate JNK activity, an effector of the PCP pathway (97). Consistent with this, in *Xenopus*, overexpression of mNkd disrupts convergent extension (97).

It is clear that non-canonical Wnt signaling plays an essential role in gastrulation in establishing the bipolar cell morphology necessary for mediolateral intercalation and convergent extension. It is also clear that these pathways act in parallel, but many questions still remain. For example, how is the orientation of cell polarity determined? It has been shown that positional information of the anterior-posterior axis that is established by graded activin-like signaling is important for this process (98), but how it is coordinated with Wnt signaling is unknown. Recent evidence also suggests that all Wnt downstream signaling may overlap. The zebrafish homolog of *Drosophila nemo*, nemo-like kinase (*Nlk*), can positively influence cell fate and patterning decisions, and convergent extension (99). In concert with *wnt8*, *Nlk* can downregulate two *tcf3* homologs that repress target genes of Wnt/β-catenin signaling, whereas, with wnt11, *Nlk* enhances prechordal plate cell migration and convergent extension (99). In addition, signal transducer and activator of transcription 3 (Stat3), which is activated in the organizer by the Wnt/β-catenin pathway, is required for the convergence of paraxial cells and the anterior migration of the dorsal mesendoderm in zebrafish embryos (100, 101). Thus, the future challenge is to ascertain not only how the various Wnt signaling pathways are regulated in coordination with one another, but also how they are integrated with signals from other factors to control the tissue specification and cell movements necessary for an embryo to gastrulate.

IV. ANALYZING GASTRULATION IN *XENOPUS*

A. The Assays: General Considerations

The two major advantages in studying gastrulation in the *Xenopus* embryo are that the morphological details of the movements are extremely well described and that there are a wide variety of single-cell and explant systems available to examine different types of cell behavior. The most commonly used assays are provided for some of the more specialized ones. There are four essential considerations for the successful use of these assays. First, always culture sibling embryos in parallel to the experiment, because once a cell or tissue is removed from the embryo, or the embryo starts to develop abnormally, it is impossible to identify its developmental stage. Second, to prevent the tissues sticking to the bottom of the dish, dissections should be carried on a bed of 1% agarose (for example, see Reference 102). Third, to prevent damage to the tissues during dissection, appropriate tools are important. Mounted eyebrow knives and hair loops are particularly useful; however, for animal cap assays, sharpened tungsten needles work well (see Reference 102). Finally, the culture medium is critical. Certain explants may thrive in one media yet perform poorly in another. Thus, it is important to choose appropriately from the wide variety of media formulations available for the culture of *Xenopus* embryos, explants, and single cells (102, 103). The addition of antibiotics to reduce contamination is also recommended (102, 103).

B. Intact Embryos

Methods for obtaining, culturing, devitellinizing, and scoring embryos can be found in Reference 102. Due to their opaque nature, and without the availability of specialized equipment—for example, see the beautiful magnetic resonance imaging (MRI) analysis of *Xenopus* gastrulation (104, 105)—only the superficial movements of gastrulation can be observed in living embryos. Fixed embryos, however, can be used for a variety of techniques. For example, to confirm that it is the process of gastrulation being examined, *in situ* hybridization is essential to analyze normal tissue specification (106). In addition, scanning electron microscopy (SEM) is useful to obtain a snapshot of particular cell activities such as cell shape or the degree of polarity (for example, see Reference 36).

Recently, an automated method for analyzing whole embryos in three dimensions has been described, called surface imaging microscopy (SIM; see Reference 107–109). SIM generates three-dimensional reconstructions of embryo samples at high resolution and high contrast, and has been used to visualize gastrulation in *Xenopus* (108). Briefly, embryos are fixed; stained, for example, with propidium iodide (109); and embedded. Embryo images are then captured serially, directly from the block face. The images are later assembled using freely available software (107) into a three-dimensional volume. This software may also be used to analyze two-dimensional images of the outside of the embryo, correcting for the curvature of the embryos' surface (107). However, it is the variety of the *ex vivo* assays that make *Xenopus* such an attractive model for this work.

C. Mesendoderm Migration

There are several approaches available for examining mesendoderm migration, and the most appropriate assay, as with all assays, depends on what you want to examine. For example, to determine whether cells are capable of any form of motility, the simplest system is to use dispersed cells plated on a fibronectin-coated dish (110). This single-cell assay can also reveal information on the type of cytoplasmic protrusions being produced. However, it is rather limited, and the observed cell behavior may bear little resemblance to that normally seen *in vivo* or in other assays (for example, see References 47 and 111). To examine factors important for directed migration, an assay initially developed for observation of single cells (40) and later refined by Rudi Winklbauer provides an excellent method (23). Alternatively, the mechanics of the mesendoderm tissue movements can be examined using the large explant system developed by Lance Davidson (4, 112).

Winklbauer directed migration assay (23, 36, 40, 41; Figure 13.4): This assay is used to examine the directed migration of the leading edge of the mesendoderm. This tissue normally migrates across the fibronectin fibrils of the blastocoel roof extracellular matrix. In *Xenopus,* these cells move as a sheet (Figure 13.2), and thus, explants are used and not dispersed cells. If plated on a fibronectin-coated dish, an explant moves in a random direction. However, when plated on the natural matrix transferred to a dish, an explant moves in its normal direction, toward the animal pole. Cells are visibly polarized on this matrix. They are unipolar, extending protrusions only in the direction of migration, and underlap neighboring cells in direction of migration (see Figure 13.2). This underlapping is less pronounced in explants migrating on fibronectin-coated dishes.

To prepare the matrix, a strip of blastocoel roof spanning the entire animal hemisphere of the embryo is dissected at stage 10 in Modified Barth's Solution (MBS). The explant is then placed blastocoel-side down in a new tissue culture dish in MBS (Figure 13.4a). Greiner tissue culture dishes are best for this assay (Greiner Bio-One Inc., Longwood, FL). The blastocoel roof explant is weighted down with a coverslip supported with silicone grease, directly placing the extracellular matrix and the tissue culture dish in apposition. The position and orientation of the tissue is marked on the dish by scratching the tissue culture plastic to demarcate the outline of the explant. In addition, the position of the animal pole and blastopore lip should

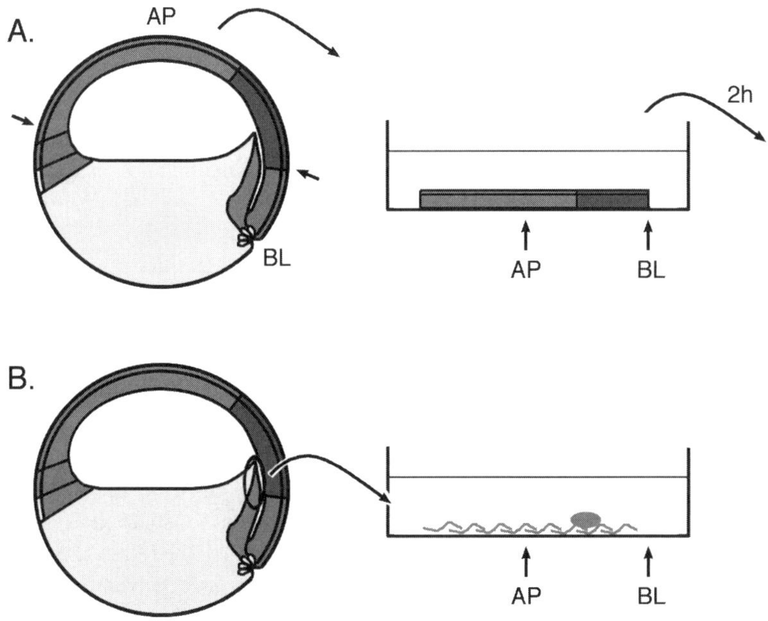

FIGURE 13.4 Winklbauer-directed mesendoderm migration assay. (A) The blastocoel roof matrix is made by placing a strip of presumptive ectoderm blastocoel-side down on a Griener petri dish for 2 hours. The position of the animal pole (AP) and blastopore lip (BL) is marked on the dish. (B) An explant containing the leading-edge mesendoderm is placed on the deposited matrix, and its direction of movement is scored.

be marked. The matrix adheres to the dish and is transferred to the plastic. After 2–3 hours, the tissue is removed by gentle aspiration (stage 11). The matrix is blocked for nonspecific binding by incubation in 50 mg/ml bovine serum albumin (BSA) for 30 minutes. In order to ensure that a polarized matrix is deposited, the substrate must be made during the gastrula stages (22); if not, fibronectin fibrils may be deposited but they will not support directed cell migration.

A leading-edge explant, containing approximately 200 cells and including the prospective head mesoderm, is dissected at stage 10.5 in MBS (113) and placed on the blastocoel roof matrix between the blastopore lip and animal pole. The explant is gently held down with a coverslip bridge supported on silicone grease. A photograph is immediately taken, and then another is taken 1 hour later for analysis of directed migration. Be sure to include some of the marked outline in the photograph as a landmark for measurements. After completion of the experiment, remove the explant by gentle aspiration. The matrix should then be examined to confirm that a contiguous fibronectin fibril network was correctly deposited, as the fibrils are necessary for directed migration (23). This can be done by immunohistochemistry using an antibody to fibronectin.

To observe cell motility in single cells, a similar prospective head mesoderm explant can be dissociated in Ca^{2+}- and Mg^{2+}-free MBS containing 0.1 mM EDTA (103, 111). These cells can be plated either on the blastocoel roof matrix, described above, or on a fibronectin-coated dish or coverslip. To prepare the fibronectin matrix, a pool of 100 mg/ml fibronectin in MBS is incubated in a marked region of a dish or coverslip for 2 hours at room temperature. After washing with MBS, the substrate is blocked for nonspecific binding as described above for the blastocoel roof matrix. Note that the dispersed cells do not recognize the polarized matrix and migrate randomly (111). Single head mesoderm cells cultured in this way should have a bipolar or tripolar morphology with lamelliform protrusions protruding from opposite sides of the cell (111).

Potential pitfalls and tips: Explants for this assay can be dissected from stage 10; however, they are easiest to obtain at stage 10.5 because the leading edge is distinct. As with all explant assays, the most critical issue is the dissection of the appropriate region of tissue. The only way to really test for this is to use the tissue in the assay and determine that it behaves appropriately, in this case migration towards the animal pole. However, an earlier injection of mRNA encoding green fluorescent protein (GFP) targeted to the dorsoanterior mesoderm (11) can be used to help guide the dissection. In this case, at stage 10 the localization of the GFP can be compared to the position of the blastopore lip and then again after tissue isolation to make sure the dissected region is labeled. This is particularly useful if other substances such as mRNA or morpholino oligonucleotides are simultaneously introduced into the embryo. Equally important is the application of an appropriate amount of pressure to the coverslip holding the explant in place. Too little and the explant may become dislodged, too much and the explant will not move. Again, the only way to test this is practice. In addition, it helps if the coverslip is coated with BSA (5%, 30 minutes) to prevent the explant adhering to it. Finally, coverslips used for the fibronectin matrix should be cleaned with a brief wash in dilute acid-alcohol followed by flaming (4).

Davidson large explant system for mesendoderm migration (4, 112): This assay is used to investigate the mechanics of mesendoderm migration and utilizes a wide (180° minimum) explant centered on the midline of the dorsoanterior marginal zone. This explant exhibits consistent patterns of gene expression that parallel those in the whole embryo, in addition to undergoing stereotypical cell movements. When it is plated onto a fibronectin-coated dish, the mesendoderm moves away from the explant as an intact sheet. Cells within the migrating mesendoderm display monopolar protrusive activity and radial intercalation, as they would normally *in vivo*. In addition, later movements such as those of the prospective brain tissues and the lateral involuting marginal zone can be examined (112).

The onset of mesendoderm migration occurs prior to any external cue that gastrulation has begun, which cannot be seen until stage 10 (see above). Thus, embryos must be opened to determine this pregastrula stage (4). Explants are prepared in modified Danilchik's solution (often referred to as Danilchik's For Amy or DFA; 113) supplemented with antibiotics (114). The explants are dissected by making two incisions, 90° from either side of the dorsoanterior midline. The vegetal endoderm and any tissues that have undergone internal involution are removed. A few rows of sub-blastoporal endoderm are left attached to prevent small wounds to the leading edge disrupting cell movements. The explants are placed internal-side down on a fibronectin-coated dish or coverslip, and held in place by a coverslip bridge supported by silicone grease for monitoring by time lapse micrography.

The original paper describing this explant also contains two modified dissections to further examine mesendoderm mechanics, the so-called "cap-less" and "donut" explants (4).

Potential pitfalls and tips*:* Similar issues to those described above for the directed migration assay are applicable here. However, bear in mind that the expansion of the mesendoderm through cell migration should not be confused with convergent extension, as it is a different cellular mechanism (see above).

D. Tissue Separation

Following involution, mesoderm cells display a separation behavior that allows them to migrate across the surface of the overlying ectoderm (see above). If this separation did not occur, these cells would sink into the blastocoel roof. This behavior can be analyzed in an *in vitro* assay in which test tissue is placed on the inner surface of a blastocoel roof explanted at stage 10+ to 10.5. The blastocoel roofs are dissected in MBS and prevented from rolling up by applying a sliver of BSA-saturated (5%, 30 minutes) coverslip resting on silicone grease. Test tissue is then monitored to determine whether it remained separated from or integrated into the blastocoel roof explant. Potential test tissue might include, for example, involuted mesoderm or cell aggregates of treated animal cap cells (53, 115).

E. Convergent Extension

The cell behaviors and mechanics of convergent extension in *Xenopus* are best seen in cultured explants. As described above, different types of polarized cell behaviors drive convergent extension in the IMZ and prospective neural tissue, and various explant assays have been designed for their analysis, the most famous of which are the "Keller explants." There are essentially three types of explants named after Ray Keller: the Keller sandwich, the open-faced Keller explant, and the shaved Keller explant (Figure 13.5). Each explant is a variation on a theme but with a distinct use.

The Keller sandwich (Figure 13.5a, b; 3, 14, 15): This explant is particularly valuable for examining the full extent of convergent extension in all future dorsal tissues, enabling comparison of the extension of the prospective notochord, somitic mesoderm, and neural plate. It has further uses to examine planar signaling (113, 116) but it is not appropriate for analyzing individual cell behaviors. The sandwich is made by combining two identical dorsoanterior marginal zone explants, in the same animal-vegetal orientation, and placed with their inner surfaces together (Figure 13.5b). The sandwich includes the IMZ, part of the animal cap, the prospective neural tissue, and superficial epithelium. The explants are dissected at stage 10 by making two incisions, 45° from either side of the dorsoanterior midline (Figure 13.5a). The animal cap is then cut and the flap of tissue is pulled vegetally, allowing the deeper involuted region, which includes the future head mesoderm cells, to be teased away.

The explant is then released from the embryo by cutting along the bottle cells (Figure 13.5a). Immediately following dissection, the two halves of the sandwich should be assembled (Figure 13.5b) and held together by coverslip bridges supported by silicone grease to allow healing. After 15–20 minutes, the sandwich should be

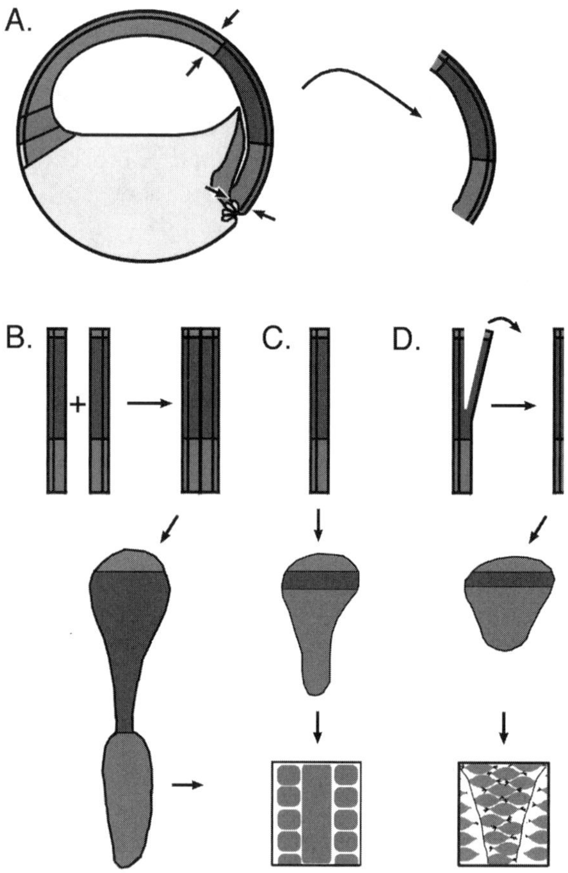

FIGURE 13.5 Keller explants. (A) The basis of all Keller explants is a dorsoanterior explant that includes the IMZ, part of the animal cap, the prospective neural tissue, and superficial epithelium. (B) The Keller sandwich comprises two of these explants in the same animal-vegetal orientation, with the deep tissues apposed. Both the prospective mesoderm and neural ectoderm undergo convergent extension, and differentiated notochord and somites can be observed. (C) The open-faced Keller explant comprises one explant held flat using a coverslip bridge. Only the prospective mesoderm undergoes convergent extension, and differentiated notochord and somites can be observed. (D) The shaved Keller explant requires the removal of the deepest layers of the tissue. There is little extension of this tissue, but mediolateral intercalation behavior can be observed.

healed enough to move into a fresh agarose-coated dish. The explants are dissected and cultured in DFA (113) supplemented with antibiotics (114). Similar explants can also be used to examine cell movements around the entire circumference of the embryo (3).

Potential pitfalls and tips: When using this sandwich explant for the examination of planar signaling, it is particularly important to ensure that the prospective head mesoderm is completely removed, which can otherwise crawl up in between the two halves of the sandwich and participate in vertical signaling. One way to control for this is to label the B and C tiers of the embryo at the 32-cell stage with different lineage tracers (for example, see Reference 117).

Open-faced Keller explants (Figure 13.5c; 118): This explant is useful for observing convergent extension of only the mesoderm because the prospective neurectoderm does not extend. In addition, the deep mesoderm that is visible at the surface of the explant does not undergo mediolateral intercalation behavior until later stages of development (27, 119), as it is organized from the superficial IMZ layers inwards. This explant is prepared in the same way as the Keller sandwich except that only one explant is used, and it must be held flat under a coverslip supported by silicone grease to prevent it from rolling up.

Mediolateral intercalation behavior but not convergent extension can be observed in a variation of this explant in which 180° of the embryo centered on dorsoanterior marginal zone is dissected. These explants are plated deep mesoderm down on fibronectin-coated glass or plastic (112).

Potential pitfalls and tips*:* It is essential to make the explants wide enough to include both prospective notochord and somite tissue because chordamesoderm explants do not extend that well. A minimum width of 90° centered on the dorsoanterior midline is required. In addition, it is worth noting that cells in these explants undergo radial intercalation, and thus tracking cells over time can be somewhat limited. However, the result of mediolateral intercalation can be monitored when explants are dissected from embryos that have been injected into one cell at the two-cell stage with a lineage tracer (for example, see Reference 90).

Shaved Keller explants (Figure 13.5d; 118): This explant is best suited for the observation of the mediolateral intercalation behavior of the cells closest to the epithelial layer. It is not useful for examining convergent extension. It is made in the same way as an open-faced explant, except that the layers of deep mesoderm cells are gently shaved away with an eyebrow knife (Figure 13.5d).

Potential pitfalls and tips*:* When shaving this explant, the more cell layers that can be removed the better, otherwise radial intercalation will remove them from the plane of view. Thus, when you think you have shaved off enough, keep going! It is also important to note that mediolateral intercalation behavior does not begin until stage 10.5 and does not reach the center of this type of explant until stage 10.75 (118). The use of GFP with a membrane localization signal is also particularly useful in tracking individual cell behaviors in this explant.

In addition to the Keller explants described above, this laboratory has designed assays to examine almost every cell behavior in the *Xenopus* embryo. The following is a partial list.

The deep cell explant (118): This explant allows the observation of cell movement within the deep layers of mesoderm. This dissection is similar to that for the open-faced explant, with two exceptions. First, the superficial endoderm is removed, and second, the mesoderm must be straightened out, as it has partially involuted at the time of dissection. The explant is made at stage 10.5 to 11, by which time radial intercalation, which is mostly over by stage 10.5 in this region, has thinned the tissue such that almost all cells are visible and the overlying epithelial layer is no longer required to organize mediolateral intercalation.

The deep neural explants (29, 30): Two versions of this explant are used to examine convergent extension of the prospective neural plate, one that contains the mesoderm and one that does not. The explants are made at stage 11.5 to 12 in DFA. First, the neural epithelium is removed from the embryo. The deep neural tissue, the mesoderm, and the endodermal epithelium are then excised by cutting into the animal cap at the boundary of the involuting tissue and the blastocoel cavity, and slightly animal to the blastopore, above the estimated limit of involution (30). The lateral width of the explant is such that all tissues fated to become neural are included (5, 6). The endodermal epithelium with or without the deep mesoderm is then sheared off, and the explant is cultured under a coverslip bridge supported with silicon grease. Be warned, these explants are not for the faint of heart!

Three notochord explant (13): This assay provides an alternative way to examine mediolateral intercalation. Three notochord anlage are grafted together side by side, with the center and side notochords differently labeled. When three, wild-type notochord anlage are grafted together, the cells intercalate mediolaterally to form one notochord, and elongate. These explants are dissected at stage 10 (13). They are prepared and cultured in a similar way to open-faced Keller explants except that only the prospective chordamesoderm is dissected and the epithelium is removed (13).

F. Animal Cap Assays

Animal cap explants are useful because they are composed of pluripotent cells. When removed from the embryo at the blastula stage and cultured intact in a balanced salts solution, animal cap explants round up and differentiate into "atypical epidermis." When this tissue is exposed to certain growth factors such as FGF or the TGFβ family member activin, the tissue differentiates into mesoderm and neural tissues (for example, see References 65 and 120). These induced animal caps also undergo convergent extension movements that commence at the same time as those in sibling embryos cultured in parallel (121). They have been used extensively to screen for molecules involved in mesoderm induction and patterning (for review, see Reference 122; see also Reference 98) and to examine mesoderm cell motility. However, although they are simple to prepare (see Reference 102), their use for gastrulation studies is limited.

Potential pitfalls and tips: The competence of the animal cap to respond to factors changes over time, and thus it is important to cut explants at the appropriate time and in a timely fashion (see Reference 102). In addition, factors should be added to the animal caps before they are allowed to round up. Uninduced animal caps should be cultured in parallel to test animal cap dissection because any inclusion of prospective mesoderm tissues in the explants can lead to false positives. Explant size can also impact the results of the experiments, with larger animal caps containing a more heterogeneous population of cells. In addition to being studied as intact tissues, animal cap explants can be dissociated in Ca^{2+} and Mg^{2+} free medium and studied as dispersed cells. These dispersed cultures can be used to study tissue induction (for example, see Reference 123) or cell motility when plated on a fibronectin-coated dish (for example, see Reference 124).

V. MANIPULATING THE MOLECULES

In *Xenopus*, there are several options for manipulating growth factors and components of their signaling pathways; however, currently there is no way to remove a gene. Although this is somewhat limiting, rapidly advancing transgenic technology coupled with an exquisitely detailed description of the cell movements and the wide variety of explant assays make *Xenopus* embryos an excellent model for understanding the molecular basis of gastrulation. Techniques for altering the molecular makeup of the embryo are summarized briefly below with references indicating examples of their use.

Gain-of-function studies in which levels of genes and their products are increased have been used extensively to identify candidate molecules for a role in gastrulation. Several methods are available in *Xenopus* including the creation of transgenic animals (for example, see References 68 and 125–133), expression of hormone-inducible proteins (for example, see Reference 134), addition of proteins such as growth factors to the culture medium (for example, see Reference 123), and microinjection of DNA, mRNA, or protein (for methods, see Reference 102). Wherever possible, it is important to track the introduced molecule, preferably by using an epitope tag. Co-injection of a lineage tracer, however, will give an approximate indication of which cells inherited the injected material.

Each technique has advantages and disadvantages. For example, whereas microinjection of plasmid DNA can be used to temporally control the expression of the gene of interest, the level and distribution of expression is often mosaic. In contrast, microinjection of mRNA allows high levels of expression with a relatively even distribution; however, the developmental stages most useful for microinjection (1- to 32-cell stage) are at least 6 hours before the onset of gastrulation (9). Because mRNAs can be immediately translated, this timing is often inconsistent with the normal period of expression *in vivo*.

The disadvantage of being unable to control the timing of translation of injected mRNAs can be circumvented if the mRNA codes for a protein that can be regulated artificially. For example, a useful method to study receptor tyrosine kinase (RTK) overexpression has been developed that uses the ligands of immunophilins to act as "chemical inducers of dimerization" (CIDs; 135). Chimera RTKs have been generated in which the intracellular domain is fused to a dimerization domain and a myristoylation signal from v-Src to target the complex to the membrane (for example, see References 136 and 137). These chimera RTKs can be introduced into the embryo as microinjected mRNA, but the protein is not activated until the addition of a membrane-permeable drug. By combining this technique with mutation of specific tyrosine residues in the intracellular domain of the receptor, downstream signaling through particular receptor-associated effectors can be isolated (136). This method has been used *in vivo* to study PDGFR function during gastrulation (136). The advantage of this system is that the drug can be added at a specific time in development; however, for robust receptor activation, the drug must be injected into the blastocoel cavity. It is also difficult to control the period of receptor activation.

A new technique with enticing possibilities for the rapid and reversible regulation of specific proteins is described in Chapter 14 in this volume by Liu, Gestwicki, and Crabtree. The approach involves destabilizing a protein with a peptide tag that imparts instability in the absence of a drug, for example, rapamycin, which is fully membrane permeable. Reversibility is achieved using a second drug, FK506m, that displaces rapamycin. *Xenopus* embryos can be cultured in rapamycin with no disruption of development (Melanie Van Stry, personal communication), and thus, this method suggests an excellent way to manipulate protein activity specifically at the gastrula stages.

In gain-of-function experiments, intact embryos are scored for gastrulation defects, which may also be misleading. Gastrulation is a highly regulated process that is easily disrupted indirectly, for example, by interference with a critical earlier event such as tissue specification. Thus, it is worth repeating that obtaining a gastrulation defect does not necessarily mean that there is a defect in gastrulation. This consideration applies to all assays in which levels of molecules are altered and is particularly important when dealing with embryo explants because analysis of cell behavior is very time consuming and should not be undertaken in vain. Thus, the use of appropriate controls, as with all experiments, is vital. A nice description of controls is included in Reference 102. Coupled with other approaches such as loss-of-function experiments and *in situ* hybridization, however, overexpression can be a useful tool.

In *Xenopus*, loss-of-function studies rely on interfering with the function of a molecule in a dominant negative fashion or suppressing its expression. This might involve the use of small molecule inhibitors that act directly on protein activity (for example, see References 36 and 138), RNAi (for example, see Reference 87), antibodies (for example, see Reference 139), dominant interfering proteins (or dominant negatives; for example, see References 36, 37, and 63), or specifically modified antisense oligonucleotides (for example, see References 140–142). The activity of transcription factors can also be manipulated by reversing their normal activity, for example, by adding a strong repressor domain from another protein, such as engrailed from *Drosophila*, or activation domain such as VP16 from cytomegalovirus (for example, see Reference 143). As with gain-of-function methods, each technique has strengths and weaknesses, and for methods involving microinjection of RNA or DNA, it is important to include an epitope tag, fluorescent tag, or lineage tracer.

Dominant negatives can be used to study a wide range of molecules from growth factors and their receptors to signaling effectors to transcription factors. Specificity is an issue, however, because dominant negatives can interfere with molecules in the same family. This can be addressed by coexpression of the wild-type gene product, which should negate the effects of the dominant negative, as may overexpression of downstream signaling effectors. It should be considered, however, that the effects of the dominant negative may not be specific.

Morpholino oligonucleotides (morpholinos) can be used to knock down protein production by blocking either the translation initiation complex or the nuclear splicing machinery. Their advantage is that an antibody can be used to confirm that the protein is absent. In addition, maternal or zygotic mRNAs can be targeted (140). Similar rescue experiments to those used for dominant negatives provide useful controls; however, there is no perfect control morpholino oligonucleotide. The company-designed control morpholino is known to be nontoxic to the embryo and provides a control for injection technique. Randomized sequences that contain the same base complement as the targeted morpholino or 5-base mismatch morpholinos can have adverse effects on development, perhaps due to interference with another sequence. Although the use of two different morpholinos targeted against the same molecule may provide some reassurance that the effect is specific, it is better to use two loss-of-function techniques such as dominant negative and morpholino, if possible.

VI. THREE POPULAR VERTEBRATE MODELS AND THE STUDY OF GASTRULATION

The power of teleost embryos as a model for studying the cellular mechanisms of gastrulation was initially illustrated by the pioneering work of John P. Trinkaus in his analysis of epiboly in *Fundulus* (144). Zebrafish have now become one of the best models in which to study gastrulation in part due to the rapid escalation in genetic technology (for review, see Reference 145). Like *Xenopus*, zebrafish embryos are accessible throughout their development and are relatively simple to obtain and culture. They develop rapidly and are large enough for microinjection and cell transfer experiments (for example, see Reference 146). An ever-increasing number of mutants with cell movement defects are also being identified (access the Internet-based Zebrafish Information Network (ZFIN) for up-to-date information). In addition, the morphological details of gastrulation cell movements are being painstakingly characterized (for example, see References 76, 147, and 148). The major advantage of these embryos, however, is that they are transparent, and thus, cell motility can be readily observed in intact embryos. This not only allows the study of specific cell rearrangements, but also importantly enables the analysis of the relationship of different movements with one another.

Chick embryos also provide a well-characterized model system for the analysis of gastrulation (for review, see Reference 149). These embryos are accessible during the gastrula stages and are amenable to microdissection and the implantation of coated beads (for example, see Reference 150). In addition, although transgenic technologies are not readily available, the use of viruses to manipulate gene expression is well established (for example, see Reference 151). In contrast, the mouse has extremely powerful genetics but the embryos are relatively inaccessible, making it difficult to analyze cell motilities in living tissues (for review, see Reference 152). They are also small, making microdissections difficult for all but the most skilled (for example, see Reference 153).

VII. FUTURE PERSPECTIVES

This is an exciting period for the study of gastrulation. New tools and technologies are constantly becoming available, with improved techniques for visualization and gene manipulation. The main purpose of this chapter is to provide methods and tips for analyzing specific gastrulation cell movements in *Xenopus* embryos. It is important, however, to compare these findings to other embryos. For example, epithelial to mesenchymal transition (EMT) is less important for *Xenopus* gastrulation than it is in other vertebrate embryos such as those of the zebrafish, chicken, and mouse (for review, see Reference 62). In addition, signaling pathways may not play identical roles in every embryo. Examples were given to illustrate our current understanding of the molecular mechanisms of frog gastrulation; however, many issues remain. For example, how diverse signaling pathways are integrated to control cell movement and how cell movement pathways are coordinated with those regulating other processes such as cell division, tissue specification and patterning, and axis formation must be resolved. In other words, our future challenge is to determine how gene expression is translated into morphogenesis.

ACKNOWLEDGMENTS

I would like to take this opportunity to thank Chris Ford, Ray Keller, Mark Mercola, Dave Nagajski, Jim Smith, and Rudi Winklbauer, who sparked my initial interest in development and gastrulation and supported me in my obsession. I would also like to thank Doug Sypeck, Connie Lane, and Evelyn Houliston for their unfailing encouragement. Grateful thanks also go to Lance Davidson, Connie Lane, Marina Malikova, Amy K. Sater, Erin Smith, John Wallingford, and Rudi Winklbauer for critical reading of the manuscript and helpful discussions, and to Amy K. Sater and Malcolm Whitman for their amazing patience.

REFERENCES

1. Stern, C.D. (Ed.), (2004). *Gastrulation: From Cells to Embryos.* Cold Spring Harbor Laboratory Press: Cold Spring Harbor, NY.
2. Keller, R., (1986). The cellular basis of amphibian gastrulation, in *The Cellular Basis of Morphogenesis,* L.W. Browder (Ed.). Plenum Press: New York. pp. 241–387.
3. Keller, R. and Danilchik, M., (1988). Regional expression, pattern and timing of convergence and extension during gastrulation of Xenopus laevis. *Development*, 103: 193–209.
4. Davidson, L.A., Hoffstrom, B.G., Keller, R. and DeSimone, D.W., (2002). Mesendoderm extension and mantle closure in Xenopus laevis gastrulation: combined roles for integrin alpha(5)beta(1), fibronectin, and tissue geometry. *Dev Biol*, 242: 109–29.
5. Keller, R.E., (1975). Vital dye mapping of the gastrula and neurula of Xenopus laevis. I. Prospective areas and morphogenetic movements of the superficial layer. *Dev Biol*, 42: 222–41.
6. Keller, R.E., (1976). Vital dye mapping of the gastrula and neurula of Xenopus laevis II. Prospective areas and morphogenetic movements of the deep layer. *Devl. Biol.*, 51: 118–137.
7. Hardin, J. and Keller, R., (1988). The behaviour and function of bottle cells during gastrulation of Xenopus laevis. *Development*, 103: 211–30.
8. Keller, R.E., (1981). An experimental analysis of the role of bottle cells and the deep marginal zone in gastrulation of Xenopus laevis. *J Exp Zool*, 216: 81–101.
9. Nieuwkoop, P.D. and Faber, J., (1967). *Normal Table of* Xenopus laevis *(Daudin).* 2nd ed. North-Holland Publishing Company: Amsterdam.

10. Winklbauer, R. and Schurfeld, M., (1999). Vegetal rotation, a new gastrulation movement involved in the internalization of the mesoderm and endoderm in Xenopus. *Development*, 126: 3703–13.
11. Lane, M.C. and Sheets, M.D., (2002). Rethinking axial patterning in amphibians. *Dev Dyn*, 225: 434–47.
12. Marsden, M. and DeSimone, D.W., (2001). Regulation of cell polarity, radial intercalation and epiboly in Xenopus: novel roles for integrin and fibronectin. *Development*, 128: 3635–47.
13. Goto, T., Davidson, L., Asashima, M. and Keller, R., (2005). Planar cell polarity genes regulate polarized extracellular matrix deposition during frog gastrulation. *Curr Biol*, 15: 787–93.
14. Keller, R.E., Danilchik, M., Gimlich, R. and Shih, J., (1985). The function and mechanism of convergent extension during gastrulation of Xenopus laevis. *J Embryol Exp Morphol*, 89 Suppl: 185–209.
15. Keller, R.E., Danilchik, M., Gimlich, R. and Shih, J., (1985). Convergent extension by cell intercalation during gastrulation of Xenopus laevis, in *Molecular Determinants of Animal Form,* G.M. Edelman (Ed.). Academic Press: San Diego. pp. 111–141.
16. Keller, R.E., (1984). The cellular basis of gastrulation in Xenopus laevis: Active postinvolution convergence and extension by mediolateral interdigitation. *Am. Zool.*, 24: 589–603.
17. Keller, R.E., (1978). Time-lapse cinemicrographic analysis of superficial cell behavior during and prior to gastrulation in Xenopus laevis. *J. Morphol.*, 157: 223–248.
18. Keller, R.E., (1980). The cellular basis of epiboly: an SEM study of deep-cell rearrangement during gastrulation in Xenopus laevis. *J Embryol Exp Morphol*, 60: 201–34.
19. Boucaut, J.-C., Darribère, T., Shi, D.L., Boulekbache, H., Yamada, K.M. and Thiery, J.P., (1985). Evidence for the role of fibronectin in amphibian gastrulation. *J. Embryol. Exp. Morph.*, 89 (Suppl.): 211–227.
20. Nakatsuji, N. and Johnson, K.E., (1983). Comparative study of extracellular fibrils on the ectodermal layer in gastrulae of five amphibian species. *J. Cell Sci.*, 59: 61–70.
21. Lee, G., Hynes, R.O. and Kirschner, M., (1984). Temporal and spatial regulation of fibronectin in early Xenopus development. *Cell*, 36: 729–740.
22. Nagel, M. and Winklbauer, R., (1999). Establishment of substratum polarity in the blastocoel roof of the Xenopus embryo. *Development*, 126: 1975–84.
23. Winklbauer, R. and Nagel, M., (1991). Directional mesoderm cell migration in the Xenopus gastrula. *Dev. Biol.*, 148: 573–589.
24. Keller, R.E. and Shook, D., (2004). Gastrulation in amphibians, in *Gastrulation: From Cells to Embryos,* C.D. Stern (Ed.). Cold Spring Harbor Laboratory Press: Cold Spring Harbor, NY. pp. 171–203.
25. Keller, R. and Tibbetts, P., (1989). Mediolateral cell intercalation in the dorsal, axial mesoderm of Xenopus laevis. *Dev Biol*, 131: 539–49.
26. Keller, R.E., Shih, J. and Wilson, P.A., (1989). Morphological polarity of intercalating deep mesodermal cells in the organizer of Xenopus laevis gastrulae, in *Proceedings of the 47th Annual Meeting of the Electron Microscopy Society of America,* G.W. Bailey (Ed.). San Francisco Press: San Francisco. pp. 840–841.
27. Keller, R., Cooper, M.S., Danilchik, M., Tibbetts, P. and Wilson, P.A., (1989). Cell intercalation during notochord development in Xenopus laevis. *J. Exp. Zool.*, 251: 134–154.

28. Elul, T., Koehl, M.A. and Keller, R.E., (1998). Patterning of morphogenetic cell behaviors in neural ectoderm of Xenopus laevis. *Ann N Y Acad Sci*, 857: 248–51.
29. Elul, T., Koehl, M.A. and Keller, R., (1997). Cellular mechanism underlying neural convergent extension in Xenopus laevis embryos. *Dev Biol*, 191: 243–58.
30. Elul, T. and Keller, R., (2000). Monopolar protrusive activity: a new morphogenic cell behavior in the neural plate dependent on vertical interactions with the mesoderm in Xenopus. *Dev Biol*, 224: 3–19.
31. Ezin, A.M., Skoglund, P. and Keller, R., (2003). The midline (notochord and notoplate) patterns the cell motility underlying convergence and extension of the Xenopus neural plate. *Dev Biol*, 256: 100–114.
32. Shih, J. and Keller, R., (1992). Patterns of cell motility in the organizer and dorsal mesoderm of Xenopus laevis. *Development*, 116: 915–930.
33. Yokota, C., Kofron, M., Zuck, M., Houston, D.W., Isaacs, H., Asashima, M., Wylie, C.C. and Heasman, J., (2003). A novel role for a nodal-related protein; Xnr3 regulates convergent extension movements via the FGF receptor. *Development*, 130: 2199–212.
34. Sivak, J.M., Petersen, L.F. and Amaya, E., (2005). FGF signal interpretation is directed by Sprouty and Spred proteins during mesoderm formation. *Dev Cell*, 8: 689–701.
35. Kwan, K.M. and Kirschner, M.W., (2003). Xbra functions as a switch between cell migration and convergent extension in the Xenopus gastrula. *Development*, 130: 1961–72.
36. Nagel, M., Tahinci, E., Symes, K. and Winklbauer, R., (2004). Guidance of mesoderm cell migration in the Xenopus gastrula requires PDGF signaling. *Development*, 131: 2727–2736.
37. Ataliotis, P., Symes, K., Chou, M.M., Ho, L. and Mercola, M., (1995). PDGF signalling is required for gastrulation of Xenopus laevis. *Development*, 121: 3099–3110.
38. Mercola, M., Melton, D.A. and Stiles, C.D., (1988). Platelet-derived growth factor A chain is maternally encoded in Xenopus embryos. *Science*, 241: 1223–1225.
39. Jones, S.D., Ho, L., Smith, J.C., Yordan, C., Stiles, C.D. and Mercola, M., (1993). The Xenopus platelet-derived growth factor a receptor: cDNA cloning and demonstration that mesoderm induction establishes the lineage-specific pattern of ligand and receptor expression. *Dev. Gen.*, 14: 185–193.
40. Nakatsuji, N. and Johnson, K.E., (1983). Conditioning of a culture substratum by the ectodermal layer promotes attachment and oriented locomotion by amphibian gastrula mesodermal cells. *J. Cell Sci.*, 59: 43–60.
41. Shi, D.-L., Darribere, T., Johnson, K.E. and Boucaut, J.-C., (1989). Initiation of mesodermal cell migration and spreading relative to gastrulation in the urodele amphibian Pleurodeles waltl. *Development*, 105: 351–363.
42. Liu, L., Korzh, V., Balasubramaniyan, N.V., Ekker, M. and Ge, R., (2002). Platelet-derived growth factor A (pdgf-a) expression during zebrafish embryonic development. *Dev Genes Evol*, 212: 298–301.
43. Liu, L., Chong, S.-W., Balasubramaniyan, N.V., Korzh, V. and Ge, R., (2002). Platelet-derived growth factor receptor alpha (pdgfr-a) gene in zebrafish embryonic development. *Mech Dev*, 116: 227–230.
44. Montero, J.A., Kilian, B., Chan, J., Bayliss, P.E. and Heisenberg, C.P., (2003). Phosphoinositide 3-kinase is required for process outgrowth and cell polarization of gastrulating mesendodermal cells. *Curr Biol*, 13: 1279–89.
45. Ulrich, F., Concha, M.L., Heid, P.J., Voss, E., Witzel, S., Roehl, H., Tada, M., Wilson, S.W., Adams, R.J., Soll, D.R. and Heisenberg, C.P., (2003). Slb/Wnt11 controls hypoblast cell migration and morphogenesis at the onset of zebrafish gastrulation. *Development*, 130: 5375–84.

46. Etienne-Manneville, S. and Hall, A., (2002). Rho GTPases in cell biology. *Nature*, 420: 629–35.
47. Symes, K. and Mercola, M., (1996). Embryonic mesoderm cells spread in response to PDGF and signalling by PI3 kinase. *Proc. Natl. Acad. Sci. USA*, 93: 9641–9644.
48. Ren, R., Nagel, M., Tahinci, E., Winklbauer, R. and Symes, K., (2006). Migrating anterior mesoderm cells and intercalating trunk mesoderm cells have distinct responses to Rho and Rac during Xenopus gastrulation. *Dev Dyn*, in press.
49. Teckchandani, A.M., Feshchenko, E.A. and Tsygankov, A.Y., (2001). c-Cbl facilitates fibronectin matrix production by v-Abl-transformed NIH3T3 cells via activation of small GTPases. *Oncogene*, 20: 1739–55.
50. Niehrs, C., Keller, R., Cho, K.W.Y. and De Robertis, E.M., (1993). The homeobox gene *goosecoid* controls cell migration in Xenopus embryos. *Cell*, 72: 491–503.
51. Kramer, K.L., Barnette, J.E. and Yost, H.J., (2002). PKCgamma regulates syndecan-2 inside-out signaling during xenopus left-right development. *Cell*, 111: 981–90.
52. Kramer, K.L. and Yost, H.J., (2002). Ectodermal syndecan-2 mediates left-right axis formation in migrating mesoderm as a cell-nonautonomous Vg1 cofactor. *Dev Cell*, 2: 115–24.
53. Wacker, S., Grimm, K., Joos, T. and Winklbauer, R., (2000). Development and control of tissue separation at gastrulation in Xenopus. *Dev Biol*, 224: 428–39.
54. Winklbauer, R., Medina, A., Swain, R.K. and Steinbeisser, H., (2001). Frizzled-7 signalling controls tissue separation during Xenopus gastrulation. *Nature*, 413: 856–60.
55. Medina, A., Swain, R.K., Kuerner, K.M. and Steinbeisser, H., (2004). Xenopus paraxial protocadherin has signaling functions and is involved in tissue separation. *Embo J*, 23: 3249–58.
56. Hukriede, N.A., Tsang, T.E., Habas, R., Khoo, P.L., Steiner, K., Weeks, D.L., Tam, P.P. and Dawid, I.B., (2003). Conserved requirement of Lim1 function for cell movements during gastrulation. *Dev Cell*, 4: 83–94.
57. Kane, D. and Adams, R., (2002). Life at the edge: epiboly and involution in the zebrafish. *Results Probl Cell Differ*, 40: 117–35.
58. Hasegawa, K. and Kinoshita, T., (2000). Xoom is required for epibolic movement of animal ectodermal cells in Xenopus laevis gastrulation. *Dev Growth Differ*, 42: 337–46.
59. Hasegawa, K., Sakurai, N. and Kinoshita, T., (2001). Xoom is maternally stored and functions as a transmembrane protein for gastrulation movement in Xenopus embryos. *Dev Growth Differ*, 43: 25–31.
60. Bruce, A.E., Howley, C., Dixon Fox, M. and Ho, R.K., (2005). T-box gene eomesodermin and the homeobox-containing Mix/Bix gene mtx2 regulate epiboly movements in the zebrafish. *Dev Dyn*, 233: 105–14.
61. Smith, J.C., (2004). Role of T-box genes during gastrulation, in *Gastrulation: From Cells to Embryos,* C.D. Stern (Ed.). Cold Spring Harbor Laboratory Press: Cold Spring Harbor, NY. pp. 571–589.
62. Münsterberg, A.E. and Wheeler, G., (2005). Cell movements during early vertebrate morphogenesis, in *Cell Signaling and Growth Factors in Development,* K. Unsicker and K. Krieglstein (Eds.). Wiley-VCH: Weinheim. pp. 107–140.
63. Amaya, E., Musci, T.J. and Kirschner, M., (1991). Expression of a dominant negative mutant of the FGF receptor disrupts mesoderm formation in Xenopus embryos. *Cell*, 66: 257–270.
64. Amaya, E., Stein, P.A., Musci, T.J. and Kirschner, M.W., (1993). FGF signalling in the early specification of mesoderm in Xenopus. *Development*, 118: 477–487.

65. Kimelman, D. and Kirschner, M., (1987). Synergistic induction of mesoderm by FGF and TGF-beta and the identification of an mRNA coding for FGF in the early Xenopus embryo. *Cell*, 51: 869–877.
66. Slack, J.M.W., Darlington, B.G., Heath, J.K. and Godsave, S.F., (1987). Mesoderm induction in early Xenopus embryos by heparin-binding growth factors. *Nature* (London). 326: 197–200.
67. Isaacs, H.V., Pownall, M.E. and Slack, J.M., (1994). eFGF regulates Xbra expression during Xenopus gastrulation. *Embo J*, 13: 4469–81.
68. Kroll, K.L. and Amaya, E., (1996). Transgenic Xenopus embryos from sperm nuclear transplantation reveal FGF signaling requirements during gastrulation. *Development*, 122: 3173–3183.
69. Nutt, S.L., Dingwell, K.S., Holt, C.E. and Amaya, E., (2001). Xenopus Sprouty2 inhibits FGF-mediated gastrulation movements but does not affect mesoderm induction and patterning. *Genes Dev*, 15: 1152–66.
70. Kinoshita, N., Iioka, H., Miyakoshi, A. and Ueno, N., (2003). PKC delta is essential for Dishevelled function in a noncanonical Wnt pathway that regulates Xenopus convergent extension movements. *Genes Dev*, 17: 1663–76.
71. Wallingford, J.B., Ewald, A.J., Harland, R.M. and Fraser, S.E., (2001). Calcium signaling during convergent extension in Xenopus. *Curr Biol*, 11: 652–61.
72. Chung, H.A., Hyodo-Miura, J., Nagamune, T. and Ueno, N., (2005). FGF signal regulates gastrulation cell movements and morphology through its target NRH. *Dev Biol*, 282: 95–110.
73. Sasai, N., Nakazawa, Y., Haraguchi, T. and Sasai, Y., (2004). The neurotrophin-receptor-related protein NRH1 is essential for convergent extension movements. *Nat Cell Biol*, 6: 741–8.
74. McKendry, R., Hsu, S.C., Harland, R.M. and Grosschedl, R., (1997). LEF-1/TCF proteins mediate wnt-inducible transcription from the Xenopus nodal-related 3 promoter. *Dev Biol*, 192: 420–31.
75. Munoz-Sanjuan, I. and Brivanlou, A.H., (2004). Modulation of BMP signaling during vertebrate gastrulation, in *Gastrulation: From Cells to Embryos,* C.D. Stern (Ed.). Cold Spring Harbor Laboratory Press: Cold Spring Harbor, NY. pp. 475–489.
76. Myers, D.C., Sepich, D.S. and Solnica-Krezel, L., (2002). Bmp activity gradient regulates convergent extension during zebrafish gastrulation. *Dev Biol*, 243: 81–98.
77. Heisenberg, C.P., Tada, M., Rauch, G.J., Saude, L., Concha, M.L., Geisler, R., Stemple, D.L., Smith, J.C. and Wilson, S.W., (2000). Silberblick/Wnt11 mediates convergent extension movements during zebrafish gastrulation. *Nature*, 405: 76–81.
78. Moon, R.T., Campbell, R.M., Christian, J.L., McGrew, L.L., Shih, J. and Fraser, S., (1993). Xwnt-5A: a maternal Wnt that affects morphogenetic movements after overexpression in embryos of Xenopus laevis. *Development*, 119: 97–111.
79. Wallingford, J.B., Fraser, S.E. and Harland, R.M., (2002). Convergent extension. The molecular control of polarized cell movement during embryonic development. *Dev Cell*, 2: 695–706.
80. Houston, D.W. and Wylie, C., (2004). The role of Wnts in gastrulation, in *Gastrulation: From Cells to Embryos,* C.D. Stern (Ed.). Cold Spring Harbor Laboratory Press: Cold Spring Harbor, NY. pp. 521–537.
81. Wallingford, J.B., (2005). Vertebrate gastrulation: polarity genes control the matrix. *Curr Biol*, 15: R414–6.
82. Veeman, M.T., Axelrod, J.D. and Moon, R.T., (2003). A second canon. Functions and mechanisms of beta-catenin-independent Wnt signaling. *Dev Cell*, 5: 367–77.

83. Strutt, D., (2003). Frizzled signalling and cell polarisation in Drosophila and vertebrates. *Development*, 130: 4501–13.
84. Tada, M. and Smith, J.C., (2000). Xwnt11 is a target of Xenopus Brachyury: regulation of gastrulation movements via Dishevelled, but not through the canonical Wnt pathway. *Development*, 127: 2227–38.
85. Topczewski, J., Sepich, D.S., Myers, D.C., Walker, C., Amores, A., Lele, Z., Hammerschmidt, M., Postlethwait, J. and Solnica-Krezel, L., (2001). The zebrafish glypican knypek controls cell polarity during gastrulation movements of convergent extension. *Dev Cell*, 1: 251–64.
86. Wallingford, J.B., Rowning, B.A., Vogeli, K.M., Rothbacher, U., Fraser, S.E. and Harland, R.M., (2000). Dishevelled controls cell polarity during Xenopus gastrulation. *Nature*, 405: 81–5.
87. Habas, R., Kato, Y. and He, X., (2001). Wnt/Frizzled activation of Rho regulates vertebrate gastrulation and requires a novel Formin homology protein Daam1. *Cell*, 107: 843–54.
88. Habas, R., Dawid, I.B. and He, X., (2003). Coactivation of Rac and Rho by Wnt/Frizzled signaling is required for vertebrate gastrulation. *Gen. Dev.*, 17: 295–309.
89. Kim, G.H. and Han, J.K., (2005). JNK and ROKalpha function in the noncanonical Wnt/RhoA signaling pathway to regulate Xenopus convergent extension movements. *Dev Dyn*, 232: 958–68.
90. Tahinci, E. and Symes, K., (2003). Distinct functions of Rho and Rac are required for convergent extension during Xenopus gastrulation. *Dev Biol*, 259: 318–335.
91. Unterseher, F., Hefele, J.A., Giehl, K., De Robertis, E.M., Wedlich, D. and Schambony, A., (2004). Paraxial protocadherin coordinates cell polarity during convergent extension via Rho A and JNK. *Embo J*, 23: 3259–69.
92. Bakkers, J., Kramer, C., Pothof, J., Quaedvlieg, N.E., Spaink, H.P. and Hammerschmidt, M., (2004). Has2 is required upstream of Rac1 to govern dorsal migration of lateral cells during zebrafish gastrulation. *Development*, 131: 525–37.
93. Goto, T. and Keller, R., (2002). The planar cell polarity gene strabismus regulates convergence and extension and neural fold closure in Xenopus. *Dev Biol*, 247: 165–81.
94. Malbon, C.C., Wang, H. and Moon, R.T., (2001). Wnt signaling and heterotrimeric G-proteins: strange bedfellows or a classic romance? *Biochem Biophys Res Commun*, 287: 589–93.
95. Penzo-Mendez, A., Umbhauer, M., Djiane, A., Boucaut, J.C. and Riou, J.F., (2003). Activation of Gbetagamma signaling downstream of Wnt-11/Xfz7 regulates Cdc42 activity during Xenopus gastrulation. *Dev Biol*, 257: 302–14.
96. Lin, F., Sepich, D.S., Chen, S., Topczewski, J., Yin, C., Solnica-Krezel, L. and Hamm, H., (2005). Essential roles of Gα12/13 signaling in distinct cell behaviors driving zebrafish convergence and extension gastrulation movements. *J Cell Biol*, 169: 777–87.
97. Yan, D., Wallingford, J.B., Sun, T.Q., Nelson, A.M., Sakanaka, C., Reinhard, C., Harland, R.M., Fantl, W.J. and Williams, L.T., (2001). Cell autonomous regulation of multiple Dishevelled-dependent pathways by mammalian Nkd. *Proc Natl Acad Sci U S A*, 98: 3802–7.
98. Ninomiya, H., Elinson, R.P. and Winklbauer, R., (2004). Antero-posterior tissue polarity links mesoderm convergent extension to axial patterning. *Nature*, 430: 364–7.
99. Thorpe, C.J. and Moon, R.T., (2004). Nemo-like kinase is an essential co-activator of Wnt signaling during early zebrafish development. *Development*, 131: 2899–909.

100. Yamashita, S., Miyagi, C., Carmany-Rampey, A., Shimizu, T., Fujii, R., Schier, A.F. and Hirano, T., (2002). Stat3 controls cell movements during zebrafish gastrulation. *Dev Cell*, 2: 363–75.
101. Yamashita, S., Miyagi, C., Fukada, T., Kagara, N., Che, Y.S. and Hirano, T., (2004). Zinc transporter LIVI controls epithelial-mesenchymal transition in zebrafish gastrula organizer. *Nature*, 429: 298–302.
102. Sive, H.L., Grainger, R.M. and Harland, R.M., (1998). *Early Development of* Xenopus laevis. *A Laboratory Manual.* Cold Spring Harbor Laboratory Press: Cold Spring Harbor, NY.
103. Peng, H.B., (1991). Solutions and protocols, in Xenopus laevis: *Practical Uses in Cell and Molecular Biology,* B.K. Kay and H.B. Peng (Eds.). Academic Press: San Diego. pp. 657–662.
104. Huber, M.M., Staubli, A.B., Kustedjo, K., Gray, M.H., Shih, J., Fraser, S.E., Jacobs, R.E. and Meade, T.J., (1998). Fluorescently detectable magnetic resonance imaging agents. *Bioconjug Chem*, 9: 242–9.
105. Jacobs, R.E. and Fraser, S.E., (1994). Magnetic resonance microscopy of embryonic cell lineages and movements. *Science*, 263: 681–4.
106. Harland, R.M., (1991). In situ hybridization: an improved whole mount method for Xenopus embryos, in Xenopus laevis: *Practical Uses in Cell and Molecular Biology,* B.K. Kay and H.B. Peng (Eds.). Academic Press: San Diego. pp. 685–695.
107. Tyszka, J.M., Ewald, A.J., Wallingford, J.B. and Fraser, S.E., (2005). New tools for visualization and analysis of morphogenesis in spherical embryos. *Dev Dyn*, 234: 974–83.
108. Ewald, A.J., Peyrot, S.M., Tyszka, J.M., Fraser, S.E. and Wallingford, J.B., (2004). Regional requirements for Dishevelled signaling during Xenopus gastrulation: separable effects on blastopore closure, mesendoderm internalization and archenteron formation. *Development*, 131: 6195–209.
109. Ewald, A.J., McBride, H., Reddington, M., Fraser, S.E. and Kerschmann, R., (2002). Surface imaging microscopy, an automated method for visualizing whole embryo samples in three dimensions at high resolution. *Dev Dyn*, 225: 369–75.
110. Nakatsuji, N. and Johnson, K.E., (1982). Cell locomotion in vitro by Xenopus laevis gastrula mesodermal cells. *Cell Motil*, 2: 149–61.
111. Winklbauer, R., Selchow, A., Nagel, M., Stoltz, C. and Angres, B., (1991). Mesoderm cell migration in the Xenopus gastrula, in *Gastrulation: Movements, Patterns, and Molecules,* R. Keller, W.H. Clark and F. Griffin (Eds.). Plenum Press: New York. pp. 147–168.
112. Davidson, L.A., Keller, R. and DeSimone, D., (2004). Patterning and tissue movements in a novel explant preparation of the marginal zone of Xenopus laevis. *Gene Expr Patterns*, 4: 457–66.
113. Sater, A.K., Steinhardt, R.A. and Keller, R., (1993). Induction of neuronal differentiation by planar signals in Xenopus embryos. *Dev Dyn*, 197: 268–80.
114. Poznanski, A. and Keller, R., (1997). The role of planar and early vertical signaling in patterning the expression of Hoxb-1 in Xenopus. *Dev Biol*, 184: 351–66.
115. Winklbauer, R. and Keller, R.E., (1996). Fibronectin, mesoderm migration, and gastrulation in Xenopus. *Dev Biol*, 177: 413–26.
116. Doniach, T., Phillips, C.R. and Gerhart, J.C., (1992). Planar induction of anteroposterior pattern in the developing central nervous system of Xenopus laevis. *Science*, 257: 542–545.
117. Vodicka, M.A. and Gerhart, J.C., (1995). Blastomere derivation and domains of gene expression in the Spemann Organizer of Xenopus laevis. *Development*, 121: 3505–18.

118. Keller, R., Shih, J. and Domingo, C., (1992). The patterning and functioning of protrusive activity during convergence and extension of the Xenopus organizer. *Development suppl.*: 81–91.
119. Wilson, P., Oster, G. and Keller, R., (1989). Cell rearrangement and segmentation in Xenopus: direct observation of cultured explants. *Development*, 105: 155–166.
120. Smith, J.C., (1987). A mesoderm-inducing factor is produced by a Xenopus cell line. *Development*, 99: 3–14.
121. Symes, K. and Smith, J.C., (1987). Gastrulation movements provide an early marker of mesoderm induction in Xenopus laevis. *Development*, 101: 339–349.
122. Kessler, D.S., (2004). Activin and Vg1 and the search for embryonic inducers, in *Gastrulation: From Cells to Embryos,* C.D. Stern (Ed.). Cold Spring Harbor Laboratory Press: Cold Spring Harbor, NY. pp. 505–520.
123. Green, J.B.A. and Smith, J.C., (1990). Graded changes in dose of a Xenopus activin A homologue elicit stepwise transitions in embryonic cell fate. *Nature*, 347: 391–394.
124. Smith, J.C., Symes, K., Hynes, R.O. and DeSimone, D., (1990). Mesoderm induction and the control of gastrulation in Xenopus laevis: the roles of fibronectin and integrins. *Development*, 108: 229–238.
125. Amaya, E., (2005). Xenomics. *Genome Res*, 15: 1683–91.
126. Chae, J., Zimmerman, L.B. and Grainger, R.M., (2002). Inducible control of tissue-specific transgene expression in Xenopus tropicalis transgenic lines. *Mech Dev*, 117: 235–41.
127. Bronchain, O.J., Hartley, K.O. and Amaya, E., (1999). A gene trap approach in Xenopus. *Curr Biol*, 9: 1195–8.
128. Amaya, E. and Kroll, K.L., (1999). A method for generating transgenic frog embryos. *Methods Mol Biol*, 97: 393–414.
129. Allen, B.G. and Weeks, D.L., (2005). Transgenic Xenopus laevis embryos can be generated using phiC31 integrase. *Nat Methods*, 2: 975–9.
130. Kelly, L.E., Davy, B.E., Berbari, N.F., Robinson, M.L. and El-Hodiri, H.M., (2005). Recombineered Xenopus tropicalis BAC expresses a GFP reporter under the control of Arx transcriptional regulatory elements in transgenic Xenopus laevis embryos. *Genesis*, 41: 185–91.
131. Sakamaki, K., Takagi, C., Yoshino, J., Yokota, H., Nakamura, S., Kominami, K., Hyodo, A., Takamune, K., Yuge, M. and Ueno, N., (2005). Transgenic frogs expressing the highly fluorescent protein venus under the control of a strong mammalian promoter suitable for monitoring living cells. *Dev Dyn*, 233: 562–9.
132. Ryffel, G.U., Werdien, D., Turan, G., Gerhards, A., Goosses, S. and Senkel, S., (2003). Tagging muscle cell lineages in development and tail regeneration using Cre recombinase in transgenic Xenopus. *Nucleic Acids Res*, 31: e44.
133. Hartley, K.O., Nutt, S.L. and Amaya, E., (2002). Targeted gene expression in transgenic Xenopus using the binary Gal4-UAS system. *Proc Natl Acad Sci U S A*, 99: 1377–82.
134. Kolm, P.J. and Sive, H.L., (1995). Efficient hormone-inducible protein function in Xenopus laevis. *Dev. Biol.*, 171: 267–272.
135. Spencer, D.M., Wandless, T.J., Schreiber, S.L. and Crabtree, G.R., (1993). Controlling signal transduction with synthetic ligands. *Science*, 262: 1019–1024.
136. Van Stry, M., Kazlauskas, A., Schreiber, S.L. and Symes, K., (2005). Distinct effectors of PDGFRa signaling are required for cell survival during embryogenesis. *Proc Natl Acad Sci U S A*, 102: 8233–8238.
137. Yang, J., Symes, K., Mercola, M. and Schreiber, S.L., (1998). Small-molecule control of insulin and PDGF receptor signaling and the role of membrane attachment. *Curr. Biol.*, 8: 11–18.

138. Schreiber, S.L., (2000). Target-oriented and diversity-oriented organic synthesis in drug discovery. *Science*, 287: 1964–9.
139. Ramos, J.W. and DeSimone, D.W., (1996). Xenopus embryonic cell adhesion to fibronectin: position-specific activation of RGD/synergy site-dependent migratory behavior at gastrulation. *J Cell Biol*, 134: 227–40.
140. Heasman, J., (2002). Morpholino oligos: making sense of antisense? *Dev Biol*, 243: 209–14.
141. Heasman, J., Kofron, M. and Wylie, C., (2000). Beta-catenin signaling activity dissected in the early Xenopus embryo: a novel antisense approach. *Dev Biol*, 222: 124–34.
142. Dagle, J.M. and Weeks, D.L., (2001). Oligonucleotide-based strategies to reduce gene expression. *Differentiation*, 69: 75–82.
143. Conlon, F.L. and Smith, J.C., (1999). Interference with brachyury function inhibits convergent extension, causes apoptosis, and reveals separate requirements in the FGF and activin signalling pathways. *Dev Biol*, 213: 85–100.
144. Trinkaus, J.P., (1951). A study of the mechanism of epiboly in the egg of Fundulus heteroclitus. *J. Exp. Zool.*, 118: 268–320.
145. Kane, D.A. and Warga, R.M., (2004). Teleost gastrulation, in *Gastrulation: From Cells to Embryos,* C.D. Stern (Ed.). Cold Spring Harbor Laboratory Press: Cold Spring Harbor, NY. pp. 157–169.
146. Ho, R.K. and Kimmel, C.B., (1993). Commitment of cell fate in the early zebrafish embryo. *Science*, 261: 109–11.
147. Glickman, N.S., Kimmel, C.B., Jones, M.A. and Adams, R.J., (2003). Shaping the zebrafish notochord. *Development*, 130: 873–87.
148. Montero, J.A., Carvalho, L., Wilsch-Brauninger, M., Kilian, B., Mustafa, C. and Heisenberg, C.P., (2005). Shield formation at the onset of zebrafish gastrulation. *Development*, 132: 1187–98.
149. Stern, C.D., (2004). Gastrulation in the chick, in *Gastrulation: From Cells to Embryos*, C.D. Stern (Ed.). Cold Spring Harbor Laboratory Press: Cold Spring Harbor, NY. pp. 219–232.
150. Yang, X., Dormann, D., Munsterberg, A.E. and Weijer, C.J., (2002). Cell movement patterns during gastrulation in the chick are controlled by positive and negative chemotaxis mediated by FGF4 and FGF8. *Dev Cell*, 3: 425–37.
151. Iba, H., (2000). Gene transfer into chicken embryos by retrovirus vectors. *Dev Growth Differ*, 42: 213–8.
152. Tam, P.P.L. and Gad, J.M., (2004). Gastrulation in the mouse, in *Gastrulation: From Cells to Embryos,* C.D. Stern (Ed.). Cold Spring Harbor Laboratory Press: Cold Spring Harbor, NY. pp. 233–262.
153. Beddington, R.S., (1994). Induction of a second neural axis by the mouse node. *Development*, 120: 613–620.

14 Bringing Small Molecule Regulation of Protein Activity to Developmental Systems

Karen J. Liu, Jason E. Gestwicki, and Gerald R. Crabtree

CONTENTS

I. SUMMARY

Rapid and reversible regulation of protein activity, as with temperature-sensitive alleles, allows a practical fusion of genetics and biochemistry and is essential to distinguish coincidence from causality. We have developed a method of making conditional protein alleles that permits the rapid and reversible regulation of specific proteins. One application of this approach involves destabilizing a protein with a peptide tag that confers instability in the absence of a drug. We describe the uses of this approach and its application to studies of signal transduction and development in *Xenopus* and mice.

II. INTRODUCTION

The study of signal transduction processes in development is necessarily complicated by the precise spatial and temporal coordination required to properly pattern an embryo. Biologists have learned much about signaling in the embryo by perturbing different pathways and observing the phenotypic and molecular changes resulting from these manipulations. Commonly, genetic methods such as null mutations are used to disrupt protein function. However, *in vivo* analyses are difficult, because one must evaluate the response of many tissues involved in a developmental process. These tissues are necessarily signaling to each other, thus making it difficult to distinguish immediate molecular and cell biological changes from later secondary effects. Many of these responses, such as cell shape changes and tissue-tissue interactions, are difficult to recapitulate *in vitro* and are best studied in whole organisms.

Methods to specifically regulate protein function, such as temperature-sensitive alleles, Cre/LoxP, morpholino oligonucleotides, RNA interference (RNAi), and small molecule inhibitors, ideally provide conditional control of timing and tissue-specificity of gene activation or repression (Fire, Xu et al. 1998; Nagy 2000). Each of these methods allows one to examine specific periods of gene function during biological processes; however, each has strengths and limitations. We will briefly discuss these features and highlight the characteristics required in ideal methods to study signaling and development.

Temperature-sensitive alleles, which produce a mutant phenotype at a nonpermissive temperature and a wild-type phenotype at a permissive temperature, can be used to limit gene function to specific temporal windows (Figure 14.1). In nonvertebrate model organisms, this has been well exploited, most notably in studies of cell-cycle regulation in yeast, as well as in studies of *sevenless* and *ras* signaling in *Drosophila* (Hartwell, Culotti et al. 1970; Mullins and Rubin 1991; Simon, Bowtell et al. 1991; Murray and Hunt 1993). Unfortunately, these approaches are not feasible in vertebrates because generation of temperature-sensitive alleles is not straightforward; in mammals, there is the additional complication that the tolerable thermal range is small. Also, mammals autoregulate their internal temperature regardless of environmental temperature.

In mice, the primary approach to studying gene function has been the use of homologous recombination in embryonic stem (ES) cells to delete genes of interest (Capecchi 1989; Thompson, Clarke et al. 1989). This method provides great specificity in the production of mutant alleles; however, many genes have multiple roles in diverse tissues during development. Thus, the manifestation of early mutant phenotypes often prevents the study of later developmental programs. Conditional knockouts, most commonly using the Cre/*loxP* system, in which a tissue-specific Cre recombinase acts on a *loxP*-flanked allele, provide more spatial and temporal specificity. Currently, the number of reliable Cre strains and the slow kinetics of recombination limit this system. As in knockout mice, Cre-mediated conditional

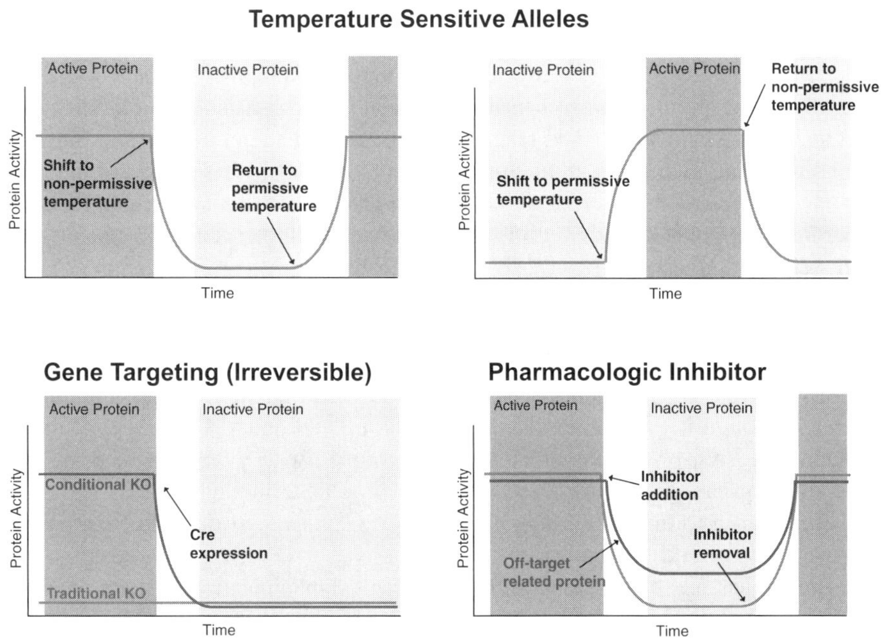

FIGURE 14.1 Regulating protein function during development. Traditionally, temperature-sensitive alleles have worked well when defining windows of protein function during development. Recent advances in mammalian genetics, including knockouts produced via homologous recombination and conditional alleles, have provided great specificity in mutant analyses. However, these methods are not reversible. Pharmacophores can be used to define windows of time in which protein activity is permitted. Addition of an activator (or agonist) is shown in the schematic, but inhibitors provide similar possibilities. A unique feature of pharmacological intervention is that the activation/inhibition is reversible. This reversal can either be passive (e.g., removal of media and replacement with drug-free media) or active (e.g., addition of a competitive inhibitor of the original pharmacophore). The two drugs determine the limits of the window.

mutagenesis results in irreversible loss of function mutations that can manifest as complex pleiotropic phenotypes.

A complementary approach, using transcriptional transactivation or repression (for example, "Tet-on" and "Tet-off" systems), provides drug-dependent control of gene transcription (Gossen and Bujard 1992). Because this method is reversible, it is often used to regulate the expression of transgenes. Finally, the recent use of RNA interference and antisense morpholino oligonucleotides (AMOs) in developmental systems has dramatically increased our ability to study loss of function phenotypes in a spatially and temporally controlled manner (Heasman, Kofron et al. 2000; Kunath, Gish et al. 2003). Unfortunately, these methods again act at the level of transcription and translation, resulting in slow kinetics due to potentially slow rates of transcription and perdurance of the protein of interest.

Function-blocking small molecules, which can act directly on protein activity and can be administered at specific time points, are powerful tools for studying signal transduction (Schreiber 2000). For example, use of the drug cyclopamine, which inhibits Hedgehog signaling by binding to the Smoothened receptor, has helped to define the roles of Hedgehog signaling in development (Chen, Taipale et al. 2002). Many of these results have since been confirmed using specific agonists and antagonists of Smoothened (Frank-Kamenetsky, Zhang et al. 2002). Similarly, the use of lithium chloride in a number of model organisms has helped to pinpoint the window of activity required for activated β-catenin in the establishment of early embryonic axes (Cooke and Smith 1988; Ghiglione, Lhomond et al. 1993; Schneider, Steinbeisser et al. 1996; Logan, Miller et al. 1999).

Administration of lithium chloride at the 32- to 64-cell stage in *Xenopus* appears to mimic the activity of the Wnt signaling pathway by inhibiting glycogen synthase kinase-3 (GSK-3) (Smith and Harland 1991). This inhibition results in global activation of β-catenin transcriptional activity and subsequent hyper-dorsalization of the embryo. Though it is possible that lithium acts on additional signaling molecules, including inositol phosphates, this approach coupled with molecular analyses has been quite revealing (Harwood 2005). However, to date, inhibitors of GSK-3 have not been shown to mimic a double knockout of GSK3, a test that has been passed for cyclosporine and FK506.

Our lab has for some time used pharmacological manipulation to study the roles of calcineurin/NFAT signaling in development. The immunosuppressants cyclosporine A and FK506 are small molecules that are permeable to the placenta and inhibit calcineurin function in the embryo within 15 minutes of administration to pregnant mice; mutant phenotypes can be faithfully recapitulated with these drugs (Graef, Chen et al. 2001; Graef, Wang et al. 2003; Neilson, Winslow et al. 2004). Treatment with these inhibitors at nanomolar concentrations has allowed us to precisely define the timing of developmental requirements for calcineurin activity (Graef, Chen et al. 2001; Graef, Wang et al. 2003; Chang, Neilson et al. 2004). In addition, antagonists permit rapid reversal of their activity in organisms from yeast to mammals (Biggar and Crabtree 2000; Biggar and Crabtree 2001). Thus, specific drugs are extremely powerful tools, allowing the study of later developmental phenotypes while bypassing the early embryonic requirements of signaling. The specificity and usefulness of these drugs gave rise to our efforts to combine chemical approaches with versatile molecular systems to rapidly and specifically target a diverse range of proteins.

Chemical genetic approaches toward pharmacological manipulation have become popular, and the hope is that identification of chemical regulators in developmental assays may be useful not only for biological studies but also for therapeutic use (reviewed in (Bishop, Ubersax et al. 2000; Schreiber 2003; Yeh and Crews 2003)). However, limitations, such as difficulty in identification of target genes, can prevent accurate assessment of molecule specificity. In addition, drug discovery approaches are most readily applied to enzymes, as the active site provides a clear target for inhibition. Many proteins, such as transcription factors or structural elements, cannot be regulated in this fashion. As a result, these methods are often not generalizable for the developmental biologist. Recent attempts to identify new biologically active small molecules through chemical genetics screens have yielded some new compounds that perturb development (Peterson, Link et al. 2000; Frank-Kamenetsky, Zhang et al. 2002; Armstrong, Yuan et al. 2004; Peterson, Shaw et al.

2004; Liu, Wu et al. 2005). Unfortunately, the number is still small, and unlike genetic mutants, the identification and validation of specific drug targets is not straightforward and ultimately depends upon the ability of the drug to reproduce critical early aspects of the mutant phenotype. This approach is not generalizable, and each drug/target combination must be individually validated.

Despite the limitations of chemical approaches, small molecules possess many advantages. For example, drug action is often reversible, and inhibitory activity is tunable by controlling the concentration of the drug. Unlike nucleotides or proteins, small molecules can be membrane permeable and have favorable pharmacokinetics in whole organisms. Due to these advantages, the pursuit of ideal drug-based methods that minimize limitations is a useful goal.

An ideal method for making chemically conditional alleles should possess several key features. Small molecule-sensitive conditional alleles must be highly specific, applicable to a wide range of targets, and regulated rapidly and reversibly. Regulation of target proteins should be rapid and reversible. In a method developed in the Shokat lab, kinases were mutated in such a way that they become sensitive to a chemical inhibitor but retain full activity in the absence of drug (Bishop, Ubersax et al. 2000). Recently, knock-in mice produced by the Ginty and Shokat labs using these techniques demonstrated specific pharmacological control of Trk receptor tyrosine kinases (Chen, Ye et al. 2005). This method may be broadly applicable to protein kinases. In another approach, proteins fused to steroid-binding domains can be held dormant until addition of hormone (Levy, Johnston et al. 1999; Picard 2000). Yet another method fuses proteins to an engineered dihydrofolate reductase (DHFR), resulting in a short-lived protein. Addition of methotrexate inhibits DHFR-mediated degradation, allowing conditional control of protein levels (Levy, Johnston et al. 1999).

In addition to the methods outlined above, a number of small molecule techniques take advantage of inducible protein-protein interactions to regulate protein function. These strategies are primarily “drug-on” approaches, in which addition of drug activates the protein. Several of these techniques are outlined in Figure 14.2. The Crabtree and Schreiber labs have for some years been developing methods for regulating protein function based on the concept that induced proximity regulates many processes including receptor function, GTPase action, vesicle fusion, transcription, and chromatin remodeling. These methods are inspired by the observation that some natural products obtain specificity and efficacy by inducing the dimerization of two cellular proteins and that induced proximity or recruitment is perhaps the most commonly used biologic regulatory mechanism. The conditional dimerization system that we, and others, have co-opted uses cell-permeable small molecules with surfaces that bind two proteins and induce their proximity. Examples include FK506, FK1012, rapamycin, coumermycin, and a wide range of other molecules.

More recently, we have developed a strategy that combines a drug-on approach, in which we use a small molecule to maintain stability of a target protein (Stankunas, Bayle et al. 2003), with a drug-off system, in which addition of a second drug allows either destabilization or relocalization of our target protein. Using this system, we have developed a general method of making fast-acting, small molecule-sensitive protein alleles. When fused to an 89 amino acid tag, FRB^{PLF}, proteins are rapidly destabilized. In the presence of rapamycin analogues (rapalogues), FRB^{PLF}-tagged

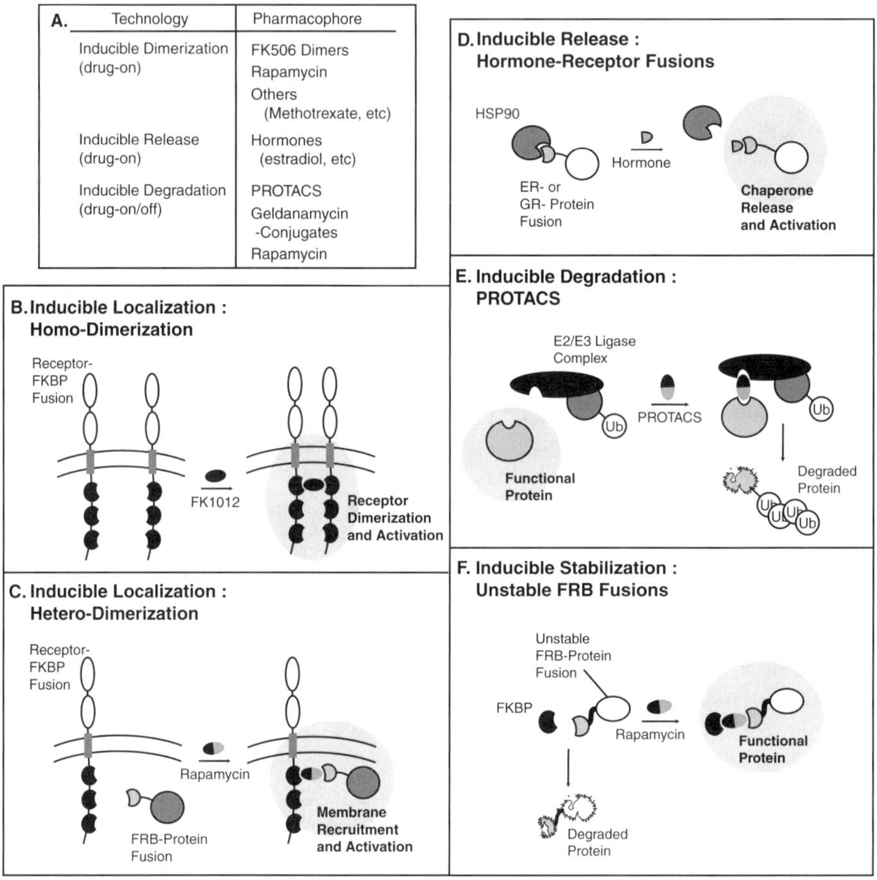

FIGURE 14.2 Strategies for controlling protein function. (A) Examples of strategies for manipulating protein function with small molecules are shown. These examples possess two key features: (1) *generality*, the drug (or class of drugs) can be used to control many different proteins, and (2) *speed*, the drug controls protein function after translation. These approaches bypass requirements for new protein synthesis or degradation of pre-existing pools. (B) Drug-dependent inducible dimerization systems using derivatives of FK506 have been used to study receptor signaling (Yap, Brieher et al. 1997; Yang, Symes et al. 1998; Pownall, Welm et al. 2003). (C) Heterodimerization using rapamycin has been applied toward studies of dimerization states of ErbB receptors and to make drug-dependent chimeric proteins (Muthuswamy, Gilman et al. 1999; de Graffenried, Laughlin et al. 2004). Chimeric methotrexate-based ligands are also useful for studying protein-protein interactions (Althoff and Cornish 2002). (D) Hormone inducible release of proteins can be used to study transcriptional regulators during development (Kolm and Sive 1995), whereas recent studies have demonstrated the utility of (E) inducible degradation of proteins via recruitment of a ubiquitination complex (Kolm and Sive 1995; Sakamoto, Kim et al. 2001; Schneekloth, Fonseca et al. 2004) and (F) inducible stabilization using unstable Frb fusions (Stankunas, Bayle et al. 2003).

proteins dimerize with endogenous FKBPs. This interaction stabilizes the fusion proteins and restores both protein levels and activity. Because our technique exploits a dimerization mechanism, we can also regulate protein localization and function using induced proximity, allowing fine manipulation of the system.

Among other applications, chemical dimerization has been used to regulate cell membrane receptors (Spencer, Wandless et al. 1993; Pruschy, Spencer et al. 1994), death inducers (Spencer, Belshaw et al. 1996; Belshaw, Spencer et al. 1996), exchange factors (Holsinger, Spencer et al. 1995), GTPases (Castellano, Montcourrier et al. 1999), and transcription factors (Ho, Biggar et al. 1996; Rivera, Clackson et al. 1996; Belshaw, Ho et al. 1996).

In this review, we set out to describe the parameters of our system. Specifically, we will focus on one example in depth: the expression of a chemically regulated luciferase protein in *Xenopus laevis*. This example will illustrate the boundaries and versatility of the system for studying signaling.

III. RAPAMYCIN-DEPENDENT DIMERIZATION

Rapamycin is a natural product isolated from *Streptomyces hygroscopicus* that possesses potent antibacterial, antitumor, and immunosupressive activity (reviewed in (Sehgal 1998)). Rapamycin is fully membrane permeable, has a prolonged serum half-life in humans and rodents, and displays remarkable efficacy and selectivity. Some of these advantages are derived from rapamycin's high affinity (Kd ~1 nM) for the cytosolic FK506-binding protein 12 (FKBP12) (Harding, Galat et al. 1989; Siekierka, Hung et al. 1989). Additionally, the drug-FKBP12 complex develops affinity for the FK506-Rapamycin Binding (FRB) domain of Target of rapamycin (Tor) (Chen, Zheng et al. 1995; Liberles, Diver et al. 1997). Tor is a member of the PI3-kinase superfamily and it functions in cell size regulation and nutrient sensing. Inhibition of Tor signaling by rapamycin leads to G1-to-S arrest (reviewed in (Jacinto and Hall 2003)). Interestingly, FKBP12 has little affinity for FRB in the absence of drug. Rapamycin, however, rapidly triggers the formation of the ternary complex (FKBP12-rapamycin-FRB). This assembly mechanism via recruitment of two endogenous cellular proteins provides rapamycin with potency and selectivity.

In the ternary complex, one structural face of rapamycin binds to FKBP, whereas the opposite side recognizes a cleft between two helices of FRB (Choi, Chen et al. 1996). The FRB domain is an 89 amino acid, 4-helical bundle, and portions of helices 1 and 4 make direct contact with rapamycin (Marianayagam, Khan et al. 2002). When the FRB-contacting region of rapamycin is chemically modified, affinity for FRB is lost. Likewise, mutations in the drug-contact region of FRB can abolish rapamycin binding. However, when mutations are paired with appropriate rapamycin modifications, new drug-protein binding partners can be generated (Liberles, Diver et al. 1997) (Figure 14.3b). One example of this "bump-hole" strategy employs a chemical analog of rapamycin, C20-methallylrapamycin (MaRap) (Stankunas, Bayle et al. 2003). MaRap binds FRB mutants in which accommodating amino acid replacements have been made at three positions. Specifically, lysine 2095, threonine 2098, and tryptophan 2101 are replaced with proline, leucine, and phenylalanine, respectively. This mutant is named FRB^{PLF} to reflect those three amino acid replacements. MaRap has no affinity for wild-type FRB and thus minimal immunosuppressive activity; however, this analog rapidly forms a complex with FRB^{PLF}.

Rapamycin-FRB pairs have been explored as research tools in applications that exploit the natural binding mode of the drug. The use of chemical inducers of dimerization to regulate protein localization and therefore function has been reviewed elsewhere (Klemm, Schreiber et al. 1998). Briefly, one example that has been described is a means to control transcriptional activation. In this three-hybrid system, a fusion protein is expressed that contains FKBP12 and a GAL4 DNA-binding domain (Ho, Biggar et al. 1996). Alone, this construct does not drive expression from a GAL4 promoter because it lacks a transcriptional activation domain. Transcription is dependent on recruitment of FRB and VP16 fusion by rapamycin. In this system, transcriptional levels are placed under control of rapamycin.

Recent attempts to generate new FRB mutants with enhanced specificity for rapamycin analogs led to a fortuitous discovery that some mutations at position 2098 confer drug-dependent stability (Stankunas, Bayle et al. 2003; Gestwicki, Stankunas et al. 2005). Experiments *in vitro* revealed that mutations in that position resulted in FRB domains that are thermodynamically unstable. In cultured cells and in mice, fusion proteins containing this mutant FRB^{PLF} domain are rapidly degraded. As mentioned previously, amino acid 2098 in FRB lies at the rapamycin contact region. Consequently, rapamycin can serve as a "chemical chaperone" to restore FRB^{PLF} stability. Although the mechanism of stability control is not known, it appears that solvent-exposed hydrophobic side chains in the unbound FRB^{PLF} are strongly disfavored. Rapamycin can protect these side chains in the ternary complex. Thus, unstable mutant FRB^{PLF} is degraded in the absence of rapamycin, but protein levels can be restored by drug addition. Competitive inhibitors of the FKBP-rapamycin interaction, such as FK506m, trigger FKBP release from the ternary structure. This event decreases the affinity of rapamycin for FRB^{PLF} and destabilizes the domain. Thus, the two drugs, rapamycin and FK506m (Figure 14.3a), can be used to control the stability of FRB (Stankunas, Bayle et al. 2003).

Instability in the FRB domain can lead to degradation of FRB^{PLF}-protein fusions. We have generated fusions to a variety of intracellular proteins, including GSK3β, Pax6, luciferase, GFP, glutathione transferase, calcineurin B1, CREB, and HNF-1 (Stankunas, Bayle et al. 2003). In all cases, fusion levels are lower in the absence of rapamycin and can be restored by drug. The general success of this approach suggests that inducible stabilization may have broad applications for regulating protein function. However, the sample size of available fusions remains small, and many protein structural classes have not been adequately studied.

Additionally, for some targets, such as calcineurin B1, protein levels can be enhanced by rapamycin addition, but corresponding functional increases are not seen (Feng Chen, Gerald R. Crabtree, unpublished observations). The reasons for this are not yet clear, but might involve failure of fusion proteins to properly assemble into multiprotein signaling complexes. Because of these concerns, it has become important to test unstable FRB^{PLF}-protein fusions and document useful experimental parameters for each target. In this review, we will focus on the requirements for

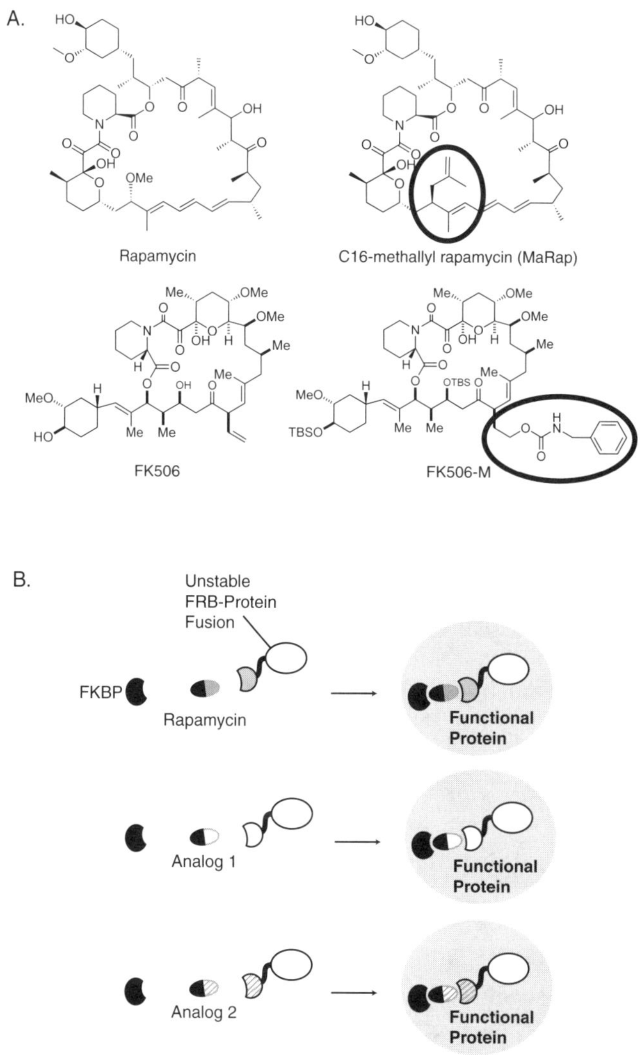

FIGURE 14.3 Chemical analogs provide specificity and reversibility, allowing use of orthogonal systems. (A) Structures of rapamycin, the rapamycin analog C20-MaRap, FK506, and FK506m. Rapamycin is a small membrane-permeable drug. Modified rapamycin analogues, such as C20-methallylrapamycin, shown here, contain bulky chemical "bumps" that allow specific binding to mutant FRB domains. The immunosuppressant FK506 binds with high affinity to FKBP12 and induces its dimerization with calcineurin, disrupting FKBP12 interaction with the FRB domain. FK506m, a derivative of FK506, binds only to FKBP-12 and inhibits rapamycin binding to FKBP-12. (B) Different drug specificities allow use of orthogonal systems. Identification of additional FRB domains with varying specificities and stabilities will be extremely useful. Using a library of rapalogues that bind with differing affinities to a panel of FRB domains allows us to use different drugs to regulate stability and proximity of a suite of FRB-tagged proteins.

experimental and practical analysis of new targets and the potential of this technique for studying signal transduction and development.

When testing a new protein target for use in inducible stabilization applications, we recommend that experiments in a cultured cell system be performed prior to advancing to model organism systems. Our standard procedure is to generate unstable FRB fusions for transfection into at least two cell lines, such as COS1 cells and mouse embryonic fibroblasts (MEFs). Overexpression of protein, either under the control of a strong promoter or in a cell line that is efficiently transfected, sometimes obscures the apparent stability of a tagged protein. This is likely due to overexpression artifacts that result from impaired ability of the cellular proteolytic machinery to fully clear unstable, but abundant, fusions. In cultured cells, rapamycin (5–10 nM) is added to half the wells. One day later, cells are harvested, total proteins separated by SDS-PAGE, and Western blots performed with anti-FRB antibodies. Ideally, a functional assay (e.g., kinase activity, reporter assay, fluorescence, or other phenotypic outcome) is also included. Additionally, both C- and N-terminal FRB fusions are routinely generated. We have commonly observed differences between fusions at either end.

Finally, the kinetics of degradation must be examined. This is commonly performed using translation inhibition by cycloheximide to compare degradation kinetics of the parent protein and the FRB-fusion. In our hands, targets such as GSK3β are lost within 30–60 minutes if rapamycin is not added (Stankunas, Bayle et al. 2003). FK506m also can be added to determine if stability is reversible. Understanding these properties (apparent stability, fusion function, and degradation kinetics) in cultured cells not only facilitates the design of biological inquiries using these cells but also serves as a benchmark for model organism experiments.

IV. APPLYING FRB^{PLF} INSTABILITY TO DEVELOPMENTAL SYSTEMS

Because we are interested in signal transduction during development, we decided to apply conditional stabilization and dimerization to problems in vertebrate embryogenesis. We have used both overexpression and replacement strategies in amphibians and knock-in approaches in mice. This section describes our application of the FRB^{PLF}-rapamycin-FKBP dimerization system to *Xenopus laevis*.

Xenopus has long been a favored model organism of the developmental biologist. Embryos are abundant and accessible, readily synchronized, and easy to manipulate. Exogenous DNA, RNA, proteins, or chemicals can simply be injected into the *Xenopus* embryo and targeted to specific tissue types at different stages. These manipulations can also be combined with loss-of-function antisense morpholino oligonucleotides (Heasman, Kofron et al. 2000; Sive, Grainger et al. 2000).

Overexpression studies in *Xenopus* have advantages and limitations. In many cases, one injects DNA or RNA at the early cleavage stages and observes the phenotypic changes at later stages. Therefore, *Xenopus* embryos have been, for some time, an ideal *in vivo* reaction vessel. Expression of a molecule of interest in *Xenopus* can often suggest signaling interactions (i.e., how and where a molecule can act).

However, because the embryo is undergoing rapid molecular, morphological, and cell biological changes during these time periods, it is often difficult to distinguish secondary results from immediate effects. Furthermore, the endogenous concentrations of active proteins are often difficult to determine and thus impossible to compare to overexpressed protein. Thus, we surmised that rapamycin-mediated regulation would be particularly useful in this system, permitting fine control of protein expression.

A. Toxicity and Pharmacology

In order to use our FRB^{PLF} system, we set out to determine the toxicity and pharmacokinetics of rapamycin in *Xenopus*. The distribution, metabolism, and excretion of a drug may be different in each organism. As a result, effective use of any new drug requires knowledge of its behavior *in vivo*. As described in the previous section, much is known about the mechanism of rapamycin action and the dimerization of FRB domains with FKBPs. However, administration of rapamycin to mammalian embryos is teratogenic and mimics a Tor mutant phenotype, indicating that rapamycin has a single target in the mouse (Hentges, Sirry et al. 2001). The latter is, of course, essential if the small molecule regulator is to be used as a monitor of the activity of a single gene and not a collection of off-target effects. Therefore, we were concerned about potential toxic effects of rapamycin to *Xenopus* embryos.

To assess this, we incubated embryos beginning at 5 hours after fertilization with doses of rapamycin ranging from 0.1 nM to 2 μM (Figure 14.4). Embryos incubated for up to 1 week in rapamycin displayed no overt phenotypic effects except at the highest doses; these phenotypes were due to solvent effects. We also verified the proper patterning of embryonic tissues using mRNA *in situ* hybridization for several molecular markers (data not shown). It is possible that *Xenopus* embryos are not affected by rapamycin-mediated G1 arrest because of the nature of the early embryonic cell cycle, which consists of alternating S- and M- phases (Gerhart 1980).

We then wanted to determine the pharmacokinetics of rapamycin in *Xenopus*. The lack of toxicity of rapamycin could be due to impermeability of the vitelline membrane, inability to pass into the cells, or rapid clearance by the embryo's metabolic pathways. We used two methods to assess the bioavailability of drug after dosing. First, we used an activity-based assay (Figure 14.4a), in which rapamycin is extracted from embryos using ethyl acetate and remaining activity is quantified. Extracted rapamycin is applied to cultured cells carrying a three-component rapamycin-dependent transcriptional switch (outlined in the previous section and in Figure 14.4a). In the presence of rapamycin, FKBP and FRB dimerize, recruiting the VP16 activation domain to a secreted alkaline phosphatase (SEAP) reporter driven by GAL4-UAS. Second, we examined extracted rapamycin samples for liquid chromatography/mass spectrometry (LC/MS), which would reveal both the parent drug and metabolic breakdown products.

We found that when incubated in *Xenopus* lysates, about 70% of the rapamycin activity was recoverable 10 hours after addition of drug. In comparison, only about 40% of drug activity was retained from *C. elegans* lysates, whereas in COS1 lysates, approximately 80% activity was recovered (Figure 14.4b). Analyses *in vivo* showed

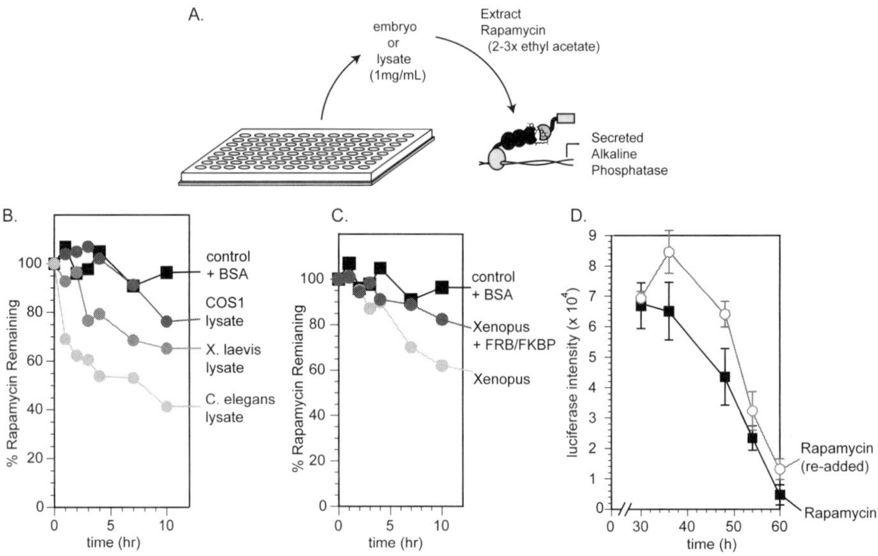

FIGURE 14.4 Rapamycin pharamacokinetics in *Xenopus laevis*. (A) Schematic depicting the approach for determining rapamycin pharmacology in organisms and lysates. (B) Kinetic profile for rapamycin degradation in lysates from three organisms. Lysates were freshly prepared by mechanical disruption in lysis buffer: 10 mM phosphate-buffered saline (PBS) with 5% SDS. Total protein was determined by Bradford assay and normalized to 1 mg/mL with lysis buffer. BSA (1 mg/mL) is included in the control to prevent drug adhesion to the plastic well plates. At time zero, 1 mL of lysate was treated with rapamycin (100 nM). At each time, 100 μL of lysate was removed and extracted with 2–3 volumes of ethyl acetate. The organic layers were pooled and crude purified through ~0.1 g silica gel. This material was concentrated by evaporation and the resulting film resuspended in 20 μL of ethanol. One half of this material (10 μL) was subject to LC/MS-MS. The remaining material was twofold serially diluted 10 times in sterile phosphate buffered saline. The dilution series was tested for activity in a bioassay. The bioassay for rapamycin and derivatives involves COS1 cells transiently transfected with three constructs (as previously described): a GAL4 DNA-binding domain fused to FKBP12, FRB fused to a VP16, and a GAL4-reporter upstream of secreted alkaline phosphatase. Controls for both the LC/MS-MS and bioassay were extracted/handled in the same manner as the experimental samples. Additional controls suggest that approximately 60–80% of the rapamycin can be extracted in this manner. (C) Partial protection of degradation can be achieved by overexpression of FRB- and FKBP-fusions. Extraction was performed from whole embryos treated at stage 5 with 100 nM rapamycin. Each data point represents a single embryo, washed twice with PBS, resuspended in 100 μL lysis buffer, and extracted two- to threefold with ethyl acetate (as above). No attempt to control for total protein level was performed in the experiments using whole embryos. (D) Readdition of rapamycin to *Xenopus* embryos injected with 0.1 ng Luc-PLF mRNA partially prolongs expression of luciferase fusion. Drug was readded every 24 hours to a final concentration of 100 nM.

that co-injection of FRB and FKBP mRNAs resulted in increased recoverable rapamycin activity, suggesting protection of the drug when bound to FRB/FKBP heterodimers (Figure 14.4c).

To determine the parameters of stabilization in *Xenopus*, we injected mRNA encoding FRB^{PLF}-tagged luciferase as a protein activity assay. Luciferase activity is easily quantifiable, allowing us to vary our culture conditions and rapidly assess protein activity over the course of several days. In all the experiments shown, 0.5 pg of FRB^{PLF}-luciferase mRNA was injected into the embryo at the one- or two-cell stage. Injected embryos were allowed to recover and incubated with or without rapamycin. Rapamycin was added at the indicated concentrations 5 hours after fertilization (approximately stage 6). We found that luciferase activity could be stabilized for several days, with peak activity occurring at about 24 hours. Readdition of rapamycin at approximately 24 hours resulted in additional stabilization of luciferase activity (Figure 14.4d).

Once we had determined the pharmacokinetics of rapamycin, we set out to characterize FRB^{PLF}-mediated protein destabilization in *Xenopus* embryos. Again using luciferase as an example, we established the kinetics of protein stabilization and activity, and assessed the effects of varying the dose of mRNA, drug, time of addition, and reversibility of the system. Finally, because *Xenopus* is a poikilotherm, we assessed the effect of varying the temperature of the embryos on protein stabilization and mapped the amount of protein activity versus protein stabilization and embryonic stages.

B. Dose-Dependent Protein Stabilization

After incubation with rapamycin, embryos retain high levels of active FRB^{PLF}-fused luciferase protein, whereas untreated embryos show decreased FRB^{PLF}-luciferase (Figure 14.5b). This is confirmed by activity assays, and the stabilization is dose dependent. Addition of the range of doses of rapamycin (from 0.1 nM to 100 nM) shown in Figure 14.5c resulted in increasing levels of luciferase activity. As before, we observe a peak of activity at 24 hours. The fold induction of activity increases over time, again in a dose-dependent fashion (Figure 14.5d). The inset depicts the fold induction at 72 hours as a function of rapamycin concentration.

These data also suggest that varying the dose of drug can produce differing concentrations of proteins. This provides an additional level of control over expression systems such as inducible transgenes, in which changing amounts of drug will not necessarily be reflected in the output of protein activity. The difference between the EC50 of luciferase stabilization and the observed Kd of rapamycin may reflect the previously observed degradation or decreased bioavailability of rapamycin *in vivo*.

C. Reversibility

As outlined in our original goals, in order to determine the window of protein requirement for specific developmental processes, we must be able to reverse the drug effects. This will allow us to differentiate between immediate effects of protein manipulation from later effects, which may result from residual protein activity. The

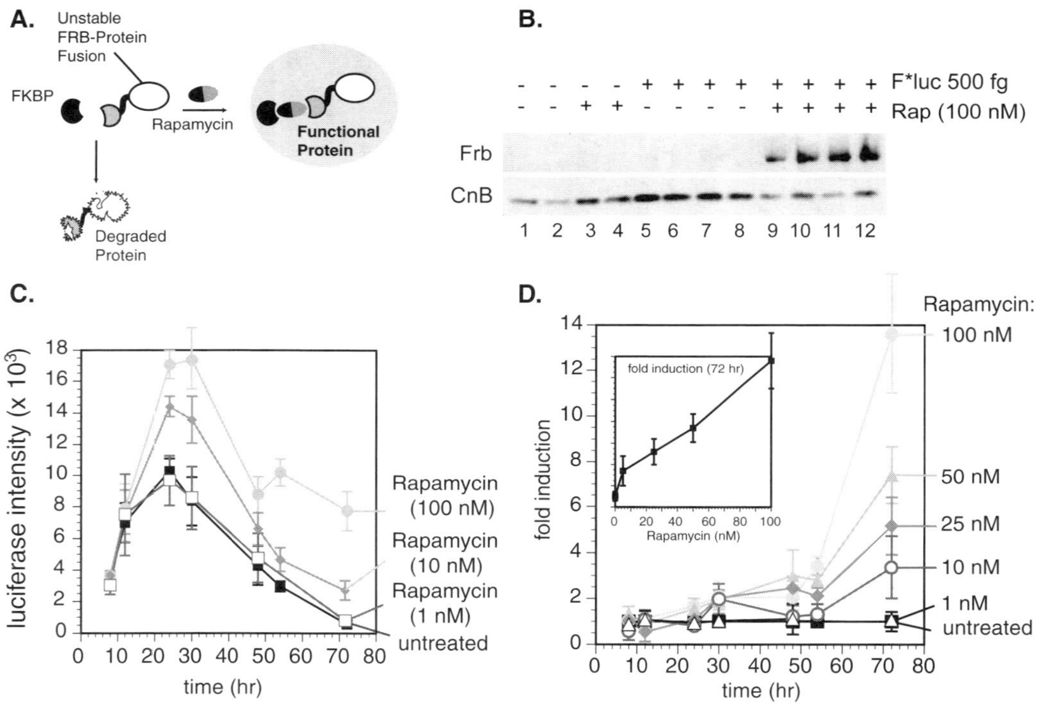

FIGURE 14.5 Dose-dependent control over protein levels. (A) Rapamycin control over FRB-fusion protein stability is shown schematically. (B) Western blots. (C) Time course of luciferase activity for embryos treated with a range of concentrations of rapamycin, as indicated. At each time point, individual embryos were lysed and 0.5 embryos assayed for luciferase activity. (D) The data from part B is shown plotted as fold induction of luciferase activity compared to the untreated control. The inset shows the fold induction at 72 hours as a function of rapamycin concentration. The high apparent functional affinity of rapamycin is likely due to partial degradation of drug (see Figure 14.3).

immunosuppressant FK506 binds with high affinity to FKBP12 and induces its dimerization with calcineurin, disrupting FKBP12 interaction with the FRB domain (Ho, Clipstone et al. 1996). FK506m, a derivative of FK506, binds only to FKBP-12 and inhibits rapamycin binding to FKBP-12 (structures shown in Figure 14.3a). FK506m efficiently eliminates rapamycin-mediated stabilization and thus rapidly eliminates protein activity (outlined in Figure 14.5a) (Spencer, Wandless et al. 1993). Thus, addition of FK506m rapidly reverses the stabilization of Frb^{PLF}-luciferase (Figure 14.6b).

D. Temperature of Embryos

We also determined the effect of temperature on drug potency. Again, this can be assessed using our luciferase assays; however, it must be kept in mind that we are also looking at changes in RNA availability and protein translation and stability. In this case, the results can be charted in two ways: drug response according to time and drug response according to developmental staging. These kinds of experiments will be broadly applicable to other poikilothermic model organisms. In particular, emerging systems such as medaka fish (*Oryzias latipes*) may be particularly well-suited to exploit temperature sensitivity of drug binding (Winkler, Loosli et al. 2000). In many cases, a physiological temperature for drug activity is assumed (e.g., 37°C for humans), whereas the temperature of the target organism is different and, in some cases, as in *Xenopus*, highly variable.

Figure 14.7 demonstrates the effect of temperature on the variability of drug activity. Embryos were injected at the two-cell stage with 0.5 ng of FRB^{PLF}-luciferase mRNA. 100 nM rapamycin was added at 5 hours post fertilization, and embryos were then incubated at a range of temperatures. In Figure 14.7, we show three representative temperatures (15°, 20°, and 28°C). Stabilization of luciferase activity is highest at 28°, and in the absence of drug, activity drops rapidly, presumably due to more rapid mRNA degradation and protein turnover. At 20°, we see a peak of stabilization at 24 to 30 hours with a steady increase in the fold stabilized activity. At 15°, both basal protein activity and stability remain over a much longer period of time, likely due again to slower mRNA degradation and protein turnover. When mapped against embryonic stages, we find that basal activity of FRB^{PLF}-luciferase is similar at all temperatures tested, whereas rapamycin treatment produces a marked temperature-dependent increase in activity (Figure 14.7).

E. Addition of Rapamycin at Specific Times

Finally, rapamycin was added at specific times during the experiment. We found that in each case, addition of rapamycin stabilized the activity of the FRB^{PLF}-luciferase, resulting in a longer period of peak activity (Figure 14.8). Addition of rapamycin at later time points may be particularly useful when studying more advanced developmental processes, particularly because addition of rapamycin appears to protect the FRB^{PLF}-tagged construct from degradation. Furthermore, this data suggests that rapamycin accessibility in the *Xenopus* embryo does not diminish during the first week of development, allowing us to exploit FRB^{PLF}-FKBP dimerization in later stages of development.

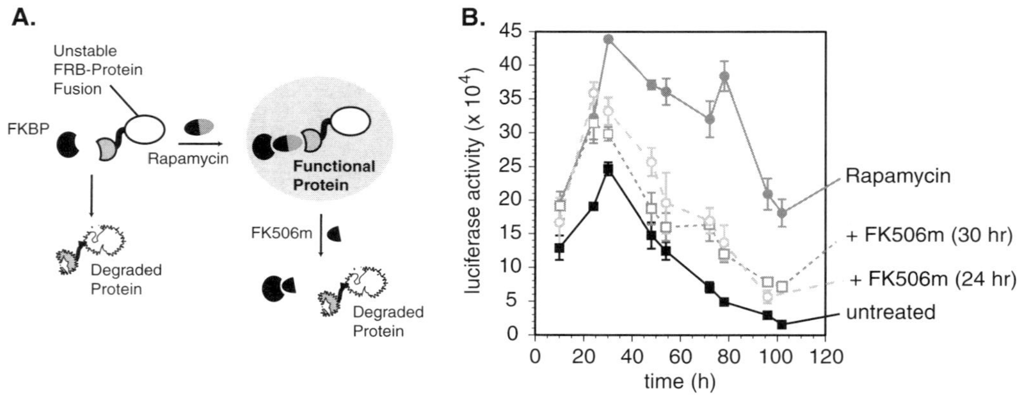

FIGURE 14.6 Reversal of stability. (A) Rapamycin control over FRB-fusion protein stability and reversal of stability by the competitor drug, FK506m, is shown schematically. (B) Kinetic profile of luciferase activity for embryos treated with rapamycin at 3 hours past fertility and then with FK506m at the time points indicated.

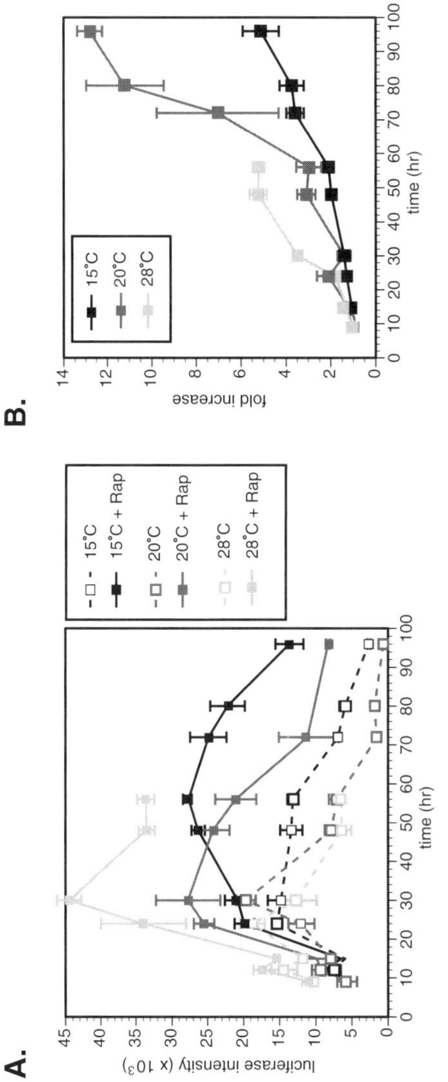

FIGURE 14.7 Induction of luciferase varies with temperature. (A) The kinetic profile of lucerifase activity is plotted versus the storage temperature. (B) The data from part A are converted to fold induction compared with the untreated control.

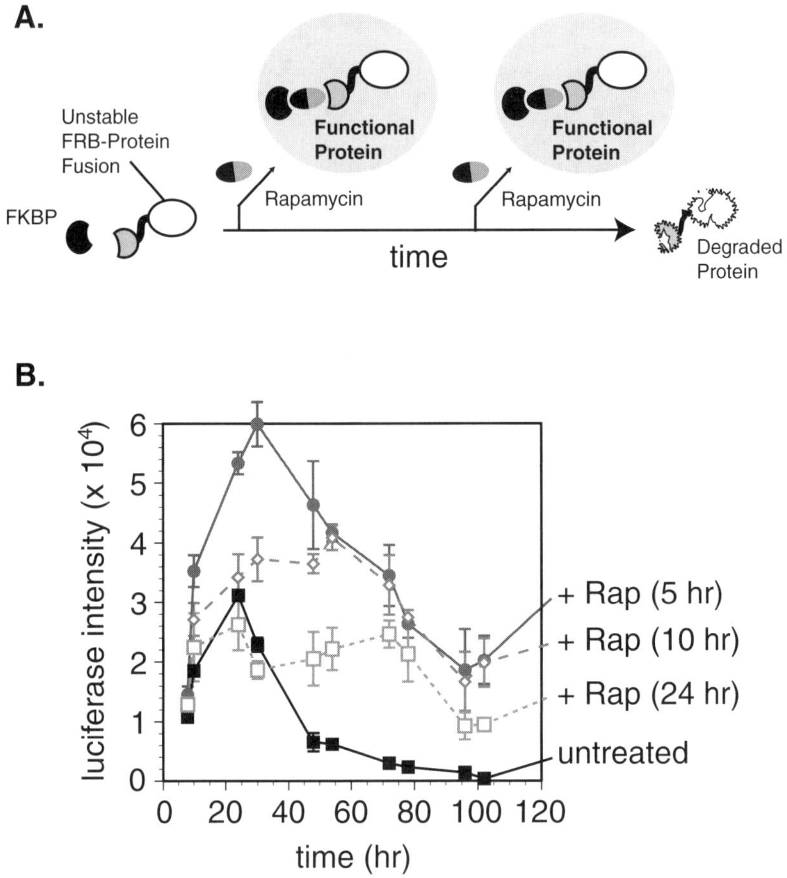

FIGURE 14.8 Addition of rapamycin at distinct time points. (A) Schematic time course of drug addition experiment. At specific times during the experiment, rapamycin was added to stabilize the luciferase-FRB fusion and enhance activity. (B) Kinetic plot of luciferase activity demonstrates that luciferase activity is stabilized at the point rapamycin is added.

F. Considerations

We recommend that any researcher moving this FRB^{PLF}-mediated destabilization into other organisms should begin with an assessment of the response of their model system to the drug. Rapamycin may be teratogenic to the organism of interest, as it is in mammals. The half-life of rapamycin in different culture media will need to be assessed. Is the drug able to enter the animal? What methods should be used to efficiently ensure accessibility to the drug, such as feeding, soaking, or injection? If the drug is indeed entering the animal, is it being metabolized? What is the half-life of the drug in the animal? This, again, can be determined as we outlined above, using an activity assay, such as the SEAP-based transcriptional reporter assay and LC/MS to determine degradation byproducts. Preliminary data suggests unfavorable pharmacokinetics of rapamycin in *C. elegans* (Figure 14.4b), whereas rapamycin does appear to enter *Drosophila* S2 cells (M. Povelones). Further studies are needed in these systems.

From the point of view of the biologist, this will be useful information for any drug that may be tested. First, degradation byproducts will likely have different pharmacologic activity than the original compound. Second, identifying the degradation byproducts of any applied drug will tell us about the mechanism of degradation and perhaps lead to a method for bypassing the degradation machinery.

G. Expressing Fusion Proteins

Varying the dose of mRNA injected will obviously vary the amount of exogenous protein produced in the embryo. The appropriate dose of mRNA used should be determined empirically for appropriate expression of each target protein, and if possible, the activity should be quantitated. For most overexpressed proteins, there will be a certain amount of background activity, so a dose should be found that produces protein activity only in the presence of drug while producing negligible activity in the absence of drug. Again, FRB^{PLF}-luciferase can be used to easily quantitate the amount of protein activity present. The use of this FRB^{PLF}-luciferase in other emerging systems will be very helpful for the researcher trying to understand the experimental conditions. Clearly defining the experimental conditions with an activity assay will also help scientists understand how their organism is responding to the system and help in the later interpretation of data.

Finally, in this review, we have outlined our strategy of using RNA overexpression for our target gene of interest. Although this approach bypasses problems of DNA injection or transfection (rates of transcription adding another level of complexity), results will of course be complicated by the presence of the endogenous protein. Replacement strategies, such as knock-ins or depletions coupled with replacement by expression of DNA/mRNA/protein, will allow us to study endogenous signaling functions *in vivo*. The choice of strategy will necessarily change the parameters and manipulability of stabilization.

V. FUTURE DIRECTIONS

The use of FRB^{PLF}-mediated instability in mammalian systems is currently hampered by the teratogenic properties of rapamycin. Embryos from pregnant mice treated with rapamycin fail to develop a telencephalon, phenocopying loss-of-function mutants in Tor/Frap (Hentges, Sirry et al. 2001). To avoid the teratogenic side effects of rapamycin, an analog, C20-methallylrapamycin, may be useful. Because MaRap contains a bulky chemical substitution at its FRB binding interface, it no longer interacts with the wild-type FRB domain. The compensatory hole created by the three-point mutations in FRB^{PLF} restores binding specificity. Thus, MaRap can interact with FRB^{PLF} but not with wild-type FRB (Stankunas, Bayle et al. 2003).

Inducible stabilization is ideally suited to mammalian tissue culture and explant systems, where we can exploit genetic knock-in strategies to study protein function under the control of the endogenous promoter. Toward this end, our lab has produced FRB^{PLF}-tagged knock-ins of two different proteins, GSK3β and Pax-6, allowing study of the endogenous proteins in a rapalogue-dependent fashion (Bayle, Grimley et al., 2006; Stankunas, Bayle et al. 2003). Because our system exploits FRB heterodimerization with FKBPs, our lab has also generated a mouse line in which a nuclear export sequence (NES) has been fused to endogenous FKBP-12 (exporter mice). FKBP-12 is highly expressed and ubiquitous. In mice carrying the exporter allele, FRB^{PLF}-tagged proteins are conditionally exported from the nucleus upon addition of rapamycin or rapamycin analogues (Bayle, Grimley et al., 2006). The FKBP-12 knockout is embryonic lethal; the fact that mice carrying FKBP-12 fused to a nuclear export sequence are viable and fertile with no apparent abnormalities leads us to believe that this modification does not alter the wild-type function of the protein (Shou, Aghdasi et al. 1998).

MaRap stabilization of FRB^{PLF} has proven to be quite useful in culture or *ex utero*; however, the *in vivo* pharmacokinetics are poor (Stankunas, Bayle et al. 2003). Alternative strategies include searching for rapamycin analogues with improved pharmacokinetics that retain specific binding to FRB^{PLF} (and not to endogenous Tor) (JHB et al., unpublished data). In addition, identification of additional FRB domains with varying specificities and stabilities would be extremely useful (JEG et al., unpublished data). Both these approaches are being actively pursued in our lab. We now have a library of rapalogues that bind with differing affinities to a panel of FRB domains, allowing us to use different drugs to regulate stability and proximity of a suite of FRB-tagged proteins (JHB et al., unpublished data).

We are also interested in further studying FRB^{PLF}-mediated instability. The original FRB^{PLF} mutations were selected to provide binding specificity to MaRap, and the resulting instability was a fortuitous side effect. As a result, little is known about the origins and limitations of FRB^{PLF}-mediated instability. Thus, our lab has mutagenized the FRB domain and tested the effects of these mutations on the folding energy of fusion. We found that few mutations provided increased stability over the wild-type FRB domain (FRB^{KTW}), and very few were found that were more unstable than FRB^{PLF} (JEG, unpublished data). However, these studies, coupled with the identification of additional rapalogues outlined above, have provided us with an expanded toolbox for chemical regulation of a series of fusion proteins.

VII. MODELING SIGNAL TRANSDUCTION: DEVELOPMENTAL CIRCUITS IN BIOLOGICAL SYSTEMS

The vast quantity of genetic data that has become available in the past few years has taught us a great deal about developmental biology, but it has also thrown into stark relief the tremendous complexity of interactions required for successful development of multicellular organisms. Genetic circuits, coupled with adaptable transcriptional modules, likely confer much of the specificity underlying different developmental design plans. However, in a complex developmental process, the encoded genes are interpreted as myriad intercellular and intracellular interactions. As a result, we must understand not just the genetic programming underlying development but also the molecular connections resulting from this programming, including post-translational modifications, signaling interactions, and localization of proteins.

Understanding how these systems fit together to create a functioning multicellular organism will require new biological and technological innovations. Using the systems outlined here, we are greatly expanding our ability to regulate proteins of interest during development. The Frb-dimerization system is specific and efficient, and bypasses the need for screening of chemical libraries. By combining this system with gene-trapping or transposon-mediated insertional techniques, we can rapidly recover rapamycin-regulatable protein alleles. These alleles can be screened in parallel using various cellular assays. The ultimate goal is to use these protein alleles *in vivo* to identify the critical timing, dose, and localization required for specific protein function in developmental events.

ACKNOWLEDGMENTS

Many thanks to Marc Dionne and Joe Arron for critical reading of the manuscript, Kryn Stankunas and Hank Bayle for exchanging ideas and unpublished data, Julie Baker and Will Talbot for reagents and support, and Jean-Pierre Saint-Jeannet and Ryuichi Nishinakamura for *Xenopus* FKBP plasmids. KJL is funded by NIH T-90DK070103, JEG is a fellow of the Helen Hay Whitney Foundation, and GRC is an investigator of the Howard Hughes Medical Institute.

REFERENCES

Althoff, E. A. and V. W. Cornish (2002). "A bacterial small-molecule three-hybrid system." *Angew Chem Int Ed Engl* 41(13): 2327–30.

Armstrong, J. I., S. Yuan, et al. (2004). "Identification of inhibitors of auxin transcriptional activation by means of chemical genetics in Arabidopsis." *Proceedings of the National Academy of Sciences of the United States of America* 101(41): 14978–83.

Bayle, J. H., J. S. Grimley, et al. (2006). "Rapamycin analogs with differential binding specificity permit orthogonal control of protein activity." *Chem Biol* 13(1): 99–107.

Belshaw, P. J., S. N. Ho, et al. (1996). "Controlling protein association and subcellular localization with a synthetic ligand that induces heterodimerization of proteins." *Proceedings of the National Academy of Sciences of the United States of America* 93: 4604–07.

Belshaw, P. J., D. M. Spencer, et al. (1996). "Controlling programmed cell death with a cyclophilin-cyclosporin-based chemical inducer of dimerization." *Chem Biol* 3(9): 731–38.

Biggar, S. R. and G. R. Crabtree (2000). "Chemically regulated transcription factors reveal the persistence of repressor-resistant transcription after disrupting activator function." *J Biol Chem* 275(33): 25381–90.

Biggar, S. R. and G. R. Crabtree (2001). "Cell signaling can direct either binary or graded transcriptional responses." *Embo J* 20(12): 3167–76.

Bishop, A. C., J. A. Ubersax, et al. (2000). "A chemical switch for inhibitor-sensitive alleles of any protein kinase." *Nature* 407(6802): 395–401.

Capecchi, M. R. (1989). "Altering the genome by homologous recombination." *Science* 244(4910): 1288–92.

Castellano, F., P. Montcourrier, et al. (1999). "Inducible recruitment of Cdc42 or WASP to a cell-surface receptor triggers actin polymerization and filopodium formation." *Curr Biol* 9(7): 351–60.

Chang, C. P., J. R. Neilson, et al. (2004). "A field of myocardial-endocardial NFAT signaling underlies heart valve morphogenesis." *Cell* 118(5): 649–63.

Chen, J., X. F. Zheng, et al. (1995). "Identification of an 11-kDa FKBP12-rapamycin-binding domain within the 289-kDa FKBP12-rapamycin-associated protein and characterization of a critical serine residue." *Proceedings of the National Academy of Sciences of the United States of America* 92: 4947–51.

Chen, J. K., J. Taipale, et al. (2002). "Inhibition of Hedgehog signaling by direct binding of cyclopamine to Smoothened." *Genes & Development* 16(21): 2743–48.

Chen, X., H. Ye, et al. (2005). "A chemical-genetic approach to studying neurotrophin signaling." *Neuron* 46(1): 13–21.

Choi, J., J. Chen, et al. (1996). "Structure of the FKBP12-rapamycin complex interacting with the binding domain of human FRAP." *Science* 273: 239–42.

Cooke, J. and E. J. Smith (1988). "The restrictive effect of early exposure to lithium upon body pattern in Xenopus development, studied by quantitative anatomy and immunofluorescence." *Development (Cambridge, England)* 102(1): 85–99.

de Graffenried, C. L., S. T. Laughlin, et al. (2004). "A small-molecule switch for Golgi sulfotransferases." *Proceedings of the National Academy of Sciences of the United States of America* 101(48): 16715–20.

Fire, A., S. Xu, et al. (1998). "Potent and specific genetic interference by double-stranded RNA in Caenorhabditis elegans." *Nature* 391(6669): 806–11.

Frank-Kamenetsky, M., X. M. Zhang, et al. (2002). "Small-molecule modulators of Hedgehog signaling: identification and characterization of Smoothened agonists and antagonists." *Journal of Biology (Online)* 1(2): 10.

Gerhart, J. G. (1980). Mechanisms regulating pattern formation in the amphibian egg and early embryo. In *Biological Regulation and Development*. R. F. Goldberger. New York, Plenum Publishing. 2: 133–316.

Gestwicki, J. E., K. Stankunas, et al. (2005). Expanding the repertoire of conditional protein alleles: rapamycin-binding tags that provide reversible instability. *Submitted.*

Ghiglione, C., G. Lhomond, et al. (1993). "Cell-autonomous expression and position-dependent repression by Li+ of two zygotic genes during sea urchin early development." *The EMBO Journal* 12(1): 87–96.

Gossen, M. and H. Bujard (1992). "Tight control of gene expression in mammalian cells by tetracycline-responsive promoters." *Proceedings of the National Academy of Sciences of the United States of America* 89(12): 5547–51.

Graef, I. A., F. Chen, et al. (2001). "Signals transduced by Ca(2+)/calcineurin and NFATc3/c4 pattern the developing vasculature." *Cell* 105(7): 863–75.

Graef, I. A., F. Wang, et al. (2003). "Neurotrophins and netrins require calcineurin/NFAT signaling to stimulate outgrowth of embryonic axons." *Cell* 113(5): 657–70.

Harding, W. M., A. Galat, et al. (1989). "A receptor for the immunosuppressant FK506 is a cis-trans peptidyl-prolyl isomerase." *Nature* 341: 758–60.

Hartwell, L. H., J. Culotti, et al. (1970). "Genetic control of the cell-division cycle in yeast. I. Detection of mutants." *Proceedings of the National Academy of Sciences of the United States of America* 66(2): 352–59.

Harwood, A. J. (2005). "Lithium and bipolar mood disorder: the inositol-depletion hypothesis revisited." *Molecular Psychiatry* 10: 117–126.

Heasman, J., M. Kofron, et al. (2000). "Beta-catenin signaling activity dissected in the early Xenopus embryo: a novel antisense approach." *Developmental Biology* 222(1): 124–34.

Hentges, K. E., B. Sirry, et al. (2001). "FRAP/mTOR is required for proliferation and patterning during embryonic development in the mouse." *Proceedings of the National Academy of Sciences of the United States of America* 98(24): 13796–801.

Ho, S., N. Clipstone, et al. (1996). "The mechanism of action of cyclosporin A and FK506." *Clinical Immunology and Immunopathology* 80(3) Pt 2: S40–45.

Ho, S. N., S. R. Biggar, et al. (1996). "Dimeric ligands define a role for transcriptional activation domains in reinitiation." *Nature* 382: 822–26.

Holsinger, L. J., D. M. Spencer, et al. (1995). "Signal transduction in T lymphocytes using a conditional allele of Sos." *Proceedings of the National Academy of Sciences of the United States of America* 92(21): 9810–14.

Jacinto, E. and M. N. Hall (2003). "Tor signalling in bugs, brain and brawn." *Nat Rev Mol Cell Biol* 4(2): 117–26.

Klemm, J. D., S. L. Schreiber, et al. (1998). "Dimerization as a regulatory mechanism in signal transduction." *Annu Rev Immunol* 16: 569–92.

Kolm, P. J. and H. L. Sive (1995). "Efficient hormone-inducible protein function in Xenopus laevis." *Dev Biol* 171(1): 267–72.

Kunath, T., G. Gish, et al. (2003). "Transgenic RNA interference in ES cell-derived embryos recapitulates a genetic null phenotype." *Nature Biotechnology* 21(5): 559–61.

Levy, F., J. A. Johnston, et al. (1999). "Analysis of a conditional degradation signal in yeast and mammalian cells." *European Journal of Biochemistry/FEBS* 259(1–2): 244–52.

Liberles, S. D., S. T. Diver, et al. (1997). "Inducible gene expression and protein translocation using nontoxic ligands identified by a mammalian three-hybrid screen." *Proceedings of the National Academy of Sciences of the United States of America* 94(15): 7825–30.

Liu, J., X. Wu, et al. (2005). "A small-molecule agonist of the Wnt signaling pathway." *Angewandte Chemie (international ed. in English)* 44(13): 1987–90.

Logan, C. Y., J. R. Miller, et al. (1999). "Nuclear beta-catenin is required to specify vegetal cell fates in the sea urchin embryo." *Development (Cambridge, England)* 126(2): 345–57.

Marianayagam, N. J., F. Khan, et al. (2002). "Fast folding of a four-helical bundle protein." *Journal of the American Chemical Society* 124(33): 9744–50.

Mullins, M. C. and G. M. Rubin (1991). "Isolation of temperature-sensitive mutations of the tyrosine kinase receptor sevenless (sev) in Drosophila and their use in determining its time of action." *Proceedings of the National Academy of Sciences of the United States of America* 88(21): 9387–91.

Murray, A. and T. Hunt (1993). *The Cell Cycle: An Introduction*. New York, Oxford University Press.

Muthuswamy, S. K., M. Gilman, et al. (1999). "Controlled dimerization of ErbB receptors provides evidence for differential signaling by homo- and heterodimers." *Mol Cell Biol* 19(10): 6845–57.

Nagy, A. (2000). "Cre recombinase: the universal reagent for genome tailoring." *Genesis* (New York) 26(2): 99–109.

Neilson, J. R., M. M. Winslow, et al. (2004). "Calcineurin B1 is essential for positive but not negative selection during thymocyte development." *Immunity* 20(3): 255–66.

Peterson, R. T., B. A. Link, et al. (2000). "Small molecule developmental screens reveal the logic and timing of vertebrate development." *Proceedings of the National Academy of Sciences of the United States of America* 97(24): 12965–69.

Peterson, R. T., S. Y. Shaw, et al. (2004). "Chemical suppression of a genetic mutation in a zebrafish model of aortic coarctation." *Nature Biotechnology* 22(5): 595–99.

Picard, D. (2000). "Posttranslational regulation of proteins by fusions to steroid-binding domains." *Methods in Enzymology* 327: 385–401.

Pownall, M. E., B. E. Welm, et al. (2003). "An inducible system for the study of FGF signalling in early amphibian development." *Dev Biol* 256(1): 89–99.

Pruschy, M. N., D. M. Spencer, et al. (1994). "Mechanistic studies of a signaling pathway activated by the organic dimerizer FK1012." *Chem Biol* 1(3): 163–72.

Rivera, V. M., T. Clackson, et al. (1996). "A humanized system for pharmacologic control of gene expression." *Nature Medicine* 2: 1028–32.

Sakamoto, K. M., K. B. Kim, et al. (2001). "Protacs: chimeric molecules that target proteins to the Skp1-Cullin-F box complex for ubiquitination and degradation." *Proceedings of the National Academy of Sciences of the United States of America* 98(15): 8554–59.

Schneekloth, J. S., Jr., F. N. Fonseca, et al. (2004). "Chemical genetic control of protein levels: selective in vivo targeted degradation." *J Am Chem Soc* 126(12): 3748–54.

Schneider, S., H. Steinbeisser, et al. (1996). "Beta-catenin translocation into nuclei demarcates the dorsalizing centers in frog and fish embryos." *Mechanisms of Development* 57(2): 191–98.

Schreiber, S. L. (2000). "Target-oriented and diversity-oriented organic synthesis in drug discovery." *Science* 287(Mar 17): 1964–69.

Schreiber, S. L. (2003). "The small-molecule approach to biology." *Chemical & Engineering News* 81(9): 51–61.

Sehgal, S. N. (1998). "Rapamune (RAPA, rapamycin, sirolimus): mechanism of action immunosuppressive effect results from blockade of signal transduction and inhibition of cell cycle progression." *Clinical Biochemistry* 31(5): 335–40.

Shou, W., B. Aghdasi, et al. (1998). "Cardiac defects and altered ryanodine receptor function in mice lacking FKBP12." *Nature* 391(6666): 489–92.

Siekierka, J. J., S. H. Y. Hung, et al. (1989). "A cytosolic binding protein for the immunosuppressant FK506 has peptidyl-prolyl isomerase activity but is distinct from cyclophilin." *Nature* 341: 755–57.

Simon, M. A., D. D. Bowtell, et al. (1991). "Ras1 and a putative guanine nucleotide exchange factor perform crucial steps in signaling by the sevenless protein tyrosine kinase." *Cell* 67(4): 701–16.

Sive, H. L., R. M. Grainger, et al. (2000). *Early Development of Xenopus laevis: A Laboratory Manual.* New York, Cold Spring Harbor Laboratory Press.

Smith, W. C. and R. M. Harland (1991). "Injected Xwnt-8 RNA acts early in Xenopus embryos to promote formation of a vegetal dorsalizing center." *Cell* 67(4): 753–65.

Spencer, D. M., P. J. Belshaw, et al. (1996). "Functional analysis of Fas signaling in vivo using synthetic inducers of dimerization." *Curr Biol* 6: 839–47.

Spencer, D. M., T. J. Wandless, et al. (1993). "Controlling signal transduction with synthetic ligands." *Science* 262(5136): 1019–24.

Stankunas, K., J. H. Bayle, et al. (2003). "Conditional protein alleles using knockin mice and a chemical inducer of dimerization." *Molecular Cell* 12(6): 1615–24.

Thompson, S., A. R. Clarke, et al. (1989). "Germ line transmission and expression of a corrected HPRT gene produced by gene targeting in embryonic stem cells." *Cell* 56(2): 313–21.

Winkler, S., F. Loosli, et al. (2000). "The conditional medaka mutation eyeless uncouples patterning and morphogenesis of the eye." *Development (Cambridge, England)* 127(9): 1911–19.

Yang, J., K. Symes, et al. (1998). "Small-molecule control of insulin and PDGF receptor signaling and the role of membrane attachment." *Curr Biol* 8(1): 11–8.

Yap, A. S., W. M. Brieher, et al. (1997). "Lateral clustering of the adhesive ectodomain: a fundamental determinant of cadherin function." *Curr Biol* 7(5): 308–15.

Yeh, J. R. and C. M. Crews (2003). Chemical genetics: adding to the developmental biology toolbox. *Dev Cell* 5: 11–9.

15 Systems Analysis of Signaling Pathways

Gregory R. Hoffman, Kevin Brown, Adrian Salic, and Ethan Lee

CONTENTS

I. INTRODUCTION: WHY DO WE NEED SYSTEMS BIOLOGY?

A remarkable result from the sequencing of the human genome is the realization that a surprisingly small number of distinct signaling pathways are responsible for controlling a wide range of cellular behaviors [1, 2]. In contrast, the list of cellular components associated with any particular signaling pathway, as well as the interactions connecting those components, has expanded dramatically through both the rapid progress of traditional experimental investigations and the recent advent of high-throughput techniques [3–5]. Thus, the experimentalist studying signal transduction is faced with complexity on two levels: (1) the sheer number of components that are involved in a particular signaling cascade [6, 7] and (2) the contextual specificity of signaling pathways [8]. The first level, component complexity, presents difficulties in interpreting how perturbations of a single component affect the global behavior of the signaling network. The second level, contextual specificity, refers to the observation that the same signaling pathway is continuously reused in a variety of contexts to control different aspects of cellular behavior, but often with subtle modifications that can affect its dynamic behavior. Similarly, individual cellular components often have multiple functions and participate in more than one distinct signaling pathway, depending on the cellular context. In the face of these challenges, it is apparent that previous descriptions of signaling pathways as linear cascades that relay and amplify information no longer provide an adequate framework to interpret and explain experimental observations. Rather, signal transduction occurs through highly branched networks involving feedback and crosstalk between distinct modules to produce specific outputs.

For the experimentalist attempting to use intuition to understand a signaling network, the complexities described above are a challenge to the development of hypotheses regarding the global behavior of a signal transduction network and the role that a particular component plays in determining signal output. Systems biology has emerged in response to these challenges and is an attempt to provide a formal framework to the systems-level behavior of biological processes. Many authors have offered their own definitions of systems biology [9–12], but for the purposes of this chapter, we will discuss systems biology of signaling pathways in the context of computational models that apply mathematical formalisms to describe the dynamic behavior of signaling networks. The goal of such computational methods is both to synthesize existing experimental knowledge in order to make quantitative models of signaling pathways and to use those models to develop experimentally testable hypotheses.

The term "systems" is borrowed from the engineering discipline of systems analysis, a field that endeavors to understand the behavior of large assemblies of interrelated elements, ranging from electrical circuits to traffic patterns [13]. The application of systems-level analysis has provided important insights into biological problems, particularly in the area of metabolism, where theoretical work has played a fundamental role in explaining the design and regulation of metabolic pathways [14, 15]. The recent surge of interest in systems biology as it relates to signal transduction reflects the potential of computational modeling to provide similar insights into signal transduction networks. In contrast to the traditionally reductionist fields of biochemistry and molecular biology, the field of embryology naturally takes a global view in the study of how signal transduction pathways influence the development of the entire organism and, as such, is uniquely positioned to take advantage of the opportunities provided by systems biology.

In fact, embryology arguably witnessed some of the earliest attempts to apply mathematical formalism to biology, with the classic Turing models of developmental patterning [16]. These models offered phenomenological descriptions of pattern formation arising from a slowly diffusing morphogen in combination with a rapidly diffusing inhibitor. Since that time, experimental biologists have accumulated a wealth of molecular data about many signal transduction pathways involved in cell and developmental biology. Application of computational methods will help guide experimental work by providing testable hypotheses regarding the underlying mechanisms of signal transduction networks.

The goal of this chapter is not to provide a recipe for the computational modeling of signaling pathways. Just as each signaling pathway requires specialized experimental techniques, each pathway will also require a unique set of modeling tools or, at the very least, variations on techniques that have proven successful in the past. Rather, we hope to provide an understanding of the biological insights that systems-level analysis can (and cannot) provide and offer examples of experimental techniques and subsequent data generation that are essential to a successful modeling experiment. The scope of this chapter is limited to a few signaling pathways, including Notch, Wnt, and MAPK, which provide important examples of the relationship between computational modeling and experimental embryology. Whenever possible, the reader is directed to additional references in which other signaling pathways have been analyzed with similar approaches.

II. PART 1: MODELING OF SIGNALING NETWORKS

In building mathematical descriptions of signaling networks, one important feature of any given model lies at the level of complexity of the model itself. Every mathematical model represents some level of simplification of the actual biological process being described, and each decision about what to include and what to exclude from the mathematical analysis is crucial to the interpretation of the results. There are two extremes when deciding how much biological detail to include in a model, and the level of detail that is chosen depends on the question being asked. The first approach is to reduce the system to the smallest number of essential parts. We will discuss models that include just a few essential components of a signaling network and yet are deeply informative as to the fundamental behavior of the system. The predictions made by these simple models are generally qualitative rather than quantitative, suggesting that the pathway will generate oscillatory behavior, for example, rather than correctly determining the amplitude or period of these oscillations. These minimal models sometimes have the added benefit of being analytically tractable or, at least, lend themselves to a more complete analysis than a large-scale simulation.

The second approach makes an attempt to capture most of the known components of a network. These "comprehensive" models aim to be quantitative in their predictions and often reveal global properties of the network that emerge from a particular combination of small signaling modules. We will present examples of such comprehensive models that are beginning to provide reliable quantitative predictions about the behavior of signaling networks. These models are generally too large to be amenable to any analysis other than simulation, but they have the great benefit of being more directly comparable to the experimental situation and can be used to generate a large number of experimentally testable predictions. We will also describe new modeling approaches that use statistical methods to generate models that are predictive even in the face of missing pathway components, unknown or poorly measured rate parameters, and other experimental limitations.

A useful analogy to clarify the distinction between these two levels of mathematical abstraction is to model the question: How does an airplane fly? One model is a small set of hydrodynamic equations describing how the flow of air over an airfoil produces lift, the upward force that keeps the plane in the air. These equations describe the qualitative behavior of the system. Another answer to this question can be found in the detailed mathematical simulation of a 747 used by Boeing to test how changes in the design of the plane will affect its flight performance. This model results in quantitative and predictive information about the system. Both mathematical descriptions answer important questions, but the kinds of queries one can present to the model and the subsequent answers to those queries are very different.

A. Minimal Models

1. Introduction to Signaling Motifs

Before undertaking any systems-level analysis of a signal transduction network, it is important to realize that complex behaviors can arise from modules comprising very few components. Learning to recognize these signaling modules (or signaling motifs) and understanding their properties can provide important insights into the behavior of a larger network. Traditionally, signaling motifs are recognized based on a functional characterization of signaling pathways, and the examples of feedback loops and three-kinase cascades discussed below can be recognized intuitively due to their obvious functional and structural organization. More recently, graph-theoretical methods have been developed to systematically identify statistically over-represented patterns in large sets of protein-protein or genetic interaction data (for review, see Reference 17). These methods have revealed the existence of a number of simple, repeated network motifs [18, 19] and have been used recently to show that certain motifs arise dynamically during the process of signal transduction [20].

In addition, Bayesian methods that allow one to infer network connections from interaction data are continually being refined and improved [21, 22]. Much like the recognition of a structurally conserved kinase domain in the primary sequence of a protein suggests that the protein may possess certain functional characteristics, the recognition of a topologically conserved signaling module within a signaling network implies that the network may exhibit certain behaviors associated with that module. In many situations, the phenomenological behavior of a large system can be described in general terms by reducing the system to a minimal set of components, and this reduction can be helpful in experimentally probing the role of these motifs in a given biological process.

a. Feedback Loops

A familiar topological feature, or motif, found in many growth factor signaling pathways is the feedback loop. Usually, these loops are imbedded in large networks, but often the general behavior of a feedback system can be understood by reducing the system to its simplest form, a two-component loop, which accounts for the oscillatory behavior observed in a variety of biological processes. The first mathematical analysis of a minimal negative feedback loop in biochemical systems was presented in the context of metabolic enzymes by Goodwin in 1965 [23]. The basic architecture of a minimal feedback loop includes an activating component that stimulates the activity of a second component. The second component then feeds back on the first to inhibit its function. A mathematical description of the dynamics of this minimal negative feedback loop reveals that the system behaves as a simple harmonic oscillator. Depending on the parameters governing the interaction between the two components, such a feedback loop is capable of producing sustained oscillations, oscillations that decay over time (damped oscillations), or a non-oscillatory sustained activation. Models based on this basic mechanism have been used to explain a range of oscillatory behaviors in cell biology, including oscillations of intracellular calcium [24], the cell cycle [25], and circadian rhythms [26]. For an excellent discussion of the history of modeling oscillatory behavior in biology, see Reference 27.

The minimal feedback model can explain the oscillatory behavior of the segmentation clock, responsible for periodic somite generation during development of the vertebrate body axis. The vertebrate body plan comprises a series of segments arranged along the anterior-posterior axis that develop into repeated structures, such as the vertebrae, ribs, and associated skeletal muscles, in the adult organism. These segments are derived from somites, which are produced by the presomatic mesoderm at regular intervals during embryonic development. The temporal interval governing somite formation varies between species, ranging from 30 minutes in zebrafish to 8 hours in humans [28]. The regular period of somite formation leads to the suggestion that this event is controlled by a timing mechanism referred to as the segmentation clock [29]. Palmeirim and colleagues made an important observation regarding the molecular mechanisms driving this clock when they observed oscillatory expression of the mRNA for the basic helix-loop-helix (bHLH) gene, c-hairy1 [30]. C-hairy1, the chicken homologue of the *Drosophila* hairy gene, is conserved across vertebrates, and the two murine homologues are referred to as Hes1 and Hes7, whereas the zebrafish homologues are known as Her1 and Her7. Each of these homologues undergoes periodic oscillations in their mRNA and protein levels, and mutations or knockdown of these genes lead to severe defects in somitogenesis.

A minimal mathematical model describing oscillations of the mouse segmentation clock was recently proposed, based on the experimental observation that the Hes1 protein can block production of its own mRNA by binding to and inhibiting transcription from the *hes1* promoter. Both Hes1 protein and mRNA are unstable; thus, once transcription is blocked, Hes1 is degraded and its levels rapidly drop to a point where inhibition of the promoter is relieved, allowing another round of protein synthesis to begin. A mathematical description of this auto-regulatory loop relies on the concentrations of just two components, the Hes1 protein and *hes1* mRNA [31]. An essential requirement for such a two-component system to exhibit sustained oscillations is the existence of a time delay in the negative feedback loop. Monk [31] and Lewis [32] independently proposed that the time required for transcription and translation provides the necessary temporal delay to produce sustained oscillations in protein levels of the *hairy*-related genes (and thus eliminating the need to include additional unknown intermediates).

Lewis' simulations of the zebrafish segmentation clock were based on a similar two-component model and demonstrated that given the short half-life of Her7 protein and *her7* mRNA, the period of oscillations is largely controlled by the rate of transcription and translation [32]. Using published values for these rates, Lewis demonstrated that Her7 oscillations occur with a period similar to that observed during zebrafish somitogenesis, providing support for Her7 oscillations in driving the segmentation clock. Thus, the complex behavior of periodic somite production can be explained by a simple transcriptional loop involving a single gene product.

It is satisfying that the mathematical analysis of the Hes1 and Her7 oscillations is consistent with many experimental observations regarding the segmentation clock. However, a number of other genes involved in somitogenesis also exhibit oscillatory behavior, and others have offered alternative models in which a negative feedback loop involving lunatic fringe [33] or axin [34] drives the clock. The existence of multiple feedback loops capable of generating the same oscillatory behavior in the segmentation clock highlights the risk of attributing the behavior of a complex system to one particular network motif. There is significant potential that an alternative mechanism not included in the minimal model may be responsible for the observed dynamics. One should not be satisfied simply with agreement between experiment and theory; the true power of such models is their ability to predict previously unappreciated biology that can be tested experimentally.

Support for *hairy*-related genes as the basis of the segmentation clock comes from an experiment by Hirata et al. designed to test one such prediction from the minimal feedback model using the mouse homologue, Hes7 [35]. Specifically, the theoretical model suggests that a critical parameter for the maintenance of sustained oscillations is the short half-life of the Hes7 protein. If Hes7 degradation is artificially altered to be faster than the time delay introduced by transcription and translation, the system will oscillate with a period controlled by the length of the time delay, suggesting that Hes7 is not the segmentation clock. If Hes7 turnover is altered to be slower than the time required to transcribe and translate additional Hes7, the system will fail, producing damped oscillations and leading to a loss of organizing information over time.

In an elegant experimental test of the theoretical work, Hirata et al. [35] identified lysine residues in the Hes7 protein critical for its rapid degradation and made transgenic mice expressing Hes7 with a mutation in one such residue, K14R. In these Hes7 mutant mice, where the Hes7 degradation is slower than that of the wild-type protein, the early somites were well formed but later segments became increasingly disorganized as the information in the Hes7 feedback loop decayed due to damped oscillations. Importantly, oscillations in lunatic fringe and axin mRNA expression were maintained even in the disorganized segments, arguing against their role in organizing the clock. It should be made clear that the minimal Hes7 model does not fully recapitulate all the details of segment formation during development, but provides the basis for cell autonomous oscillations. The lunatic fringe feedback loop, through the Notch pathway, likely plays an important role in synchronizing the oscillations between neighboring cells, whereas Wnt signaling most likely plays a role in coordinating somitogenesis with global patterning events [36, 37]. More comprehensive models will be required to completely understand the events of somitogenesis, but the example of the Hes7 feedback loop illustrates how relatively simple theoretical models of small signaling motifs can have an important impact on experimental embryology, and it forms an excellent foundation on which to base a more comprehensive model.

Similar negative feedback loops can account for important properties of the signaling networks involving NFκB/IκB [38] or p53/MDM2 [39], and similar computational models of these systems have been developed recently. Recognizing the potential oscillatory properties of a negative feedback loop in the topology of the p53/MDM2 network led Alon and colleagues to predict that p53 and MDM2 protein levels will oscillate following a sufficiently strong DNA damage event, a prediction confirmed by subsequent experimentation [40]. Importantly, in the case of p53 signaling, many of the parameters describing the system are unknown or difficult to quantify, such as the amount of DNA damage induced by a particular dose of irradiation, making the development of quantitative models difficult. However, predicting the qualitative behavior of the system based on the minimal feedback loop has led to a number of important experimental observations. In particular, in studying the properties of this feedback system, Alon and colleagues observed that individual cells expressing GFP-tagged p53 oscillate with the same amplitude for different numbers of cycles [40]. Thus, the damped oscillations observed at the population level are the result of an average of individual asynchronous cells, rather than decaying oscillations of single cells in synchrony.

b. Three-Kinase Cascades

A second example of a motif commonly encountered in growth factor signaling pathways is a three-kinase cascade, such as the canonical mitogen activated kinase (MAPK) cascade involving Raf, MEK, and ERK. The MAPK cascade has been the subject of extensive theoretical work, recently reviewed by Kolch et al. [41], and provides another clear example of how models involving a minimal number of components can reveal important insights into the underlying design principles of a signaling network. More generally, the three kinases that comprise the cascade are referred to as a MAP kinase kinase kinase (MAPKKK), a MAP kinase kinase (MAPKK), and the MAP kinase (MAPK) itself, which lies at the end of the pathway. This three-tiered architecture is also found in other mammalian kinase pathways, such as the p38 and JNK pathways, and in five distinct MAPK pathways in yeast [42]. The conserved architecture of the MAPK pathway raises the question as to why the three-kinase design is so well preserved through evolution.

MAPK activation during maturation of *Xenopus* oocytes has provided a useful experimental system for understanding the design principles of the MAPK cascade. During maturation, oocytes reenter the meiotic cell cycle from a G_2 arrest, where the cells again arrest as mature eggs at metaphase of meiosis II awaiting fertilization. Exit from the G_2 arrest requires progesterone-mediated activation of the MAPK cascade, which leads to formation of the Cdk2/cyclinB complex and cell cycle entry [43, 44]. The MAPKKK in the oocyte is Mos, which accumulates in response to progesterone stimulation, leading in turn to the sequential activation of MEK and ERK. An important insight into signaling properties of the MAPK cascade came when Huang and Ferrell used recombinant, active Mos to stimulate concentrated cellular extracts from *Xenopus* oocytes and observed dramatic "switch-like" activation of ERK [45]. In the parlance of systems biology, ERK activation is said to be "ultrasensitive," meaning the response of the system is more sensitive to an activating stimulus than a simple linear response and is similar to that of a cooperative enzyme (i.e., more Hill-like than Michaelis-Menten). An intuitive explanation often proffered for the evolutionary conservation of kinase cascades is their potential for signal amplification, specifically that the activation of a few molecules of Mos could give rise to a large amount of active ERK. The results from Huang et al. suggest an alternative explanation; namely, that the cascade acts as a threshold detector, where ERK only becomes active once a sufficient quantity of active Mos has accumulated. Below this threshold, ERK is always off, allowing the system to ignore noisy or weak activating signals and hence avoid improper meiotic reentry.

How does the MAPK pathway generate the observed ultrasensitive response? In order to understand the switch-like behavior of the three-tiered kinase cascade, Huang and Ferrell developed a minimal mathematical description of the ERK-MAPK pathway, which included the phosphorylation events leading from Mos to ERK and allowed for inactivation of each enzyme by phosphatases [45]. The central mechanistic feature of this model rests on the observation that activation of ERK requires phosphorylation of both the threonine and tyrosine residues in the TxY motif of its activation loop [46]. The mathematical model makes the assumption that these two phosphorylation events occur in a distributive manner, meaning that a single interaction between MEK and ERK phosphorylates either the tyrosine or the threonine residue in the activation loop but not both. In order to become activated, a singly phosphorylated ERK species must subsequently interact with another molecule of MEK, allowing the second phosphorylation to take place.

The consequence of this distributive phosphorylation mechanism is that the activation of ERK is proportional to the square of the active MEK concentration. MEK itself also requires dual phosphorylation of its activation loop to become activated, and simulations based on mathematical models of two successive distributive phosphorylation events reproduce the sharp switch-like activation of ERK observed experimentally. The modeling effort of Huang and Ferrell provides theoretical support for the proposal that three-kinase cascades have evolved to produce switch-like responses [45].

c. *Bistability*

When intact *Xenopus* oocytes are treated with progesterone and the response of individual oocytes is recorded, the MAPK response is even more switch-like than was observed in the oocyte extracts, with all of the ERK in an individual oocyte simultaneously switching from the inactive, unphosphorylated state to the active, doubly phosphorylated state. The unphosphorylated and phosphorylated forms of ERK are never observed together in the same oocyte [47]. This extreme form of switch-like behavior is referred to as bistability and describes a system that toggles between the inactive and active state without the presence of a stable intermediate. A characteristic of bistable systems is the capacity to generate irreversible state transitions, a feature that provides an attractive explanation for a wide range of biological phenomena that involve irreversible changes in cell fate, such as cell cycle entry, cell differentiation, and the lysis-lysogeny switch in *E. coli* (for a summary of bistability, see Reference 48).

The ultrasensitivity described by the minimal MAPK model outlined above does not account for the irreversibility of MAPK activation observed experimentally for intact oocytes, as the model predicts that removal of the stimulus allows ERK to return to its inactive state. Following the initial observation that oocyte maturation involves a bistable MAPK response, several authors proposed theoretical models of this system, describing a variety of mechanisms capable of producing bistable behavior in the MAPK pathway. These models are enlightening, illustrating the many ways in which relatively simple biochemical mechanisms can be used to produce complex bistable behavior (reviewed in Reference 49). At the same time, the variety of theoretical explanations for bistability in the MAPK cascade challenges the experimentalist to identify which mechanism is at work in this particular context.

A common mechanism for achieving bistability in signaling pathways is through positive feedback. In *Xenopus* oocytes, Mos-stimulated ERK activity stimulates the accumulation of additional Mos by increasing Mos protein synthesis and stabilizing Mos protein levels through direct phosphorylation, a classic feedback loop. Ferrell and Machleder [47] postulated that this feedback loop contributed to the bistability observed experimentally in oocyte maturation, and they generated a mathematical model of oocyte maturation that hinged on the activation of ERK driving the production of additional Mos. Indeed, their model showed that this feedback loop is capable of producing irreversible, bistable activation of the MAPK pathway. A model by Xiong et al. made a number of important experimental predictions, most notably that breaking the proposed feedback loop should eliminate the bistability and allow for reversibility of MAPK activation. In elegant experimental confirmation of their theoretical work, the authors generated oocytes expressing an inducible form of Raf (ΔRaf:ER), which can be directly activated by estradiol to artificially induce oocyte maturation in the absence of progesterone treatment [94].

As predicted by the feedback model, activation of Raf in the oocytes produced a bistable response, leading to switch-like activation of ERK that was maintained even after Raf activity was reduced by washing away the estradiol. By blocking ERK-dependent Mos production with cycloheximide treatment to break the feedback loop, ERK activity was reversible and returned to basal levels upon inactivation of Raf following removal of estradiol. This work on MAPK signaling in oocytes suggests that the biochemical features of the three-kinase cascade are responsible for generating switch-like changes in cell fate and provides a compelling example of how theoretical analysis coupled with careful experimentation can provide deep insights into a biological process.

B. Comprehensive Models

1. Introduction to Emergent Properties

Signaling modules provide the building blocks from which larger signaling networks are assembled, and these modules can be thought of as performing specific information processing tasks within the network [50]. The addition of a positive feedback circuit into the minimal model of a three-tiered kinase cascade described above is an example of a common progression in computational modeling of signal transduction pathways. The basic features of a network are first described by a minimal model, subsequent experimental observations lead to the introduction of additional details to the model, and further modeling and experimentation then capture the full behavior of the system. In many situations, the qualitative features of a complex network cannot be attributed to the behavior of a single motif but rather emerge from the collective behavior of the network as a whole. The idea of "emergent properties," a term borrowed from physics to describe behaviors of a system that cannot be predicted from the properties of its individual (atomic) components, was first introduced into the biological literature by Bhalla and Iyengar, who used a simulation of growth factor activation of MAPK signaling in mammalian cells to demonstrate that important signal processing features of a pathway can "emerge" from the combination of multiple feedback mechanisms [51]. Specifically, in their model of EGF-receptor signaling, a feedback loop involving PKC phosphorylation of Raf results in a bistable system, capable of maintaining sustained ERK activation even after removal of the initial EGF stimulus.

A second feedback loop involving the induction of MAP-kinase phosphatase 1 (MKP1) by ERK, which subsequently feeds back to dephosphorylate and inactivate ERK, stabilizes the system by maintaining a constant level of ERK phosphorylation under conditions of variable or noisy stimulation. Similar to the examples of minimal models outlined above, computational modeling of larger systems can provide a powerful tool for the experimentalist by predicting important qualitative features of the signaling network and pointing to specific modules, such as feedback loops, to account for the qualitative behavior of the network such as noise filtering and bistability.

To what extent can conclusions based on qualitative models of signaling pathways be generalized to other cellular contexts? Does the conserved architecture of the MAPK pathway always imply conserved functional features, or has the pathway been adapted by evolution to exhibit different systems-level properties in different cellular contexts? The MAPK cascade is a good example of context-sensitive complexity and warns us against using models specific for one biological system to describe other, even closely related, systems. In both the yeast mating response [52] and growth factor signaling pathways in mammalian cells [53, 54], the MAPK cascade does not work as it does in *Xenopus* oocytes. In each of these systems, experimental evidence demonstrates that the response of ERK to stimuli is still switch-like, but the eventual activation level is not a simple all-or-nothing response but rather continuously varies with the input signal level. Hence, the nonlinearity of activation is preserved, but the positive feedback that creates bistability appears to be missing. This highlights the need, as with all experimental results, to carefully consider whether models and inferences built from one's favorite experimental system are really appropriate for a new system. Although evolution has certainly reused useful strategies, it has definitely not followed a one-size-fits-all path.

2. Comprehensive Models

In many cases, the cellular response to a signaling network can only be understood based on quantitative features of the pathway, where cells interpret the amplitude or duration of a particular signal to achieve a specific cellular outcome. The MAPK pathway again serves as a valuable example. In the rat neuronal PC12 cell line, activation of the MAPK cascade is able to induce cell cycle entry or terminal differentiation depending on quantitative differences in the duration of ERK activation [55]. Specifically, NGF treatment of PC12 cells results in sustained ERK kinase activity and leads to differentiation, whereas EGF stimulation, acting through many of the same cellular components, results in transient ERK activity and proliferation. It is also apparent that quantitative aspects of the MAPK pathway, such as the concentration of proteins in the signaling network, influence signal specificity.

In one particularly striking experimental example, mice expressing an allelic series of Grb2 mutants exhibit distinct developmental defects in tissue morphogenesis and cell proliferation based on quantitative differences in the activity of this important adaptor protein [56]. These quantitative aspects of signal transduction cannot be explained by minimal models that provide phenomenological descriptions of signaling events. Moreover, it is often the case that qualitative features of the network are highly dependent on the parameters of the system, such as component concentrations and reaction rates. For example, the feedback loops in the MAPK cascade described above are similar in architecture to the minimal feedback loop that gives rise to oscillations in the *hairy*-related proteins during somitogenesis.

Although oscillations of ERK activity are theoretically possible in response to EGF stimulation and are experimentally observed in some settings [57], Bhalla and Iyengar found that the parameters of the network in mammalian cells are such that, within physiologically relevant boundaries, the system only produces the sustained activation observed experimentally [51]. Clearly, understanding many biological processes will require a modeling strategy that fully accounts for the complexity of components in a signaling pathway and provides a theoretical framework to understand the quantitative behavior of these systems. In parallel, experimentalists must develop more quantitative techniques for both measuring important model parameters and monitoring signaling pathways in order to test and refine model predictions. It is in this regard that experimental systems like *Xenopus* oocyte extracts, which lie somewhere between *in vitro* and *in vivo* systems, are serving as a powerful tool to obtain the kinds of data that can appropriately inform and constrain mathematical models.

Comprehensive models inevitably involve a large number of parameters, many of which are often poorly measured. As described elsewhere in this volume, extract-based biochemical systems have been widely used in embryology to study the regulation of important signaling pathways controlling development, as well as to study the cell cycle. Recently, Salic and colleagues successfully reconstituted the cytosolic aspects of the canonical Wnt signaling pathway in *Xenopus* egg extracts [58]. In the absence of a Wnt signal, β-catenin forms a complex with two scaffolding proteins, axin and APC. In this complex, β-catenin is phosphorylated by GSK3, an event that targets β-catenin for degradation.

Stimulation of the Wnt pathway activates disheveled (Dsh) to inhibit β-catenin phosphorylation and stabilize levels of β-catenin. Addition of recombinant, active Dsh to egg extracts reproduces this regulation, leading to stabilization of β-catenin.

Using immunodepletion of key components of the pathway and addition of recombinant proteins, Salic et al. reported detailed dose-response profiles for a number of key components of the Wnt pathway and developed a computational model of β-catenin degradation using these measurements [59]. In particular, the model suggested that the low cytosolic concentration of axin is key to efficient Wnt signaling and that control of axin stability is a critical aspect of regulating the canonical Wnt pathway. This work is notable for the experimental effort made to measure the concentration of each component of the pathway and for examining the effects of changing these concentrations on the behavior of the system.

In one of the most comprehensive models of a signaling pathway described to date, Schoeberl et al. simulated the behavior of EGF receptor activation of MAPK signaling [60], including 94 cellular compounds and 95 reaction rates, and simulations based on this model generated a number of important quantitative predictions. One particularly interesting result concerned the relative importance of surface versus internalized receptors in generating the phospho-Erk signal. Using the model, Schoeberl and colleagues found that at high ligand (EGF) concentrations, internalized receptors affect the overall MAPK signal very little because the internalization dynamics operate on a much slower time scale than Erk activation; essentially Erk has reached maximal activity before a significant number of receptors have been endocytosed. However, at low (compared to the K_d of the receptor) ligand concentrations, where Erk phosphorylation is slower, internalized receptors play a greater role in the signaling to Erk.

Although these are intriguing predictions, they have yet to be tested experimentally, and the authors of the study correctly note that internalization may be much more complicated than is represented in their model. Another interesting prediction of the model concerned the effects of receptor number. Overexpression of the EGF receptor in PC12 cells, in combination with EGF treatment, is sufficient to induce differentiation [61], suggesting a strong increase in signaling potency with receptor dose. The simulations from Schoeberl et al. showed a similar effect [60]; at high receptor number, the Erk signal is prolonged due to saturation of the endocytic machinery, and at physiological receptor numbers, inhibition of receptor internalization slows the rate of phospho-Erk signal attenuation.

These simulations demonstrate that ligand affinities for the EGF receptor, rates of receptor endocytosis, the kinetics of receptor activation, and receptor number all make specific contributions to shaping the quantitative behavior of EGF signaling dynamics. This work is also a good example of the utility of comprehensive models; because of the large number of distinct processes included (cell surface events, signaling cascades, receptor internalization and recycling, etc.), there is an opportunity to make a variety of very specific perturbations affecting one protein or part of one process in the model and compare these perturbations directly with analogous cell biological experiments. In other words, there are many more "knobs to turn" in comprehensive models than in minimal ones. That so many predictions can be made from comprehensive models presents a challenge to experimentation, because theoretical predictions can be made far more quickly than they can be tested experimentally.

3. Statistical Methods for Large Models with Unknown or Underdetermined Parameters

Even models of modest size invariably contain unknown parameters. Models of signal transduction pathways discussed in this review include many rate constants in the chemical kinetic description of such networks. In addition, initial conditions such as the basal concentrations of signaling molecules are important and often unknown. When one thinks about making comprehensive models of large scale, the problem is even more daunting; a recent article lists ~1000 parts in the epidermal growth factor receptor pathway [62]. A model of such a pathway would undoubtedly include thousands of rate constants, and obtaining a precise measurement of each parameter is beyond the reach of current experimental techniques. Recently, some authors [63–65] have used concepts and approaches borrowed from statistical mechanics and Bayesian statistics in an attempt to deal with models where many parameters are not well measured. These methods employ a cost function, which quantifies the degree of model agreement with available experimental data.

The cost function is then employed to generate an ensemble of different models, each of which has a different set of rate constants that provide an equally good fit of the model to experimental data. Thus, each unknown rate constant can be considered as the average of a large collection of parameters that all fit the experimental data. This procedure is carried out in an attempt both to generate predictions from the model that are independent of specific parameter values and, perhaps more importantly, to place confidence estimates (error bars) on model predictions, by using the same model ensemble to compute a standard deviation for quantities of interest. These techniques also highlight the danger of producing large, multiparameter models with insufficient information; despite the fact that a huge range of parameter combinations fit the observed experimental data, some model queries have such large error bars that they cannot really be considered predictions at all, as nearly any experimental outcome would fall inside the model's error bars.

Further model analysis using this statistical approach revealed some interesting properties that may be generic to all large, multiparameter models of signal transduction. The authors found that the most important degrees of freedom for generating the fit to observed data were not single rate constants but combinations of rate constants, and these combinations fall off rapidly (exponentially) in importance [63, 64]. A simple enzyme kinetic example can help explain why considering ratios of rate constants, rather than the individual constants themselves, can be important in determining signaling behavior. A simple receptor-ligand binding equation has two rate constants: k_{on} and k_{off}. If downstream processes are sensitive only to the equilibrium receptor occupancy, then K_d is the relevant parameter and not the individual on and off rates. Specifically, if we rescale both the on and off rates by the same number—any number—then the dissociation constant does not change and the behavior will be unaltered. Conversely, some combinations will be critical in generating the observed behavior. Unlike the dissociation constant example, a coordinated change in the parameters that affect a particular behavior will have a dramatic effect on the ability of the model to accurately produce the appropriate signaling behavior.

Such an analysis of NGF and EGF signaling in PC12 cells yielded some intriguing results [63, 64]. The model included the canonical Erk cascade from the EGF and NGF receptors through Sos to Erk, along with negative feedback via p90/RSK. An EGF-activated inhibitory pathway to Raf (via PI3 kinase) and a parallel NGF-activated stimulatory cascade through the small GTPase Rap1 and kinase B-Raf were also included. The model consisted of 30 chemical species and 48 rate parameters; lacking reliable measurements for more than a couple of the rate constants, all were treated as unknown and allowed to fluctuate during statistical sampling. The authors tested the effects of PI3K inhibition on Erk kinetics, reasoning that if the inhibition through Raf was strong, inhibiting PI3K in combination with EGF could result in sustained EGF activation, thereby converting EGF from a proliferative to differentiative agent.

Surprisingly, the model predicted that inhibition of PI3K would have little effect on Erk signaling, certainly not converting the response from transient (EGF-like) to sustained (NGF-like), and the error bars on the prediction of the phospho-Erk response were quite small. When the prediction was tested in PC12 cells, using the chemical inhibitor LY294002 to prevent PI3K activation, the experiment confirmed the theoretical prediction. Additionally, the model weakly predicted (with large error bars) the complementary experiment, that inhibition of Rap1 would cause NGF to produce a transient, rather than sustained, phospho-Erk signal. This experiment had been done previously but was not used to inform the model [66], and the results of the previous experimentation agree with the (weak) prediction. Together, these results give us the following picture of the network leading to Erk activation by EGF and NGF in PC12 cells: The classical cascade through Ras produces a transient pulse in response to both treatments, and NGF adds an additional positive signal through the Rap1 cascade. It is this additional boost that sustains Erk activation in response to NGF. These results also point to the importance of placing error bars on one's model predictions, just as one would place error bars on experimental data; many predictions had large error bars, about which the model cannot be definitive.

This type of statistical analysis implies that even in a large model with tens or hundreds of parameters, there are actually very few degrees of freedom that control the model's agreement with data, and they are not individual rate constants: They are combinations that are difficult or impossible to predict without carrying out a suitable mathematical analysis. In addition, the individual rate constants that participate have the potential to identify dynamical modules in a signaling network, not unlike the static modules identified by graphical techniques mentioned above. For the experimentalist, the implication of these models is that manipulation of a few rate constants in a coordinated manner will have dramatic effects on the behavior of the system, whereas other coordinated changes will effectively leave the system unchanged. Thus, these theoretical techniques not only provide experimentally testable predictions about the behavior of the signaling network, but also help guide experimental work by pointing to specific points in the network that are most likely to change the overall behavior of the system if they were altered experimentally.

C. Spatial Models

An important goal of developmental biologists is to relate the underlying molecular events in the cell to the overall form and function of an organism. The modeling efforts described above all describe the dynamic behavior of a signaling network as it changes over time. These models rest on the underlying assumption that the system is well stirred, or in other words, the dynamic behavior of the network is the same at different points in space. Obviously, differences in the behavior of signal transduction networks at different points in space are of critical importance to developmental biology and are required to explain such phenomena as the response of tissues to morphogen gradients. The fundamental difference between models that describe the changes of the system over space and time (as opposed to just time) is that such systems are described by partial differential equations (PDEs) rather than ordinary differential equations (ODEs) [67]. Simulations of partial differential equations are more technically challenging than those of ODEs, and relatively few examples of spatial models of signaling pathways exist in the literature.

One important example of spatial modeling is that of planar cell polarity (PCP) in *Drosophila* (see Reference 68, for review). Epithelial cells are polarized along the basal/apical axis and in the plane of the sheet of cells (orthogonal). This latter property of planar cell polarity generates an asymmetry along the plane of epithelium and is best exemplified (and extensively studied) in *Drosophila* in the regular arrays of actin-based body bristles and hairs as well as the in ommatidial facets of the compound eyes. An interesting aspect of PCP signaling in the *Drosophila* wing hair cells is that mutant clones of certain components of the PCP pathway affect not only polarity (i.e., orientation of bristles) in the mutant clones themselves, but also the orientation of nearby wild-type cells (referred to as domineering nonautonomy). Particularly puzzling were experimental results indicating that domineering nonautonomy was observed in only a certain subset of mutants of PCP components.

A mathematical model proposed by Amonlirdviman et al. attempts to account for the differences in the behavior of different components of the PCP pathway [69]. The authors made an initial assumption that PCP signaling occurs through a feedback mechanism in which an initial asymmetric cue results in relocalization of PCP components from a uniform distribution to a highly asymmetric one. Using a model that presupposes a set of defined interactions between just four key PCP components (Dsh, Fz, Vang, and Pk), an asymmetry generating cue, and a feedback mechanism, the authors simulated the behavior of the system such that loss of function of certain components (Fz and Vang) exhibited nonautonomy, whereas others (Dsh and Pk) exhibited autonomy effects.

It had been previously demonstrated that certain mutants of Fz exhibit autonomous cellular behavior. Given that Fz has been shown to interact with Dsh, the authors hypothesized that autonomous mutants of Fz may be defective in their interaction with Dsh, but still be able to interact with Vang in adjacent cells. A prediction of this model is that autonomous mutants of Fz may mimic the simulated behavior of Dsh mutants. In contrast, nonautonomous mutants would be expected to be unable to interact with neighboring Vang. Both of these predictions were confirmed experimentally with observations that the localization of Fz, Vang, and

Dsh proteins in the pupal wing of autonomous and nonautonomous Fz mutants correlated favorably with their predicted changes based on this mathematical model for planar cell polarity.

D. Summary

An emerging theme in modern cell biology is the integration of biological measurements with quantitative mathematical models to generate unified descriptions of complex cellular events. Computer models of signal transduction networks are increasingly being used as a tool in the attempt to achieve a fundamental understanding of the flow of information through these pathways. A significant challenge in modeling signal transduction pathways is in understanding the level of detail appropriate to answer the question under study. The examples described above illustrate how simple models that include a few key components of the pathway can be applied to address specific questions regarding the design of signaling networks, whereas more detailed models, which capture the dynamic behavior of a large number of network components, can be used to make quantitative predictions regarding signal output. In any modeling exercise, the goal should be to develop experimentally testable hypotheses about the behavior of the system. In the following section, we provide some guidance for obtaining the data necessary to develop and test computational models of signaling networks.

III. PART II: METHODS FOR DEVELOPING QUANTITATIVE MODELS OF SIGNALING PATHWAYS

A. The Importance of Quantitative Data

The preceding discussion illustrates the value of computational approaches in understanding signal transduction networks and, more importantly, in developing experimentally testable hypotheses regarding the function of these networks. As the above examples illustrate, a variety of experimental approaches can be useful in testing predictions based on computational modeling. The specific techniques used to test a hypothesis will be dictated by the particular biological problem under consideration. Developing a mathematical description of any signaling network requires three steps that are universal to any modeling exercise. The first prerequisite for any modeling study is to define the components of the network and describe their interactions. The second more challenging task is to quantify the concentrations of each component and measure the parameters that describe their interactions and enzymatic activities. The third and final step is the development of a mathematical description of the network suitable for computational work.

It is important again to distinguish here between minimal models and comprehensive models with regard to the experimental demands in both generating and testing the model. Minimal models, involving a few components, make significant simplifying assumptions. For example, the two-component model of the segmentation clock described in the previous section includes only two components, Hes7 protein and *hes7* mRNA. However, in the true biological context, the effects of Hes7 protein on the expression of *hes7* mRNA are indirect, and quantitative measurements of protein and mRNA levels are not particularly informative in this model.

On the other hand, determining the rates of transcription and translation relative to the rate of protein turnover was critical to validating the predictions of the model, illustrating that quantitative experimentation remains important even to simple idealized models. Comprehensive models aimed at predicting the dynamic behavior of the network are far more demanding with regard to precise measurement of component concentrations and kinetic parameters. In the following discussion, we will provide a brief overview of techniques that can be used to generate the requisite quantitative data for a modeling study and provide references to specific protocols for techniques that have been particularly useful in our own experimental work.

B. Establishing Network Topology: Getting the Parts List

The first step in any modeling exercise is to describe the connectivity of the signaling network. This process involves defining the biochemical species (proteins, lipids, small molecules, nucleic acids, etc.) involved at each step in the signaling cascade(s) that will be included in the model. Within signal transduction networks, the components of the network themselves transmit information through post-translational modifications, conformational changes in three-dimensional structure, and the assembly and disassembly of protein complexes, each of which must be represented in the biochemical description of the network. For many well-studied networks, simply defining the connectivity of the network can be a daunting task; for example, a recent comprehensive connection map of the EGF-receptor signaling pathway developed from a review of the current literature contained 322 distinct chemical species, including 202 individual proteins [70].

Defining the scale of the network to be modeled is an essential first step. Should a model of EGF-receptor signaling include all 322 components, or is there a smaller subnetwork that can be modeled with sufficient accuracy to provide insight into the system? A useful approach suggested by Neves and Iyengar is to begin by defining individual signaling modules within the network and then connect these modules into larger networks [71]. Most researchers will have a clear picture of the network that is appropriate for their work, and the published literature obviously provides an essential resource for defining the components of a signaling network. New sites maintained by the Alliance for Cell Signaling (http://www.signaling-gateway.org/) [72] and *Science*'s STKE (http://stke.sciencemag.org/) provide a good starting point for defining the topology of many networks. Both resources provide network maps for important signaling pathways based on data from the literature, and the sites are manually curated by experts in the field.

Many current modeling efforts rely completely on connectivity maps generated from the published literature. However, the process of discovery in biology is far from complete, and novel components central to the function of even some of the most well-studied pathways continue to be discovered [73]. The most exciting applications of systems biology in the coming years will involve the incorporation of newly discovered components into existing theoretical descriptions of a given network. By including modeling studies early in the experimental process, hypotheses can be generated and tested experimentally as part of the initial characterization of novel proteins [72]. Historically, genetic studies in *S. cerevisae*, *C. elegans*, and *Drosophila* have been used to identify essential components of a pathway, and the recent advent of RNAi technology has made genetic screens possible in a variety of tissue culture models, as well [74].

Traditional genetic approaches and high-throughput RNAi screens identify gene products that are essential to the function of a network, but additional biochemical data is required to establish their connectivity to other components of the pathway. Computational approaches have been developed to infer pathway connectivity based on indirect experimental evidence; however, we prefer to focus on experimentally validated interactions. Yeast two-hybrid analysis continues to be a powerful tool for mapping protein-protein interactions and has recently been applied on a genome-wide scale, providing useful data on the connectivity of signaling networks, and has recently been applied to a large-scale screen of the human genome [75]. *In vitro* expression cloning (IVEC) is a powerful biochemical method used to screen for novel components of signal transduction networks [76–78]. IVEC strategies rely on *in vitro* transcription of cDNA libraries, generally divided into small pools, which can then be screened in a variety of biochemical assays. IVEC screens have been particularly valuable in identifying substrates of E3 ligases and can also be used to directly identify protein-protein interactions. The availability of genome-wide cDNA collections, such as that available for *Drosophila* [79], now make it possible to systematically screen the genome using IVEC-based strategies.

Affinity purification techniques have proven value in analyzing protein-protein interactions. An important recent advance in affinity purification strategies is the application of tandem affinity purification (TAP) techniques, which use two sequential affinity purification steps to isolate a bait protein along with associated proteins [80]. The original TAP-tag comprised a protein A Ig-binding domain, a recognition sequence for the TEV protease, and a calmodulin-binding domain. Bait proteins fused to this tag are first passed over immunoglobulin beads, washed, and eluted by cleavage with the highly specific TEV protease. The recovered sample is then passed over a calmodulin column and eluted by calcium chelating agents. Following the elution, the bait protein, along with any interacting proteins, is resolved using gel electrophoresis, and specifically associated bands are identified by mass spectrometry (MS). The two-step purification ensures a high purity, with the caveat that weakly interacting proteins may be lost in this process. In principle, any pair of tags can be used in a TAP strategy, and a number of variations on the original technique have been reported.

Systematic large-scale TAP-tagging experiments have been performed to define the protein-protein interaction map in *S. cerevisae* [81, 82], and a number of protein-protein interaction databases from these experiments are available (see Reference 83 for an extensive discussion of resources for searching and analyzing protein-protein interaction databases). These comprehensive databases are valuable tools, but more focused strategies aimed at characterizing the interactions within a specific signaling network are particularly promising for defining the topology of a given network. A notable example of this type of pathway-focused proteomics approach was recently reported for the TNF-α pathway [83]. In this report, the authors used a TAP strategy to systematically identify protein-protein interactions involving 32 core components of the TNF-α signaling network following ligand stimulation. A notable aspect of this work was the use of RNAi to knock down novel components of the network identified in the TAP experiments in order to validate their role in the NFκB response to TNF-α stimulation. This study generated a large connection map, involving over 680 nonredundant proteins, and should provide an important foundation for future theoretical work on the TNF-α network.

C. Measuring Parameters: Putting Numbers on the Arrows

Simulating the dynamic behavior of a complex signaling network requires knowledge of quantitative values describing the concentration of each component, rate constants for enzymatic reactions, and binding affinities for protein-protein interactions. High-throughput omic-scale experiments are often of limited value in providing this quantitative information, as enzymatic rate constants and binding affinities for protein-protein interactions identified in a network require assays unique to each component. Thus, there is a significant gap between the rapid identification of novel network components and the comparatively protracted process of quantitative measurement of the activities and concentrations of these components [84]. Many modeling efforts rely on published literature to obtain system parameters, and some efforts have been made to compile this information in a central database, the Database of Quantitative Cellular Signaling (DOQCS) (doqcs.ncbs.res.in). System parameters available in the literature are often derived from disparate cell types and experimental conditions and should be incorporated into models with that caveat in mind, and attempts should be made to measure these parameters directly whenever possible. For large comprehensive models of signaling networks, direct measurement of each parameter is often prohibitive. The Wnt signaling pathway described above is one example where direct measurement of many parameters was accomplished.

Measuring component concentrations and the variation of these concentrations during the dynamic process of signal transduction can be achieved by relatively simple techniques. Western blotting approaches are routinely used to measure the time course of changes in the state of a protein, such as a change in protein levels or its degree of phosphorylation. In general, Western blotting techniques provide qualitative information on the relative changes in protein level. Quantitative Western blotting can be performed using purified recombinant protein to develop a standard curve that is then used to determine the absolute concentration of the component in a cell lysate. Often, simply knowing the concentration of an individual component provides significant insight into the function of a signaling pathway. Such was the case in recent studies of Wnt signaling, where quantitative Western blotting was used to measure the low concentration of axin in egg extracts relative to other components of the pathway [59].

Mass spectrometry-based approaches such as stable isotope labeling by amino-acids in culture (SILAC) and isotope coded affinity tag (ICAT) provide alternative methods for quantifying protein concentrations in a more high-throughput manner [3]. Both MS techniques allow for the relative quantification of protein levels by incorporating heavy isotopes into complex protein mixtures to provide internal standards. SILAC requires that cells are grown in media containing isotopically labeled amino acids and, therefore, may not be practical for many applications in developmental biology where such metabolic labeling strategies are not possible. ICAT incorporates the isotopic label by selective labeling of reactive-thiol groups in cellular extracts, making this technique more generally applicable.

In addition to measurements of component concentrations, computational modeling also requires values for reaction rates and binding affinities of interacting components. A variety of *in vitro* biochemical methods are available for measuring the enzymatic activity of kinases, phosphatases, GTPases, and other enzymes typically involved in growth factor signaling pathways, and specific protocols for measuring their activities are available elsewhere in this volume and in the published literature. Similarly, the binding affinities of protein-protein interactions are routinely measured by a variety of techniques, including fluorescence spectroscopy [85], isothermal titration calorimetry [86], and surface plasmon resonance measurements using the BIACORE system [87], all of which have their unique advantages and have been reviewed elsewhere. These approaches are limited by their requirement for highly purified components and the possibility that the values of the parameters measured *in vitro* may be quite different from their true value in the cell, where effects of molecular crowding and interactions with other components may dramatically influence binding events [88].

For this reason, in our own studies of the Wnt pathway, we relied on extract-based methods that are biochemically accessible and more accurately mimic the conditions of intact cells. Specifically, we assembled protein complexes using radiolabeled proteins produced by *in vitro* translation and monitored the dissociation of these complexes following dilution into egg extracts, where the concentration of cellular components approximates that of the intact oocyte [59]. Another important kinetic parameter that is often overlooked in many mathematical descriptions of signaling networks is the rate of protein turnover. Extracts have historically been a valuable biochemical system in the study of protein turnover and allowed us to obtain quantitative information on the rate of β-catenin turnover under a variety of conditions.

D. Building a Mathematical Description of Signaling Networks

Biochemical interactions are typically modeled using the equations of chemical kinetics. For the examples discussed in this review, with the exception of the PCP modeling, the dynamics of each network are modeled by a set of ordinary differential equations, which describe the changes in concentration of each component over time and assume a well-stirred system (if separate cellular compartments are considered, each compartment is well-stirred and additional ODEs describe shuttling between compartments). A discussion of the analytical and numerical techniques involved in formulating and solving systems of differential equations is beyond the scope of this chapter, and the reader is referred to a number of excellent texts on the subject [89, 90]. Primary references describing the models of Her7/Hes1 [31, 32], the MAPK cascade in *Xenopus* oocyte maturation [45], and Wnt signaling [59] each provide detailed information on the derivation of the ODEs involved in simulating these networks. A number of free software packages are available for developing ODE models of biological networks, including Genesis/KinetiKit [91], the Virtual Cell [92], and Gepasi 3 [93] (reviewed in Reference 71).

In addition, general mathematical programming software, such as Matlab, provides a flexible programming environment that can easily be applied to the mathematical modeling of biological networks. Spatial organization, of obvious importance to embryology, requires the introduction of partial differential equations to describe changes in component concentrations over space and time. Methods for solving PDEs are more challenging, and models incorporating spatial heterogeneity are often simplified by using ODE models, constrained spatially by compartmentalization and allowing for flux between compartments [38]. The truism regarding the dangers of a small amount of knowledge hold both for the theoreticians entering experimental biology and the biologists applying theoretical models to their research, and caution should be taken to seek expert advice in such interdisciplinary endeavors.

REFERENCES

1. Gerhart, J., Warkany lecture: signaling pathways in development. Teratology, 1999. 60(4): pp. 226–39.
2. Pires-daSilva, A. and R.J. Sommer, The evolution of signalling pathways in animal development. Nat Rev Genet, 2003. 4(1): pp. 39–49.
3. Aebersold, R. and M. Mann, Mass spectrometry-based proteomics. Nature, 2003. 422(6928): pp. 198–207.
4. van Steensel, B., Mapping of genetic and epigenetic regulatory networks using microarrays. Nat Genet, 2005. 37 Suppl: pp. S18–24.
5. Wheeler, D.B., A.E. Carpenter, and D.M. Sabatini, Cell microarrays and RNA interference chip away at gene function. Nat Genet, 2005. 37 Suppl: pp. S25–30.
6. Hunter, T., Signaling—2000 and beyond. Cell, 2000. 100(1): pp. 113–27.
7. Hlavacek, W.S., et al., The complexity of complexes in signal transduction. Biotechnol Bioeng, 2003. 84(7): pp. 783–94.
8. Papin, J.A., et al., Reconstruction of cellular signalling networks and analysis of their properties. Nat Rev Mol Cell Biol, 2005. 6(2): pp. 99–111.
9. Ideker, T., T. Galitski, and L. Hood, A new approach to decoding life: systems biology. Annu Rev Genomics Hum Genet, 2001. 2: pp. 343–72.
10. Aderem, A., Systems biology: its practice and challenges. Cell, 2005. 121(4): pp. 511–13.
11. Liu, E.T., Systems biology, integrative biology, predictive biology. Cell, 2005. 121(4): pp. 505–06.
12. Kirschner, M.W., The meaning of systems biology. Cell, 2005. 121(4): pp. 503–04.
13. Blanchard, B.S. and W.J. Fabrycky, *Systems engineering and analysis*. 3rd ed. 1998, Upper Saddle River, NJ: Prentice Hall.
14. Fell, D., *Understanding the control of metabolism*. 1st ed. Frontiers in metabolism, 2. 1997, London; Miami; Brookfield, VT: Portland Press;
15. Heinrich, R. and S. Schuster, *The regulation of cellular systems*. 1996, New York: Chapman & Hall.

16. Turing, A.M., The chemical basis of morphogenesis. Philos. Trans. R. Soc. Lond. [B], 1952. 237: pp. 37–72.
17. Ma'ayan, A., R.D. Blitzer, and R. Iyengar, Toward predictive models of mammalian cells. Annu Rev Biophys Biomol Struct, 2005. 34: pp. 319–49.
18. Milo, R., et al., Network motifs: simple building blocks of complex networks. Science, 2002. 298(5594): pp. 824–27.
19. Kashtan, N., et al., Topological generalizations of network motifs. Phys Rev E Stat Nonlin Soft Matter Phys, 2004. 70(3 Pt 1): p. 031909.
20. Ma'ayan, A., et al., Formation of regulatory patterns during signal propagation in a mammalian cellular network. Science, 2005. 309(5737): pp. 1078–83.
21. Barash, Y. and N. Friedman, Context-specific Bayesian clustering for gene expression data. J Comput Biol, 2002. 9(2): pp. 169–91.
22. Friedman, N., et al., Using Bayesian networks to analyze expression data. J Comput Biol, 2000. 7(3–4): pp. 601–20.
23. Goodwin, B.C., Oscillatory behavior in enzymatic control processes. Adv. Enzyme Regul., 1965. 3: pp. 425–38.
24. Goldbeter, A., G. Dupont, and M.J. Berridge, Minimal model for signal-induced Ca2+ oscillations and for their frequency encoding through protein phosphorylation. Proc Natl Acad Sci U S A, 1990. 87(4): pp. 1461–65.
25. Tyson, J.J., Modeling the cell division cycle: cdc2 and cyclin interactions. Proc Natl Acad Sci U S A, 1991. 88(16): pp. 7328–32.
26. Goldbeter, A., A model for circadian oscillations in the Drosophila period protein (PER). Proc Biol Sci, 1995. 261(1362): pp. 319–24.
27. Goldbeter, A., Computational approaches to cellular rhythms. 2002. 420(6912): pp. 238–45.
28. Pourquie, O., The vertebrate segmentation clock. J Anat, 2001. 199(Pt 1–2): pp. 169–75.
29. Bessho, Y. and R. Kageyama, Oscillations, clocks and segmentation. Curr Opin Genet Dev, 2003. 13(4): pp. 379–84.
30. Palmeirim, I., et al., Avian hairy gene expression identifies a molecular clock linked to vertebrate segmentation and somitogenesis. Cell, 1997. 91(5): pp. 639–48.
31. Monk, N.A., Oscillatory expression of Hes1, p53, and NF-kappaB driven by transcriptional time delays. Curr Biol, 2003. 13(16): pp. 1409–13.
32. Lewis, J., Autoinhibition with transcriptional delay: a simple mechanism for the zebrafish somitogenesis oscillator. Curr Biol, 2003. 13(16): pp. 1398–408.
33. Dale, J.K., et al., Periodic notch inhibition by lunatic fringe underlies the chick segmentation clock. Nature, 2003. 421(6920): pp. 275–78.
34. Aulehla, A., et al., Wnt3a plays a major role in the segmentation clock controlling somitogenesis. Dev Cell, 2003. 4(3): pp. 395–406.
35. Hirata, H., et al., Instability of Hes7 protein is crucial for the somite segmentation clock. Nat Genet, 2004. 36(7): pp. 750–54.
36. Rida, P.C., N. Le Minh, and Y.J. Jiang, A Notch feeling of somite segmentation and beyond. Dev Biol, 2004. 265(1): pp. 2–22.
37. Aulehla, A. and B.G. Herrmann, Segmentation in vertebrates: clock and gradient finally joined. Genes Dev, 2004. 18(17): pp. 2060–67.
38. Hoffmann, A., et al., The IkappaB-NF-kappaB signaling module: temporal control and selective gene activation. Science, 2002. 298(5596): pp. 1241–45.
39. Lev Bar-Or, R., et al., Generation of oscillations by the p53-Mdm2 feedback loop: a theoretical and experimental study. Proc Natl Acad Sci U S A, 2000. 97(21): pp. 11250–55.

40. Lahav, G., et al., Dynamics of the p53-Mdm2 feedback loop in individual cells. Nat Genet, 2004. 36(2): pp. 147–50.
41. Kolch, W., M. Calder, and D. Gilbert, When kinases meet mathematics: the systems biology of MAPK signalling. FEBS Lett, 2005. 579(8): pp. 1891–95.
42. Roux, P.P. and J. Blenis, ERK and p38 MAPK-activated protein kinases: a family of protein kinases with diverse biological functions. Microbiol Mol Biol Rev, 2004. 68(2): pp. 320–44.
43. Gotoh, Y. and E. Nishida, The MAP kinase cascade: its role in Xenopus oocytes, eggs and embryos. Prog Cell Cycle Res, 1995. 1: pp. 287–97.
44. Ferrell, J.E., Jr., Xenopus oocyte maturation: new lessons from a good egg. Bioessays, 1999. 21(10): pp. 833–42.
45. Huang, C.Y. and J.E. Ferrell, Jr., Ultrasensitivity in the mitogen-activated protein kinase cascade. Proc Natl Acad Sci U S A, 1996. 93(19): pp. 10078–83.
46. Ballif, B.A., et al., Quantitative phosphorylation profiling of the ERK/p90 ribosomal S6 kinase-signaling cassette and its targets, the tuberous sclerosis tumor suppressors. Proc Natl Acad Sci U S A, 2005. 102(3): pp. 667–72.
47. Ferrell, J.E., Jr. and E.M. Machleder, The biochemical basis of an all-or-none cell fate switch in Xenopus oocytes. Science, 1998. 280(5365): pp. 895–98.
48. Ferrell, J.E., Jr., Self-perpetuating states in signal transduction: positive feedback, double-negative feedback and bistability. Curr Opin Cell Biol, 2002. 14(2): pp. 140–48.
49. Ferrell, J.E. and W. Xiong, Bistability in cell signaling: How to make continuous processes discontinuous, and reversible processes irreversible. Chaos, 2001. 11(1): pp. 227–36.
50. Tyson, J.J., K.C. Chen, and B. Novak, Sniffers, buzzers, toggles and blinkers: dynamics of regulatory and signaling pathways in the cell. Curr Opin Cell Biol, 2003. 15(2): pp. 221–31.
51. Bhalla, U.S. and R. Iyengar, Emergent properties of networks of biological signaling pathways. Science, 1999. 283(5400): pp. 381–87.
52. Poritz, M.A., et al., Graded mode of transcriptional induction in yeast pheromone signalling revealed by single-cell analysis. Yeast, 2001. 18(14): pp. 1331–38.
53. Whitehurst, A., M.H. Cobb, and M.A. White, Stimulus-coupled spatial restriction of extracellular signal-regulated kinase 1/2 activity contributes to the specificity of signal-response pathways. Mol Cell Biol, 2004. 24(23): pp. 10145–50.
54. Mackeigan, J.P., et al., Graded mitogen-activated protein kinase activity precedes switch-like c-Fos induction in mammalian cells. Mol Cell Biol, 2005. 25(11): pp. 4676–82.
55. Traverse, S., et al., Sustained activation of the mitogen-activated protein (MAP) kinase cascade may be required for differentiation of PC12 cells. Comparison of the effects of nerve growth factor and epidermal growth factor. Biochem J, 1992. 288 (Pt 2): pp. 351–55.
56. Saxton, T.M., et al., Gene dosage-dependent functions for phosphotyrosine-Grb2 signaling during mammalian tissue morphogenesis. Curr Biol, 2001. 11(9): pp. 662–70.
57. Maeda, M., et al., Periodic signaling controlled by an oscillatory circuit that includes protein kinases ERK2 and PKA. Science, 2004. 304(5672): pp. 875–78.
58. Salic, A., et al., Control of beta-catenin stability: reconstitution of the cytoplasmic steps of the wnt pathway in Xenopus egg extracts. Mol Cell, 2000. 5(3): pp. 523–32.
59. Lee, E., et al., The roles of APC and Axin derived from experimental and theoretical analysis of the Wnt pathway. PLoS Biol, 2003. 1(1): pp. E10.

60. Schoeberl, B., et al., Computational modeling of the dynamics of the MAP kinase cascade activated by surface and internalized EGF receptors. Nat Biotechnol, 2002. 20(4): pp. 370–75.
61. Traverse, S., et al., EGF triggers neuronal differentiation of PC12 cells that overexpress the EGF receptor. Curr Biol, 1994. 4(8): pp. 694–701.
62. Oda, K., et al., A comprehensive pathway map of epidermal growth factor receptor signaling. 2005. 1(1): pp. msb4100014-E1–msb4100014-E17.
63. Brown, K.S., et al., The statistical mechanics of complex signaling networks: nerve growth factor signaling. Physical Biology, 2004. 1(3): pp. 184–95.
64. Brown, K.S. and J.P. Sethna, Statistical mechanical approaches to models with many poorly known parameters. Physical Review E, 2003. 68, 021904
65. Battogtokh, D., et al., An ensemble method for identifying regulatory circuits with special reference to the qa gene cluster of Neurospora crassa. Proc Natl Acad Sci U S A, 2002. 99(26): pp. 16904–09.
66. York, R.D., et al., Rap1 mediates sustained MAP kinase activation induced by nerve growth factor. Nature, 1998. 392(6676): pp. 622–26.
67. Farlow, S.J., *Partial differential equations for scientists and engineers.* 1993, New York: Dover Publications.
68. Kiefer, J.C., Planar cell polarity: heading in the right direction. Dev Dyn, 2005. 233(2): pp. 695–700.
69. Amonlirdviman, K., et al., Mathematical modeling of planar cell polarity to understand domineering nonautonomy. Science, 2005. 307(5708): pp. 423–26.
70. Oda, K., et al., A comprehensive pathway map of epidermal growth factor receptor signaling. Molecular Systems Biology, 2005. msb4100014: pp. E1–E17.
71. Neves, S.R. and R. Iyengar, Modeling of signaling networks. Bioessays, 2002. 24(12): pp. 1110–17.
72. Gilman, A.G., et al., Overview of the Alliance for Cellular Signaling. Nature, 2002. 420(6916): pp. 703–06.
73. Rajalingam, K., et al., Prohibitin is required for Ras-induced Raf-MEK-ERK activation and epithelial cell migration. Nat Cell Biol, 2005. 7(8): pp. 837–43.
74. Silva, J., et al., RNA-interference-based functional genomics in mammalian cells: reverse genetics coming of age. Oncogene, 2004. 23(51): pp. 8401–09.
75. Stelzl, U., et al., A human protein-protein interaction network: a resource for annotating the proteome. Cell, 2005. 122(6): pp. 957–68.
76. Lee, L.A., et al., Drosophila genome-scale screen for PAN GU kinase substrates identifies Mat89Bb as a cell cycle regulator. Dev Cell, 2005. 8(3): pp. 435–42.
77. Lustig, K.D., et al., Small pool expression screening: identification of genes involved in cell cycle control, apoptosis, and early development. Methods Enzymol, 1997. 283: pp. 83–99.
78. King, R.W., et al., Expression cloning in the test tube. Science, 1997. 277(5328): pp. 973–74.
79. Drysdale, R.A. and M.A. Crosby, FlyBase: genes and gene models. Nucleic Acids Res, 2005. 33(Database issue): pp. D390–95.
80. Gingras, A.C., R. Aebersold, and B. Raught, Advances in protein complex analysis using mass spectrometry. J Physiol, 2005. 563(Pt 1): pp. 11–21.
81. Gavin, A.C., et al., Functional organization of the yeast proteome by systematic analysis of protein complexes. Nature, 2002. 415(6868): pp. 141–47.
82. Ho, Y., et al., Systematic identification of protein complexes in Saccharomyces cerevisae by mass spectrometry. Nature, 2002. 415(6868): pp. 180–83.

83. Xia, Y., et al., Analyzing cellular biochemistry in terms of molecular networks. Annu Rev Biochem, 2004. 73: pp. 1051–87.
84. Bailey, J.E., Complex biology with no parameters. Nat Biotechnol, 2001. 19(6): pp. 503–04.
85. Lakowicz, J.R., *Principles of fluorescence spectroscopy.* 2nd ed. 1999, New York: Kluwer Academic/Plenum.
86. Velazquez-Campoy, A., S.A. Leavitt, and E. Freire, Characterization of protein-protein interactions by isothermal titration calorimetry. Methods Mol Biol, 2004. 261: pp. 35–54.
87. Malmqvist, M. and R. Karlsson, Biomolecular interaction analysis: affinity biosensor technologies for functional analysis of proteins. Curr Opin Chem Biol, 1997. 1(3): pp. 378–83.
88. Minton, A.P., The influence of macromolecular crowding and macromolecular confinement on biochemical reactions in physiological media. J Biol Chem, 2001. 276(14): pp. 10577–80.
89. Bower, J.M. and H. Bolouri, *Computational modeling of genetic and biochemical networks.* 2001, Cambridge, Mass.: MIT Press.
90. Strogatz, S.H., *Nonlinear dynamics and Chaos: with applications to physics, biology, chemistry, and engineering. Studies in nonlinearity.* 1994, Reading, Mass.: Addison-Wesley Pub.
91. Bhalla, U.S., Use of Kinetikit and GENESIS for modeling signaling pathways. Methods Enzymol, 2002. 345: pp. 3–23.
92. Moraru, II, et al., The virtual cell: an integrated modeling environment for experimental and computational cell biology. Ann N Y Acad Sci, 2002. 971: pp. 595–96.
93. Mendes, P., GEPASI: a software package for modelling the dynamics, steady states and control of biochemical and other systems. Comput Appl Biosci, 1993. 9(5): pp. 563–71.
94. Xiong, W. and Ferrell, J.E., A positive feedback-based bistable "memory module" that governs a cell fate decision. Nature, 2003, 426: 460–5.

Index

C

D

E

F

G

H

I

J

K

L

M

N

Q

R

S

T

U

V

W

X

Y

Z